Recalled Earlier

Microwave Circuit Design Using Linear and Nonlinear Techniques

RECEIVED

AUG 8 1990

MSU - LIBRARY

Microwave Circuit Design Using Linear and Nonlinear Techniques

GEORGE D. VENDELIN
  Consultant
ANTHONY M. PAVIO
  Texas Instruments, Inc.
ULRICH L. ROHDE
  Compact Software, Inc.

WILEY

A Wiley-Interscience Publication

JOHN WILEY & SONS

New York　•　Chichester　•　Brisbane　•　Toronto　•　Singapore

TK
7876
.V47
1990

Copyright © 1990 by John Wiley & Sons, Inc.

All rights reserved. Published simultaneously in Canada.

Reproduction or translation of any part of this work
beyond that permitted by Section 107 or 108 of the
1976 United States Copyright Act without the permission
of the copyright owner is unlawful. Requests for
permission or further information should be addressed to
the Permissions Department, John Wiley & Sons, Inc.

Library of Congress Cataloging in Publication Data:

Vendelin, George D. (George David), 1938–
   Microwave circuit design using linear and nonlinear techniques/George D.
Vendelin, Anthony M. Pavio. Ulrich L. Rohde.
      p.   cm.
   "A Wiley-Interscience publication."
   Bibliography: p.
   1. Microwave integrated circuits.   2. Microwave amplifiers.
3. Oscillators, Microwave.   4. Electronic circuit design.
I. Pavio, Anthony M.   II. Rohde, Ulrich L.   III. Title.
TK7876.V47   1990
621.381'325—dc19
   ISBN 0-471-60276-0   89-30005
                                                           CIP

Printed in the United States of America

10 9 8 7 6 5 4 3 2 1

1912 2611   10/1/90 RB

To our wives
Barbara,
Jeanne,
Meta
for their patience and understanding.

Contents

Foreword

Over the past three-to-four decades, transistors, both bipolars and FETs, have been the dominant three-terminal (two-port) solid state device of choice of designers of microwave circuits and systems.

Although many technical treatises covering various properties of these devices have appeared on the market, there has been a distinct lack of a comprehensive treatment of these devices covering all aspects ranging from their intrinsic and circuit properties to circuit design techniques for maximizing their performance in communication and radar systems—a handbook, if you will.

Fortunately, this book, written by three experts with years of industrial experience in the field of solid-state circuit design and system applications fills this void. Not only is the treatment a repository of useful design formulas and reference data for the experienced engineer, much of it appearing in print for the first time, but it also fulfills a tutorial role for the engineer just entering the exciting field of microwave integrated circuit design. While leaving some of the research details to the voluminous referenced literature, the authors have provided all of the necessary mathematics and practical design information to assist both the student and the experienced engineer to carry out the design of hybrid and monolithic integrated circuits.

This book consists of eight chapters covering topics ranging from general descriptions of linear two-ports to workstations for computer-aided design (CAD) of microwave circuits. Each chapter is adequately represented by illustrations and contains an extensive reference list and a problem section. These features of the book coupled with its tutorial flavor makes it suitable for a graduate course in microwave solid-state circuit design.

The first chapter consists of an in-depth review of the various mathematical representations of the circuit properties of linear two-ports. Starting out with low-frequency descriptions relating terminal voltages and currents, such as hybrid parameters used in the early days of bipolars, the presentation moves to the well-known y- and z-parameter representations, then to the lesser-known $ABCD$ circuit parameters which are especially useful for noise analyses of cascaded networks and CAD applications. The extension of these representations to microwave circuits by decomposition of the voltage and current variables into forward and backward traveling waves leads, naturally, to the concept of scattering (S) and transmission parameters, the latter being the wave analog of the $ABCD$ parameters. A handy reference table relating the various two-port parameter representations is included.

With these tools established, it is only natural that they be applied to transmission lines of various types and to active circuits using them. Thus such important properties as power gain, stability analysis and conjugate matching conditions are expressed in terms of S parameters. An especially useful outcome of this chapter is the derivation of the diverse formulae for defining power gain (nine ways!) of a linear two-port. These are listed in convenient tabular form for easy reference. Other useful tables contain a listing of numerical values of the various two-port parameters typical of bipolar transistors.

The second chapter takes up the all-important topic of noise in linear two-ports. Starting from the basic definition of noise figure and signal-to-noise ratio, and the method of measuring these properties, the authors proceed to the general circuit representation of the noise properties of linear two-ports taking correlation of noise sources into account. This leads naturally to the important concept of the correlation matrix, which is so useful in CAD software for the analysis of noise in cascaded networks.

With this groundwork established, such topics as the derivation of the noise figure of cascaded networks, the influence of parasitic elements on the noise figure, the methods of determining the four noise parameters, and noise circles are addressed. Noise circles are particularly useful in computer aided design of low-noise amplifiers when combined with the corresponding gain circles. Again, as in Chapter 1, useful tables summarizing the derived formulas are included.

Chapter 3 applies the techniques established in the first two chapters to a detailed analysis of the small-signal and noise properties of silicon bipolars and gallium arsenide (GaAs) FETs in terms of their equivalent circuit and noise representations. The important topics of bias circuitry design, temperature stability, and impedance matching are covered.

Again extensive tables are presented covering such diverse topics as properties of semiconductors and the equivalent circuit and noise parameter values for commercial bipolars and FETs. All told, there are 24 tables, 53 figures, and 68 references in this chapter, as well as a very complete bibliography.

The next chapter concentrates on the basics of transistor amplifier design. Such topics as unilateral and nonunilateral design approaches are addressed. As vehicles for illustrating these design approaches, the authors "walk" through the steps leading to the design of actual transistor amplifiers. Both small signal and power amplifiers are covered in some detail with such ancillary topics as distortion analysis included. Other circuit designs illustrated are low-noise and traveling wave amplifiers.

Chapter 5 takes up the topic of high-power amplifier design in more detail. Since linear two-port theory is not applicable to large-signal operation of transistors, the authors begin the chapter with the topics of device modeling and characterization for large-signal applications. These are covered in meticulous detail. Various experimental methods of large-signal characterization are described such as load-pull, along with a discussion of their limitations and difficulties of application. Large signal simulation of transistors by time domain techniques and the harmonic balance methods are discussed.

The chapter then moves on to the design of the optimum load of a power amplifier and applies the concept to single-stage and multistage cascaded and distributed power amplifier designs. Distributed amplifiers are discussed in considerable detail both from the device and the circuit points of view. Most of the power amplifier designs are Class A or AB, although a short section on Class B design is included.

Oscillators are the topic of Chapter 6. Starting with a derivation of the conditions for oscillation (Barkhausen criteria), the authors then briefly discuss the expanded Smith Chart (reflection coefficient greater than unity) and its role in oscillator design. The chapter addresses the various methods of stabilizing oscillations in oscillators, such as by series or parallel resonance circuits and solid state resonators (YIG spheres, dielectric pucks, cavities), and for tuning (varactor, coaxial lines, and YIG spheres). Experimental techniques for characterizing the properties of resonators such as resonant frequency and Q factor are also addressed.

Next, the analysis of noise and its measurement in oscillators is explained in some detail. Both thermal and $1/f$ noise sources are considered.

To round out the chapter, the reader is led through a typical oscillator design using the previously described theory. The chapter ends with a description of a nonlinear FET model useful in oscillator design and the computer-aided design of oscillators by the harmonic balance technique.

Chapter 7 addresses the difficult topic of practical mixer design; a topic that has not received adequate attention in textbooks until recently, although a large body of experience resides in industry. The authors first address diode mixers and the various circuit configurations that can be built around them, for example, single, single-balanced, and double-balanced. These topics are treated in detail.

Next, the theory and operation of FET mixers are explained. FET mixers are of interest because they can exhibit conversion gain and high dynamic range. Both single-gate and dual-gate mixers are discussed. Mixer configurations based on single devices and distributed formats are addressed. The dual-gate FET is particularly suited for broadband applications based on the traveling wave concept because of the ease with which the oscillator and signal excitations can be isolated because of the additional gate electrode. The chapter concludes with a discussion of special mixer circuits such as the image rejection or single-sideband mixer. This chapter is illustrated adequately by over 130 figures!

The last chapter concentrates on the role of computer-aided design (CAD) workstations in circuit design—a relatively new concept and the "wave of the future." Here the authors demonstrate how workstations can be used not only for circuit design and simulation, both linear and nonlinear, but also in pattern layout and in the control of measurement equipment in the laboratory. The workstation concept is ideally suited to the design of microwave integrated circuits.

The authors end the chapter with an excellent example to illustrate the power and versatility of CAD tools and the many techniques covered in the previous seven chapters, namely, the design of an anticollision radar transceiver system with all its functional blocks—low-noise and power amplifier, mixer, and oscillator, among others. A comparison is made between the simulated performance and actual measurements.

Although I have attempted to address the salient features of this book, only a personal examination will convey to the prospective reader the broad scope of its coverage. The authors are to be commended for their exhaustive treatment of the subject matter.

ROBERT A. PUCEL
Consulting Scientist
Raytheon Company

Lexington, Massachusetts
November 1989

Preface

The intention of this book is to be a complete introduction to the MIC and MMIC technology and at the same time be useful as a reference book. While leaving some technology of the research details to the referenced literature, we have provided all necessary details to enable an RF engineer to become a good microwave engineer. It is a skilled mixture of mathematical information and practical design results that will assist both the student and practical design engineer.

Although the book has a great deal of cross-fertilization, the primary chapter responsibilities were: Vendelin, Chapters 1, 3, 4; Rohde, Chapters 2, 6, 8 and Appendices D–G; and Pavio, Chapters 5 and 7.

A book of this magnitude is not completed without assistance from many fronts. The contributions of these engineers of Compact Software, Inc. throughout the book are acknowledged and greatly appreciated: Dr. Qui Zhang, Wei-Xu Huang, Tony Kwan, and Louis Perez.

The numerous contributions of Dr. Pieter L. D. Abrie, William C. Mueller, Dr. Robert A. Pucel, and Dr. Alfie N. Riddle provided valuable information and organization throughout the book. Dr. Abrie is also the author of *Complex Match*.

Numerous contributors included Dr. Behruz Rezvani, Dr. Craig P. Snapp, Dr. R. M. Malbon, Dr. Patrick Chye, Dr. Ding Day, Dr. J. A. Higgins, Dr. A. P. S. Khanna, Harry F. Cooke, and Vince Grande. Test fixture inputs were provided by Dr. Werner Schuerch of Inter-Continental Microwave. The impedance matching section on discontinuities in Chapter 3 was written by Professor G. R. Branner. Chapter 4 inputs came from Wayne Kennan, Michael Frank, and Robert Zona. Dr. Dan Raicu greatly assisted with multistage high power, wide dynamic range, and low-noise design trade-offs. Inputs on transformer feedback were provided by Hansel B. Mead of Q-Bit. A portion of the oscillator noise discussion was written by Bill Chan.

Chapter 5 utilized several MMIC design examples furnished by Texas Instruments, Inc. and was reviewed by Ralph H. Halladay of Texas Instruments. Dr. Rowan Gilmore allowed parts of his dissertation to be used in the Chapter 6 presentation of nonlinear designs. Dr. Yusuke Tajima of Raytheon Corporation supplied comments on the nonlinear oscillator approach. Chapter 7 utilizes several MIC and MMIC mixers and special circuits provided by Texas Instruments and was reviewed by Ben R. Hallford of Texas Instruments.

In Chapter 8, the information on anticollision radar was supplied by Honeywell Corporation of West Germany. Synergy Microwave Corporation engineers contributed a variety of circuits and approaches throughout the book.

Brenda Biggs, Maria Galli, and Michelle Jensen contributed great typing assistance, our editor, George Telecki, provided constant encouragement, and the copy editor, Barbara Zeiders, refined our rough manuscript.

GEORGE D. VENDELIN
ANTHONY M. PAVIO
ULRICH L. ROHDE

Saratoga, California
Dallas, Texas
Paterson, New Jersey
November 1989

Microwave Circuit Design
Using Linear and
Nonlinear Techniques

1 *S* Parameters

1.1 DEFINITIONS AND USE WITH AMPLIFIERS

The necessary tools for the design of microwave amplifiers, oscillators, and mixers are found from an understanding of transmission lines, two-port networks, and impedance-matching techniques. With a knowledge of the transistor S parameters, the design of amplifier, oscillator, and mixer circuits will be shown to be essentially an impedance-matching problem. The design tools are presented in Chapters 1, 2, 3, and 8.

The design techniques are shown schematically in Fig. 1.1. For the amplifier, the power available from the source should be delivered to the input of the transistor; this is the input-matching problem M_1, which is an idealized lossless matching circuit. The output lossless matching structure M_2 (also an idealized lossless matching circuit) should be designed to deliver the maximum power to the load. This type of amplifier, which is simultaneously conjugately matched at input and output ports, is possible only when the stability factor is sufficiently large ($k > 1$). The power gain is the maximum available gain (G_{ma}). The details of this design procedure are given in Chapters 4 and 5.

The oscillator is a similar design problem, where the stability factor must now be less than 1 ($k < 1$). Notice that the load is receiving power for either case; that is, the load would not know whether the transistor is an amplifier or an oscillator. Usually, the same transistors are used for both applications. Some type of feedback may be required to achieve $k < 1$ at the frequency of interest. The lossless input structure M_3 resonates the input port, whereas the output matching structure M_4 is usually designed to deliver the maximum power to the load. A complete discussion of the oscillator design procedure is given in Chapter 6.

The mixer design can also be viewed as an impedance-matching problem at the signal frequency, intermediate frequency, and local oscillator frequency via the three matching structures M_5, M_6, and M_7 shown in the mixer block diagram (see Fig. 1.1). A complete discussion of the mixer design is given in Chapter 7.

Finally, the most important tool of all, the computer, is discussed in Chapter 8. An up-to-date discussion of linear and nonlinear software is presented in this chapter.

1.1.1 Low-Frequency Two-Port Network

Electrical networks can be described by the number of external terminals that are available for measurement and analysis. A single terminal pair is a one-port

1

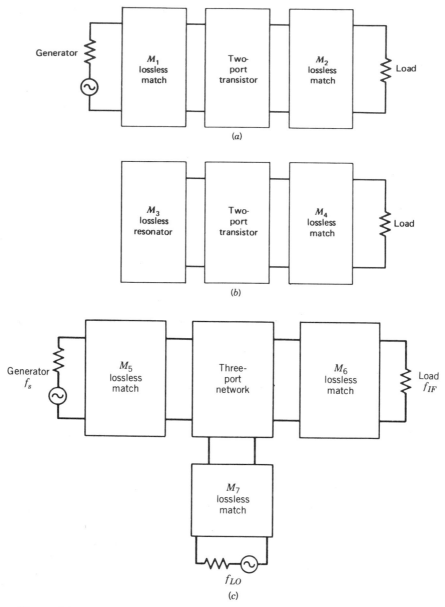

Figure 1.1 Block diagram for (*a*) amplifier, (*b*) oscillator, and (*c*) mixer designs.

network. A two-port network may be formed by either two terminal pairs or two terminals above ground (Fig. 1.2). A three-port network may be formed by three terminal pairs or three terminals above ground (e.g., the transistor three-port in Fig. 1.2).

The most commonly encountered network in circuit design is the two-port network. The two-port network is described by a set of four independent param-

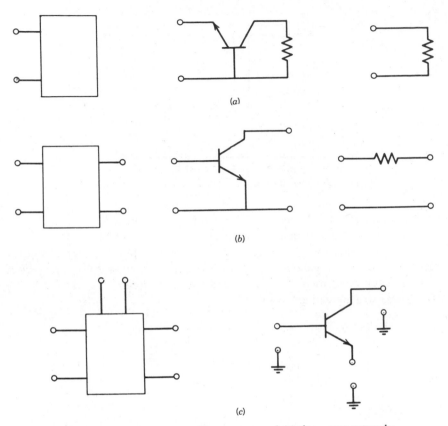

Figure 1.2   (*a*) One-port, (*b*) two-port, and (*c*) three-port networks.

eters, which can be related to the terminal voltage and current at low frequencies. Figure 1.3 shows a two-port network representation of a transistor. The commonly used two-port network parameters are the *h*, *z*, *y*, and *ABCD* parameters.

The hybrid *h* parameters were used in the early days of the transistor because this set described the physical operation of the transistor—for example, the short-circuit current gain, h_{21}. These parameters are called hybrid because the units are hybrid and are given by

$$v_1 = h_{11}i_1 + h_{12}v_2 \tag{1.1}$$

$$i_2 = h_{21}i_1 + h_{22}v_2 \tag{1.2}$$

or in matrix form,

$$\begin{bmatrix} v_1 \\ i_2 \end{bmatrix} = \begin{bmatrix} h_{11} & h_{12} \\ h_{21} & h_{22} \end{bmatrix} \begin{bmatrix} i_1 \\ v_2 \end{bmatrix} \tag{1.3}$$

The measurements of these parameters could be made accurately only for transis-

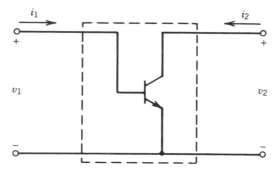

Figure 1.3 Two-port network representation for transistor.

tors operating up to 1 MHz because of the difficulties in defining a short circuit and an open circuit and maintaining a stable two-port with dc bias.

For higher-frequency, very high frequency (VHF) and ultrahigh frequency (UHF) transistors, the parameter sets preferred for measurement accuracy were the open-circuit *z* parameters to 1 MHz and the short-circuit *y* parameters to 500 MHz. These sets are defined as follows:

z parameters:

$$\begin{bmatrix} v_1 \\ v_2 \end{bmatrix} = \begin{bmatrix} z_{11} & z_{12} \\ z_{21} & z_{22} \end{bmatrix} \begin{bmatrix} i_1 \\ i_2 \end{bmatrix} \tag{1.4}$$

y parameters:

$$\begin{bmatrix} i_1 \\ i_2 \end{bmatrix} = \begin{bmatrix} y_{11} & y_{12} \\ y_{21} & y_{22} \end{bmatrix} \begin{bmatrix} v_1 \\ v_2 \end{bmatrix} \tag{1.5}$$

The algebra of two-port network calculations can be simplified by choosing the simplest two-port network description. For example, for the parallel–parallel interconnection shown in Fig. 1.4 the overall *y* parameters are easily obtained by adding the individual *y* parameters:

$$\begin{bmatrix} i_1 \\ i_2 \end{bmatrix} = \begin{bmatrix} i_1^a + i_1^b \\ i_2^a + i_2^b \end{bmatrix} = \begin{bmatrix} y_{11}^a + y_{11}^b & y_{12}^a + y_{12}^b \\ y_{21}^a + y_{21}^b & y_{22}^a + y_{22}^b \end{bmatrix} \begin{bmatrix} v_1 \\ v_2 \end{bmatrix} \tag{1.6}$$

Similarly, for the series–series interconnection, the *z* parameters are added; for the series–parallel interconnection the *h* parameters are added. When cascading two-port networks, as shown in Fig. 1.5, the chain or *ABCD* parameters will result in less algebra calculation. The *ABCD* parameters are defined as follows:

$$\begin{bmatrix} v_1 \\ i_1 \end{bmatrix} = \begin{bmatrix} A & B \\ C & D \end{bmatrix} \begin{bmatrix} v_2 \\ -i_2 \end{bmatrix} \tag{1.7}$$

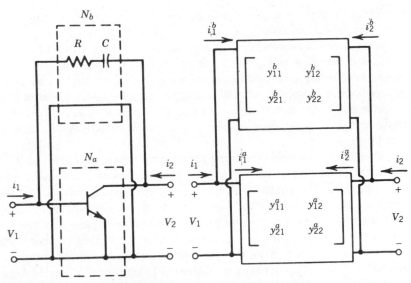

Figure 1.4 Parallel–parallel interconnection.

where the units are again hybrid. For the cascaded interconnection, the overall chain matrix is simply the product of individual chain matrices:

$$\begin{bmatrix} v_1 \\ i_1 \end{bmatrix} = \begin{bmatrix} v_1^a \\ i_1^a \end{bmatrix} = \begin{bmatrix} A^a & B^a \\ C^a & D^a \end{bmatrix} \begin{bmatrix} v_2^a \\ -i_2^a \end{bmatrix} = \begin{bmatrix} A^a & B^a \\ C^a & D^a \end{bmatrix} \begin{bmatrix} A^b & B^b \\ C^b & D^b \end{bmatrix} \begin{bmatrix} v_2^b \\ -i_2^b \end{bmatrix} \quad (1.8)$$

where the final answer is always a 2×2 matrix, regardless of the length of the cascaded network. All of these two-port parameter sets have four independent

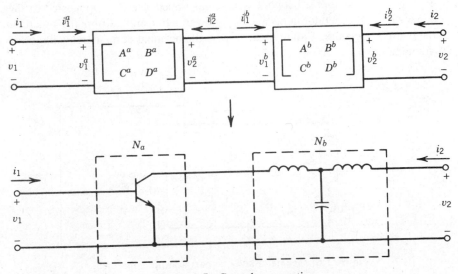

Figure 1.5 Cascade connection.

parameters, and only one set needs to be measured. Later, the relationship between the two-port parameter sets will be given (see Table 1.1).

At microwave frequencies, z, y, or h parameters are very difficult to measure. The reason is that open and short circuits to microwave signals are very difficult to realize because of lead inductance and fringing capacitance. These measurements require tuning stubs separately adjusted at each frequency to reflect open or short circuits to device terminals. For the active devices, such as the transistor and negative-resistance diode, open-circuit or short-circuit conditions will often result in oscillation.

To solve these problems, scattering parameters, which are related to incident and reflected power, rather than h, z, y, or $ABCD$ parameters, are used in microwave frequencies. The advantages of using scattering parameters are that unlike terminal voltages and currents, they are related to the incident and reflected power and do not vary in magnitude along lossless transmission lines. This means that the scattering parameters can be measured on a device located at some distance from the measuring point. Also, the S parameters are measured under impedance-matching conditions, thus avoiding unstable oscillation for active-device measurement. Therefore, we introduce S parameters based on the concept of incident and reflected power.

1.1.2 Incident and Reflected Power

The scattering parameters originally come from transmission-line theory; actually, they are the parameters of power. Scattering parameters are very useful in the design of microwave circuits and various matching networks. For easy understanding, we introduce S parameters using the concept of incident and reflected voltage and current parameters rather than transmission-line theory, where rms values of voltage and current are assumed unless stated otherwise.

In Fig. 1.6, a one-port network with input impedance Z is connected to the generator E_g with generator impedance Z_g. Power is transmitted from the generator to the one-port network. From circuit theory, the terminal current and voltage can be found as

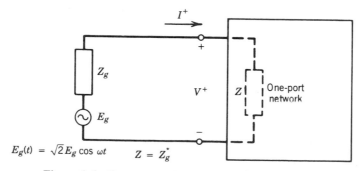

Figure 1.6 Generator and one-port network connected.

$$i = \frac{E_g}{Z_g + Z} \tag{1.9}$$

and

$$v = \frac{ZE_g}{Z_g + Z} \tag{1.10}$$

If the load impedance is equal to the conjugate of the generator impedance, namely $Z = Z_g^*$, the one-port network is conjugate matched with the generator. Under this condition, the terminal current is defined as the incident current,

$$I^+ = \frac{E_g}{Z_g^* + Z_g} = \frac{E_g}{2\,\mathrm{Re}(Z_g)} \tag{1.11}$$

and the terminal voltage is defined as the incident voltage,

$$V^+ = \frac{Z_g^* E_g}{Z_g^* + Z_g} = \frac{Z_g^* E_g}{2\,\mathrm{Re}(Z_g)} \tag{1.12}$$

The relationship between incident voltage and incident current is given by

$$V^+ = Z_g^* I^+ \tag{1.13}$$

In this case the load receives the maximum available power from the generator:

$$P_A = \mathrm{Re}(VI^*) = \frac{|E_g|^2}{4\,\mathrm{Re}(Z_g)} \tag{1.14}$$

Note that incident voltage V^+ and incident current I^+ are independent of the impedance of the one-port network; instead, they are related to the generator impedance Z_g, which is called the reference impedance of the network. It means that the incident voltage and current act as a constant for a given generator, regardless what kind of load is connected. This is very useful in practical measurement, since the incident voltage and current can be used as the measurement reference. It should be pointed out that when $\mathrm{Re}\,Z_g = 0$, the incident voltage and incident current cannot be uniquely determined since the maximum available power for a generator cannot be defined when the internal resistance of the generator is equal to zero.

In the general case, the actual terminal voltage and current are not equal to the incident voltage and current. The difference between the actual terminal voltage and the incident voltage is defined to be the reflected voltage:

$$V^- = V - V^+ \tag{1.15}$$

In the same way, the reflected current can be defined as

$$-I^- = I - I^+ \tag{1.16}$$

The direction of incident and reflected voltage and current are shown in Fig. 1.7; the minus sign in front of I^- means that the forward direction of I^- is chosen to be opposite the direction of I^+. The ratio of the reflected voltage to the incident voltage is called the voltage reflection coefficient and is given by

$$\Gamma_v = \frac{V^-}{V^+} = \frac{V}{V^+} - 1 = \frac{Z}{Z + Z_g} \frac{Z_g + Z_g^*}{Z_g^*} - 1 = \frac{Z_g(Z - Z_g^*)}{Z_g^*(Z + Z_g)} \tag{1.17}$$

The same definition applies to the current reflection coefficient:

$$\Gamma_i = \frac{I^-}{I^+} = 1 - \frac{I}{I^+} = 1 - \frac{Z_g + Z_g^*}{Z + Z_g} = \frac{Z - Z_g^*}{Z + Z_g} \tag{1.18}$$

The relationship between the voltage and current reflection coefficients is given by

$$\Gamma_v Z_g^* = \Gamma_i Z_g \tag{1.19}$$

From the equations above, we may find that the voltage and current reflection coefficients are not equal in the general case; the difference is two times the phase angle of the reference impedance. When the generator impedance is real, the voltage and current reflection coefficients are identical. This is the most practical case. At the condition $Z = Z_g^*$, the two reflection coefficients become zero. Only in this situation are the terminal voltage and current equal to the incident voltage and current. In other words, the incident power is completely absorbed by the load. This situation is what we are trying to approach in most circuit design.

The relation of reflected voltage V^- and current I^- is given by

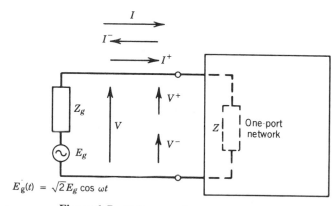

Figure 1.7   Voltages and currents at the port.

$$V^- = \Gamma_v V^+ = \Gamma_v Z_g^* I^+ = \Gamma_i Z_g I^+ = Z_g I^- \qquad (1.20)$$

Equations (1.13) and (1.20) mean that the incident voltage and current waves see the conjugate impedance of reference impedance Z_g, whereas the reflected voltage and current waves see the reference impedance Z_g itself.

Next we introduce normalized voltage waves and normalized current waves, which are directly related to the incident and reflected power. From (1.14), the incident power is defined as the maximum available power from a given generator, namely:

$$P_{inc} = \text{Re}\left[V^+ (I^+)^*\right] = \frac{|E_g|^2}{4\,\text{Re}(Z_g)} \qquad (1.21)$$

To introduce normalized incident voltage wave, the incident voltage is used to express incident power.

$$P_{inc} = \frac{|V^+|^2\,\text{Re}(Z_g)}{|Z_g^*|^2} \qquad (1.22)$$

The square root of the incident power is defined as the incident voltage wave a [with units of $(\text{watts})^{1/2}$]:

$$a = \sqrt{P_{inc}} = \frac{V^+\sqrt{\text{Re}(Z_g)}}{Z_g^*} \qquad (1.23)$$

Similarly, the normalized reflected voltage wave b is defined as

$$b = \sqrt{P_r} = \frac{V^-\sqrt{\text{Re}(Z_g)}}{Z_g} \qquad (1.24)$$

It may be helpful to remember that the waves arrive in alphabetical order: a is incident and b is reflected. Obviously, $|a|^2$ and $|b|^2$ represent the incident and reflected powers, respectively. Alternatively, the incident and reflected powers may also be expressed by incident and reflected currents as

$$P_{inc} = |I^+|^2\,\text{Re}(Z_g) \qquad (1.25)$$

and

$$P_r = |I^-|^2\,\text{Re}(Z_g) \qquad (1.26)$$

The a and b may be expressed by incident and reflected currents as

$$a = \sqrt{P_{inc}} = I^+\sqrt{\text{Re}(Z_g)} \qquad (1.27)$$

$$b = \sqrt{P_r} = I^- \sqrt{\mathrm{Re}(Z_g)} \qquad (1.28)$$

Hence a and b are also called the normalized incident and reflected current waves, or simply, normalized incident and reflected waves, since the normalized current waves and the normalized voltage waves are the same.

The terminal voltage and current are related to a and b by the following equations:

$$V = V^+ + V^- = \frac{Z_g^*}{\sqrt{\mathrm{Re}(Z_g)}} a + \frac{Z_g}{\sqrt{\mathrm{Re}(Z_g)}} b \qquad (1.29)$$

$$I = I^+ - I^- = \frac{1}{\sqrt{\mathrm{Re}(Z_g)}} a - \frac{1}{\sqrt{\mathrm{Re}(Z_g)}} b \qquad (1.30)$$

or

$$a = \frac{V + Z_g I}{2\sqrt{\mathrm{Re}(Z_g)}} \qquad (1.31)$$

$$b = \frac{V - Z_g^* I}{2\sqrt{\mathrm{Re}(Z_g)}} \qquad (1.32)$$

The dimension or units of a and b are the square root of power. They are directly related to power flow, which is why they are also often called power waves. It should be pointed out that the normalized incident wave a is relative only to the generator impedance Z_g, whereas the normalized reflected wave b is relative to both the generator impedance Z_g and the network impedance. The measurement of incident and reflected power is discussed next.

The ratio of the normalized reflection wave and the normalized incident wave is called the reflection coefficient Γ which is a complex number with magnitude and phase:

$$\Gamma = b/a \qquad (1.33)$$

We use this definition later to introduce S parameters for two-port and n-port networks.

In practice, incident and reflected powers can be measured by using a directional coupler. First, we need a basic explanation for the ideal directional coupler. Assume that there are two lines parallel and close to each other, and there is current I passing through line 1–2 in the direction shown in Fig. 1.8. The power will be coupled to line 3–4 via the mutual capacitance and inductance. The coupled currents by the mutual capacitance C_m are represented by I_3 and I_4 in line 3–4 flowing in the opposite direction. The coupled current by the mutual inductance is noted as I_L in line 3–4. From the electromagnetic induction theory (Faraday's law) the

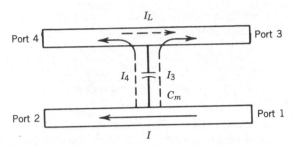

I_L

Port 4   Port 3

I_4 I_3

C_m

Port 2   Port 1

I

Figure 1.8 Physical explanation of ideal directional coupler.

direction of I_L is opposite to I. At port 3, the currents I_3 and I_L are in phase; therefore, a certain amount of power input from port 1 is coupled to port 3, whereas the currents I_4 and I_L are out of phase at port 4 and may cancel. The ideal directional coupler is designed such that I_4 and I_L cancel out and no power appears at port 4. Port 3 is usually called the coupled port and port 4 is the isolated port. Similarly, if the current is coming from port 2 to port 1, port 4 is the coupled port and port 3 is the isolated port.

An arrangement for measurement of incident and reflected power by using a directional coupler is given in Fig. 1.9. Here, as a practical example, we choose a 20-dB coupler, which means that by connecting a generator at port 1 (input power P), $0.99P$ will be incident at port 2, $0.01P$ at the coupled port 3 (or in decibels, $-10 \log 0.01 = 20$ dB), and there is no power received at the isolated port 4 under the conditions that ports 1, 2, 3, and 4 are connected by the matched load R_0, whose impedance is normally 50 Ω.

If we have an unknown load R (or Z if the load is complex) at port 2, when R is not equal to R_0, the reflected power P_r is not zero. There will be $0.99\,P_r$ reaching port 1 and $0.01P_r$ reaching port 4; there is no reflected power at port 3 according to the theory of the coupler. Finally, at port 3, we have measured power $P_3 =$

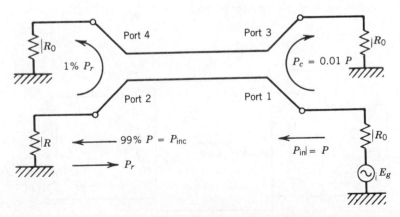

R_0

Port 4   Port 3

1% P_r R_0

$P_c = 0.01\ P$

Port 2   Port 1

R 99% $P = P_{\text{inc}}$ R_0

P_r $|P_{\text{in}}| = P$

E_g

Figure 1.9 Measurement of incident and reflected power by using directional coupler.

$0.01P$, and at port 4 we have measured power $P_4 = 0.01\,P_r$. The reflection coefficient at port 2 can be calculated as follows:

$$\Gamma_2 = \sqrt{\frac{P_r}{P_{\text{inc}}}} = \sqrt{\frac{P_r}{0.99P}} = \sqrt{\frac{1}{0.99}\frac{0.01P_r}{0.01P}} = \sqrt{\frac{P_4}{0.99P_3}} \qquad (1.34)$$

With this directional coupler and the matched conditions at ports 1, 3, and 4, we can obtain the incident power $P_{\text{inc}}\,(P_{\text{inc}} = 0.99P = 99P_3)$ and reflected power $P_r(P_r = 100P_4)$ at port 2. P_3 and P_4 can be measured at port 3 and 4. The phase conditions or angle of Γ_2 will also be measured at each frequency when the directional coupler has been calibrated for phase. A more complete discussion of network analyzer measurements is given in Refs. [1.1 to 1.6].

1.1.3   Two-Port Network S Parameters

From (1.33), $b = \Gamma a$, the reflected wave b can be expressed by the incident wave a via reflection coefficient Γ in the one-port-network case. For a two-port network, as shown in Fig. 1.10, the reflection power waves b_1 and b_2 can also be represented by the incident power waves a_1 and a_2 as follows:

$$b_1 = S_{11}a_1 + S_{12}a_2 \qquad (1.35)$$

$$b_2 = S_{21}a_1 + S_{22}a_2 \qquad (1.36)$$

or in matrix form,

$$\begin{bmatrix} b_1 \\ b_2 \end{bmatrix} = \begin{bmatrix} S_{11} & S_{12} \\ S_{21} & S_{22} \end{bmatrix} \begin{bmatrix} a_1 \\ a_2 \end{bmatrix} \qquad (1.37)$$

where incident waves a_1 and a_2 are given by

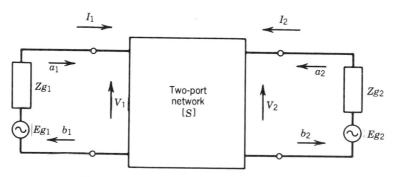

Figure 1.10   Two-port network S parameters.

$$a_1 = \frac{V_1 + Z_{g1} I_1}{2\sqrt{\mathrm{Re}(Z_{g1})}} \qquad (1.38)$$

$$a_2 = \frac{V_2 + Z_{g2} I_2}{2\sqrt{\mathrm{Re}(Z_{g2})}} \qquad (1.39)$$

S_{11}, S_{12}, and so on, are called the S parameters of the two-port network. From expression (1.37), S_{11} can be found as follows:

$$S_{11} = \left.\frac{b_1}{a_1}\right|_{a_2=0} \qquad (1.40)$$

This means that S_{11} is the reflection coefficient at port 1 of the two-port network under the condition that there is not incident power coming from port 2. In other words, S_{11} is the reflection coefficient at port 1 when port 2 is terminated in a matched load. Parameter S_{21} is found as

$$S_{21} = \left.\frac{b_2}{a_1}\right|_{a_2=0} \qquad (1.41)$$

which means that S_{21} is a transmission coefficient of the two-port network under condition $a_2 = 0$. Similarly, S_{22} and S_{12} are the reflection and transmission coefficients at port 2 of the two-port network under the condition that there is a matched termination at port 1.

From the definition it is clear that S parameters can be measured under matched conditions. Using matched resistance terminations to measure active-device S parameters, the unstable problem of active devices under open or short conditions can be avoided. Another advantage of using S parameters is that the S parameters can be measured on a two-port network or actual device located at some distance from the measuring point by using the concept of changing reference plane. This is very important, because in the measurement of a device it is impractical to attach the radio-frequency (RF) connectors to the actual device terminals because of their small size. Hence transmission lines are commonly added at the input and output ports for ease of measurement. Since transmission lines are used extensively in the design of microwave matching networks and other microwave circuits, they are reviewed briefly in the next section.

Similarly to $ABCD$ parameters, another scattering transfer parameter set, the T parameters, are introduced to simplify the calculations of cascaded two-port networks. The T matrix is defined as follows:

$$\begin{bmatrix} a_1 \\ b_1 \end{bmatrix} = \begin{bmatrix} T_{11} & T_{12} \\ T_{21} & T_{22} \end{bmatrix} \begin{bmatrix} b_2' \\ a_2' \end{bmatrix} = \begin{bmatrix} T_{11}^a & T_{12}^a \\ T_{21}^a & T_{22}^a \end{bmatrix} \begin{bmatrix} T_{11}^b & T_{12}^b \\ T_{21}^b & T_{22}^b \end{bmatrix} \begin{bmatrix} b_2' \\ a_2' \end{bmatrix} \qquad (1.42)$$

The relationship between T and S parameters is given by

$$S_{11} = T_{21}/T_{11} \tag{1.43}$$

$$S_{21} = 1/T_{11} \tag{1.44}$$

$$S_{12} = T_{22} - T_{21}T_{12}/T_{11} \tag{1.45}$$

$$S_{22} = -T_{12}/T_{11} \tag{1.46}$$

and

$$T_{11} = 1/S_{21} \tag{1.47}$$

$$T_{12} = -S_{22}/S_{21} \tag{1.48}$$

$$T_{21} = S_{11}/S_{21} \tag{1.49}$$

$$T_{22} = S_{12} - S_{11}S_{22}/S_{21} \tag{1.50}$$

Figure 1.11 shows that when two-port networks are cascaded, the overall T matrix is simply the product of the two individual T matrices. It is also desirable to have conversions among the z, y, h, and S parameters; the two-port network parameters conversions are tabulated in normalized form in Table 1.1.

For the basic two-port network feedback connections pictured in Fig. 1.12, it can be shown [1.5] that a common formula (amenable to a common subroutine) can handle all four cases. The S matrix of the resultant two-port as a function of the S matrices (S_A and S_B) of the interconnected two-ports is determined for the parallel–parallel connection as follows:

$$S_{p-p} = f(S_A, S_B) \tag{1.51a}$$

The same formula can be generalized to express the S matrix in any of the four situations indicated in Fig. 1.12.

$$S = E_k \times f(E_k \times S_A, E_k S_B) \tag{1.51b}$$

where

$$E_k = \begin{bmatrix} c & 0 \\ 0 & d \end{bmatrix} \tag{1.51c}$$

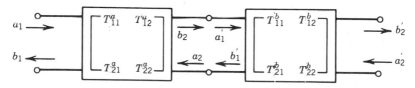

Figure 1.11 Cascaded T matrices.

and the numeric parameters c and d are related to the type of interconnection at the input and output of the composite two-port.

$$c = 1, \quad d = 1, \qquad \text{for parallel–parallel interconnection.}$$

$$c = -1, d = -1, \qquad \text{for series–series interconnection.}$$

$$c = -1, d = 1, \qquad \text{for series–parallel interconnection.}$$

$$c = 1, \quad d = -1, \qquad \text{for parallel–series interconnection.}$$

The matrix function of the two matrix variables, f, appearing in (1.51a) is

$$f(S_A, S_B) = A^{-1} \times \left[B + 4 \times C \times S_B \times (A - B \times S_B)^{-1} \times C \right] \quad (1.51d)$$

where $A = 3 \times I$, $B = S_A - I$, $C = S_A + I$, and I is the unity matrix,

$$I = \begin{bmatrix} 1 & 0 \\ 0 & 1 \end{bmatrix} \qquad (1.51e)$$

An instrument for the measurement of S parameters was introduced by Hewlett-Packard in 1965. The predecessor to this instrument was the Rohde & Schwarz Z-G diagraph, introduced in 1960 [1.6]. Today the S parameters can be measured with a variety of instruments from 1 kHz to 40 GHz using the HP8510, Wiltron 360, or EIP network analyzer. The design of amplifiers and oscillators is based on these parameter sets, for both bipolar and FET circuits. These parameters are best understood by illustrating some examples of the simple bipolar transistor model given in Fig. 1.13. A basic understanding of the parameter sets is given in Table 1.2, which lists the two-port parameters of the transistor in Fig. 1.13 versus frequency.

In Fig. 1.14 the common-emitter S parameters have been plotted up to 10 GHz, showing the capacitive terminations of the transistor because the bonding inductances have not yet been added to this simple model. From Table 1.2 at 100 MHz we see that the common emitter S_{21} is higher than the common base and common collector, which explains why most bipolar transistor amplifiers are common-emitter.

1.1.4 Three-Port Network and n-Port Network S Parameters

A three-port design may be required if a third port is available for tuning or passive termination. An example is the dual-gate GaAs metal semiconductor field-effect transistor (MESFET), which usually operates in the common-source mode with port 1 as gate 1, port 2 as the drain, and port 3 as gate 2. The three-port is identified in Fig. 1.15 on page 28.

TABLE 1.1   Conversions between Two-Port Parameters Normalized to

	S	z
S	$$\begin{bmatrix} b_1 \\ b_2 \end{bmatrix} = \begin{bmatrix} S_{11} & S_{12} \\ S_{21} & S_{22} \end{bmatrix} \begin{bmatrix} a_1 \\ a_2 \end{bmatrix}$$	$S_{11} = \dfrac{(z_{11} - 1)(z_{22} + 1) - z_{12}z_{21}}{(z_{11} + 1)(z_{22} + 1) - z_{12}z_{21}}$ $S_{12} = \dfrac{2z_{12}}{(z_{11} + 1)(z_{22} + 1) - z_{12}z_{21}}$ $S_{21} = \dfrac{2z_{21}}{(z_{11} + 1)(z_{22} + 1) - z_{12}z_{21}}$ $S_{22} = \dfrac{(z_{11} + 1)(z_{22} - 1) - z_{12}z_{21}}{(z_{11} + 1)(z_{22} + 1) - z_{12}z_{21}}$
Z	$z_{11} = \dfrac{(1 + S_{11})(1 - S_{22}) + S_{12}S_{21}}{(1 - S_{11})(1 - S_{22}) - S_{12}S_{21}}$ $z_{12} = \dfrac{2S_{12}}{(1 - S_{11})(1 - S_{22}) - S_{12}S_{21}}$ $z_{21} = \dfrac{2S_{21}}{(1 - S_{11})(1 - S_{22}) - S_{12}S_{21}}$ $z_{22} = \dfrac{(1 - S_{11})(1 + S_{22}) + S_{12}S_{21}}{(1 - S_{11})(1 - S_{22}) - S_{12}S_{21}}$	$$\begin{bmatrix} V_1 \\ V_2 \end{bmatrix} = \begin{bmatrix} z_{11} & z_{12} \\ z_{21} & z_{22} \end{bmatrix} \begin{bmatrix} I_1 \\ I_2 \end{bmatrix}$$
Y	$y_{11} = \dfrac{(1 - S_{11})(1 + S_{22}) + S_{12}S_{21}}{(1 + S_{11})(1 + S_{22}) - S_{12}S_{21}}$ $y_{12} = \dfrac{-2S_{12}}{(1 + S_{11})(1 + S_{22}) - S_{12}S_{21}}$ $y_{21} = \dfrac{-2S_{21}}{(1 + S_{11})(1 + S_{22}) - S_{12}S_{21}}$ $y_{22} = \dfrac{(1 + S_{11})(1 - S_{22}) + S_{12}S_{21}}{(1 + S_{11})(1 + S_{22}) - S_{12}S_{21}}$	$\dfrac{z_{22}}{\Delta^z} \qquad \dfrac{-z_{12}}{\Delta^z}$ $\dfrac{-z_{21}}{\Delta^z} \qquad \dfrac{z_{11}}{\Delta^z}$
H	$h_{11} = \dfrac{(1 + S_{11})(1 + S_{22}) - S_{12}S_{21}}{(1 - S_{11})(1 + S_{22}) + S_{12}S_{21}}$ $h_{12} = \dfrac{2S_{12}}{(1 - S_{11})(1 + S_{22}) + S_{12}S_{21}}$ $h_{21} = \dfrac{-2S_{21}}{(1 - S_{11})(1 + S_{22}) + S_{12}S_{21}}$ $h_{22} = \dfrac{(1 - S_{22})(1 - S_{11}) - S_{12}S_{21}}{(1 - S_{11})(1 + S_{22}) + S_{12}S_{21}}$	$\dfrac{\Delta^z}{z_{22}} \qquad \dfrac{z_{12}}{z_{22}}$ $\dfrac{-z_{21}}{z_{22}} \qquad \dfrac{1}{z_{22}}$
A	$A = \dfrac{(1 + S_{11})(1 - S_{22}) + S_{12}S_{21}}{2S_{21}}$ $B = \dfrac{(1 + S_{11})(1 + S_{22}) - S_{12}S_{21}}{2S_{12}}$ $C = \dfrac{(1 - S_{11})(1 - S_{22}) - S_{12}S_{21}}{2S_{21}}$ $D_{22} = \dfrac{(1 - S_{11})(1 + S_{22}) + S_{12}S_{21}}{2S_{21}}$	$\dfrac{z_{11}}{z_{21}} \qquad \dfrac{\Delta^z}{z_{21}}$ $\dfrac{1}{z_{21}} \qquad \dfrac{z_{22}}{z_{21}}$

$Z_0 = 1$ with $\Delta^K = K_{11}K_{22} - K_{12}K_{21}$

y	h	A	
$S_{11} = \dfrac{(1 - y_{11})(1 + y_{22}) + y_{12}y_{21}}{(1 + y_{11})(1 + y_{22}) - y_{12}y_{21}}$	$S_{11} = \dfrac{(h_{11} - 1)(h_{22} + 1) - h_{12}h_{21}}{(h_{11} + 1)(h_{22} + 1) - h_{12}h_{21}}$	$\dfrac{A + B - C - D}{A + B + C + D}$	$\dfrac{2(AD - BC)}{A + B + C + D}$
$S_{12} = \dfrac{-2y_{12}}{(1 + y_{11})(1 + y_{22}) - y_{12}y_{21}}$	$S_{12} = \dfrac{2h_{12}}{(h_{11} + 1)(h_{22} + 1) - h_{12}h_{21}}$		
$S_{21} = \dfrac{-2y_{21}}{(1 + y_{11})(1 + y_{22}) - y_{12}y_{21}}$	$S_{21} = \dfrac{-2h_{21}}{(h_{11} + 1)(h_{22} - 1) - h_{12}h_{21}}$		
$S_{22} = \dfrac{(1 + y_{11})(1 - y_{22}) + y_{12}y_{21}}{(1 + y_{11})(1 + y_{22}) - y_{12}y_{21}}$	$S_{22} = \dfrac{(1 + h_{11})(1 - h_{22}) + h_{12}h_{21}}{(h_{11} + 1)(h_{22} + 1) - h_{12}h_{21}}$	$\dfrac{2}{A + B + C + D}$	$\dfrac{-A + B - C + D}{A + B + C + D}$
$\dfrac{y_{22}}{\Delta^y} \qquad \dfrac{-y_{12}}{\Delta^y}$ $\dfrac{-y_{21}}{\Delta^y} \qquad \dfrac{y_{11}}{\Delta^y}$	$\dfrac{\Delta^h}{h_{22}} \qquad \dfrac{h_{12}}{h_{22}}$ $\dfrac{-h_{12}}{h_{22}} \qquad \dfrac{1}{h_{22}}$	$\dfrac{A}{C} \qquad \dfrac{\Delta^A}{C}$ $\dfrac{1}{C} \qquad \dfrac{D}{C}$	
$\begin{bmatrix} I_1 \\ I_2 \end{bmatrix} = \begin{bmatrix} y_{11} & y_{12} \\ y_{21} & y_{22} \end{bmatrix}\begin{bmatrix} V_1 \\ V_2 \end{bmatrix}$	$\dfrac{1}{h_{11}} \qquad \dfrac{-h_{12}}{h_{11}}$ $\dfrac{h_{21}}{h_{11}} \qquad \dfrac{\Delta^h}{h_{11}}$	$\dfrac{D}{B} \qquad \dfrac{-\Delta^A}{B}$ $\dfrac{-1}{B} \qquad \dfrac{A}{B}$	
$\dfrac{1}{y_{11}} \qquad \dfrac{-y_{12}}{y_{11}}$ $\dfrac{y_{21}}{y_{11}} \qquad \dfrac{\Delta^y}{y_{11}}$	$\begin{bmatrix} V_1 \\ I_2 \end{bmatrix} = \begin{bmatrix} h_{11} & h_{12} \\ h_{21} & h_{22} \end{bmatrix}\begin{bmatrix} I_1 \\ V_2 \end{bmatrix}$	$\dfrac{B}{D} \qquad \dfrac{\Delta^A}{D}$ $\dfrac{-1}{D} \qquad \dfrac{C}{D}$	
$\dfrac{-y_{22}}{y_{21}} \qquad \dfrac{-1}{y_{21}}$ $\dfrac{-\Delta^y}{y_{21}} \qquad \dfrac{-y_{11}}{y_{21}}$	$\dfrac{-\Delta^h}{h_{21}} \qquad \dfrac{-h_{11}}{h_{21}}$ $\dfrac{-h_{22}}{h_{21}} \qquad \dfrac{-1}{h_{21}}$	$\begin{bmatrix} V_1 \\ I_1 \end{bmatrix} = \begin{bmatrix} A & B \\ C & D \end{bmatrix}\begin{bmatrix} V_2 \\ -I_2 \end{bmatrix}$	

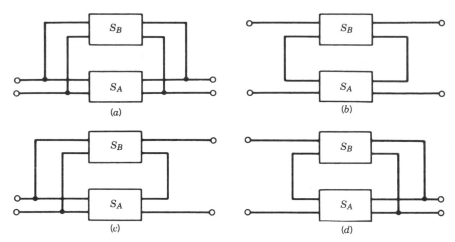

Figure 1.12 Four basic two-port feedback connections: (*a*) parallel–parallel; (*b*) series–series; (*c*) parallel–series; (*d*) series–parallel.

The transistor must be described by a three-port set of *S* parameters given by

$$b_1 = S_{11}a_1 + S_{12}a_2 + S_{13}a_3 \tag{1.52}$$

$$b_2 = S_{21}a_1 + S_{22}a_2 + S_{23}a_3 \tag{1.53}$$

$$b_3 = S_{31}a_1 + S_{32}a_2 + S_{33}a_3 \tag{1.54}$$

Since the termination at each port is given by

$$a_1 = b_1\Gamma_G \tag{1.55}$$

$$a_2 = b_2\Gamma_L \tag{1.56}$$

$$a_3 = b_3\Gamma_3 \tag{1.57}$$

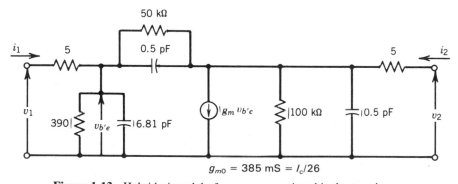

Figure 1.13 Hybrid-pi model of a common-emitter bipolar transistor.

TABLE 1.2 Two-Port Parameters of a Bipolar Transistor (Fig. 1.13) versus Frequency

Common-Emitter S Parameters

Frequency (Hz)	S_{11}/M S_{21}/M	S_{11}/DE S_{21}/DE	S_{12}/M S_{22}/M	S_{12}/DE S_{22}/DE
1	0.742	-619×10^{-9}	0.00171	8.69×10^{-6}
	32.9	180	0.961	-364×10^{-9}
1×10^3	0.742	-619×10^{-6}	0.00171	0.00869
	32.9	180	0.961	-364×10^{-6}
1×10^6	0.742	-0.619	0.00173	8.61
	32.9	180	0.961	-0.364
10×10^6	0.742	-6.19	0.00319	54.4
	32.9	177	0.960	-3.63
100×10^6	0.745	-56.8	0.0238	57.5
	29.0	151	0.854	-33.8
1×10^9	0.754	-158	0.0490	3.48
	5.99	92.7	0.308	-115
10×10^9	0.792	-176	0.0325	-48.7
	0.404	29.5	0.643	-147

Common-Emitter h Parameters

Frequency (Hz)	h_{11}/M h_{21}/M	h_{11}/DE h_{21}/DE	h_{12}/M h_{22}/M	h_{12}/DE h_{22}/DE
1	386	-1.05×10^{-6}	0.00763	7.93×10^{-6}
	147	-1.07×10^{-6}	0.00296	7.97×10^{-6}
1×10^3	386	-0.00105	0.00763	0.00793
	147	-0.00107	0.00296	0.00797
1×10^6	386	-1.05	0.00772	7.86
	147	-1.07	0.00300	7.90
10×10^6	380	-10.4	0.0140	47.0
	144	-10.6	0.00543	47.2
100×10^6	183	-60.3	0.0568	24.5
	69.4	-62.0	0.0221	25.6
1×10^9	21.4	-73.0	0.0643	1.28
	7.8	-89.5	0.0255	12.5
10×10^9	5.54	-22.1	0.0625	-13.8
	0.776	-116	0.0542	49.7

Common-Emitter z Parameters

Frequency (Hz)	z_{11}/M z_{21}/M	z_{11}/DE z_{21}/DE	z_{12}/M z_{22}/M	z_{12}/DE z_{22}/DE
1	8.86	1.29×10^{-6}	2.57	-38.2×10^{-9}
	49.5×10^3	180	338	-7.97×10^{-6}

TABLE 1.2   (*Continued*)

Common-Emitter *z* Parameters

Frequency (Hz)	z_{11}/M z_{21}/M	z_{11}/DE z_{21}/DE	z_{12}/M z_{22}/M	z_{12}/DE z_{22}/DE
1×10^3	8.86 49.5×10^3	0.00129 180	2.57 338	-38.2×10^{-6} -0.00797
1×10^6	8.89 48.9×10^3	1.25 171	2.57 334	-0.0376 -7.90
10×10^6	9.79 26.5×10^3	3.32 122	2.57 184	-0.191 -47.2
10×10^6	9.79 26.5	3.32 122	2.57 184	-0.191 -47.2
100×10^6	10.1 3.14×10^3	-0.122 92.4	2.57 45.2	-1.15 -25.6
1×10^9	9.99 308	-5.60 77.9	2.52 39.2	-11.3 -12.5
10×10^9	6.38 14.3	-18.9 14.8	1.15 18.5	-63.5 -49.7

Common-Emitter *y* Parameters

Frequency (Hz)	y_{11}/M y_{21}/M	y_{11}/DE y_{21}/DE	y_{12}/M y_{22}/M	y_{12}/DE y_{22}/DE
1	0.00259 0.380	1.05×10^{-6} -16.8×10^{-9}	19.7×10^{-6} 67.9×10^{-6}	-180 10.31×10^{-6}
1×10^3	0.00259 0.380	0.00105 -16.8	19.7×10^{-6} 67.9×10^{-6}	-180 0.0103
1×10^6	0.00259 0.380	1.05 -0.0168	20.0×10^{-6} 69.0×10^{-6}	-171 10.2
10×10^6	0.00263 0.380	10.4 -0.168	36.8×10^{-6} 140×10^{-6}	-123 60.9
100×10^6	0.00547 0.379	60.3 -1.68	310×10^{-6} 0.00123	-95.2 85.8
1×10^9	0.0468 0.368	73.0 -16.5	0.00301 0.0119	-106 80.0
10×10^9	0.180 0.140	22.1 -93.4	0.0113 0.062	-172 52.9

Common-Base *S* Parameters

Frequency (Hz)	S_{11}/M S_{21}/M	S_{11}/DE S_{21}/DE	S_{12}/M S_{22}/M	S_{12}/DE S_{22}/DE
1	0.900 1.89	180×10^{-9} -17.5×10^{-9}	239×10^{-6} 0.998	10.9×10^{-6} -20.7×10^{-9}

TABLE 1.2 *(Continued)*

Common-Base S Parameters

Frequency (Hz)	S_{11}/M S_{21}/M	S_{11}/DE S_{21}/DE	S_{12}/M S_{22}/M	S_{12}/DE S_{22}/DE
1×10^3	0.900 1.89	180 -17.5×10^{-6}	239×10^{-6} 0.998	0.0109 -20.7×10^{-6}
1×10^6	0.900 1.89	180 -0.0175	244×10^{-6} 0.998	10.7 -0.0207
10×10^6	0.900 1.89	180 -0.175	513×10^{-6} 0.998	62.1 -0.207
100×10^6	0.900 1.89	180 -1.75	0.00455 0.998	85.9 -2.07
1×10^9	0.887 1.86	177 -17.6	0.0453 1.01	78.9 -20.7
10×10^9	0.764 0.301	-177 -123	0.174 0.854	10.1 -139

Common-Base h Parameters

Frequency (Hz)	h_{11}/M h_{21}/M	h_{11}/DE h_{21}/DE	h_{12}/M h_{22}/M	h_{12}/DE h_{22}/DE
1	2.62 0.993	8.17×10^{-9} 180	126×10^{-6} 20.1×10^{-6}	10.9×10^{-6} 9.03×10^{-6}
1×10^3	2.62 0.993	8.17×10^{-6} 180	126×10^{-6} 20.1×10^{-6}	0.0109 0.00903
1×10^6	2.62 0.993	0.00817 180	128×10^{-6} 20.3×10^{-6}	10.7 8.96
10×10^6	2.62 0.993	0.0817 180	270×10^{-6} 37.4×10^{-6}	62.2 57.6
100×10^6	2.62 0.993	0.817 179	0.00240 317×10^{-6}	87.0 86.8
1×10^9	2.69 0.988	7.88 173	0.0241 0.0032	89.8 93.5
10×10^9	5.35 0.507	7.34 115	0.294 0.0523	68.5 79.2

Common-Base z Parameters

Frequency (Hz)	z_{11}/M z_{21}/M	z_{11}/DE z_{21}/DE	z_{12}/M z_{22}/M	z_{12}/DE z_{22}/DE
1	8.86 49.5×10^3	1.29×10^{-6} -9.04×10^{-6}	6.29 49.8×10^3	1.83×10^{-6} -9.03×10^{-6}
1×10^3	8.86 49.5×10^3	0.00129 -0.00904	6.29 49.8×10^3	0.00183 -0.00903

TABLE 1.2 *(Continued)*

Common-Base *z* Parameters

Frequency (Hz)	z_{11}/M z_{21}/M	z_{11}/DE z_{21}/DE	z_{12}/M z_{22}/M	z_{12}/DE z_{22}/DE
1×10^6	8.89 48.9×10^3	1.25 -8.96	6.32 49.2×10^3	1.78 -8.96
10×10^6	9.79 26.5×10^3	3.32 -57.7	7.22 26.7×10^3	4.56 -57.6
100×10^6	10.1 3.14×10^3	-0.122 87.4	7.56 3.16×10^3	0.227 -86.8
1×10^9	9.99 307	-5.60 -100	7.49 311	-3.70 -93.5
10×10^9	6.38 9.69	-18.9 -144	5.61 19.1	-10.6 -79.2

Common-Base *y* Parameters

Frequency (Hz)	y_{11}/M y_{21}/M	y_{11}/DE y_{21}/DE	y_{12}/M y_{22}/M	y_{12}/DE y_{22}/DE
1	0.382 0.380	-8.17×10^{-9} 180	48.2×10^{-6} 67.9×10^{-6}	-180 10.3×10^{-6}
1×10^3	0.382 0.380	-8.17×10^{-6} 180	48.2×10^{-6} 67.9×10^{-6}	-180 0.0103
1×10^6	0.382 0.380	-0.00817 180	49.1×10^{-6} 69.0×10^{-6}	-169 10.2
10×10^6	0.382 0.380	-0.0817 180	103×10^{-6} 140×10^{-6}	-118 60.9
100×10^6	0.382 0.379	-0.817 179	915×10^{-6} 0.00123	-93.8 85.8
1×10^9	0.372 0.367	-7.88 165	0.00895 0.0119	-98.1 80.0
10×10^9	0.187 0.095	-7.34 108	0.548 0.0623	-119 52.9

Common-Collector *S* Parameters

Frequency (Hz)	S_{11}/M S_{21}/M	S_{11}/DE S_{21}/DE	S_{12}/M S_{22}/M	S_{12}/DE S_{22}/DE
1	0.985 1.89	-30.9×10^{-9} -17.1×10^{-9}	0.0127 0.888	983×10^{-9} 180
1×10^3	0.985 1.89	-30.9×10^{-6} -17.1×10^{-6}	0.0127 0.888	983×10^{-6} 180
1×10^6	0.985 1.89	-0.0309 -0.0171	0.0127 0.888	0.983 180

TABLE 1.2 *(Continued)*

Common-Collector *S* Parameters

Frequency (Hz)	S_{11}/M S_{21}/M	S_{11}/DE S_{21}/DE	S_{12}/M S_{22}/M	S_{12}/DE S_{22}/DE
10×10^6	0.985 1.89	-0.309 -0.171	0.0129 0.888	9.73 180
100×10^6	0.985 1.89	-3.09 -1.71	0.0256 0.888	58.1 179
1×10^9	0.979 1.81	-30.3 -16.7	0.214 0.832	65.5 166
10×10^9	0.777 0.477	-146 -68.4	0.464 0.550	-21.1 -156

Common-Collector *h* Parameters

Frequency (Hz)	h_{11}/M h_{21}/M	h_{11}/DE h_{21}/DE	h_{12}/M h_{22}/M	h_{12}/DE h_{22}/DE
1	386 148	-1.05×10^{-6} 180	0.992 0.00296	-61.0×10^{-9} 7.97×10^{-6}
1×10^3	386 148	-0.00105 180	0.992 0.00296	-61.0×10^{-6} 0.00797
1×10^6	386 148	-1.05 179	0.992 0.00300	-0.0609 7.90
10×10^6	380 145	-10.4 170	0.991 0.00543	-0.590 47.2
100×10^6	183 69.9	-60.3 119	0.949 0.0221	-1.42 25.6
1×10^9	21.4 7.94	-73.0 97.7	0.936 0.0255	-0.0881 12.5
10×10^9	5.54 0.966	-22.1 134	0.939 0.0542	0.906 49.7

Common-Collector *z* Parameters

Frequency (Hz)	z_{11}/M z_{21}/M	z_{11}/DE z_{21}/DE	z_{12}/M z_{22}/M	z_{12}/DE z_{22}/DE
1	49.8×10^3 49.8×10^3	-9.03×10^{-5} -9.03×10^{-6}	335 338	-8.03×10^{-6} -7.97×10^{-6}
1×10^3	49.8×10^3 49.8×10^3	-0.00903 -0.00903	335 338	-0.00803 -0.00797
1×10^6	49.2×10^3 49.2×10^3	-8.96 -8.96	331 334	-7.96 -7.90
10×10^6	26.7×10^3 26.7×10^3	-57.6 -57.6	182 184	-47.7 -47.2

TABLE 1.2 *(Continued)*

Common-Collector *z* Parameters

Frequency (Hz)	z_{11}/M z_{21}/M	z_{11}/DE z_{21}/DE	z_{12}/M z_{22}/M	z_{12}/DE z_{22}/DE
100×10^6	3.16×10^3 3.16×10^3	-86.8 -86.9	42.8 45.2	-27.0 -25.6
1×10^9	311 311	-93.5 -94.8	36.7 39.2	-12.6 -12.5
10×10^9	19.1 17.8	-79.2 -96.2	17.3 18.5	-48.8 -49.7

Common-Collector *y* Parameters

Frequency (Hz)	y_{11}/M y_{21}/M	y_{11}/DE y_{21}/DE	y_{12}/M y_{22}/M	y_{12}/DE y_{22}/DE
1	0.00259 0.382	1.05×10^{-6} 180	0.00257 0.382	-180 -8.17×10^{-9}
1×10^3	0.00259 0.382	0.00105 180	0.00257 0.382	-180 -8.17×10^{-6}
1×10^6	0.00259 0.382	1.05 180	0.00257 0.382	-179 -0.00817
10×10^6	0.00263 0.382	10.4 180	0.00261 0.382	-170 -0.0817
100×10^6	0.00547 0.382	60.3 179	0.00518 0.382	-121 -0.817
1×10^9	0.468 0.372	73.0 171	0.0438 0.372	-107 -7.88
10×10^9	0.180 0.174	22.1 156	0.170 0.187	-157 -7.34

a straightforward calculation will give

$$S'_{11} = \frac{b_1}{a_1} = S_{11} + S_{12}\frac{S_{21}\Gamma_L + \Gamma_L\Gamma_3(S_{23}S_{31} - S_{21}S_{33})}{(1 - S_{22}\Gamma_L)(1 - S_{33}\Gamma_3) - S_{23}S_{32}\Gamma_3\Gamma_L}$$
$$+ S_{13}\frac{S_{31}\Gamma_3 + \Gamma_L\Gamma_3(S_{32}S_{21} - S_{31}S_{22})}{(1 - S_{22}\Gamma_L)(1 - S_{33}\Gamma_3) - S_{23}S_{32}\Gamma_3\Gamma_L} \qquad (1.58)$$

$$S'_{22} = \frac{b_2}{a_2} = S_{22} + S_{21}\frac{S_{12}\Gamma_G + \Gamma_G\Gamma_3(S_{13}S_{32} - S_{33}S_{12})}{(1 - S_{11}\Gamma_G)(1 - S_{33}\Gamma_3) - S_{13}S_{31}\Gamma_3\Gamma_G}$$
$$+ S_{23}\frac{S_{32}\Gamma_3 + \Gamma_G\Gamma_3(S_{31}S_{12} - S_{11}S_{32})}{(1 - S_{11}\Gamma_G)(1 - S_{33}\Gamma_3) - S_{13}S_{31}\Gamma_3\Gamma_G} \qquad (1.59)$$

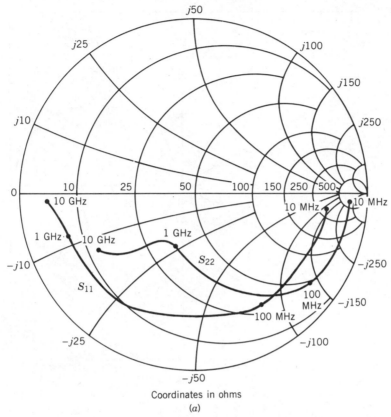

Figure 1.14 Polar plot of common-emitter S parameters versus frequency (a) S_{11} and S_{22}, on Smith chart coordinates ($Z_0 = 50 \ \Omega$): Coordinates are in ohms. (b) S_{12} on polar coordinates. (c) S_{21} on polar coordinates.

$$S'_{33} = \frac{b_3}{a_3} = S_{33} + S_{31} \frac{S_{13}\Gamma_G + \Gamma_L\Gamma_G(S_{12}S_{23} - S_{22}S_{13})}{(1 - S_{11}\Gamma_G)(1 - S_{22}\Gamma_L) - S_{12}S_{21}\Gamma_L\Gamma_G}$$

$$+ S_{32} \frac{S_{23}\Gamma_L + \Gamma_L\Gamma_G(S_{21}S_{13} - S_{11}S_{23})}{(1 - S_{11}\Gamma_G)(1 - S_{22}\Gamma_L) - S_{12}S_{21}\Gamma_L\Gamma_G} \tag{1.60}$$

Because of the mathematical complexity, the three-port is usually reduced to a two-port by terminating port 3. For dual-gate amplifiers an RF short circuit at port 3 is often chosen for a two-port stability factor greater than unity,

$$\Gamma_3 = -1 \tag{1.61}$$

and for a simple RF realization with a capacitor. The two-port S parameters of this network can be derived from (1.52)–(1.54). Alternatively, an RF short could be

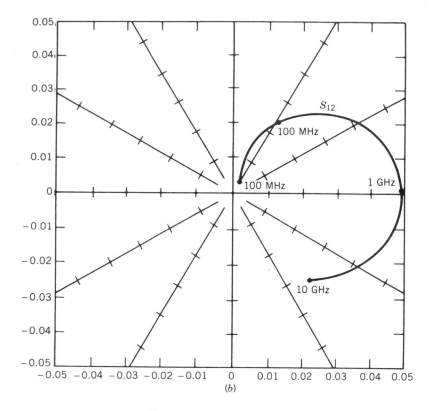

Figure 1.14 (*Continued*)

connected to gate 2 and the two-port *S* parameters of the transistor could be measured.

The effect of the third-port termination can be investigated if $\Gamma_L = 0$. This reduces (1.58) to

$$S'_{11} = S_{11} + \frac{S_{13}S_{31}\Gamma_3}{1 - S_{33}\Gamma_3} \qquad (1.62)$$

Similarly, if $\Gamma_G = 0$, (1.59) reduces to

$$S'_{22} = S_{22} + \frac{S_{23}S_{32}\Gamma_3}{1 - S_{33}\Gamma_3} \qquad (1.63)$$

In the more general case of arbitrary choice of Γ_G, Γ_L, and Γ_3, a large number of interesting amplifier and oscillator designs are possible.

For an *n*-port network with port number $i = 1, 2, \ldots, n$, as shown in Fig. 1.16, the power waves may be defined as follows:

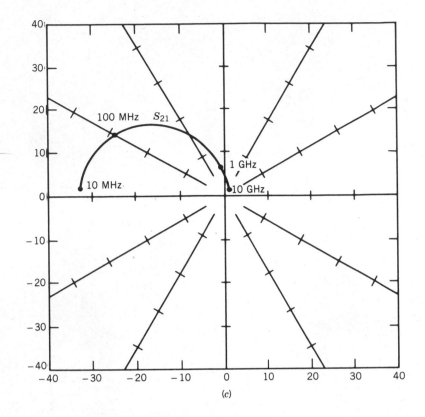

(c)

Figure 1.14   (*Continued*)

$$a_i = \frac{V_i + Z_{gi} I_i}{2\sqrt{\text{Re}(Z_{gi})}} \tag{1.64}$$

$$b_i = \frac{V_i - Z_{gi}^* I_i}{2\sqrt{\text{Re}(Z_{gi})}} \tag{1.65}$$

where V_i and I_i are the terminal voltage and current at port i, respectively, and Z_{gi} is the complex impedance connected to port i. For a linear network the reflection power waves b_i are linearly related to the incident power a_i as given by

$$[B] = [S][A] \tag{1.66}$$

where $[A] = [a_1, a_2, a_3, \ldots, a_{n-1}, a_n]^T$ is the incident power wave vector, and $[B] = [b_1, b_2, b_3, \ldots, b_{n-1}, b_n]^T$ is the reflected power wave vector. The superscript T indicates a transpose operation.

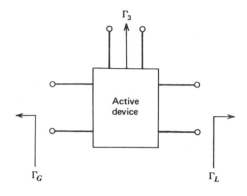

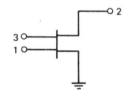

Figure 1.15   Three-port networks.

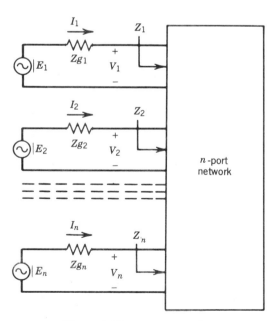

Figure 1.16   *n*-port network.

$$[S] = \begin{bmatrix} S_{11} & S_{12} & S_{13} & \cdots & S_{1n} \\ S_{21} & S_{22} & S_{23} & \cdots & S_{2n} \\ \vdots & \vdots & \vdots & \vdots & \vdots \\ S_{n1} & S_{n2} & S_{n3} & \cdots & S_{nn} \end{bmatrix} \tag{1.67}$$

is the S matrix of the n-port network. Each element of $[S]$ depends on the property of the network and the exterior connected impedances as well. This matrix, called the generalized scattering matrix, relates the reflected waves with the incident waves defined from the exterior complex generator impedances. When Z_{gi} is a real number and equals the characteristic impedance of the transmission line connected to port i, this generalized scattering matrix becomes the common scattering matrix.

1.2 TRANSMISSION LINES

Transmission lines are the basic passive elements in microwave circuit design. They can be used to transmit microwave signals from one point to another; or to construct directional couplers, filters, power dividers, and so on. Transmission lines are also used extensively for network matching in the design of microwave amplifiers, oscillators, and mixers. Some actual transmission lines are shown in Fig. 1.17; the symbol for a transmission line is shown in Fig. 1.18.

1.2.1 Basic Equations

At low frequencies, where the length of a transmission line is much smaller than the signal wavelength, the voltage and current along a transmission line can be considered as constants. At high frequency, the signal wavelength is comparable to the physical dimension. The voltage and current vary along the line because of the effects of the distributed inductance and capacitance of the lines. A circuit model for an incremental length of line can be used to describe the distributed character of the transmission line and is shown in Fig. 1.19.

For sinusoidal excitation, the time variation factor $e^{j\omega t}$ is dropped for simplicity in the following discussion. The voltage and current along a transmission line are functions of position and satisfy the differential equations

$$\frac{dV}{dx} = IZ \tag{1.68}$$

$$\frac{dI}{dx} = VY \tag{1.69}$$

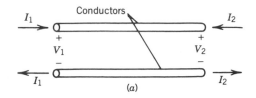

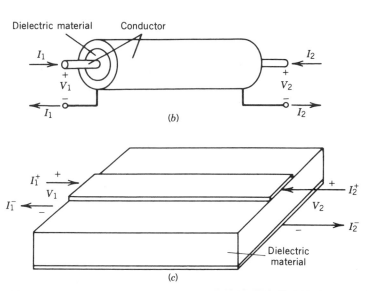

Figure 1.17   Some actual transmission lines: (*a*) parallel line; (*b*) coaxial line; (*c*) microstrip line.

where $Z = r + j\omega l$ and $Y = g + j\omega c$ are the series impedance and shunt suscep-tance per unit length of the transmission line. The parameters l, c, r, and g are the inductance, capacitance, resistance, and conductance per unit length of the trans-mission line. The wave equations are obtained by eliminating I or V:

$$\frac{dV^2}{dx^2} - \gamma^2 V = 0 \qquad (1.70)$$

$$\frac{dI^2}{dx^2} - \gamma^2 I = 0 \qquad (1.71)$$

where γ is the complex propagation constant and is given by

$$\gamma = \sqrt{ZY} = \sqrt{(r + j\omega l)(g + j\omega c)} = \alpha + j\beta \qquad (1.72)$$

The attenuation constant α is given in nepers per unit length and the phase constant β, in radians per unit length. The general solutions of (1.70) and (1.71) are

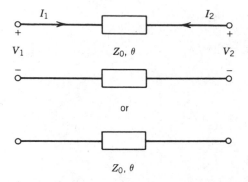

Figure 1.18 Transmission-line symbols.

$$V = Ae^{-\gamma x} + Be^{\gamma x} \tag{1.73}$$

$$I = \frac{A}{Z_0} e^{-\gamma x} - \frac{B}{Z_0} e^{\gamma x} \tag{1.74}$$

where

$$Z_0 = \sqrt{\frac{r + j\omega l}{g + j\omega c}} = \frac{V^+}{I^+} = \frac{V^-}{I^-} \tag{1.75}$$

is the complex characteristic impedance of the transmission line.

Equations (1.73) and (1.74) represent the voltage and current along the transmission line as a superposition of waves traveling in opposite directions. The term $Ae^{-\gamma x} = V^+$ represents the incident voltage wave and $Be^{\gamma x} = V^-$, the reflected voltage wave. At microwave frequencies, r and g are usually much smaller than ωl and ωc, and can be neglected. In this case the amplitudes of the voltage and current do not change with position and the transmission line is called lossless. For a lossless transmission line the characteristic impedance is

$$Z_0 = \sqrt{l/c} \tag{1.76}$$

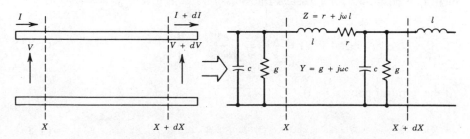

Figure 1.19 Distributed transmission-line model.

and is real; the phase constant is expressed by

$$\beta = \omega \sqrt{lc} = 2\pi / \lambda_g \tag{1.77}$$

The phase velocity is given by

$$v_p = 1/\sqrt{lc} = \omega/\beta \tag{1.78}$$

and the group velocity is

$$v_g = d\omega/d\beta \tag{1.79}$$

The group delay of a signal is given by

$$\tau = \frac{d\theta}{df} = \frac{d\beta}{df} d = \frac{2\pi}{v_g} d \tag{1.80}$$

For a signal with no dispersion,

$$v_p = v_g = c/\sqrt{\epsilon_{eff}} \tag{1.81}$$

where ϵ_{eff} is the effective dielectric constant of the line and c is the velocity of light. For example, the group delay at 1 GHz for a 10-cm length of air line is calculated as follows:

$$v_g = c = 3 \times 10^{10} \text{ cm/s}$$

$$\tau = \frac{2\pi \times 10}{3 \times 10^{10}} = 2.09 \times 10^{-9} \text{ s} = 2.09 \text{ ns}$$

Similar to (1.17), the ratio of the reflected voltage wave and the incident voltage wave is defined as the reflection coefficient. Figure 1.20 represents a transmission line of characteristic impedance Z_0 terminated in a load Z_L. The reflection coefficient Γ at position x for a lossless transmission line is given by

$$\Gamma(x) = \frac{V^-}{V^+} = \frac{Be^{j\beta x}}{Ae^{-j\beta x}} = \frac{B}{A} e^{j2\beta x} \tag{1.82}$$

At the terminal port ($x = 0$)

$$\Gamma_L = B/A \tag{1.83}$$

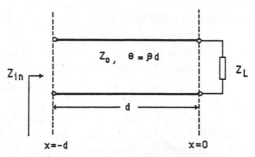

Figure 1.20 Transmission line terminated in a load Z_L.

where Γ_L is called the load reflection coefficient. At the input port $(x = -d)$,

$$\Gamma = \frac{B}{A} e^{-j2\beta d} = \frac{B}{A} e^{-j2\theta} = \Gamma_L e^{-j2\theta} \qquad (1.84)$$

where $\theta = \beta d$ is called the electrical length of the transmission line. The input impedance of the transmission line is given by

$$Z_{\text{in}} = \frac{V}{I} = Z_0 \frac{e^{j\theta} + \Gamma_L e^{-j\theta}}{e^{j\theta} - \Gamma_L e^{-j\theta}} \qquad (1.85)$$

At $x = 0$ the input impedance is equal to Z_L and (1.85) becomes

$$Z_L = Z_0 \frac{1 + \Gamma_L}{1 - \Gamma_L} \qquad (1.86)$$

or

$$\Gamma_L = \frac{Z_L - Z_0}{Z_L + Z_0} \qquad (1.87)$$

Substituting (1.87) into (1.85) gives

$$Z_{\text{in}} = Z_0 \frac{Z_L + jZ_0 \tan \beta d}{Z_0 + jZ_L \tan \beta d} \qquad (1.88)$$

Equations (1.87) and (1.88) are the most useful formulas in the design of transmission-line networks. From these equations, the following important conclusions are obtained immediately:

1. When load impedance Z_L equals Z_0, the characteristic impedance of the transmission line, the load reflection coefficient $\Gamma_L = 0$, and the input

impedance equals the characteristic impedance of the transmission line, namely $Z_{in} = Z_0$.

2. For an open transmission line $\Gamma_L = 1$, $Z_{in} = -jZ_0 \cot \theta$. Under the condition $\theta < \pi/2$, the behavior of the input impedance is like that of a capacitance. Hence a short open-circuit transmission line can be used as a capacitance element.

3. For a shorted transmission line $\Gamma_L = -1$, $Z_{in} = jZ_0 \tan \theta$. Under the condition $\theta < \pi/2$, the behavior of the input impedance is like that of an inductance. Hence a short short-circuit transmission line can be used as an inductance element.

4. When electrical length $\theta = \pi/2$ (of physical length $l = \lambda/4$), the transmission line is called a quarter-wave transformer. The quarter-wave transformer has the following important property:

$$Z_{in} = Z_0^2/Z_L \tag{1.89}$$

Equation (1.89) shows that a real impedance can be changed by a factor Z_0^2 if a quarter-wavelength transmission line is used. This property is often used as a microwave matching technique.

The most useful two-port matrices for a matched lossless transmission line with electrical length θ are the S matrix and the $ABCD$ matrix. They are given as follows:
S matrix:

$$\begin{bmatrix} b_1 \\ b_2 \end{bmatrix} = \begin{bmatrix} 0 & e^{-j\theta} \\ e^{-j\theta} & 0 \end{bmatrix} \begin{bmatrix} a_1 \\ a_2 \end{bmatrix} \tag{1.90}$$

$ABCD$ matrix:

$$\begin{bmatrix} v_1 \\ i_1 \end{bmatrix} = \begin{bmatrix} \cos \theta & jZ_0 \sin \theta \\ j \sin \theta/Z_0 & \cos \theta \end{bmatrix} \begin{bmatrix} v_2 \\ -i_2 \end{bmatrix} \tag{1.91}$$

Another useful concept in transmission-line theory is that of the voltage standing-wave ratio (VSWR). The two waves traveling in opposite directions in the transmission line form a standing-wave pattern. In some places these two waves are in phase and produce the maximum amplitude $V_{max} = |A| + |B|$; in other places the two waves are out of phase and the standing wave has a minimum amplitude, $V_{min} = |A| - |B|$. The ratio of V_{max} to V_{min} is called the voltage standing-wave ratio (VSWR). The relationship between VSWR and the reflection coefficient is as follows:

$$VSWR = \frac{V_{max}(x)}{V_{min}(x)} = \frac{1 + |\Gamma_L|}{1 - |\Gamma_L|} \qquad (1.92)$$

or

$$|\Gamma_L| = \frac{VSWR - 1}{VSWR + 1} \qquad (1.93)$$

The spacing between two adjacent minimum values (or maximum values) equals the half-wavelength; hence the standing-wave pattern can be used to measure the signal frequency.

1.2.2 Shift in Reference Planes

As we mentioned before, in actual measurement it is impractical to attach the RF connectors to the actual device terminals because of their small size. Usually, the 50-Ω transmission lines are used in between the actual measuring point and the device terminals are shown in Figure 1.21. The S matrices of the 50-Ω transmission lines are represented by S_1 and S_2:

$$S_1 = \begin{bmatrix} 0 & e^{-j\theta_1} \\ e^{-j\theta_1} & 0 \end{bmatrix} \qquad (1.94)$$

$$S_2 = \begin{bmatrix} 0 & e^{-j\theta_2} \\ e^{-j\theta_2} & 0 \end{bmatrix} \qquad (1.95)$$

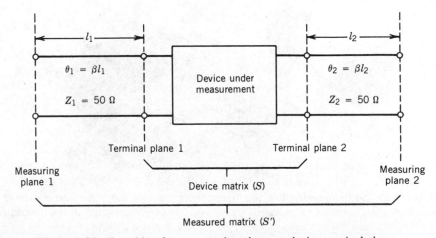

Figure 1.21 Transition from measuring planes to device terminal planes.

The *S* matrix of the device can be represented by S_1 and S_2 and the measured matrix S' as follows:

$$\begin{bmatrix} S_{11} & S_{12} \\ S_{21} & S_{22} \end{bmatrix} = \begin{bmatrix} S'_{11}e^{j2\theta_1} & S'_{12}e^{j(\theta_1+\theta_2)} \\ S'_{21}e^{j(\theta_2+\theta_1)} & S'_{22}e^{j2\theta_2} \end{bmatrix} \tag{1.97}$$

Some discussion of this result is appropriate for the application of *S*-parameter data. Usually, the device under test (DUT) is embedded in a 50-Ω test fixture using either coaxial or microstrip transmission lines. The simplest correction to find the actual device data is the reference plane correction given by (1.97); more sophisticated deembedding methods are necessary for frequencies above 6 GHz. These techniques include through-delay [1.7], through-reflect-line [1.8–1.12], and many variations of these calibration procedures.

A typical microstrip-line test fixture is shown in Fig. 1.22 for a packaged transistor. This test fixture uses through-delay calibration for deembedding the effect of the transition from the data. The substrate is alumina ($\epsilon_r = 10$) with a thickness of 10, 15, or 25 mils. The substrates use a constant length of 330 mils, giving a through length of 660 mils. At present the delay lines are 1.160, 0.760,

Figure 1.22   50-Ω microstrip-line test fixture with through and delay sections for *S*-parameter measurements of packaged transistors. (Photo courtesy of Inter-Continental Microwave, Santa Clara, Calif.)

and 0.730 in. for a full-frequency coverage of 400 MHz to 26 GHz. These lengths have been selected to give a phase shift of about 20° to 160° over the 0.4–4 GHz, 2–18 GHz, and 4–26 GHz bands, respectively. This will give about 360° rotation on the Smith chart near the upper frequency of each band. All the Avantek microstrip packaged transistors and chips, which are described in Chapter 3, have been measured with this type of test fixture.

1.2.3 Microstrip Lines

One of the important transmission lines is the microstrip line, used extensively in microwave integrated circuit designs because they are easily fabricated and connect easily to active devices. The basic microstrip-line geometry and field pattern are shown in Fig. 1.23. This transmission line may be visualized as a distorted coaxial line, with the top strip as the center conductor and the ground plane transformed to a flat plane. The electrical behavior is determined by the distributed inductance and capacitance of the line (l and c). The important physical and electrical parameters are defined in Table 1.3 for microstrip line. The important physical parameters are line width w, substrate thickness h, and substrate dielectric constant ϵ_r. The important electrical parameters for microstrip line design are the characteristic impedance Z_0, the guide wavelength λ_g, and the attenuation constant α. In the following we use an effective dielectric constant to express these parameters.

The transmission mode is assumed to be quasi-TEM, that is, both the electric field and magnetic field in the direction of propagation are negligible compared to the TEM fields. The electrical properties of the microstrip line can be calculated by finding the effective dielectric constant of the transmission line. A portion of the electric field lines is in the dielectric, and a few fringe electric field lines occur in the air. The distribution of the electric field lines is described by the filling factor q, which approaches unity when all the electric field lines are in the substrate. The

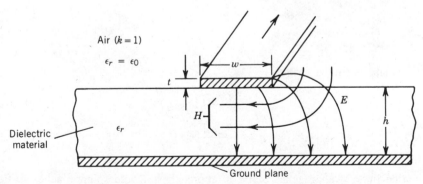

Figure 1.23 Microstrip-line geometry (w = width; h = height).

TABLE 1.3 Microstrip-Line Parameters

Physical		Electrical	
w	line width	Z_0	characteristic impedance
h	substrate thickness	α	attenuation constant
ϵ_r	substrate dielectric constant		(nepers/length)
t	metal thickness	β	phase constant (radians/length)
σ_m	metal conductivity	v_{ph}	phase velocity
σ_d	dielectric conductivity	ϵ_r'	effective dielectric constant
		λ_g	guide wavelength
		q	filling factor
		r	resistance per unit length
		l	inductance per unit length
		g	conductivity per unit length
		c	capacitance per unit length

effective or microstrip dielectric constant ϵ_r' is related to the filling factor q by [1.13]

$$\epsilon_r' = q\epsilon_r + (1 - q)\epsilon_0 \qquad (1.98)$$

$$q = \frac{\epsilon_r' - \epsilon_0}{\epsilon_r - \epsilon_0} \leq 1 \qquad (1.99)$$

If we define a dimensionless dielectric constant by

$$k = \epsilon_r/\epsilon_0 \qquad (1.100)$$

$$k' = \epsilon_r'/\epsilon_0 \qquad (1.101)$$

these equations can be written

$$k' = qk + (1 - q) \qquad (1.102)$$

$$q = \frac{k' - 1}{k - 1} \leq 1 \qquad (1.103)$$

For wide lines, q approaches unity, ϵ_r' approaches ϵ_r, and k' approaches k. For narrow lines, q approaches $\frac{1}{2}$, ϵ_r' is the average of ϵ_r and ϵ_0, and k' is the average of k and unity.

The effective dielectric constant reduces the phase velocity and reduces the characteristic impedance according to the following relations:

$$v_{ph} = \frac{1}{\sqrt{lc}} = \frac{c}{\sqrt{k'}} \qquad (1.104)$$

$$Z_0 = \sqrt{\frac{l}{c}} = \frac{Z_{0a}}{\sqrt{k'}} \qquad (1.105)$$

where c is the velocity of light and Z_{0a} is the characteristic impedance for an air dielectric. The velocity of the wave is reduced by the square root of the effective dielectric constant, but not as slowly as if all the wave is in the dielectric. The characteristic impedance is reduced by the increased capacitance of the line compared to the air line. The guide wavelength is also reduced according to the relations

$$\lambda_0 = \frac{c}{f} \qquad (1.106)$$

$$\lambda_g = \frac{v_{ph}}{f} = \frac{c}{f\sqrt{k'}} \qquad (1.107)$$

There are no simple expressions for k' and Z_0, but some approximation formulas can be found in the literature [1.14]. The solutions for the microstrip-line characteristic impedance and effective dielectric constant are given in Figs. 1.24 and 1.25. These are basic design curves for microstrip-line low-loss passive circuits. Also, Compact Software, EEsof, and other microwave software companies have software for various transmission lines, including microstrip-line design.

Another important parameter for the design of microstrip-line circuits is the attenuation factor. The effect of losses can be found at high frequencies by expanding (1.72) as follows (assuming that $rg \ll \omega^2 lc$ for low-loss lines):

$$\alpha + j\beta = \sqrt{(rg - \omega^2 lc) + j\omega(lg + rc)}$$

$$\simeq j\omega\sqrt{lc}\left[1 + \frac{j\omega(lg + rc)}{-\omega^2 lc}\right]^{1/2} \qquad (1.108)$$

Using the binomial expansion theorem,

$$(1 + x)^{1/2} = 1 + \frac{1}{2}x + \frac{1/2(-1/2)}{2!}x^2 + \cdots \qquad (1.109)$$

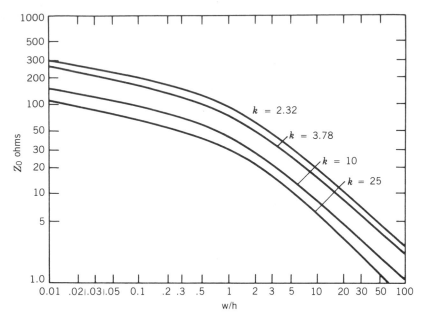

Figure 1.24 Microstrip-line characteristic impedance versus width/height (w/h) and dielectric constant.

gives

$$\alpha + j\beta \simeq j\omega\sqrt{lc}\left[1 + \frac{1}{2}\frac{j\omega(lg + rc)}{-\omega^2 lc} - \cdots\right]$$

$$= j\omega\sqrt{lc} + \frac{lg + rc}{2\sqrt{lc}} \tag{1.110}$$

Thus the attenuation constant consists of two terms that can be associated with conductor losses and dielectric losses (in units of nepers per length):

$$\alpha = \alpha_c + \alpha_d = \frac{r}{2}\sqrt{\frac{c}{l}} + \frac{g}{2}\sqrt{\frac{l}{c}}$$

$$= \frac{r}{2Z_0} + \frac{gZ_0}{2} \tag{1.111}$$

The conductor losses are usually dominant for narrow lines. In units of decibels per unit length, they can be calculated approximately by the formula

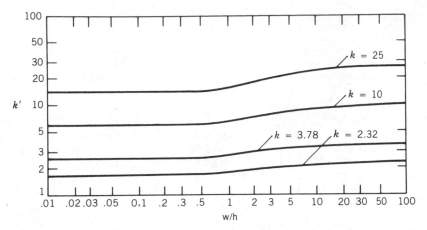

Figure 1.25 Effective dielectric constant versus width/height (w/h) and dielectric constant.

$$\alpha_c = \frac{8.68\sqrt{\pi\mu\rho}}{Z_0 w}\sqrt{f} \qquad (1.112)$$

where μ is permeability and ρ is the resistivity of the conductor. The approximation formula for dielectric loss in units of decibels per unit length is

$$\alpha_d = \frac{4.34q}{\sqrt{k'}}\sqrt{\frac{\mu_0}{\epsilon_0}}\,\sigma_d \qquad (1.113)$$

where σ_d is the conductivity of the dielectric substrate. These data are often given by the loss tangent of the material, where

$$\tan\delta_d = \frac{\sigma_d}{\omega\epsilon_r} \qquad (1.114)$$

is a parameter normally assumed to be a constant with frequency and is the loss tangent of the substrate. It is usually small except for semiconductor substrates such as silicon [1.15]. In practice, the measured attenuation of microstrip line is higher than predicted by (1.112) because of the surface roughness, which increases the surface resistance of the metal [1.16].

The important substrate materials for microstrip line are summarized in Table 1.4. The most common choice is alumina because of its low cost, but surface roughness is also an important practical consideration.

TABLE 1.4 Microstrip-Line Dielectrics

Material	$k = \epsilon_r/\epsilon_0$	Comment
Alumina Al_2O_3	9.6–10	Low cost
Sapphire single crystal, Al_2O_3	9.9	Expensive and anisotropic
	11.6	
Fused silica (quartz)	3.78	Fragile
Beryllium oxide, BeO	6.0	High thermal conductivity
Duroid	2.32	Low cost
Teflon–fiberglass	2.56	Low cost
GaAs ($\rho = 10^7\ \Omega \cdot$ cm)	12.8	Fragile
Silicon ($\rho = 10^3\ \Omega \cdot$ cm)	11.7	Low resistivity

The Q of microstrip line is calculated from

$$Q = \beta/2\alpha \tag{1.115}$$

Using 1 np = 8.686 dB gives

$$Q = \frac{2\pi}{\lambda_g}\frac{1}{2\alpha} = \frac{8.686\pi}{\alpha\lambda_g}\ \text{dB}$$

$$= \frac{27.3}{\alpha}\frac{\text{dB}}{\lambda_g} \tag{1.116}$$

Thus the Q of microstrip-line resonators will increase with frequency, since α in decibels per guide wavelength is decreasing. The dielectric loss and radiation losses should also be included for completeness. The dielectric loss is normally negligible except for semiconductor substrates such as silicon, in which the conductivity of the substrate is significant. The Q of an open-circuited resonator is thus modified to include conductor loss, dielectric loss, and radiation loss by [1.17–1.19]

$$\frac{1}{Q_t} = \frac{1}{Q_c} + \frac{1}{Q_d} + \frac{1}{Q_r} \tag{1.117}$$

The following example illustrates the use of microstrip lines in matching networks. A more complete discussion of matching is given in Section 3.4.

Example 1.1: Microstrip Matching Design input- and output-matching networks for the amplifier shown in Fig. 1.26, where the reflection coefficients for match in a 50-Ω system are $\Gamma_G = 0.3 \angle -50°$ and $\Gamma_L = 0.4 \angle 170°$.

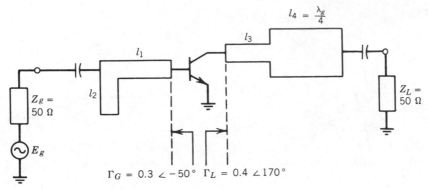

Figure 1.26 Amplifier schematic (Example 1.1).

SOLUTION

INPUT-MATCHING NETWORK DESIGN. To design the input-matching network, we locate Γ_G (point A) on the Y Smith chart shown in Fig. 1.27a. A transmission line of length $l_1 = 0.222\lambda_g (0.072 + 0.15 = 0.222)$ is obtained by moving from A to C (in the direction toward the load), and then using an open-circuited shunt stub of length $l_2 = 0.092\lambda_g$ to meet the match point. The characteristic impedance of the input network microstrip lines is 50 Ω.

OUTPUT-MATCHING NETWORK DESIGN. We locate Γ_L (point A) in the Z Smith chart in Fig. 1.27b. We move from point A toward the load along the constant Γ circle, and meet the axis of line $jX/Z_0 = 0$ at point B; a transmission line l_3 of length $0.014\lambda_g$ is thus obtained. At point B, $R/Z_0 = 0.45$, we have the characteristic impedance 50 Ω, so

$$R = 50(0.45) = 22.5$$

Choosing a quarter-wave transformer to transform 22.5 to 50, we should determine its characteristic impedance Z:

$$Z = \sqrt{50(22.5)} = 33.6 \ \Omega$$

This matching problem can also be solved by matching Γ_G^* and Γ_L^* to the center of the Smith chart by the techniques discussed in Chapter 3 (see Section 3.4). If a conjugate power match is created at the transistor terminal planes and lossless elements are used for matching, a conjugate power match also exists at the generator and load ports.

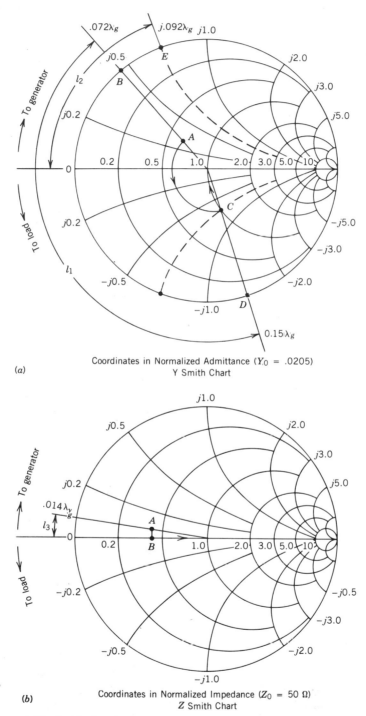

Coordinates in Normalized Admittance (Y_0 = .0205)
Y Smith Chart

(a)

Coordinates in Normalized Impedance (Z_0 = 50 Ω)
Z Smith Chart

(b)

Figure 1.27   Smith chart impedance matching: (*a*) input match; (*b*) output match.

1.3 USE OF *S* PARAMETERS WITH AMPLIFIERS

1.3.1 Stability

The amplifier design problem (Fig. 1.1) is very simple when the two-port transistor is unconditionally stable. In this case a simultaneous conjugate match at the input and output ports will deliver the maximum power to the load and therefore will produce the highest power gain.

First, we should obtain the expression for input and output reflection coefficients, Γ_1 and Γ_2 in Fig. 1.28 by setting $Z_{01} = Z_{02} = Z_0$ and $l_1 = l_2 = 0$ for

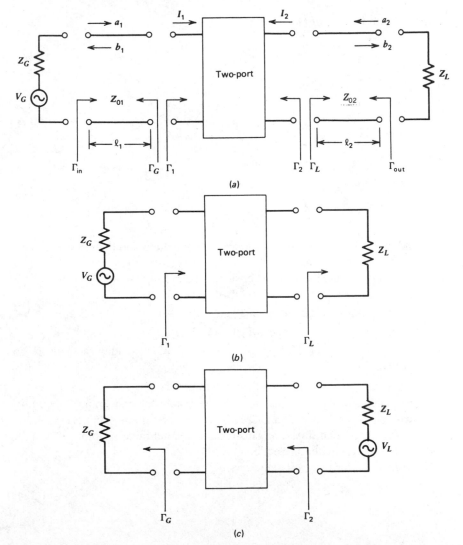

Figure 1.28 Description of a two-port: (*a*) two-port with generator at port 1 and load at port 2; (*b*) measurement of Γ_1; (*c*) measurement of Γ_2.

convenience. There is only a phase shift associated with the length of the input and output lossless transmission lines, and this can be dropped temporarily. With the use of (1.37) and

$$\Gamma_L = a_2/b_2 \tag{1.118}$$

the input reflection coefficient Γ_1 is

$$\Gamma_1 = \frac{b_1}{a_1} = S_{11} + \frac{S_{12}S_{21}\Gamma_L}{1 - S_{22}\Gamma_L} = S_{11}' \tag{1.119}$$

Similarly, with the use of (1.37) and

$$\Gamma_G = a_1/b_1 \tag{1.120}$$

the output reflection coefficient Γ_2 is

$$\Gamma_2 = \frac{b_2}{a_2} = S_{22} + \frac{S_{12}S_{21}\Gamma_L}{1 - S_{11}\Gamma_G} = S_{22}' \tag{1.121}$$

When $\Gamma_L = \Gamma_G = 0$, these reduce to

$$\Gamma_1 = S_{11} \tag{1.122}$$

$$\Gamma_2 = S_{22} \tag{1.123}$$

The input and output reflection coefficients are functions of all four *S* parameters and the terminations Γ_L and Γ_G.

The question of stability can be considered from three points of view:

1. In the Γ_L plane, what values of Γ_L give $|S_{11}'| > 1$?
2. In the S_{11}' plane, where does $|\Gamma_L| = 1$ plot?
3. If $(S_{11}')^* = \Gamma_G$ and $(S_{22}')^* = \Gamma_L$, the resistors terminating the network are positive.

All of these points of view will lead to the same conclusion concerning the stability of the two-port. If conditions 1 and 2 both give unconditional stability, the necessary and sufficient conditions are satisfied for unconditional stability. The third condition will be discussed in Section 1.3.2.

The conditions for two-port stability are

$$|S_{11}'| < 1 \tag{1.124}$$

$$|S_{22}'| < 1 \tag{1.125}$$

for all possible load terminations with a positive real part (i.e., positive resistors). If either of the conditions in (1.124) and (1.125) can be violated, the two-port is only conditionally stable. The condition $|S'_{11}| > 1$ or $|S'_{22}| > 1$ is equivalent to a negative resistance at that port, which is needed for oscillator design.

From previous considerations [Equations (1.119) and (1.121)] the values of S'_{11} and S'_{22} are

$$S'_{11} = S_{11} + \frac{S_{12}S_{21}\Gamma_L}{1 - S_{22}\Gamma_L} = \frac{S_{11} - D\Gamma_L}{1 - S_{22}\Gamma_L} \qquad (1.126)$$

$$S'_{22} = S_{22} + \frac{S_{12}S_{21}\Gamma_G}{1 - S_{11}\Gamma_G} = \frac{S_{22} - D\Gamma_G}{1 - S_{11}\Gamma_G} \qquad (1.127)$$

$$D = S_{11}S_{22} - S_{12}S_{21} \qquad (1.128)$$

For unconditional stability, all values of Γ_L must ensure that the input reflection coefficient S'_{11} is less than unity, and all values of Γ_G must ensure that the output reflection coefficient S'_{22} is less than unity.

Consider the mapping of $|S'_{11}| = 1$ in the Γ_L plane, as shown in Fig. 1.29. Using the following substitutions in (1.126) will give the conditions for stability:

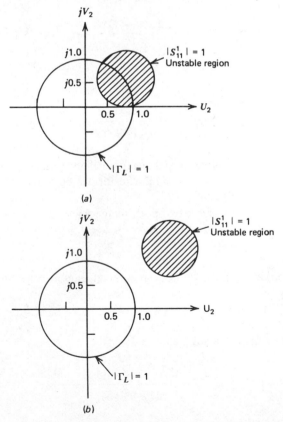

(a)

(b)

Figure 1.29 Stability viewed in the Γ_L plane: (a) conditionally stable; (b) unconditionally stable.

$$S_{11} = S_{11R} + jS_{11I} \tag{1.129}$$

$$S_{22} = S_{22R} + jS_{22I} \tag{1.130}$$

$$D = D_R + jD_I \tag{1.131}$$

$$\Gamma_L = U_2 + jV_2 \tag{1.132}$$

From (1.124) and (1.126) the boundary for stability is

$$|S_{11} - D\Gamma_L| < |1 - S_{22}\Gamma_L| \tag{1.133}$$

Substituting (1.129)–(1.132) and squaring both sides gives the following circle:

$$(U_2 - U_{2G})^2 + (V_2 - V_{2G})^2 = r_G^2 \tag{1.134}$$

where the center is

$$C_G = U_{2G} + jV_{2G}$$

$$= \frac{(S_{22} - DS_{11}^*)^*}{|S_{22}|^2 - |D|^2} \tag{1.135}$$

and the radius is

$$r_G = \frac{|S_{12}S_{21}|}{||S_{22}|^2 - |D|^2|} \tag{1.136}$$

The inside of this circle is the unstable region if the point $U_2 + jV_2 = 0$ gives stability, that is, if $|S_{11}'| < 1$. A similar result can be found in the Γ_G plane for $S_{22}' = 1$ by using (1.125) and (1.127).

For an unconditionally stable two-port, the geometry of Fig. 1.28b is satisfied:

$$|C_G| - r_G > 1 \tag{1.137}$$

which finally produces the result (a derivation appears in Appendix A)

$$1 - |S_{11}|^2 - |S_{22}|^2 + |D|^2 > 2|S_{12}||S_{21}| \tag{1.138}$$

If we define the stability factor k as

$$k = \frac{1 - |S_{11}|^2 - |S_{22}|^2 + |D|^2}{2|S_{12}||S_{21}|} \tag{1.139}$$

a stability factor greater than unity is required for unconditional stability.

For the second view of unconditional stability, plot the circle $\Gamma_L = 1$ in the

S'_{11} plane, as shown in Fig. 1.30. This again produces a circle with radius and center given by

$$C_L = S_{11} + \frac{S_{12}S_{21}S_{22}^*}{1 - |S_{22}|^2} \qquad (1.140)$$

$$r_L = \frac{|S_{12}S_{21}|}{1 - |S_{22}|^2} \qquad (1.141)$$

Since C_L could be zero,

$$r_L < 1 \qquad (1.142)$$

is required for stability.

$$|S_{12}S_{21}| < 1 - |S_{22}|^2 \qquad (1.143)$$

and similarly for the $\Gamma_G = 1$ circle in the S'_{22} plane:

$$|S_{12}S_{21}| < 1 - |S_{11}|^2 \qquad (1.144)$$

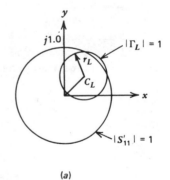

Figure 1.30   Stability viewed in the S'_{11} plane: (a) conditionally stable; (b) unconditionally stable.

In summary, the conditions (necessary and sufficient) for unconditional stability are

$$k = \frac{1 - |S_{11}|^2 - |S_{22}|^2 + |D|^2}{2\,|S_{12}|\,|S_{21}|} > 1 \qquad (1.145)$$

$$|S_{12}S_{21}| < 1 - |S_{11}|^2 \qquad (1.146)$$

$$|S_{12}S_{21}| < 1 - |S_{22}|^2 \qquad (1.147)$$

Since normally $|D| < 1$, one can then show that $k > 1$ is sufficient to guarantee unconditional stability. Otherwise, three conditions are required (necessary and sufficient) for absolute unconditional stability.

The first step in amplifier or oscillator design is to determine the stability factor versus frequency. For the regions where $k < 1$, the stability circles are plotted in the Γ_L plane and in the Γ_G plane. The matching circuits in Fig. 1.1 must be designed to avoid the unstable regions given by stability circles. Normally, the low-frequency range of the transistor will give $k < 1$, so the low-frequency value of Γ_L and Γ_G must be carefully chosen. A short-circuit termination at low frequencies will usually guarantee low-frequency stability.

1.3.2 Simultaneous Conjugate Match

Before deriving the relationships for simultaneous conjugate match, it will be useful to review the general conditions of conjugate power match between a generator and a load. Consider a generator described by Γ_G and a load described by Γ_L connected to a transmission line of Z_0 characteristic impedance (Fig. 1.31). For maximum power transfer one must satisfy both

$$Z_L = Z_G^* \qquad (1.148)$$

and

$$Z_G = Z_0 \qquad (1.149)$$

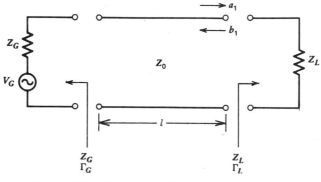

Figure 1.31 Generator delivering power to a load.

Equation (1.148) is well-known conjugate impedance matching, and (1.149) expresses the requirement for the generator to deliver all of its available power to the transmission line. If (1.148) is satisfied, then (for $l = 0$)

$$\Gamma_L = \Gamma_G^* \tag{1.150}$$

is an equivalent way of expressing maximum power transfer. Alternatively, if a conjugate power match exists, then at any point on a transmission line the generator and load reflection coefficients are conjugate. The conjugate power match for a lossless line can be expressed by

$$\Gamma_G = (\Gamma_L e^{-j2\beta l})^* \tag{1.151}$$

The mismatch loss is an important concept in the design of amplifiers. Normally, the generator is matched to the transmission line, so that $\Gamma_G = 0$. In this case the conjugate mismatch loss is defined by the ratio of available power from the generator to the power delivered to the load:

$$M_c = \frac{P_A}{P_L} = \frac{P_A}{P_A - P_{\text{ref}}} = \frac{1}{1 - |\Gamma_L|^2} \geq 1 \tag{1.152}$$

For the more general case of a reflecting generator and a reflecting load, we must know the available power from the generator. If the generator were connected to a nonreflecting load, the amplitude of the wave emitting from the generator would be b_G. In the next section we show that

$$a_1 = b_G + b_1 \Gamma_G \tag{1.153}$$

The net power delivered to the load is

$$P_L = |a_1|^2 - |b_1|^2 = |a_1|^2 (1 - |\Gamma_L|^2) \tag{1.154}$$

Since $b_1 = a_1 \Gamma_L$ in Fig. 1.30, (1.153) gives

$$a_1 = b_G + a_1 \Gamma_L \Gamma_G \tag{1.155}$$

$$a_1 = \frac{b_G}{1 - \Gamma_L \Gamma_G} \tag{1.156}$$

Thus the power to the load may be written

$$P_L = \frac{|b_G|^2 (1 - |\Gamma_L|^2)}{|1 - \Gamma_L \Gamma_G|^2} \tag{1.157}$$

which is a function of b_G, Γ_L, and Γ_G. For a maximum power transfer (1.150) is satisfied, which is now substituted in (1.157) to give the available power from the generator:

$$P_A = \frac{|b_G|^2 (1 - |\Gamma_G^*|^2)}{|1 - \Gamma_G^* \Gamma_G|^2} = \frac{|b_G|^2}{1 - |\Gamma_G|^2} \tag{1.158}$$

Thus the conjugate mismatch loss in the general case is found from (1.157) and (1.158):

$$M_c = \frac{P_A}{P_L} = \frac{|1 - \Gamma_L \Gamma_G|^2}{(1 - |\Gamma_L|^2)(1 - |\Gamma_G|^2)} \geq 1 \tag{1.159}$$

The power delivered to the load may be calculated for this expression if Γ_L, Γ_G, and P_A are known.

Turning from one-port conjugate mismatch loss, the requirements for simultaneous conjugate power match of a two-port will now be derived. For a two-port with the stability factor (k) greater than unity, it is possible to simultaneously conjugately match the two-port to produce the maximum possible gain (G_{ma}). The conditions for simultaneous conjugate match are

$$S_{11}' = S_{11} + \frac{S_{12} S_{21} \Gamma_L}{1 - S_{22} \Gamma_L} = \Gamma_G^* \tag{1.160}$$

and

$$S_{22}' = S_{22} + \frac{S_{12} S_{21} \Gamma_G}{1 - S_{11} \Gamma_G} = \Gamma_L^* \tag{1.161}$$

which become

$$(1 - \Gamma_L S_{22})(S_{11} - \Gamma_G^*) + \Gamma_L S_{12} S_{21} = 0 \tag{1.162}$$

and

$$(1 - \Gamma_G S_{11})(S_{22} - \Gamma_L^*) + \Gamma_G S_{12} S_{21} = 0 \tag{1.163}$$

From (1.162) and (1.163) we find that

$$\Gamma_L = \frac{\Gamma_G^* - S_{11}}{\Gamma_G^* S_{22} - D} \tag{1.164}$$

$$\Gamma_G = \frac{\Gamma_L^* - S_{22}}{\Gamma_L^* S_{11} - D} \tag{1.165}$$

Substituting (1.163) gives

$$\Gamma_G^2 - \Gamma_G \frac{B_1}{C_1} + \frac{C_1^*}{C_1} = 0 \tag{1.166}$$

TABLE 1.5 Four Cases of (1.169)

$B_1 > 0$ Normal Condition		$B_1 < 0$	
Case 1	Case 2	Case 3	Case 4
$\dfrac{B_1}{2\lvert C_1\rvert} > 1$	$\dfrac{B_1}{2\lvert C_1\rvert} < 1$	$\dfrac{B_1}{2\lvert C_1\rvert} > 1$	$\dfrac{B_1}{2\lvert C_1\rvert} < 1$
$k > 1$	$k < 1$	$k > 1$	$k < 1$
Useful solution	$\lvert\Gamma_{Gm}\rvert = 1$	Potentially unstable	$\lvert\Gamma_{Gm}\rvert = 1$
given in	not useful	even though	not useful
(1.170)		$k > 1$	

$$C_1 = S_{11} - DS_{22}^* \tag{1.167}$$

$$B_1 = 1 - \lvert S_{22}\rvert^2 + \lvert S_{11}\rvert^2 - \lvert D\rvert^2 \tag{1.168}$$

The solution for (1.166) is

$$\Gamma_{Gm} = \frac{B_1}{2C_1} \pm \frac{1}{2}\sqrt{\left(\frac{B_1}{C_1}\right)^2 - 4\frac{C_1^*}{C_1}}$$

$$= \frac{C_1^*}{\lvert C_1\rvert}\left[\frac{B_1}{2\lvert C_1\rvert} \pm \sqrt{\frac{B_1^2}{\lvert 2C_1\rvert^2} - 1}\right] \tag{1.169}$$

In examining (1.169) there are four cases to consider, as shown in Table 1.5. Since it can be shown that $B_1/2\lvert C_1\rvert > 1$ corresponds to $k > 1$, this is the only case that can produce a useful solution. This statement is proved in Appendix B.

Also, $B_1 < 0$ can occur only if the two-port is potentially unstable, that is, if (1.145)–(1.147) are not satisfied. Thus the only useful solution from Table 1.5 is case 1, which gives

$$\Gamma_{Gm} = \frac{C_1^*}{\lvert C_1\rvert}\left[\frac{B_1}{2\lvert C_1\rvert} - \sqrt{\frac{B_1^2}{\lvert 2C_1\rvert^2} - 1}\right] \tag{1.170}$$

Notice that for very small S_{12} and small D, $C_1 \simeq S_{11}$.

$$\Gamma_{Gm}\big|_{S_{12}=0} \simeq \frac{S_{11}^*}{\lvert C_1\rvert}\left[\frac{B_1}{2\lvert C_1\rvert} - \sqrt{\frac{B_1^2}{\lvert 2C_1\rvert^2} - 1}\right] \tag{1.171}$$

which shows that the optimum value of Γ_{Gm} may be near S_{11}^*. Note that the angle is essentially the angle of S_{11}^*. A similar result for the load termination is

$$\Gamma_{Lm} = \frac{C_2^*}{\lvert C_2\rvert}\left(\frac{B_2}{2\lvert C_2\rvert} - \sqrt{\frac{B_2^2}{\lvert 2C_2\rvert^2} - 1}\right) \tag{1.172}$$

$$C_2 = S_{22} - DS_{11}^* \qquad (1.173)$$

$$B_2 = 1 - |S_{11}|^2 + |S_{22}|^2 - |D|^2 \qquad (1.174)$$

$$\Gamma_{Lm}\big|_{S_{12}=0} \approx \frac{S_{22}^*}{|C_2|}\left[\frac{B_2}{2|C_2|} - \sqrt{\frac{B_2^2}{|2C_2|^2} - 1}\right] \qquad (1.175)$$

Useful solutions are possible only when

$$\frac{B_1}{|2C_1|} > 1 \qquad (1.176)$$

$$\frac{B_2}{|2C_2|} > 1 \qquad (1.177)$$

which implies that $k > 1$. A proof appears in Appendix B.

1.3.3 Power Gains

When we discuss the power gains of the two-port network, it is important to give the representation of the generator. When a generator or source of power is connected to the two-port, the generator emits a wave b_G if a nonreflecting load is connected ($\Gamma_1 = 0$). In the general case where the load is not matched, consider Fig. 1.32. The first wave incident on the two-port is b_G, which is reflected as $b_G\Gamma_1$, which is reflected as $b_G\Gamma_1\Gamma_G$, and so on. Thus the sum of the reflected waves coming toward the generator is

$$b_1 = b_G\Gamma_1[1 + \Gamma_1\Gamma_G + (\Gamma_1\Gamma_G)^2 + \cdots]$$
$$= \frac{b_G\Gamma_1}{1 - \Gamma_1\Gamma_G} \qquad (1.178)$$

Since $\Gamma_1 = b_1/a_1$, (1.178) becomes

$$b_1 = \frac{b_G b_1}{a_1 - \Gamma_G b_1} \qquad (1.179)$$

$$a_1 = b_G + b_1\Gamma_G \qquad (1.180)$$

This is an important relationship in *S*-parameter analysis. The incident wave on the two-port a_1 is not equal to b_G unless the load is nonreflecting, which would give $b_1 = 0$.

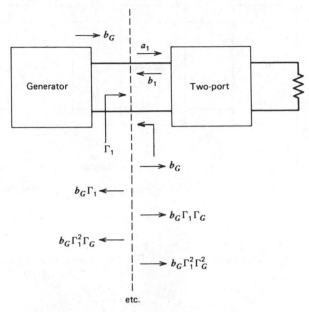

Figure 1.32 Wave reflections at generator port.

An alternative expression for a_1 is given as follows. Since $\Gamma_1 = b_1/a_1$, (1.180) gives

$$a_1 = b_G + \Gamma_1\Gamma_G a_1 \qquad (1.181)$$

$$a_1 = \frac{b_G}{1 - \Gamma_1\Gamma_G} \qquad (1.182)$$

Now we discuss the power gain definitions.

Several definitions of power gains are important in the design of amplifiers. Consider the power flow from generator to load shown in Fig. 1.33, where M_1 and M_2 are lossless matching networks. Three power gains may be defined. The transducer power gain is the most significant, since it shows the insertion effect of the total amplifier. If the amplifier is omitted, $P_A = P_L$ (for a conjugate generator and load impedance). The transducer power gain is defined by

$$G_T = G_T(\Gamma_G, \Gamma_L, S) = \frac{P_L}{P_A} \qquad (1.183)$$

where S is the S-parameter matrix of the two-port. The available power gain is defined by

$$G_A = G_A(\Gamma_G, S) = \frac{P_{\text{avo}}}{P_A} \qquad (1.184)$$

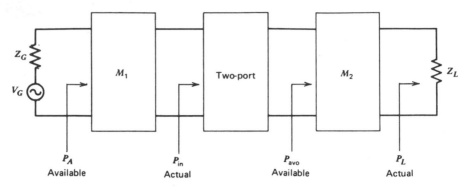

Figure 1.33 Available power and actual power for a two-port connected to a generator and a load.

which is not a function of the load match Γ_L. The power gain is defined by

$$G = G(\Gamma_L, S) = \frac{P_L}{P_{in}} \qquad (1.185)$$

which is not a function of the generator match Γ_G. The transducer gain can only approach the available power gain and the power gain:

$$G_T \leq G_A \qquad \text{where equality occurs for a } \Gamma_L \text{ conjugate match}$$
$$G_T \leq G \qquad \text{where equality occurs for a } \Gamma_G \text{ conjugate match}$$

For a simultaneous conjugate match at both ports, the transducer power gain is the maximum available gain G_{ma} given by (1.217). This is possible only for $k > 1$.

The transducer power gain is derived below. From the definition, G_T gives

$$G_T(\Gamma_G, \Gamma_L, S) = \frac{|b_2|^2}{|b_G|^2} \left(1 - |\Gamma_L|^2\right) \left(1 - |\Gamma_G|^2\right) \qquad (1.186)$$

Using (1.35), (1.36), and (1.182) gives

$$\frac{b_2}{a_1} = \frac{S_{21}}{1 - S_{22}\Gamma_L} \qquad (1.187)$$

$$\frac{a_1}{b_G} = \frac{1}{1 - \Gamma_1\Gamma_G} \qquad (1.188)$$

$$G_T = \frac{\left(1 - |\Gamma_L|^2\right)|S_{21}|^2 \left(1 - |\Gamma_G|^2\right)}{|1 - S_{22}\Gamma_L|^2 |1 - \Gamma_1\Gamma_G|^2} \qquad (1.189)$$

$$\Gamma_1 = S'_{11} = S_{11} + \frac{S_{12} S_{21} \Gamma_L}{1 - S_{22} \Gamma_L} \qquad (1.190)$$

$$G_T = \frac{(1 - |\Gamma_L|^2) |S_{21}|^2 (1 - |\Gamma_G|^2)}{|1 - S_{22} \Gamma_L|^2 |1 - S_{11} \Gamma_G - (S_{12} S_{21} \Gamma_L \Gamma_G)/(1 - S_{22} \Gamma_L)|^2}$$

$$= \frac{(1 - |\Gamma_L|^2) |S_{21}|^2 (1 - |\Gamma_G|^2)}{|(1 - S_{22} \Gamma_L)(1 - S_{11} \Gamma_G) - S_{12} S_{21} \Gamma_L \Gamma_G|^2} \qquad (1.191)$$

This is the exact solution for transducer power gain as a function of Γ_G, Γ_L, and all four S parameters. Comparing (1.189) and (1.191), we can also express G_T as

$$G_T = \frac{(1 - |\Gamma_L|^2) |S_{21}|^2 (1 - |\Gamma_G|^2)}{|1 - \Gamma_2 \Gamma_L|^2 |1 - S_{11} \Gamma_G|^2} \qquad (1.192)$$

$$\Gamma_2 = S'_{22} = S_{22} + \frac{S_{12} S_{21} \Gamma_G}{1 - S_{11} \Gamma_G} \qquad (1.193)$$

The available power gain follows by substituting $\Gamma_L = \Gamma_2^*$ in (1.192). This gives

$$G_A = G_A(\Gamma_G, S) = \frac{(1 - |\Gamma_G|^2) |S_{21}|^2}{|1 - S_{11} \Gamma_G|^2 (1 - |S'_{22}|^2)} \qquad (1.194)$$

which is a function of Γ_G and all four S parameters. The power gain follows by substituting $\Gamma_G = \Gamma_1^*$ in (1.186). This gives

$$G = G(\Gamma_L, S) = \frac{(1 - |\Gamma_L|^2) |S_{21}|^2}{|1 - S_{22} \Gamma_L|^2 (1 - |S'_{11}|^2)} \qquad (1.195)$$

which is a function of Γ_L and all four S parameters.

A useful approximation for transducer power gain is to assume that $S_{12} = 0$ in (1.191) to give G_{TU}, the unilateral transducer power gain:

$$G_{TU} = \frac{(1 - |\Gamma_L|^2) |S_{21}|^2 (1 - |\Gamma_G|^2)}{|1 - S_{22} \Gamma_L|^2 |1 - S_{11} \Gamma_G|^2} \qquad (1.196)$$

The effect of Γ_L mismatch and Γ_G mismatch can be clearly seen in this equation. For an input and output match, $\Gamma_L = S_{22}^*$ and $\Gamma_G = S_{11}^*$, this reduces to

$$G_{TU\text{max}} = \frac{|S_{21}|^2}{(1 - |S_{22}|^2)(1 - |S_{11}|^2)} \qquad (1.197)$$

which is a useful approximation for the maximum gain with simultaneous conjugate match. This form is simple to calculate and shows the importance of input and output mismatch, which is inherent at the input and output ports. If the input transistor mismatch is

$$T_1 = \frac{1}{1 - |S_{11}|^2} \tag{1.198}$$

and the output transistor mismatch is

$$T_2 = \frac{1}{1 - |S_{22}|^2} \tag{1.199}$$

the maximum unilateral transducer power gain is

$$G_{\text{TUmax}} = T_1 |S_{21}|^2 T_2 \tag{1.200}$$

For a two-port that is conditionally stable ($k < 1$), the maximum achievable gain is infinite, but a useful figure of merit is the maximum stable gain G_{ms}, given by

$$G_{ms} = \frac{|S_{21}|}{|S_{12}|} = \frac{|y_{21}|}{|y_{12}|} = \cdots \tag{1.201}$$

This is the gain that can be achieved by resistively loading the two-port such that $k = 1$ and then simultaneously conjugately matching the input and output ports. For conditionally stable two-ports, the maximum stable gain is a helpful limit in power gain that can be approached as the input and output mismatch is reduced. If a simultaneous conjugate match is attempted, the two-port will oscillate if $k < 1$.

A variation of the maximum stable gain is the conjugate stable gain proposed by Rosemarin [1.20]:

$$G_{pm} = 2k \frac{|S_{21}|}{|S_{12}|} \tag{1.202}$$

which may be larger than G_{ms}. The G_{pm} is the maximum stable gain when the input is conjugately matched with lossless elements.

The maximum available gain can be derived by considering the power gain as a function of the Γ_L termination. If the two-port is unconditionally stable, this power gain can be maximized by proper choice of the load termination. Using (1.195) for power gain and expressing S_{11}' as

$$S_{11}' = S_{11} + \frac{S_{12}S_{21}\Gamma_L}{1 - S_{22}\Gamma_L} = \frac{S_{11} - \Gamma_L D}{1 - S_{22}\Gamma_L} \tag{1.203}$$

gives

$$G(\Gamma_L, S) = \frac{|S_{21}|^2 (1 - |\Gamma_L|^2)}{|1 - S_{22}\Gamma_L|^2 - |S_{11} - \Gamma_L D|^2} \qquad (1.204)$$

Expanding the denominator gives

$$G(\Gamma_L, S) = \frac{|S_{21}|^2 (1 - |\Gamma_L|^2)}{(1 - |S_{11}|^2) + |\Gamma_L|^2 (|S_{22}|^2 - |D|^2) - 2\,\mathrm{Re}(\Gamma_L C_2)} \qquad (1.205)$$

$$C_2 = S_{22} - DS_{11}^* \qquad (1.206)$$

The power gain is a function of Γ_L and is expressed as

$$G(\Gamma_L, S) = |S_{21}|^2 g_2 \qquad (1.207)$$

$$\Gamma_L = U_2 + jV_2 \qquad (1.208)$$

$$\mathrm{Re}(\Gamma_L C_2) = \mathrm{Re}(U_2 + jV_2)[\mathrm{Re}(C_2) + j\,\mathrm{Im}(C_2)]$$

$$= U_2\,\mathrm{Re}(C_2) - V_2\,\mathrm{Im}(C_2) \qquad (1.209)$$

Therefore,

$$g_2 = \frac{|1 - U_2^2 - V_2^2|}{|1 - |S_{11}|^2 + (U_2^2 + V_2^2)(|S_{22}|^2 - |D|^2) - 2U_2\,\mathrm{Re}(C_2) + 2V_2\,\mathrm{Im}(C_2)|} \qquad (1.210)$$

This can be expressed as a circle in the Γ_L plane in the form

$$(U_2 - U_{2c})^2 + (V_2 - V_{2c})^2 = \rho_{2c}^2 \qquad (1.211)$$

Expanding (1.210) gives

$$U_2^2 + V_2^2 + 2\,\frac{U_2\,\mathrm{Re}(C_2 g_2)}{-1 - |S_{22}|g_2 + |D|g_2} - 2\,\frac{V_2\,\mathrm{Im}(C_2 g_2)}{-1 - |S_{22}|g_2 + |D|g_2}$$

$$= \frac{-1 + g_2 - |S_{11}|^2 g_2}{-1 - |S_{22}|^2 g_2 + |D|^2 g_2} \qquad (1.212)$$

Adding the terms

$$\frac{\mathrm{Re}(C_2^2 g_2^2) + \mathrm{Im}(C_2^2 g_2^2)}{|-1 - |S_{22}|^2 g_2 + |D|^2 g_2|^2}$$

to both sides of (1.212) gives the radius of the circle, where the following identity is also used:

$$\text{Re}(C_2^2) + \text{Im}(C_2^2) = |S_{22} - DS_{11}^*|^2$$
$$= |S_{12}S_{21}|^2 + (1 - |S_{11}|^2)(|S_{22}|^2 - |D|^2) \qquad (1.213)$$

Regrouping terms as in (1.211) gives

$$\rho_{2c}^2 = \left| \frac{-1 + g_2 - |S_{11}|^2 g_2}{-1 - |S_{22}|^2 g_2 + |D|^2 g_2} + \frac{|S_{22} - DS_{11}^*|^2}{|-1 - |S_{22}|^2 g_2 + |D|^2 g_2|^2} \right|$$

$$= \frac{\left| 1 + g_2\{|S_{22}|^2 + |S_{11}|^2 - |D|^2 - 1\} + g_2^2\{-|S_{22}|^2 + |D|^2 + |S_{11}|^2|S_{22}|^2 - |S_{11}|^2|D|^2 + |S_{12}S_{21}|^2 + |S_{22}|^2 - |D|^2 - |S_{11}|^2|S_{22}|^2 + |S_{11}|^2|D|^2\} \right|}{\left| -1 - |S_{22}|^2 g_2 + |D|^2 g_2 \right|^2}$$

$$= \frac{1 - 2k|S_{12}S_{21}| g_2 + |S_{12}S_{21}|^2 g_2^2}{|-1 - |S_{22}|^2 g_2 + |D|^2 g_2|^2} \qquad (1.214)$$

To maximize the gain, set $\rho_{2c} = 0$ and solve for g_{20}:

$$g_{20}^2 - \frac{2k}{|S_{12}S_{21}|} g_{20} + \frac{1}{|S_{12}S_{21}|^2} = 0 \qquad (1.215)$$

$$g_{20} = \frac{1}{|S_{12}S_{21}|} \left(k \pm \sqrt{k^2 - 1} \right) \qquad (1.216)$$

Thus the maximum available gain is

$$G_{ma} = \left| \frac{S_{21}}{S_{12}} \right| \left(k \pm \sqrt{k^2 - 1} \right) \qquad (1.217)$$

where $k > 1$. The sign in (1.217) is usually negative, since B_1 is usually positive. B_1 was given by (1.168). Although G_{ma} is called the maximum available gain, it is also the maximum power gain of (1.185) and the maximum transducer power gain of (1.191).

One additional power gain is the unilateral power gain [1.21]. This is the maximum available power gain when the two-port has been simultaneously conjugately matched and the feedback parameter has been neutralized to zero. The conditions are indicated in Fig. 1.34 for unilateral power gain. Notice that for $S_{12} = 0$, the stability factor is infinite, but G_{ma} does exist [see (1.217)]. From the general considerations of passivity of a two-port, the network must have $U > 1$

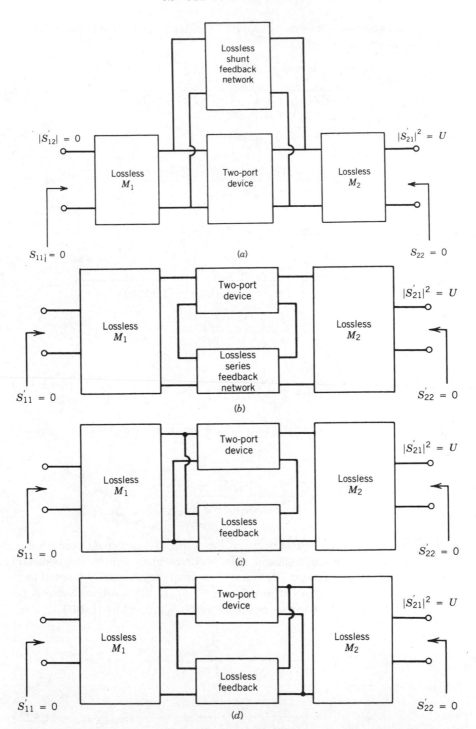

Figure 1.34 Conditions for unilateral gain: (*a*) shunt feedback; (*b*) series feedback; (*c*) parallel–series feedback; (*d*) series–parallel feedback.

for net power flowing out of the two-port. This may be expressed in y or z parameters of the two-port device as

$$U = \frac{|z_{21} - z_{12}|^2}{4\,\mathrm{Re}(z_{11})\,\mathrm{Re}(z_{22}) - 4\,\mathrm{Re}(z_{12})\,\mathrm{Re}(z_{21})} \qquad (1.218)$$

Since the stability factor k is also expressed as

$$k = \frac{2\,\mathrm{Re}(z_{11})\,\mathrm{Re}(z_{22}) - \mathrm{Re}(z_{12}z_{21})}{|z_{12}z_{21}|} \qquad (1.219)$$

the unilateral gain may be expressed as

$$\begin{aligned}
U &= \frac{|z_{21} - z_{12}|^2}{2k\,|z_{12}z_{21}| - 2\,\mathrm{Re}(z_{12})\,\mathrm{Re}(z_{21}) - 2\,\mathrm{Im}(z_{12})\,\mathrm{Im}(z_{21})} \\[2mm]
&= \frac{|z_{21}/z_{12} - 1|^2}{2k\,|z_{21}/z_{12}| - 2\,\mathrm{Re}(z_{12}^* z_{21})/|z_{12}|^2} \\[2mm]
&= \frac{|z_{21}/z_{12} - 1|^2}{2k\,|z_{21}/z_{12}| - 2\,\mathrm{Re}(z_{21}/z_{12})} \qquad (1.220)
\end{aligned}$$

Now using the conversion formula given in Table 1.1, we find that the unilateral gain is

$$U = \frac{|S_{21}/S_{12} - 1|^2}{2k\,|S_{21}/S_{12}| - 2\,\mathrm{Re}(S_{21}/S_{12})} \qquad (1.221)$$

This gain is the highest possible gain that the active two-port could ever achieve. The frequency where the unilateral gain becomes unity defines the boundary between an active and a passive network. This frequency is usually referred to as f_{max}, the maximum frequency of oscillation. If S_{12} of the two-port device is set equal to zero in (1.221), one may again derive $G_{TU\,max}$ given by (1.197).

$$U = \frac{|S_{21} - S_{12}|^2}{2k\,|S_{21}\,S_{12}| - 2\,\mathrm{Re}(S_{21}\,S_{12}^*)}$$

$$U\big|_{S_{12}=0} = \frac{|S_{21}|^2}{[1 - |S_{11}|^2][1 - |S_{22}|^2]} = G_{TU\,max} \qquad (1.222)$$

All of the power gains defined in this chapter can be expressed in decibels. A summary of the power gains appears in Table 1.6. In decibels,

$$U_{dB} = 10 \log U \tag{1.223}$$

However, for the case of a transducer gain given by S_{21}, this is a voltage gain, so it must be squared to give the transducer power gain:

$$G_{TdB} = \mid S_{21} \mid^2_{dB} = 10 \log \mid S_{21} \mid^2$$

$$= 20 \log \mid S_{21} \mid \tag{1.224}$$

The power gains for a typical common-source GaAs MESFET are plotted versus frequency in Fig. 1.35. This figure indicates that the largest gain is always U and that G_{ma} will exist only for $k > 1$.

TABLE 1.6   Nine Power Gains

Transducer power gain in 50-Ω system	$G_T = \mid S_{21} \mid^2$
Transducer power gain for arbitrary Γ_G and Γ_L	$G_T = \dfrac{(1 - \mid \Gamma_G \mid^2) \mid S_{21} \mid^2 (1 - \mid \Gamma_L \mid^2)}{\mid (1 - S_{11} \Gamma_G)(1 - S_{22} \Gamma_L) - S_{12} S_{21} \Gamma_G \Gamma_L \mid^2}$
Unilateral transducer power gain	$G_{TU} = \dfrac{\mid S_{21} \mid^2 (1 - \mid \Gamma_G \mid^2)(1 - \mid \Gamma_L \mid^2)}{\mid 1 - S_{11} \Gamma_G \mid^2 \mid 1 - S_{22} \Gamma_L \mid^2}$
Power gain with input conjugate matched	$G = \dfrac{\mid S_{21} \mid^2 (1 - \mid \Gamma_L \mid^2)}{\mid 1 - S_{22} \Gamma_L \mid^2 (1 - \mid S_{11}' \mid^2)} = \dfrac{\mid S_{21} \mid^2}{1 - \mid S_{11} \mid^2}$ (for $\Gamma_L = 0$)
Available power gain with output conjugate matched	$G_A = \dfrac{\mid S_{21} \mid^2 (1 - \mid \Gamma_G \mid^2)}{\mid 1 - S_{11} \Gamma_G \mid^2 (1 - \mid S_{22}' \mid^2)} = \dfrac{\mid S_{21} \mid^2}{1 - \mid S_{22} \mid^2}$ (for $\Gamma_G = 0$)
Maximum available power gain	$G_{ma} = \left| \dfrac{S_{21}}{S_{12}} \right| (k - \sqrt{k^2 - 1})$
Maximum unilateral transducer power gain	$G_{TUmax} = \dfrac{\mid S_{21} \mid^2}{(1 - \mid S_{11} \mid^2)(1 - \mid S_{22} \mid^2)}$
Maximum stable power gain	$G_{ms} = \dfrac{\mid S_{21} \mid}{\mid S_{12} \mid}$
Unilateral power gain	$U = \dfrac{1/2 \mid S_{21}/S_{12} - 1 \mid^2}{k \mid S_{21}/S_{12} \mid - \text{Re}(S_{21}/S_{12})}$

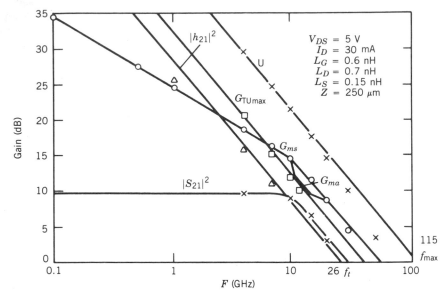

Figure 1.35 Power gains for a GaAs MESFET versus frequency. (AT-10600 from Avantek.)

REFERENCES

1.1 R. A. Hackborn, "An Automatic Network Analyzer System," *Microwave Journal*, May 1968.

1.2 "S-Parameters, Circuit Analysis and Design," *Hewlett-Packard Application Note 95*, September 1968.

1.3 "S-Parameter Design," *Hewlett-Packard Application Note 154*, April 1972.

1.4 "Automating the HP8410B Microwave Network Analyzer," *Hewlett-Packard Application Note 221A*, June 1980.

1.5 P. Bodharamik, L. Besser, and R. W. Newcomb, "Two Scattering Matrix Programs for Active Circuit Analysis," *IEEE Transactions on Circuit Theory*, Vol. CT-18, November 1971, pp. 610–619.

1.6 Rohde & Schwarz, *Instruction Manual for Z-G Diagraph, Models ZGU and ZDU*, 1963.

1.7 "Measurement and Modelling of GaAs FET Chips," *Avantek Application Note*, October 1983.

1.8 N. R. Franzen and R. A. Speciale, "A New Procedure for System Calibration and Error Removal in Automated *S*-Parameter Measurement," *Proceedings of the 5th European Microwave Conference*, September 1975.

1.9 "Specifying Calibration Standards for Use with the HP 8510 Network Analyzer," *Hewlett-Packard Product Note 8510-5*, March 1986.

1.10 H. F. Cooke, "Test and Measurement Notebook: Universal Fixture for Transistor Chip and Microwave Amplifier Measurement," *Microwave Systems News and Communications Technology*, Vol. 17, No. 3, March 1987, pp. 64–70.

1.11 "Applying the HP 8510B TRL Calibration for Non-coaxial Measurements," *Hewlett-Packard Product Note 8510-8*, October 1987.

1.12 J. Curran, "Applying TRL Calibration for Non-coaxial Measurements," *Microwave Systems News and Communications Technology*, Vol. 18, No. 3, March 1988, pp. 91–98.

1.13 H. A. Wheeler, "Transmission-Line Properties of Parallel Strips Separated by a Dielectric Sheet," *IEEE Transactions on Microwave Theory and Techniques*, March 1965, pp. 172–185.

1.14 M. V. Schneider, "Dielectric Loss in Integrated Microwave Circuits," *Bell System Technical Journal*, Vol. 48, No. 7, September 1969, pp. 2325–2332.

1.15 T. M. Hyltin, "Microstrip Transmission on Semiconductor Dielectrics," *IEEE Transactions on Microwave Theory and Techniques*, No. 6, November 1965, pp. 777–781.

1.16 R. A. Pucel, D. J. Masse, and C. P. Hartwig, "Losses in Microstrip," *IEEE Transactions on Microwave Theory and Techniques*, Vol. MTT-16, June 1968, pp. 342–350; and correcting pp. *IEEE Transactions on Microwave Theory and Techniques*, Vol. MTT-16, December 1968, p. 1064.

1.17 E. J. Denlinger, "Radiation from Microstrip Resonators," *IEEE Transactions on Microwave Theory and Techniques*, Vol. MTT-17, No. 4, April 1969, pp. 235–236.

1.18 E. J. Denlinger, "Frequency Dependent Solution for Microstrip Transmission Lines," *IEEE Transactions on Microwave Theory and Techniques*, Vol. MTT-19, No. 1, January 1971, pp. 30–39.

1.19 E. Belohoubek and E. J. Denlinger, "Loss Considerations for Microstrip Resonators," *IEEE Transactions on Microwave Theory and Techniques*, Vol. MTT-23, June 1975, pp. 522–526.

1.20 D. Rosemarin, "A Design Cure for Unstable Transistors," *Microwave & RF*, March 1983, pp. 94–95.

1.21 S. J. Mason, "Power Gain in Feedback Amplifiers," *IRE Transactions on Circuit Theory*, Vol. CT-1, June 1954, pp. 20–25.

BIBLIOGRAPHY

Da Silva, E. F., and M. K. McPhun, "Repeatability of Computer-Corrected Network Analyzer Measurements of Reflection Coefficients," *Electronic Letters*, Vol. 14, No. 25, December 1978, pp. 832–834.

Fitzpatrick, J., "Error Models for Systems Measurements," *Microwave Journal*, May 1978, pp. 63–66.

Gupta, K. C., R. Garg, and I. J. Bahl, *Microstrip Lines and Slotlines*, Artech House, Dedham, Mass., 1979.

Hand, B. P., "Developing Accuracy Specifications for Automatic Network Analyzer Systems," *Hewlett-Packard Journal*, February 1970, pp. 16–19.

Khanna, A. P. S., "Three Port *S*-Parameters Ease GaAs FET Designing," *Microwave & RF*, Vol. 24, November 1985, pp. 81–84.

Schneider, M. V., "Microstrip Dispersion," *Proceedings of the IEEE*, Vol. 60, January 1972, pp. 144–146.

Vendelin, G. D., "Comments on Microstrip vs. Balanced Stripline," *Proc. IEEE*, October 1969, pp. 1778–1780.

Vendelin, G. D., "Limitations on Stripline *Q*," *Microwave Journal*, May 1970, pp. 63–69.

PROBLEMS

1.1. At $f = 4$ GHz calculate the S parameters ($Z_0 = 50\ \Omega$), z parameters, y parameters, h parameters, and $ABCD$ parameters for the following networks.

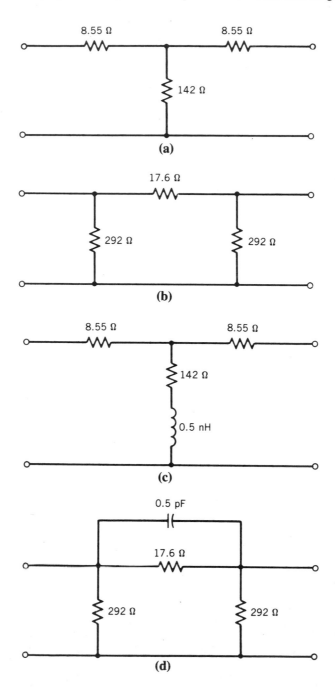

(a)

(b)

(c)

(d)

1.2. Show that the S parameters of a series and shunt element are give as follows:

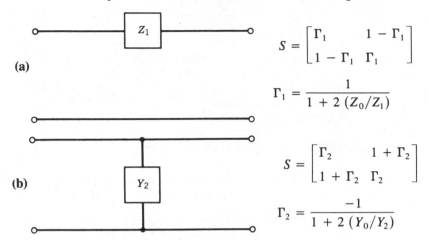

(a)

$$S = \begin{bmatrix} \Gamma_1 & 1 - \Gamma_1 \\ 1 - \Gamma_1 & \Gamma_1 \end{bmatrix}$$

$$\Gamma_1 = \frac{1}{1 + 2(Z_0/Z_1)}$$

(b)

$$S = \begin{bmatrix} \Gamma_2 & 1 + \Gamma_2 \\ 1 + \Gamma_2 & \Gamma_2 \end{bmatrix}$$

$$\Gamma_2 = \frac{-1}{1 + 2(Y_0/Y_2)}$$

1.3. A GaAs MESFET (AT-8251) has the following common-source chip S parameters at $f = 4$ GHz:

$$S_{11} = 0.82 \angle -94°$$

$V_{DS} = 5$ V $\qquad S_{21} = 4.48 \angle 110°$

$I_D = 50$ mA $\qquad S_{12} = 0.068 \angle 52°$

$$S_{22} = 0.46 \angle -20°$$

Calculate k and the following power gains in decibels: $|S_{21}|^2$, G_{ma}, G_{ms}, G_{TUmax}, and U. Estimate f_{max} and f_t (f_t is where $|h_{21}|^2 = 1$).

1.4. Give the three-port S parameters and y parameters for the transistor in Problem 1.3. Calculate the common-gate and common-drain two-port S parameters. What is the stability factor k for these two-ports? Make a power gain table for each of the three two-ports and calculate in decibels: $|S_{21}|^2$, G_{ma}, G_{ms}, G_{TUmax}, and U. Estimate f_{max} and f_t for each of the three two-ports.

1.5. Find the three-port S parameters for the following networks ($Z_0 = 50\ \Omega$).

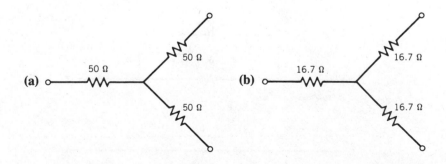

1.6. Show that the *S* parameters of an ideal transformer are given by

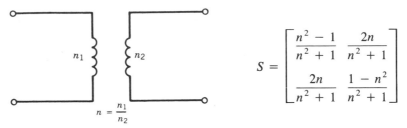

$$S = \begin{bmatrix} \dfrac{n^2 - 1}{n^2 + 1} & \dfrac{2n}{n^2 + 1} \\ \dfrac{2n}{n^2 + 1} & \dfrac{1 - n^2}{n^2 + 1} \end{bmatrix}$$

$$n = \frac{n_1}{n_2}$$

1.7. Find the *ABCD* matrix for the circuits in Problem 1.2.

1.8. Using *Z* parameters, find the new *S* parameters of the following FET circuit at 4 GHz (from Problem 1.3) for $L_s = 0.3$ nH and 1.0 nH.

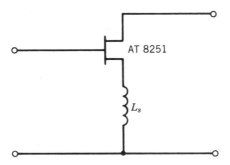

1.9. Find the *z* parameters of this network:

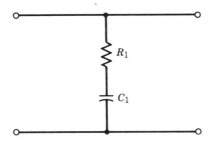

1.10. Find the *y* parameters of this network:

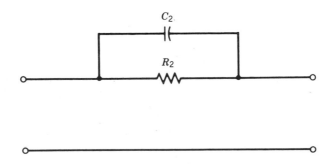

1.11. For the following two-port network, calculate the S parameters (see Problem 1.2); calculate the available power from the generator P_A; and calculate the power dissipated in R_g, R_1, and R_L. Show that $\Sigma P = 0$. Assume that $E_g = 1$ V and $Z_0 = 50\ \Omega$. (*Note:* $P_{GEN} < P_A$.)

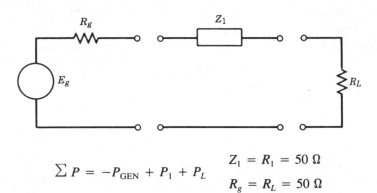

$$\Sigma P = -P_{GEN} + P_1 + P_L$$

$$Z_1 = R_1 = 50\ \Omega$$
$$R_g = R_L = 50\ \Omega$$

What is the value for R_1 for $P_L = P_A$?

1.12. Show that the equation for unilateral gain can be given by

$$U = \frac{\frac{1}{2}\left|\sqrt{S_{21}/S_{12}} - \sqrt{S_{12}/S_{21}}\right|^2}{k - \cos\theta} \qquad \text{where} \quad \theta = \frac{\angle S_{21}}{\angle S_{12}}$$

1.13. Verify that the stability factor k given in (1.219) is equivalent to the stability factor k given by (1.139).

1.14. At $f = 4$ GHz find element values that will make the following transistor (AT-10600 at $V_{DS} = 3$ V, $I_{DS} = 10$ mA) a two-port with $G_{ma} = G_{ms}$ and $k = 1.00$.

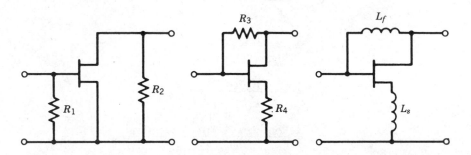

$$S = \begin{bmatrix} 0.93\ \angle -37° & 0.067\ \angle 74° \\ 2.15\ \angle 142° & 0.64\ \angle -9° \end{bmatrix}$$

Calculate U, G_{ms}, and k for the two-port.

1.15. For a lossless two-port, show that $k = 1.0$; also show that

$$|S_{12}|^2 + |S_{22}|^2 = 1$$

$$|S_{11}| = |S_{22}|$$

$$|S_{21}| = |S_{12}|$$

1.16. For a change in reference planes, show that the new S parameters are

$$\begin{bmatrix} S'_{11} & S'_{12} \\ S'_{21} & S'_{22} \end{bmatrix} = \begin{bmatrix} S_{11}e^{-j2\theta_1} & S_{12}e^{-j(\theta_1+\theta_2)} \\ S_{21}e^{-j(\theta_1+\theta_2)} & S_{22}e^{-j2\theta_2} \end{bmatrix}$$

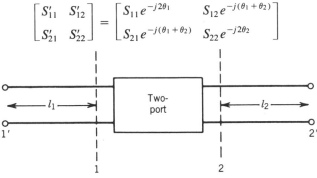

1.17. Derive the scattering parameters of two cascaded networks M and N.

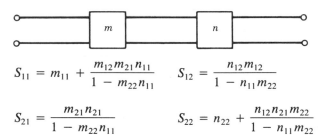

$$S_{11} = m_{11} + \frac{m_{12}m_{21}n_{11}}{1 - m_{22}n_{11}} \qquad S_{12} = \frac{n_{12}m_{12}}{1 - n_{11}m_{22}}$$

$$S_{21} = \frac{m_{21}n_{21}}{1 - m_{22}n_{11}} \qquad S_{22} = n_{22} + \frac{n_{12}n_{21}m_{22}}{1 - n_{11}m_{22}}$$

1.18. For a general cascade of two-ports, show that the y parameters are given by

$$y_{11} = \frac{y_{11A}(y_{22A} + y_{11B}) - y_{21A}y_{12A}}{y_{22A} + y_{11B}}$$

$$y_{12} = -\frac{y_{12A}y_{12B}}{y_{22A} + y_{11B}}$$

$$y_{21} = -\frac{y_{21A}y_{21B}}{y_{22A} + y_{11B}}$$

$$y_{22} = \frac{y_{22B}(y_{22A} + y_{11B}) - y_{21B}y_{12B}}{y_{22A} + y_{11B}}$$

2 Noise in Linear Two-Ports

2.0 INTRODUCTION

In Chapter 1 we learned that a linear two-port requires four complex S-parameters to describe the gain. In this chapter we find that four additional parameters are required to describe the noise; many equivalent representations are described.

Even when a two-port is linear, the output waveform may differ from the input, because of the failure to transmit all spectral components with equal gain (or attenuation) and delay. By careful design of the two-port, or by limitation of the bandwidth of the input waveform, such distortions can largely be avoided. However, noise generated within the two-port can still change the waveform of the output signal. In a linear passive two-port, noise arises only from the losses in the two-port; thermodynamic considerations indicate that such losses result in the random changes that we call noise. When the two-port contains active devices, such as transistors, there are other noise mechanisms that are present. A very important consideration in a system is the amount of noise that it adds to the transmitted signal. This is often judged by the ratio of the output signal power to the output noise power (S/N). The ratio of signal plus noise power to noise power $[(S + N)/N]$ is generally easier to measure, and approaches S/N when the signal is large.

In the evaluation of a two-port it is important to know the amount of noise added to a signal passing through it. An important parameter for expressing this characteristic is the noise factor. The signal energy coming from a generator or antenna is amplified or attenuated in passing from the input to the output of a two-port, as is the noise that accompanies the input signal energy. A system generally includes a cascade of two-port networks which constitute one overall two-port which amplifies the signal to a high-enough power level for its intended use. The noise factor of a system is defined as the ratio of signal-to-noise ratios available at input and output:

$$F = \frac{(S/N)_{\text{input}}}{(S/N)_{\text{output}}} \geq 1 \qquad (2.1)$$

The noise figure (or factor) of a receiver is an easily measured quantity that describes the signal-to-noise ratio reduction of that receiver.

When this ratio of powers is converted to decibels, it is generally referred to as the noise figure rather than noise factor. Various conventions are used to distinguish the symbols used for noise factor and noise symbol. Here we use F to represent the noise factor and NF to represent the noise figure, although the terms are usually used interchangeably.

For an amplifier with the power gain G, the noise factor can be rearranged as

$$F = \frac{S_i/N_i}{GS_i/G(N_i + N_a)} \tag{2.2}$$

where N_a is the additional noise power added by the amplifier referred to the input. This can be computed to be

$$F = 1 + N_a/N_i \tag{2.3}$$

The noise factor is often replaced by the noise figure (NF), which is defined in decibels as

$$NF = 10 \log_{10} F \tag{2.4}$$

In applications such as satellite receivers the noise factor becomes such a small number that it is inconvenient to work with. Many people have adopted the use of an equivalent noise temperature for a circuit to remedy this situation. Since the thermal noise power available from a resistor at temperature T_e is

$$N = kT_e B \tag{2.5}$$

where k is Boltzmann's constant (1.38×10^{-23} J/K), T_e is the effective temperature in kelvin, and B is the bandwidth in hertz. The equation above may be used to associate an effective noise temperature with circuits containing more than just thermal noise sources. This allows (2.3) to be written as

$$F = 1 + \frac{kT_e B}{kT_0 B} = 1 + \frac{T_e}{T_0} \tag{2.6}$$

where T_e is the effective noise temperature of the circuit and T_0 is the temperature of the generator resistor in kelvin. The noise temperature T_e now characterizes our circuit noise contribution and can be directly related to the noise factor.

Assuming a reference noise temperature of 290 K ($-273 + 290 = 17°C$) let us determine the noise temperature of the system with a noise factor of 2.6 (4.15 dB):

$$T_e = (2.6 - 1)(290) = 464 \text{ K}$$

This temperature T_e should not be confused with the environmental operating temperature T_0. It is quite common to operate low-noise amplifiers with T_e below 100 K at an ambient temperature of 290 K.

2.1 SIGNAL-TO-NOISE RATIO

Let us consider the signal-to-noise ratio of power delivered from a generator to a load as shown in Fig. 2.1. The signal power delivered to the input is given by

$$S_{\text{in}} = P_{\text{in}} = \frac{E_g^2 \, \text{Re}(Z_{\text{in}})}{|Z_g + Z_{\text{in}}|^2} \qquad (2.7)$$

where E_g is the rms voltage of the input signal supplied to the system, and the noise power supplied to the input is expressed by

$$N_{\text{in}} = \frac{\overline{v_n^2} \, \text{Re}(Z_{\text{in}})}{|Z_g + Z_{\text{in}}|^2} \qquad (2.8)$$

where the noise power at the input is provided by the noise energy of the real part of Z_g. The input impedance Z of the system in the form $Z = R_{\text{in}} + jX_{\text{in}}$ is assumed to be complex.

The Johnson noise of a resistor [here $\text{Re}(Z_g)$] is given by the mean-square voltage

$$\overline{v_n^2} = 4kTRB \qquad (2.9)$$

with k Boltzmann's constant $= 1.38 \times 10^{-23}$ J/K, T is the absolute temperature of the resistor, and B the bandwidth, is sufficiently small that the resistive component of impedance does not change. The available signal power from the generator

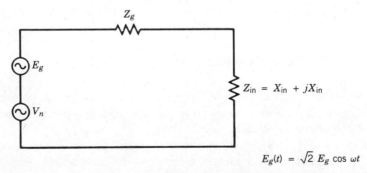

$$E_g(t) = \sqrt{2} \, E_g \cos \omega t$$

Figure 2.1 Combination of signal and noise voltages supplied to a complex termination.

has a lower limit, even if the signal is attenuated by the highest possible attenuation. The generator resistor acts as a Johnson noise generator, its power being

$$P_A = \frac{4kTRB}{4R} = kTB \qquad (2.10)$$

with k Boltzmann's constant, T the absolute temperature, and B the bandwidth. This power is the maximum available output power.

For an ambient temperature of 290 K, $kT = 4 \times 10^{-21}$ W/Hz. This expression is also given as $kT = -204$ dBW/Hz $= -174$ dBm/Hz $= -114$ dBm/MHz. We can combine (2.7) to (2.9) to obtain

$$\left(\frac{S}{N}\right)_{in} = \frac{E_g^2}{4kT \, \mathrm{Re}(Z_g)B} \qquad (2.11)$$

This is the value of S/N contributed by the generator, which does not include the noise generated by the load, in this case $\mathrm{Re}(Z_{in})$, which would need to be included in the measurement of the total S/N across the input impedance.

A critical parameter is the noise bandwidth, B_n, which is defined as the equivalent bandwidth, as shown in Fig. 2.2. For reasons of group delay correction, most practical filters have round rather than sharp corners. The noise figure measurements shown later can be used to determine the "integrated" bandwidth, which is B_n.

An active system such as a combination of amplifiers and mixers will add noise to the input signals, and the noise factor that describes this is defined as the S/N ratio at the input to the S/N ratio at the output, which is always greater than unity [2.1]. In practice, a certain minimum signal-to-noise ratio is required for operation. For example, in a communication system such a minimum is required for intelligible transmission, either voice or data. For high-performance TV reception, to provide a picture noise-free to the eye, a typical requirement is for a 60-dB S/N. In the case of a TV system, a large dynamic range is required, as well as a very large bandwidth to reproduce all colors truthfully and all shades from high-intensity white to black. Good systems will have 8-MHz bandwidth or more.

2.2 NOISE FIGURE MEASUREMENTS

Some of the noise equations are based on mathematical models and physics. To understand some of these expressions, it is useful to look at a practical case of a system with amplifiers which is to be evaluated.

Let us look at Fig. 2.3, which consists of a signal generator, the system or device under test (DUT), and a selective receiver with a build-in root-mean-square (rms) voltmeter to determine the signal and the noise voltage. It is necessary that

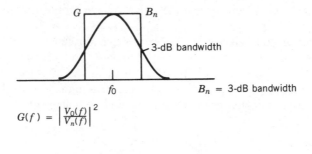

$$G(f) = \left| \frac{V_0(f)}{V_n(f)} \right|^2$$

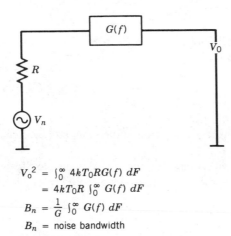

$$V_0{}^2 = \int_0^\infty 4kT_0RG(f)\, dF$$
$$= 4kT_0R \int_0^\infty G(f)\, dF$$
$$B_n = \frac{1}{G} \int_0^\infty G(f)\, dF$$
$$B_n = \text{noise bandwidth}$$

Figure 2.2 Graphical and mathematical explanation of the noise bandwidth from a comparison of the Gaussian-shaped bandwidth to the rectangular filter response.

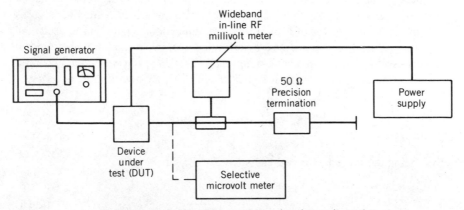

Figure 2.3 Test setup to measure signal-to-noise ratio.

the system have enough gain so that the noise voltage supplied by the generator will be indicated [2.2].

If we assume that our selective receiver is a video noise meter calibrated in rms voltage levels, we can perform two measurements. With an input termination connected to the TV system (typically, 75 Ω for cable TV, 50 Ω for satellite TV), the noise receiver/meter will read a value for proper termination which can easily be calculated. Since one-half of the mean-square noise voltage appears across the input,

$$v_{in} = \frac{v_n}{2} = \frac{\sqrt{4kTRB}}{2} \qquad (2.12)$$

With $B = 10$ MHz, $T = 290$ K (T is always expressed in absolute temperature; $T_0 = -273$ C), and $k = 1.38 \times 10^{-23}$ J/K, then for $R = 75$ Ω,

$$v_{in} = \frac{v_n}{2} = 1.73 \ \mu V \qquad (2.13)$$

where the rms noise voltage has been referred to the input port. We can verify this with our first measurement.

Now we increase the input voltage of the signal generator to a value that indicates a 60-dB S/N ratio at the output port. This should be about

$$E_g = \frac{v_n}{2} \sqrt{F} \times 1000 = 1.73 \sqrt{F} \qquad mV$$

where F is the noise factor of the receiver. For a receiver noise factor of 10 we would obtain $E_g = 5.48$ mV (rms value). If the noise energy equivalent to a noise factor of F is assumed, we need $\sqrt{F}$ times more voltage. For a 60-dB ratio, this means that $E_g = 1000 \times (v_n/2) \times \sqrt{F}$.

As they are done here, over a power range of 60 dB, the measurements can be performed over such a wide range only if special equipment is available. In cases where the internal detector of a piece of communications equipment is used, the signal-to-noise ratio measurements are performed over much smaller power ranges.

Let us assume that for the above-mentioned case ($F = 10$) we find a S/N ratio of 10 dB at the output for an input signal of 5.47 μV. By rewriting (2.6) as

$$E_g = \frac{v_n}{2} \sqrt{F} = \sqrt{kTRBF} \qquad (2.14)$$

with F being the noise factor, we can solve for F with

$$F = \frac{P_s}{P_n} = \frac{E_g^2/R}{kTB} \qquad (2.15)$$

While the input power from the thermal energy of the input termination resistor was $kTB = 4 \times 10^{-4}$ W, the input power required for the 10-dB S/N ratio was

$$P_s = \frac{(5.47 \times 10^{-6})^2}{75} = 3.98 \times 10^{-13} \text{ W} \qquad (2.16)$$

The noise factor is defined as the ratio P_s/P_n:

$$F = \frac{3.98 \times 10^{-13}}{4 \times 10^{-14}} = 10 \qquad (2.17)$$

which is the proof.

This method is used more frequently at the 3-dB point, or double the input power if the dynamic range of the detector is small or only a linear indicator is available. Because of hum and other pickup, this is not an easy measurement. Using a signal generator is very expensive because in a laboratory or production environment a wide frequency range requires several generators.

Another method is the use of a wideband noise generator. Modern gas discharge diodes or avalanche diodes are available which provide essentially white noise energy over a large frequency range. These microwave diodes typically have an output of 30 dB above kT when switched on and kT when switched off. To provide good matching at microwave frequencies, a 15-dB attenuator is cascaded. This means that the noise power of the source in the ON condition is about 15 dB above kT.

In the early 1960s, low-cost noise figure test equipment was built around vacuum diodes whose operating range was limited to 1200 MHz due to the resonate effects of the structure. Today the automatic noise gain analyzer offered currently by Hewlett-Packard and Eaton/AIL uses calibrated solid-state noise sources up to 26.5 GHz. It appears that the upper frequency limit has to do with matching, and the lower-frequency limit with $1/f$ noise.

2.3 NOISY TWO-PORT DESCRIPTION

Based on the convention by Rothe and Dahlke [2.3], any linear two-port can be in the form shown in Fig. 2.4. This general case of a noisy two-port can be redrawn showing noise sources at the input and at the output. Figure 2.4b shows this in admittance form and Fig. 2.4c in impedance form. The internal noise sources are assumed to produce very small currents and voltages, and we assume that linear two-port equations are valid. From the set of equations (1.5) and (1.4) of Chapter 1, we can describe the general case. The internal noise contributions have been expressed by using external noise sources:

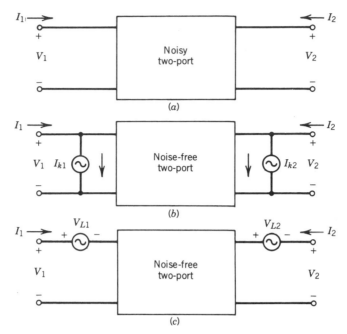

Figure 2.4 Noisy linear two-ports: (*a*) general form; (*b*) admittance form; (*c*) impedance form.

$$I_1 = y_{11}V_1 + y_{12}V_2 + I_{K1}$$
$$I_2 = y_{21}V_1 + y_{22}V_2 + I_{K2}$$

$$(2.18)$$

$$V_1 = z_{11}I_1 + z_{12}I_2 + V_{L1}$$
$$V_2 = z_{21}I_1 + z_{22}I_2 + V_{L2}$$

$$(2.19)$$

where the external noise sources are I_{K1}, I_{K2}, V_{L1}, and V_{L2}.

Since we want to describe our noisy circuit in terms of the noise figure, the *ABCD*-matrix description will be more convenient since it refers both noise sources to the input of the two-port [2.4.]. This representation is given below (note the change in direction of I_2):

$$V_1 = AV_2 + BI_2 + V_A$$
$$I_1 = CV_2 + DI_2 + I_A$$

$$(2.20)$$

where V_A and I_A are the external noise sources.

It is important to remember that all of these matrix representations are interrelated. For example, the noise sources for the *ABCD*-matrix description can be obtained from the *z*-matrix representation shown in (2.19). This transformation is

$$V_A = -\frac{I_{K2}}{y_{21}} = V_{L1} - \frac{V_{L2}z_{11}}{z_{21}} \qquad (2.21)$$

$$I_A = I_{K1} - \frac{I_{K2}y_{11}}{y_{21}} = -\frac{V_{L2}}{z_{21}} \qquad (2.22)$$

The *ABCD* representation is particularly useful based on the fact that it allows us to define a noise temperature for the two-port referenced to its input. The two-port itself (shown in Fig. 2.5) is assumed to be noise free.

In the past, z and y parameters have been used, but in microwave applications it has become common to use *S*-parameter definitions. This is shown in Fig. 2.6. The previous equations can be rewritten in their new form using S parameters:

$$\begin{bmatrix} b_1 \\ b_2 \end{bmatrix} = \begin{bmatrix} S_{11} & S_{12} \\ S_{21} & S_{22} \end{bmatrix} \begin{bmatrix} a_1 \\ a_2 \end{bmatrix} + \begin{bmatrix} b_{n1} \\ b_{n2} \end{bmatrix} \qquad (2.23)$$

There are different physical origins for the various sources of noise. Typically, thermal noise is generated by resistances and loss in the circuit or transistor, whereas shot noise is generated by current flowing through semiconductor junctions and vacuum tubes. Since these many sources of noise are represented by only two noise sources at the device input, the two equivalent input noise sources are often a complicated combination of the circuit internal noise sources. Often, some fraction of V_A and I_A is related to the same noise source. This means that V_A and I_A are not independent in general. Before we can use V_A and I_A to calculate the noise figure of the two-port, we must calculate the correlation between the V_A and I_A shown in Fig. 2.5.

The noise source V_A represents all of the device noise referred to the input when the generator impedance is zero; that is, the input is short circuited. The noise source I_A represents all of the device noise referred to the input when the generator admittance is zero; that is, the input is open circuited.

The correlation of these two noise sources considerably complicates the analysis. By defining a correlation admittance, we can simplify the mathematics and get some physical intuition for the relationship between noise figure and generator admittance. Since some fraction of I_A will be correlated with V_A, we split I_A into correlated and uncorrelated parts as follows:

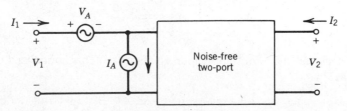

Figure 2.5 Chain matrix form of linear noisy two-ports.

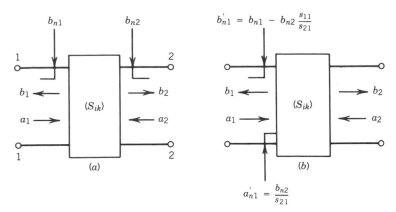

Figure 2.6   S-parameter form of linear noisy two-ports.

$$I_A = I_n + I_u \qquad (2.24)$$

I_u is the part of I_A uncorrelated with V_A. Since I_n is correlated with V_A, we can say that I_n is proportional to V_A and the constant of proportionality is the correlation admittance.

$$I_n = Y_{\text{cor}}V_A \qquad (2.25)$$

This leads us to

$$I_A = Y_{\text{cor}}V_A + I_u \qquad (2.26)$$

The following derivation of noise figure will use the correlation admittance. Y_{cor} is not a physical component located somewhere in the circuit. Y_{cor} is a complex number derived by correlating the random variables I_A and V_A. To calculate Y_{cor}, we multiply each side of (2.26) by V_A^* and average the result. This gives

$$\overline{V_A^* I_A} = Y_{\text{cor}} \overline{V_A^2} \qquad (2.27)$$

where the I_u term averaged to zero since it was uncorrelated with V_A. The correlation admittance is thus given by

$$Y_{\text{cor}} = \frac{\overline{V_A^* I_A}}{\overline{V_A^2}} \qquad (2.28)$$

Often, people use the term "correlation coefficient." This normalized quantity is defined as

$$c = \frac{\overline{V_A^* I_A}}{\sqrt{\overline{V_A^2 I_A^2}}} = Y_{\text{cor}}\sqrt{\frac{\overline{V_A^2}}{\overline{I_A^2}}} \qquad (2.29)$$

Note that the dual of this admittance description is the impedance description. Thus the impedance representation has the same equations as above with Y replaced by Z, I replaced by V, and V replaced by I.

V_A and I_A represent internal noise sources in the form of a voltage source acting in series with the input voltage and a source of current flowing in parallel with the input current. This representation conveniently leads to the four noise parameters needed to describe the noise performance of the two-port. Again using the Nyquist formula the open-circuit voltage of a resistor at the temperature T is

$$\overline{V_A^2} = 4kTRB \qquad (2.30)$$

This voltage is a mean-square fluctuation (or spectral density). It is the method used to calculate the noise density. We could also define a noise equivalent resistance for a noise voltage as

$$R_n = \frac{\overline{V_A^2}}{4kTB} \qquad (2.31)$$

The resistor R_n is not a physical resistor but can be used to simulate different portions of the noise equivalent circuit.

In a similar manner a mean-square current fluctuation can be represented in terms of an equivalent noise conductance G_n, which is defined by

$$G_n = \frac{\overline{I_A^2}}{4kTB} \qquad (2.32)$$

and

$$G_u = \frac{\overline{I_u^2}}{4kTB} \qquad (2.33)$$

for the case of the uncorrelated noise component. The input generator to the two-port has a similar contribution:

$$G_G = \frac{\overline{I_G^2}}{4kTB} \qquad (2.34)$$

with Y_G being the generator admittance and G_G being the real part. With the definition of F above, we can write

$$F = 1 + \left| \frac{I_A + Y_G E_A}{I_G} \right|^2 \qquad (2.35)$$

The use of the voltage V_A and the current I_A has allowed us to combine all the effects of the internal noise sources.

We can use the previously defined (2.28) correlation admittance, $Y_{cor} = G_{cor} + B_{cor}$, to simplify (2.35). First, we determine the total noise current:

$$\overline{I_n^2} = 4kT(Y_{cor}R_n + G_n)B \tag{2.36}$$

where R_n and G_n are as defined in (2.31) and (2.32). The noise factor can now be determined.

$$F = 1 + \frac{G_u}{G_g} + \frac{R_n}{G_g}\left[(G_G + G_{cor})^2 + (B_G + B_{cor})^2\right] \tag{2.37}$$

$$F = 1 + \frac{R_u}{R_g} + \frac{G_n}{R_g}\left[(R_G + R_{cor})^2 + (X_G + X_{cor})^2\right] \tag{2.38}$$

The noise factor is a function of various elements, and the optimum impedance for best noise figure can be determined by minimizing F with respect to generator reactance and resistance. This gives*

$$R_{0n} = \sqrt{\frac{R_n}{G_n} + R_{cor}^2} \tag{2.39}$$

$$X_{0n} = -X_{cor} \tag{2.40}$$

and

$$F_{min} = 1 + 2G_n R_{cor} + 2\sqrt{R_u G_n + (G_n R_{cor})^2} \tag{2.41}$$

At this point we see that the optimum condition for minimum noise figure is not a conjugate power match at the input port. We can explain this by recognizing that the noise source V_A and I_A represent all the two-port noise, not just the thermal noise of the input port. We should observe that the optimum generator susceptance, $-X_{cor}$, will minimize the noise contribution of the two noise generators.

In rearranging for conversion to S parameters, we write

$$F = F_{min} + \frac{G_n}{R_G}\left|Z_G - Z_{0n}\right|^2 \tag{2.42}$$

$$F = F_{min} + \frac{R_n}{G_G}\left|Y_G - Y_{0n}\right|^2 \tag{2.43}$$

From the definition of the reflection coefficient,

$$\Gamma_G = \frac{Y_0 - Y_G}{Y_0 + Y_G} \tag{2.44}$$

*In order to distinguish between optimum noise and optimum power, we have introduced the conventional O_n instead of the more familiar abbreviation opt.

and with

$$r_n = \frac{R_n}{Z_0} \qquad (2.45)$$

the normalized equivalent noise resistance

$$F = F_{\min} + \frac{4r_n |\Gamma_G - \Gamma_{0n}|^2}{(1 - |\Gamma_G|^2)|1 + \Gamma_{0n}|^2} \qquad (2.46)$$

$$r_n = (F_{50} - F_{\min}) \frac{|1 + \Gamma_{0n}|^2}{4|\Gamma_{0n}|^2} \qquad (2.47)$$

$$\Gamma_{0n} = \frac{Z_{0n} - Z_0}{Z_{0n} + Z_0} \qquad (2.48)$$

The noise performance of any linear two-port can now be determined if the values of the four noise parameters, $F_{\min}$, $r_n = R_n/50$, and Γ_{0n} are known.

2.4 NOISE FIGURE OF CASCADED NETWORKS

In a system with many circuits connected in cascade (Fig. 2.7), we must consider the contributions of the various circuits. In considering the equivalent noise resistor R_n in series with the input circuit,

$$F = \frac{R_G + R_n}{R_G} \qquad (2.49)$$

$$F = 1 + \frac{R_n}{R_G} \qquad (2.50)$$

The excess noise added by the circuit is R_n/R_G.

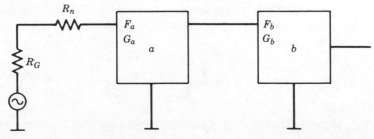

Figure 2.7 Cascaded noisy two-ports with the noise figures F_a and F_b and the gain figure G_a and G_b.

In considering two cascaded circuits a and b, by definition the available noise at the output of b is

$$N_{ab} = F_{ab} G_{ab} kTB \qquad (2.51)$$

with B the equivalent noise bandwidth in which the noise is measured. The total available gain G is the product of the individual available gains, so

$$N_{ab} = F_{ab} G_a G_b kTB \qquad (2.52)$$

The available noise from network a at the output of network b is

$$N_{a/b} = N_a G_b = F_a G_a G_b kTB \qquad (2.53)$$

The available noise added by network b (its excess noise) is

$$N_{b/b} = (F_b - 1) G_b kTB \qquad (2.54)$$

The total available noise N_{ab} is the sum of the available noise contributed by the two networks:

$$N_{ab} = N_{a/b} + N_{b/b} = F_a G_a G_b kTB + (F_b - 1) G_b kTB$$

$$= \left(F_a + \frac{F_b - 1}{G_a} \right) G_a G_b kTB \qquad (2.55)$$

$$F_{ab} = F_a + \frac{F_b - 1}{G_a} \qquad (2.56)$$

For any number of circuits, this can be extended to be

$$F = F_1 + \frac{F_2 - 1}{G_1} + \frac{F_3 - 1}{G_1 G_2} + \frac{F_4 - 1}{G_1 G_2 G_3} + \cdots \qquad (2.57)$$

When considering a long chain of cascaded amplifiers there will be a minimum noise figure achievable for this chain. This is a figure of merit and was proposed by Haus and Adler [2.5]. It is calculated by rearranging (2.57). If all stages are designed for a minimum noise figure, we find that for an infinite chain of identical stages,

$$(F_{tot})_{min} = (F_{min} - 1) + \frac{F_{min} - 1}{G_A} + \frac{F_{min} - 1}{G_A^2} + \cdots + 1 \qquad (2.58)$$

where F_{min} is the minimum noise figure for each stage and G_A is the available power gain of the identical stages. Using

$$\frac{1}{1 - X} = 1 + X + X^2 + \cdots \tag{2.59}$$

we find a quantity $(F_{\text{tot}} - 1)$, which is defined as noise measure M. The minimum noise measure

$$(F_{\text{tot}})_{\text{min}} - 1 = \frac{F_{\text{min}} - 1}{1 - 1/G_A} = M_{\text{min}} \tag{2.60}$$

refers to the noise of an infinite chain of optimum-tuned, low-noise stages, so it represents a lower limit on the noise of an amplifier.

The minimum noise measure M_{min} is an invariant parameter and is not affected by feedback. It is somewhat similar to a gain–bandwidth product, in its use as a system invariant. The minimum noise measure is achieved when the amplifier is tuned for the available power gain and $\Gamma_G = \Gamma_{0n}$, given by (2.48).

2.5 INFLUENCE OF EXTERNAL PARASITIC ELEMENTS

Mounting an active two-port such as a transistor usually adds stray capacitance and lead inductance to the device, as shown in Fig. 2.8. These external components, consisting of transmission lines and parasitic reactances modify the noise parameters and the gain. Some researchers have published the results of these parasitic effects and have made manual computations or used some limited computer programs.

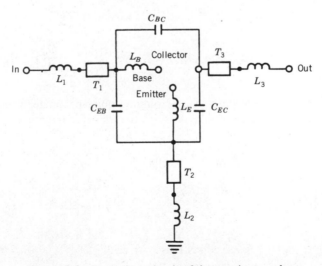

Figure 2.8   Equivalent circuit of the transistor package.

In a paper by Fukui [2.6] an attempt was made to determine the necessary equations, but the formulas are too involved even for pocket calculators (see Appendix F). A more generic study by Iversen [2.7] is also very involved because of the various matrix manipulations, and is more suitable for a computer. Besser's paper in the IEEE MTT-S in 1975 [2.8] and Vendelin's paper [2.9] in the same issue have shown for the first time some practical results using computers and even optimization methods, using an early version of Compact. The intention of these investigations was to find feedback that modifies the device noise and scattering parameters such that a noise match could also provide a low-input VSWR. It can be seen from these discussions that some feedback, besides resulting in some gain reduction, may improve the noise matching at the input for a limited frequency range. The derivation of these matrix methods is presented in Section 2.7.

A more recent paper by Suter [2.10] based on a report by Hartmann and Strutt [2.11] has given a simple transformation starting from the S parameters and the noise parameters from common-source (or common-emitter) measurements. The noise parameters for the "packaged" device are calculated. This means that the parameters for the "new" device, including the common-gate (or common-base) case, are calculated. The equations are device independent. They are valid for any active two-port.

A transformation matrix, n, may be used to combine the noise sources of the various circuit configurations. The transformation matrix parameters are given in Table 2.1 for series feedback, shunt feedback, and the common-gate (base) case, which will be important for oscillator analysis. The transformation matrix gives the new four noise parameters as follows:

$$R'_n = R_n \left| n_{11} + n_{12} Y_{\text{cor}} \right|^2 + G_n \left| n_{12} \right|^2 \qquad (2.61)$$

$$G'_n = \frac{G_n R_n}{R'_n} \left| n_{11} n_{22} - n_{12} n_{21} \right|^2 \qquad (2.62)$$

$$Y'_{\text{cor}} = \frac{R_n}{R'_n} \left(n_{21} + n_{22} Y_{\text{cor}} \right) \left(n^*_{11} + n^*_{12} Y^*_{\text{cor}} \right) + \frac{G_n}{R'_n} n_{22} n^*_{12} \qquad (2.63)$$

A final transformation to the more common noise-parameter format given by (2.43) is still needed [2.9]:

$$F_{\text{min}} = 1 + 2R'_n (G'_{\text{cor}} + G'_{0n}) \qquad (2.64)$$

$$R_n = R'_n \qquad (2.65)$$

$$G_{0n} = \sqrt{\frac{G'_n}{R'_n} + G'^2_{\text{cor}}} \qquad (2.66)$$

$$B_{0n} = -B'_{\text{cor}} \qquad (2.67)$$

TABLE 2.1

Series Feedback

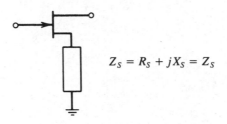

$$Z_S = R_S + jX_S = Z_S$$

$$[n] = \begin{bmatrix} n_{11} = 1 & n_{12} = Z_0 \dfrac{S_{21}M - S'_{21}N}{S_{21}C'_1 + S'_{21}C_1}\ \Omega \\[2ex] n_{21} = 0 & n_{22} = \dfrac{S_{21}C'_1}{S_{21}C'_1 + S'_{21}C_1} \end{bmatrix}$$

where

$$S'_{11} = S'_{22} = \frac{-1}{1 + 2Z_S}$$

$$S'_{12} = S'_{21} = \frac{2Z_S}{1 + 2Z_S}$$

$$M = (1 + S'_{11})(1 - S_{22}) + S'_{12}S'_{21}$$

$$N = (1 + S_{11})(1 - S_{22}) + S_{12}S_{21}$$

$$C_1 = (1 - S_{11})(1 - S_{22}) - S_{12}S_{21}$$

$$C'_1 = (1 + S'_{11})(1 - S'_{22}) - S'_{12}S'_{21}$$

$$[S]_{\text{DEVICE}} = \begin{bmatrix} S_{11} & S_{12} \\ S_{21} & S_{22} \end{bmatrix}$$

Shunt Feedback

$$Z_P = R_P + jX_P = Z_P$$

$$[n] = \begin{bmatrix} n_{11} = \dfrac{S_{21}C'_2}{S_{21}C'_2 + S'_{21}C_2} & n_{12} = 0\ \Omega \\[2ex] n_{21} = \dfrac{1}{Z_0}\dfrac{S_{21}P - S'_{21}Q}{S_{21}C'_2 + S'_{21}C_2}\ S & n_{22} = 1 \end{bmatrix}$$

where

$$S'_{11} = S'_{22} = \frac{Z_P}{2 + Z_P}$$

TABLE 2.1 (*Continued*)

$$S'_{12} = S'_{21} = \frac{2}{2 + Z_P}$$
$$P = (1 - S'_{11})(1 + S'_{22}) + S'_{12}S'_{21}$$
$$Q = (1 - S_{11})(1 + S_{22}) + S_{12}S_{21}$$
$$C_2 = (1 + S_{11})(1 + S_{22}) - S_{12}S_{21}$$
$$C'_2 = (1 + S'_{11})(1 + S'_{22}) - S'_{12}S'_{21}$$

$$[S]_{\text{DEVICE}} = \begin{bmatrix} S_{11} & S_{12} \\ S_{21} & S_{22} \end{bmatrix}$$

Common Gate

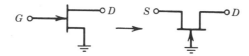

$$[n] = \begin{bmatrix} n_{11} = \dfrac{2S_{21}}{-2S_{21} + C_4} & n_{12} = 0\ \Omega \\[2em] n_{21} = \dfrac{1}{Z_0}\dfrac{C_3C_4 - 4S_{12}S_{21}}{V(-2S_{21} + C_4)}S & n_{22} = -1 \end{bmatrix}$$

where

$$V = (1 + S_{11})(1 + S_{22}) - S_{12}S_{21}$$
$$C_3 = (1 - S_{11})(1 + S_{22}) + S_{12}S_{21}$$
$$C_4 = (1 + S_{11})(1 - S_{22}) + S_{12}S_{21}$$

$$[S]_{\text{DEVICE}} = \begin{bmatrix} S_{11} & S_{12} \\ S_{21} & S_{22} \end{bmatrix} = \text{common-source } S \text{ parameters}$$

In Chapter 3 we will see examples of the different noise parameters for bipolar and field-effect transistors. Figure 2.9 shows the noise figure as a function of external feedback for a low-noise microwave bipolar transistor that is discussed in Chapter 3, the AT-41435.

2.6 NOISE CIRCLES

From Section 2.5 we see that the noise factor is a strong function of the generator admittance (or impedance) presented to the input terminals of the noisy two-port. Noise tuning is the method to change the values of the input admittance to obtain the best noise performance. There is a range of values of input reflection coefficients over which the noise figure is constant. In plotting these points of constant noise figure, we obtain the so-called noise circles, which can be drawn on the Smith chart Γ_G plane [2.12]. Using similar techniques as were used to calculate

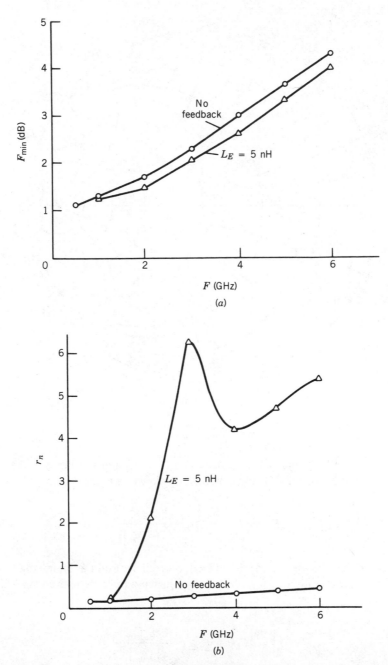

Figure 2.9 Noise parameters versus feedback for AT-41435 silicon bipolar transistor. (*a*) F_{min} for AT-41435 versus frequency and feedback; (*b*) r_n for AT-41435 versus frequency and feedback; (*c*) Γ_{0n} for AT-41435 versus frequency and feedback.

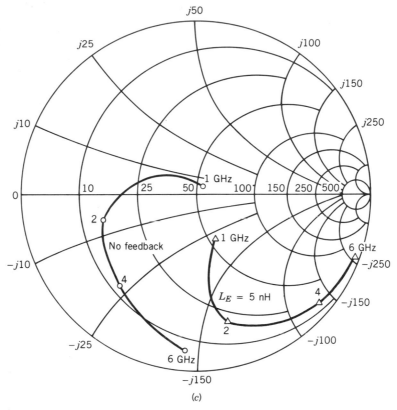

Figure 2.9 (*Continued*)

the gain circles in Chapter 1 [see (1.134) and (1.135)] and starting with the noise equation [see (2.43)] for a 50-Ω generator impedance, we find that

$$F_{50} = F_{min} + 4r_n \frac{|\Gamma_{0n}|^2}{|1 + \Gamma_{0n}|^2} \tag{2.68}$$

We want to find the position of the reflection coefficient on the Smith chart, as in the case of the gain circles, for which F = constant. First we rearrange (2.65) to read

$$r_n = (F_{50} - F_{min}) \frac{|1 + \Gamma_{0n}|^2}{4|\Gamma_{0n}|^2} \tag{2.69}$$

By introducing

$$N_i = \frac{F_i - F_{min}}{4r_n} |1 + \Gamma_{0n}|^2 \tag{2.70}$$

we can find an expression for a circle of constant noise figure as introduced by Rothe and Dahlke [2.3, 2.12]. The center for the noise circle is

$$C_i = \frac{\Gamma_{0n}}{1 + N_i} \tag{2.71}$$

and the radius

$$r_i = \frac{\sqrt{N_i^2 + N_i\left(1 - |\Gamma_{0n}|^2\right)}}{1 + N_i} \tag{2.72}$$

with the definition of N used previously. Examples of noise circles are shown in Figures 3.49 and 3.53. However, if we only consider the minimum noise figure for a given device, we will not obtain the minimum noise figure for the multistage amplifier system. This was explained when the noise measure was introduced. (See Fig. 2.10) Therefore, a better way to design the amplifier would be to use circles at constant noise measure instead of circles of constant noise figure. This was recently done by Poole and Paul [2.13]. They derived the expressions for the noise measure circles as a function of S parameters, noise parameters, and Γ_G, using

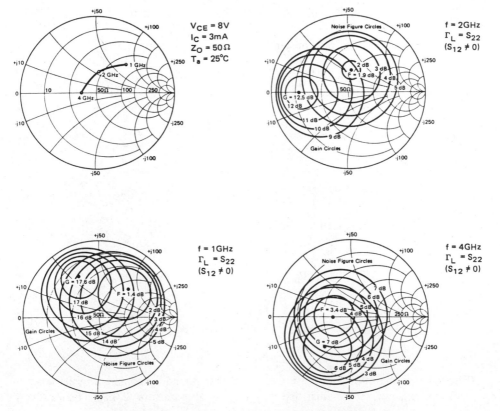

Figure 2.10 Typical noise figure circles and gain circles.

$$G_A = \frac{|S_{21}|^2(1 - |\Gamma_G|^2)}{(1 - |S_{22}|^2) + |\Gamma_G|^2(|S_{11}|^2 - |\Delta|^2) - 2\,\mathrm{Re}(\Gamma_G C_1)} \tag{2.73}$$

where

$$C_1 = S_{11} - S_{22}^* \Delta$$

$$\Delta = S_{11}S_{22} - S_{12}S_{11}$$

In terms of the generator reflection coefficient, the noise measure can be expressed as

$$M = [\,(F_{\min} - 1)(1 - |\Gamma_G|^2)\,|S_{21}|^2\,|1 + \Gamma_{0n}|^2 + 4r_n\,|S_{21}|^2\,|\Gamma_G - \Gamma_{0n}|^2\,]$$

$$\times \{\,|1 + \Gamma_{0n}|^2[\,|S_{21}|^2(1 - |\Gamma_G|^2) - (1 - |S_{22}|^2) - |\Gamma_G|^2(|S_{11}|^2 - |\Delta|^2) + 2\,\mathrm{Re}(\Gamma_G C_1)]\,\}^{-1} \tag{2.74}$$

Equation (2.74) can be shown to represent circles in the source reflection coefficient plane described by the following:

$$|\Delta_G|^2 + |\Gamma_m|^2 - \Gamma_G \Gamma_m^* - \Gamma_m^* \Gamma_m = r_m^2 \tag{2.75}$$

The centers and radii of the constant-noise-measure circles are given by

$$C_m = \frac{M|1 + \Gamma_{0n}|^2 C_1^* + 4r_n\,|S_{21}|^2 \Gamma_{0n}}{M|1 + \Gamma_{0n}|^2 P + |S_{21}|^2(4r_n - W)} \tag{2.76}$$

$$r_m = \frac{\sqrt{M^2 M_a + M M_b + M_c}}{M|1 + \Gamma_{0n}|^2 P + |S_{21}|^2(4r_n - W)} \tag{2.77}$$

where

$$P = |S_{21}|^2 + |S_{11}|^2 - |\Delta|^2$$

$$Q = |S_{21}|^2 + |S_{21}|^2 - 1$$

$$W = |1 + \Gamma_{0n}|^2(F_{\min} - 1)$$

$$M_a = |1 + \Gamma_{0n}|^4(PQ + |C_1|^2)$$

$$M_b = |1 + \Gamma_{0n}|^2\,|S_{21}|^2[\,8r_n\,\mathrm{Re}(\Gamma_{0n}C_1)$$

$$\qquad - (4r_n|\Gamma_{0n}|^2 + W)P - (W - 4r_n)Q\,]$$

$$M_c = |S_{21}|^4[\,W - 4r_n(1 - |\Gamma_{0n}|^2)\,]$$

The value of the minimum noise measure can be found by considering the noise measure circle of zero radius; that is, set r_m equal to zero in (2.77). This results in

$$M_{\min} = \frac{-M_b \pm \sqrt{M_b^2 - 4M_a M_c}}{2M_a} \tag{2.78}$$

Equation (2.78) yields the same value of $M_{\min}$ as would have been obtained by using the immittance parameter equation given by Fukui [2.12] and can, therefore, be considered as the reflection coefficient plane analog of Fukui's expression. The elimination of the need for the parameter R_{eg}, however, results in a considerable simplification compared with the earlier approach [2.12].

The value of the minimum noise measure is taken as the smallest nonnegative value of $M_{\min}$ given by (2.78). The source reflection coefficient which results in the minimum noise measure can now be obtained by employing (2.76):

$$\Gamma_{0m} = \frac{M_{\min}\left|1 + \Gamma_{0n}\right|^2 C_1^* + 4r_n \left|S_{21}\right|^2 \Gamma_{0n}}{M_{\min}\left|1 + \Gamma_{0n}\right|^2 P + \left|S_{21}\right|^2 (4r_n - W)} \tag{2.79}$$

The output reflection coefficients of the device, when Γ_{0m} is presented to the input port, is given by

$$S_{22}' = \frac{S_{22} - \Delta\Gamma_{0m}}{1 - S_{11}\Gamma_{0m}} \tag{2.80}$$

which is also (1.126) of Chapter 1.

2.7 NOISE CORRELATION IN LINEAR TWO-PORTS USING CORRELATION MATRICES

In the introduction to two-port noise theory, it was indicated that noise correlation matrices form a general technique for calculating noise in n-port networks. Haus and Adler have described the theory behind this technique [2.5]. In 1976, Hillbrand and Russer published equations and transformations that aid in supplying this method to two-port CAD [2.14].

This method is useful because it forms a base from which we can rigorously calculate the noise of linear two ports combined in arbitrary ways. For many representations, the method of combining the noise parameters is as simple as that for combining the circuit element matrices. In addition, noise correlation matrices can be used to calculate the noise in linear frequency conversion circuits. The following is an introduction to this subject.

Linear, noisy two-ports can be modeled as a noise-free two-port with two additional noise sources. These noise sources must be chosen so that they add directly to the resulting vector of the representation, as shown in (2.81) and (2.82) and Fig. 2.4.

$$\begin{bmatrix} I_1 \\ I_2 \end{bmatrix} = \begin{bmatrix} y_{11} & y_{12} \\ y_{21} & y_{22} \end{bmatrix} \begin{bmatrix} V_1 \\ V_2 \end{bmatrix} + \begin{bmatrix} i_1 \\ i_2 \end{bmatrix} \tag{2.81}$$

$$\begin{bmatrix} V_1 \\ V_2 \end{bmatrix} = \begin{bmatrix} z_{11} & z_{12} \\ z_{21} & z_{22} \end{bmatrix} \begin{bmatrix} I_1 \\ I_2 \end{bmatrix} + \begin{bmatrix} v_1 \\ v_2 \end{bmatrix} \qquad (2.82)$$

where the i and v vectors indicate noise sources for the y and z representations, respectively. This two-port example can be extended to n-ports in a straightforward, obvious way.

Since the noise vector for any respresentation is a random variable, it is much more convenient to work with the noise correlation matrix. The correlation matrix gives us deterministic numbers to calculate with. The correlation matrix is formed by taking the mean value of the outer product of the noise vector. This is equivalent to multiplying the noise vector by its adjoint (complex conjugate transpose) and averaging the result:

$$\langle \bar{i}\,\bar{i}^{+} \rangle = \begin{bmatrix} i_1 \\ i_2 \end{bmatrix} [i_1^* \quad i_2^*] = \begin{bmatrix} \langle i_1 i_1^* \rangle & \langle i_1 i_2^* \rangle \\ \langle i_1^* i_2 \rangle & \langle i_2 i_2^* \rangle \end{bmatrix} = [C_y] \quad (2.83)$$

where the angular brackets denote the average value.

Note that the diagonal terms are the "power" spectrum of each noise source and the off-diagonal terms are complex conjugates of each other and represent the cross "power" spectrums of the noise sources. "Power" is used because these magnitude-squared quantities are proportional to power.

To use these correlation matrices in circuit analysis, we must know how to combine them and how to convert them between various representations. An example using y matrices will illustrate the method for combining two-ports and their correlation matrices. Given two matrices y and y', when we parallel them we have the same port voltages, and the terminal currents add (Fig. 2.11):

$$I_1 = y_{11}V_1 + y_{12}V_2 + y'_{11}V_1 + y'_{12}V_2 + i_1 + i'_1$$
$$I_2 = y_{21}V_1 + y_{22}V_2 + y'_{21}V_1 + y'_{22}V_2 + i_2 + i'_2 \qquad (2.84)$$

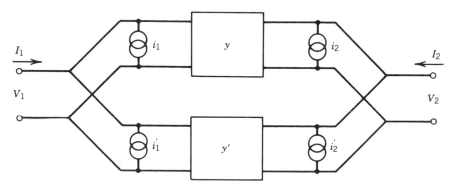

Figure 2.11   Parallel combination of two-ports using y parameters.

or

$$\begin{bmatrix} I_1 \\ I_2 \end{bmatrix} = \begin{bmatrix} y_{11} + y'_{11} & y_{12} + y'_{12} \\ y_{21} + y'_{21} & y_{22} + y'_{22} \end{bmatrix} \begin{bmatrix} V_1 \\ V_2 \end{bmatrix} + \begin{bmatrix} i_1 + i'_1 \\ i_2 + i'_2 \end{bmatrix} \tag{2.85}$$

Here we can see that the noise current vectors add just as the y parameters add. Converting the new noise vector to a correlation matrix yields

$$\langle \bar{i}_{new} \bar{i}^+_{new} \rangle = \left\langle \begin{bmatrix} i_1 + i'_1 \\ i_2 + i'_2 \end{bmatrix} [i_1^* + i_1'^* \quad i_2 i_2'^*] \right\rangle \tag{2.86}$$

$$= \begin{bmatrix} \langle i_1 i_1^* \rangle + \langle i'_1 i_1'^* \rangle & \langle i_1 i_2^* \rangle + \langle i'_1 i_2'^* \rangle \\ \langle i_2 i_1^* \rangle + \langle i'_2 i_1'^* \rangle & \langle i_2 i_2^* \rangle + \langle i'_2 i_2'^* \rangle \end{bmatrix} \tag{2.87}$$

The noise sources from different two-ports must be uncorrelated, so there are no cross products of different two-ports. By inspection (2.87) is just the addition of the correlation matrices for the individual two-ports, so

$$[C_{ynew}] = [C_y] + [C'_y] \tag{2.88}$$

The same form holds true for g, h, and z parameters, but $ABCD$ parameters have the more complicated form shown below. If

$$[A_{new}] = [A][A'] \tag{2.89}$$

then

$$[C_{Anew}] = [C_A] + [A][C_{A'}][A]^+ \tag{2.90}$$

The transformation of one representation to another is best illustrated by an example. Let us transform the correlation matrix for a Y representation to a Z representation. Starting with

$$\begin{bmatrix} I_1 \\ I_2 \end{bmatrix} = [Y] \begin{bmatrix} V_1 \\ V_2 \end{bmatrix} + \begin{bmatrix} i_1 \\ i_2 \end{bmatrix} \tag{2.91}$$

we can move the noise vector to the left side and invert y:

$$\begin{bmatrix} V_1 \\ V_2 \end{bmatrix} = [Y^{-1}] \begin{bmatrix} I_1 - i_1 \\ I_2 - i_2 \end{bmatrix} = [Y^{-1}] \begin{bmatrix} I_1 \\ I_2 \end{bmatrix} + [Y^{-1}] \begin{bmatrix} -i_1 \\ -i_2 \end{bmatrix} \tag{2.92}$$

Since $(Y)^{-1} = (Z)$, we have

$$\begin{bmatrix} V_1 \\ V_2 \end{bmatrix} = [Z] \begin{bmatrix} I_1 \\ I_2 \end{bmatrix} + [Z] \begin{bmatrix} -i_1 \\ -i_2 \end{bmatrix} \tag{2.93}$$

so

$$\begin{bmatrix} v_1 \\ v_2 \end{bmatrix} = [Z] \begin{bmatrix} -i_1 \\ -i_2 \end{bmatrix} = [T_{yz}] \begin{bmatrix} -i_1 \\ -i_2 \end{bmatrix} \qquad (2.94)$$

where the signs of i_1 and i_2 are superfluous since they will cancel when the correlation matrix is formed. Here the transformation of the Y noise current vector to the Z noise voltage vector is done simply by multiplying by (Z). Other transformations are shown in Table 2.2.

To form the noise correlation matrix, we again form the mean of the outer product:

$$\langle vv^+ \rangle = \begin{bmatrix} \langle v_1 v_1^* \rangle & \langle v_1 v_2^* \rangle \\ \langle v_1^* v_2 \rangle & \langle v_2 v_2^* \rangle \end{bmatrix} = [Z] \left\langle \begin{bmatrix} i_1 \\ i_2 \end{bmatrix} [i_1^* \ i_2^*] \right\rangle [Z]^+ \qquad (2.95)$$

or

$$[C_z] = [Z][C_y][Z]^+ \qquad (2.96)$$

where

$$v^+ = [i_1^* \ i_2^*][Z]^+$$

This is called a congruence transformation. The key to all of these derivations is the construction of the correlation matrix from the noise vector, as shown in (2.87).

These correlation matrices may easily be derived from the circuit matrices of passive circuits with only thermal noise sources. For example,

$$[C_z] = 2kT\Delta f \operatorname{Re}([Z]) \quad \text{and} \qquad (2.97)$$

$$[C_y] = 2kT\Delta f \operatorname{Re}([Y]) \qquad (2.98)$$

TABLE 2.2 Noise Matrix $T_{\alpha\beta}$ Transformation

		Original Form (α Form)					
		Y		Z		A	
Resulting Form (β Form)	Y	1	0	y_{11}	y_{12}	$-y_{11}$	1
		0	1	y_{21}	y_{22}	$-y_{21}$	0
	Z	z_{11}	z_{12}	1	0	1	$-z_{11}$
		z_{21}	z_{22}	0	1	0	$-z_{21}$
	A	0	A_{12}	1	$-A_{11}$	1	0
		1	A_{22}	0	$-A_{21}$	0	1

The $2kT$ factor comes from the double-sided spectrum of thermal noise. The correlation matrix for the $ABCD$ matrix may be related to the noise figure, as shown by Hillbrand and Russer [2.14]. We have

$$F = 1 + \frac{\overline{Y}[C_a]\overline{Y}^+}{2kT\,\mathrm{Re}(Y_G)} \tag{2.99}$$

where

$$\overline{Y} = \begin{bmatrix} Y_G \\ 1 \end{bmatrix}$$

The $ABCD$ correlation matrix can be written in terms of the noise-figure parameters as

$$[C_a] = 2kT \begin{bmatrix} R_n & \dfrac{F_0 - 1}{2} - R_n Y_{0n}^* \\ \dfrac{F_0 - 1}{2} - R_n Y_{0n} & R_n \left| Y_{0n} \right|^2 \end{bmatrix} \tag{2.100}$$

The noise correlation matrix method forms an easy and rigorous technique for handling noise in networks. This technique allows us to calculate the total noise for complicated networks by combining the noise matrices of subcircuits. It should be remembered that although noise correlation matrices apply to n-port networks, noise-figure calculations apply only to pairs of ports. The parameters of the C_a matrix can be used to give the noise parameters:

$$Y_{0n} = \sqrt{\frac{C_{ii*}}{C_{uu*}} - \left[\mathrm{Im}\left(\frac{C_{ui*}}{C_{uu*}}\right) \right]^2} + j\,\mathrm{Im}\left(\frac{C_{ui*}}{C_{uu*}}\right) \tag{2.101}$$

$$F_0 = 1 + \frac{C_{ui*} + C_{uu*}Y_{0n}^*}{kT} \tag{2.102}$$

$$R_n = C_{uu*} \tag{2.103}$$

2.8 NOISE FIGURE TEST EQUIPMENT

Figure 2.12 shows the block diagram of a noise test setup. It includes the noise source and the other components. The metering unit has a special detector which is linear and over a certain dynamic range measures linear power. The tunable receiver covers a wide frequency range (e.g., 10 to 1800 MHz) and controls the noise source. The receiver is a double-conversion superheterodyne configuration

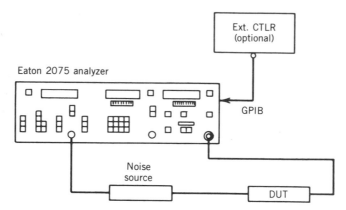

Figure 2.12 Noise figure measurement.

with sufficient image rejection to avoid double-sideband noise measurements that would give the wrong results.

These receivers are microprocessor controlled and the measurement is a two-step procedure. The first is a calibration step that measures the noise figure of the receiver system and a reference power level. Then the DUT (device under test) is inserted and the system noise figure and total output power are measured. The noise factor is calculated by

$$F_1 = F_{\text{system}} - \frac{F_2 - 1}{G_1} \tag{2.104}$$

and the gain is given by the change in output power from the reference level [2.15]. The noise of the system is calculated by measuring the total noise power with the noise source on and off. With the ENR (excess-noise ratio) known [2.15],

$$F_{\text{system}} = \frac{\text{ENR}}{Y - 1} \tag{2.105}$$

The noise bandwidth is usually set by the bandwidth of the receiver, which is assumed to be constant over the linear range. The ENR of the noise source is given by

$$\text{ENR} = \frac{T_{\text{hot}}}{T_{\text{cold}}} - 1 \tag{2.106}$$

where T_{cold} is usually room temperature (290 K). This ENR number is about 15 dB for noise sources with a 15-dB pad and 5 dB for noise sources with a 25-dB pad. Since both gain and noise were stored in the initial calibration, a noise/gain sweep can be performed.

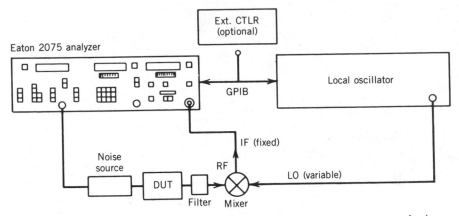

Figure 2.13 Single-sideband (SSB) noise figure measurement using an external mixer.

For frequencies above 1800 MHz we can extend the range with the help of the external signal generators, as shown in Fig. 2.13. As shown, a filter ahead of the external mixer reduces the noise energy in the image band. If the DUT has a very broad frequency range and has flat gain and noise over that range, a dB (double-sideband) measurement is possible, with the image rejection filter removed. However a SSB (single-sideband) measurement is always more accurate [2.16].

2.9 HOW TO DETERMINE THE NOISE PARAMETERS

The noise figure of a linear two-port network as a function of source admittance may be represented by

$$F = F_{\min} + \frac{R_n}{G_G} \left[(G_{0n} - G_G)^2 + (B_{0n} - B_G)^2 \right] \qquad (2.107)$$

where $G_G + jB_G$ = generator admittance presented to the input of the two-port
$\quad\;\; G_{0n} + jB_{0n}$ = generator admittance at which optimum noise figure occurs
$\quad\qquad\;\; R_n$ = empirical constant relating the sensitivity of the noise figure to generator admittance, with dimensions of resistance

It may be noted that for an arbitrary noise-figure measurement with a known generator admittance, Equation (2.107) has four unknowns, $F_{\min}$, R_n, G_{0n}, and B_{0n}. By choosing four known values of generator admittance, a set of four linear equations are formed and the solution of the four unknowns can be found [2.17, 2.18]. Equation (2.107) may be transformed to

$$F = F_{\min} + \frac{R_n |Y_{0n}|^2}{G_G} - 2R_n G_{0n} + \frac{R_n |Y_G|^2}{G_G} - 2R_n B_{0n} \frac{B_G}{G_G} \qquad (2.108)$$

or

$$F = F_{min} + \frac{R_n}{G_G} \left| Y_G - Y_{0n} \right|^2 \qquad (2.109)$$

Let

$$X_1 = F_{min} - 2R_n G_{0n}$$

$$X_2 = R_n \left| Y_{0n} \right|^2$$

$$X_3 = R_n$$

$$X_4 = R_n B_{0n}$$

Then the generalized equation may be written as

$$F_i = X_1 + \frac{1}{G_{si}} X_2 + \frac{\left| Y_{si} \right|^2}{G_{si}} X_3 - 2 \frac{G_{si}}{B_{si}} X_4 \qquad (2.110)$$

or, in matrix form,

$$[F] = [A][X] \qquad (2.111)$$

and the solution becomes

$$[X] = [A]^{-1}[F] \qquad (2.112)$$

These parameters completely characterize the noise behavior of the linear two-port network. Direct measurement of these noise parameters by this method would be possible only if the receiver on the output of the two-port were noiseless and insensitive to its input admittance. In actual practice, the receiver itself behaves as a noisy two-port network and can be characterized in the same manner. What is actually being measured is the system noise figure of the two-port and the receiver.

The two-port noise figure can, however, be calculated using the system formula (2.104). It is important to note that F_2 is assumed to be independent of the impedance of the first-stage two-port, which means that an isolator should be inserted between the first-stage two-port and the receiver. Thus it becomes apparent that to do a complete two-port noise characterization, the system noise characterization, the receiver noise characterization, and the gain of the two-port must be measured [2.19]. In addition, any losses in the input-matching networks must be carefully accounted for, because they add directly to the measured noise-figure reading [2.20].

REFERENCES

2.1   H. T. Friis, "Noise Figures of Radio Receivers," *Proceedings of the IRE*, Vol. 32, July 1944, pp. 419–422.

2.2   "IRE Standards on Methods of Measuring Noise in Linear Twoports," *Proceedings of the IRE*, Vol. 48, January 1960, pp. 60–68.

2.3   H. Rothe and W. Dahlke, "Theory of Noisy Fourpoles," *Proceedings of the IRE*, Vol. 44, June 1956, pp. 811–818.

2.4   "Representation of Noise in Linear Twoports," *Proceedings of the IRE*, Vol. 48, January 1960, pp. 69–74.

2.5   H. A. Haus and R. B. Adler, *Circuit Theory of Linear Noisy Networks*, Cambridge, Mass., MIT Press, 1959, and New York, Wiley, 1959.

2.6   H. Fukui, "The Noise Performance of Microwave Transistors," *IEEE Transactions on Electron Devices*, Vol. ED-13, March 1966, pp. 329–341.

2.7   S. Iversen, "The Effect of Feedback on Noise Figure," *Proceedings of the IEEE*, Vol. 63, March 1975, pp. 540–542.

2.8   L. Besser, "Stability Considerations of Low-Noise Transistor Amplifiers with Simultaneous Noise and Power Match," *IEEE MTT-S International Microwave Symposium Digest*, 1975, pp. 327–329.

2.9   G. D. Vendelin, "Feedback Effects on the Noise Performance of GaAs Mesfets," *IEEE MTT-S International Microwave Symposium Digest*, 1975, pp. 324–326.

2.10 W. A. Suter, "Feedback and Parasitic Effects on Noise," *Microwave Journal*, February 1983, pp. 123–129.

2.11 K. Hartman, and M. J. O. Strutt, "Changes of the Four Noise Parameters due to General Changes of Linear Two-Port Circuits," *IEEE Transactions on Electron Devices*, Vol. ED-20, No. 10, October 1973, pp. 874–877.

2.12 H. Fukui, "Available Power Gain, Noise Figure, and Noise Measure of Two-Ports and Their Graphical Representation," *IEEE Transactions on Circuit Theory*, Vol. CT-13, June 1966, pp. 137–142.

2.13 C. R. Poole and D. K. Paul, "Optimum Noise Measure Terminations for Microwave Transistor Amplifiers," *IEEE Transactions on Microwave Theory and Techniques*, Vol. MTT-33, No. 11, November 1985, pp. 1254–1257.

2.14 H. Hillbrand, and P. H. Russer, "An Efficient Method for Computer Aided Noise Analysis of Linear Amplifier Networks," *IEEE Transactions on Circuits and Systems*, Vol. CAS-23, No. 4, April 1976, pp. 235–238.

2.15 W. E. Pastori, "Topics in Noise," Eaton Electronics Instrumentation Division (EID) Seminar Notes, Los Angeles, 1983.

2.16 W. E. Pastori, "Image and Second Stage Corrections Resolve Noise Figure Measurement Confusion," *Microwave Systems News*, May 1983, pp. 67–86.

2.17 R. Q. Lane, "The Determination of Device Noise Parameters," *Proceedings of the IEEE*, Vol. 57, August 1969, pp. 1461–1462.

2.18 R. Q. Lane, "A Microwave Noise and Gain Parameter Test Set," *IEEE International Solid-State Circuits Conference*, 1978, pp. 172–173.

2.19 N. Kuhn "Accurate and Automatic Noise Figure Measurements With Standard Equipment," *IEEE MTT-S International Microwave Symposium Digest*, 1980, pp. 425–427.

2.20 E. W. Strid, "Measurement of Losses in Noise-Matching Networks," *IEEE Transactions on Microwave Theory and Techniques*, Vol. MTT-29, March 1981, pp. 247–252.

BIBLIOGRAPHY

Ambrózy, A., *Electronic Noise*, McGraw-Hill, New York, 1982.

Baechtold, W., and M. J. O. Strutt, "Noise in Microwave Transistors," *IEEE Transactions on Microwave Theory and Techniques*, Vol. MTT-16, No. 9, September 1968, pp. 578–585.

Cappy, A., "Noise Modeling and Measurement Techniques," *IEEE Transactions Microwave Theory and Techniques*, Vol. MTT-36, January 1988, pp. 1–10.

Caruso, G., and M. Sannino, "Computer-Aided Determination of Microwave Two-Port Noise Parameters," *IEEE Transactions on Microwave Theory and Techniques*, Vol. MTT-26, September 1978, pp. 639–642.

Eisenberg, J. A. "Designing Amplifiers for Optimum Noise Figure," *Microwaves*, Vol. 13, April 1974, pp. 36–41.

Fukui, H., *Low-Noise Microwave Transistors and Amplifiers*, IEEE Press, New York, 1981.

Gentili, C., *Microwave Amplifiers and Oscillators*, McGraw-Hill, New York, 1987.

Haus, H. A., and R. B. Adler, "Optimum Noise Performance of Linear Amplifiers," *Proc. IRE*, Vol. 46, August 1958, pp. 1517–1533.

Lange, J. "Noise Characterization of Linear Two-Ports in Terms of Invariant Parameters," *IEEE Journal of Solid State Circuits*, Vol. SC-2, June 1967, pp. 37–40.

Meys, R. P., "A Wave Approach to the Noise Properties of Linear Microwave Devices," *IEEE Transactions on Microwave Theory and Techniques*, Vol. MTT-26, January 1978, pp. 34–37.

Miller, C. K. S., W. C. Daywitt, and M. G. Arthur, "Noise Standards, Measurements and Receiver Noise Definitions," *Proceedings of the IEEE*, Vol. 55, June 1967.

Mitama, M., and H. Katoh, "An Improved Computational Method for Noise Parameter Measurement," *IEEE Transactions on Microwave Theory and Techniques*, Vol. MTT-27, June 1979, pp. 612–615.

Mukaihata, T. "Applications and Analysis of Noise Generation in N-Cascaded Mismatched Two-Port Networks," *IEEE Transactions on Microwave Theory and Techniques*, Vol. MTT-16, September 1968, pp. 699–708.

Mumford, W. W., and E. H. Scheibe, *Noise Performance Factors in Communications Systems*, Horizon House, Dedham, Mass., 1968.

Okean, H. S., and P. P. Lombardo, "Noise Performance of M/W and MM-Wave Receivers," *Microwave Journal*, Vol. 16, No. 1, January 1973.

Otoshi, T. Y., "The Effect of Mismatched Components on Microwave Noise-Temperature Calibrations," *IEEE Transactions on Microwave Theory and Techniques*, Vol. MTT-16, September 1968, pp. 675–686.

Penfield, P., "Wave Representation of Amplifier Noise," *IRE Transactions on Circuit Theory*, March 1962, pp. 84–86.

Sannino, M., "On the Determination of Device Noise and Gain Parameters," *Proceedings of the IEEE*, Vol. 67, September 1979, pp. 1364–1366.

van der Ziel, A., *Noise in Solid State Devices and Circuits*, Wiley, New York, 1986.

Wait, D. F., "Thermal Noise from a Passive Linear Multiport," *IEEE Transactions on Microwave Theory and Techniques*, Vol. MTT-16, September 1968, pp. 687–691.

PROBLEMS

2.1. For a passive element, show that the noise figure is given by

$$F = 1/G_A$$

where G_A is the available power gain.

For an ideal generator ($\Gamma_G = 0$), show that this reduces to

$$F = \frac{1 - |S_{22}|^2}{|S_{21}|^2}$$

What is the noise figure of an ideal 3-dB attenuator? What is the noise temperature?

2.2. Consider the following resistive network, which has a noise figure given by Problem 2.1. Find the values of R_1 and R_2 for

  (a) $F = 6$ dB

  (b) $F = 0$ dB

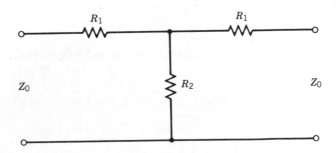

2.3. For a bandwidth of 1 MHz and room temperature (290 K), calculate the root-mean-square noise voltage V_n for the following resistors:

  (a) $R = 1$ kΩ

  (b) $R = 1$ MΩ

2.4. Using (2.41) and (2.42), derive (2.43) and (2.44).

2.5. Derive a table for the equivalence between noise temperature and noise figure for 0 to 100 K for every 10 K. Plot the Y axis as the noise figure in decibels versus the X axis as temperature in kelvin. For a noise figure of 3 dB, what is the noise temperature?

2.6. What is the minimum system noise figure for a transistor with a 1.0-dB noise figure and a 8-dB gain? What is the minimum noise temperature?

2.7. Give the S parameters and noise parameters of a GaAs FET at 12 GHz, ATF-13135:

$$S = \begin{bmatrix} 0.16 \:\underline{/37°} & 0.144 \:\underline{/-89°} \\ 2.34 \:\underline{/-84°} & 0.15 \:\underline{/46°} \end{bmatrix}$$

$$F_{min} = 1.2 \text{ dB} \qquad \Gamma_{0n} = 0.47 \:\underline{/-65°} \qquad R_n = 40 \text{ } \Omega$$

(a) Find Y_{0n} and Z_{0n}.
(b) Calculate the available gain for a noise match in the input, $G_A(\Gamma_{0n})$ in decibels.
(c) Using Table 2.1, calculate the four noise parameters for a 10-Ω source resistor.
(d) Calculate the noise-figure circles in the Γ_G plane.

2.8. For the transistor of Problem 2.7, calculate the common-gate noise parameters (see Table 2.1).

2.9. For the transistor of Problem 2.7, calculate the four noise parameters for a 200-Ω shunt resistor from gate to drain.

2.10 For a three-stage amplifier, calculate the total noise figure and noise temperature.

$$F_1 = 1.2 \text{ dB} \qquad G_{A1} = 9 \text{ dB}$$

$$F_2 = 1.4 \text{ dB} \qquad G_{A2} = 9 \text{ dB}$$

$$F_3 = 1.8 \text{ dB} \qquad G_{A3} = 9 \text{ dB}$$

2.11. For the series noise connection of Table 2.1, show that

$$N_s = \begin{bmatrix} 1 & Z_{11}^a - \dfrac{Z_{21}^a}{Z_{21}^b} Z_{11}^b \\ 0 & \dfrac{1}{1 + Z_{21}^a/Z_{21}^b} \end{bmatrix}$$

2.12. For the parallel noise connection of Table 2.1, show that

$$
N_p = \begin{bmatrix} \dfrac{1}{1 + y_{21}^a/y_{21}^b} & 0 \\[2ex] y_{11}^a - \dfrac{y_{21}^a}{y_{21}^b}\, y_{11}^b & 1 \end{bmatrix}
$$

2.13. The noise-figure data for a low-noise transistor gives $\Gamma_{0n} = 0.5\underline{/45^\circ}$ at the reference plane of the test fixture. What is the device Γ_{0n} if the test fixture has an electrical length of 1.65 cm at 10 GHz? (Given: the velocity of light $= c = 3.0 \times 10^{10}$ cm/s. Electrical length is defined as βl in the air line.)

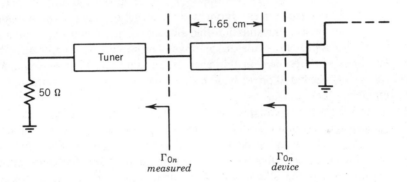

Γ_{0n}
measured

Γ_{0n}
device

2.14. Express the four noise parameters $F_{\min}$, R_n, G_{cor}, and B_{cor} in terms of the four noise parameters $F_{\min}$, R_n, G_{0n}, and B_{0n}. [For the inverse case, see (2.64) to (2.67).]

2.15. A one-stage low-noise amplifier has $F = F_{\min} = 1.4$ dB and $G = G_A = 7$ dB. By adding source feedback, the noise figure is improved until $F = 1.2$ dB and $G = 5$ dB. Show the noise measure has not changed.

3 Microwave Transistor S Parameters and Tuning Elements

3.0 INTRODUCTION

The design of modern microwave amplifiers, oscillators, and mixers requires impedance-matching techniques. From a knowledge of the active two-port S parameters, of microstrip transmission-line elements, and of impedance-matching techniques, the amplifier, oscillator, or mixer design can be completed. These additional tools are presented in this chapter. In addition, an understanding of device physics is needed to properly dc bias the transistor or diode to the required operating point.

Microwave transistors can presently be classified into four groups: silicon bipolar transistors, gallium arsenide field-effect transistors, modulation-doped gallium arsenide field-effect transistors, and heterojunction gallium arsenide bipolar transistors. Two transistors that have not made any impact to date are the gallium-arsenide bipolar and the silicon FET; the first because of no planar process technology, poor emitter efficiency (see τ_e on p. 113), and high base resistance, and the second because of poor microwave performance. Silicon was the first widely used microwave transistor because of the historical development of semiconductor technology. Gallium arsenide has become the dominant microwave transistor because of the superior semiconductor physical parameters, which are summarized in Table 3.1. The fourth column in this table refers to a heterojunction structure of AlGaAs and undoped GaAs which forms a high mobility electron gas called a modulation-doped structure or a 2 DEG (two-dimensional electron gas); the room-temperature parameters reported for these electrons are also given.

Electron mobility in GaAs n-type material with typical donor concentrations around $10^{17}/\text{cm}^3$ ranges from 4000 to 5000 cm^2/V $\cdot$ s at room temperature. The mobility in the channel at 77 K is not much higher than at room temperature. This is because of ionized impurity scattering. In undoped GaAs, electron mobility of 2 to 3×10^5 cm^2/V $\cdot$ s has been obtained at 77 K. The mobility of GaAs with feasible high electron concentrations for facilitating the fabrication of devices was found to increase through the modulation doping technique demonstrated in AlGaAs/GaAs heterojunctions, also called superlattices [3.1].

The theoretical value of $T_{\max}$ in Table 3.1 is given for an intrinsic carrier concentration of $n_i \simeq 10^{15}$ cm^{-3} [3.2, p. 19]. Other effects dominate at a much lower temperature with present technology (e.g., metal contact degradation) [3.3,

TABLE 3.1 Semiconductor Parameters at $T = 25°C$, $N \approx 10^{16}$ cm^{-3}

Parameter	GaAs	Si	Ge	GaAs 2 DEG
Electron mobility, μ_n (cm^2/V · s)	5000	1300	3800	8000
Hole mobility, μ_p (cm^2/V · s)	330	430	1800	—
Saturated drift velocity, v_s [cm/s (electrons)]	1–2×10^7	0.7×10^7	0.6×10^7	2–3×10^7
Band gap, E_g (e · V)	1.42	1.12	0.66	—
Avalanche field, E_{max} (V/cm)	4.2×10^5	3.8×10^5	2.3×10^5	—
T_{max} (theory) [°C ($N \approx 10^{15}$ cm^{-3})]	500	270	100	—
T_{max} (practical)	175	200	75	—
Thermal conductivity, σ_T, at 150°C (W/cm · °C)	0.30	1.0	0.40	0.30
σ_T at 25°C (W/cm · °C)	0.45	1.4	0.60	0.45

3.4]. Another viewpoint for maximum operating temperature is the efficiency of Si and GaAs solar cells, which decreases linearly with temperature to 0% at about 200°C for Si and 300°C for GaAs [3.2, p. 809] when the solar cells have ideal recombination currents.

The choice of transistors is usually between the silicon (Si) bipolar transistor and the GaAs MESFET. A comparison of the relative gain, noise figure, and power is given in Tables 3.2 and 3.3. It should be noted that the GaAs MESFET has lower noise, higher gain, and higher power output. The primary disadvantage is higher $1/f$ noise, which can be significant for oscillator applications.

The higher output power of the GaAs MESFET is a direct result of the higher critical field and higher saturated drift velocity. The approximate power-frequency-squared limit is given by [3.5]

$$\text{Pf}^2 \simeq \left(\frac{E_c v_s}{2\pi} \right)^2 \frac{1}{X_c} \qquad (3.1)$$

where E_c = effective electric field before avalanche breakdown
v_s = drift velocity of carriers (electrons)
X_c = device impedance level

and is a property of the material. Since the parameters E_c and v_s are higher for GaAs, the GaAs MESFET is intrinsically a higher-power device.

If we include a correction factor for the geometry of the transistor, the effective values for GaAs are

TABLE 3.2 1988 Comparison of Microwave Transistors

Parameter	GaAs MESFET				Silicon Bipolar Transistor		
	4 GHz	8 GHz	12 GHz	18 GHz	4 GHz	8 GHz	12 GHz
Gain (dB)	20	16	12	8	15	9	6
F_{min} (dB)	0.5	0.7	1.0	1.2	2.5	4.5	8
Power output (W)	25	15	8	2	6	2	0.25
Oscillator noise (1/f corner frequency)		30 MHz				10 kHz	
1981 Pf^2 (W/s^2)		10^{21}				1.5×10^{20}	
1988 Pf^2 (W/s^2)		1.8×10^{21}				1.5×10^{20}	
Theoretical Pf^2 limit (W/s^2)		5×10^{21}				5×10^{20}	

Parameter	MODFET (AlGaAs/InGaAs)[a]				HBT (AlGaAs/GaAs)[b]			
	12 GHz	18 GHz	36 GHz	60 GHz	12 GHz	18 GHz	36 GHz	60 GHz
Gain (dB)	22	16	12	8	20	16	10	7
F_{min} (dB)	0.5	0.9	1.7	2.6	4			
Power output (W)			0.15	0.10	0.4			
Oscillator noise (1/f corner frequency)		30 MHz				1 MHz		
L_g (μm) or emitter width	0.5	0.5	0.25	0.25		1.2		
Power density (W/mm)			0.9	0.5	4		1.5	
Power added efficiency (%)			38	28	48		28	

[a]From Refs. 3.8 and 3.9.
[b]From Refs. 3.10 and 3.11.

$$X_c \simeq 1\ \Omega$$

$$E_c \simeq E_{max}/4 \simeq 10^5\ \text{V/cm}$$

$$v_s \simeq v_{sat}/5 \simeq 4 \times 10^6\ \text{cm/s}$$

$$Pf^2 \simeq 5 \times 10^{21}\ \text{W/s}^2$$

In 1980, the continuous-wave (CW) performance of 10 W at 10 GHz had already been achieved, giving

$$Pf^2 = 10 \times 10^{10} \times 10^{10} = 10^{21}\ \text{W/s}^2$$

In 1988, the Class A CW performance of 8 W at 15 GHz had been reached [3.6], which is $Pf^2 = 1.8 \times 10^{21}$ W/s^2. In 1978, the CW silicon bipolar transistor had reached 1.5 W at 10 GHz [3.7]

TABLE 3.3 Scaling Parameters for Microwave Transistors

Parameter[a]	GaAs MESFET	Silicon Bipolar Transistor	MODFET	HBT
g_m/Z (mS/mm)	150–300	3000	400–500	7000
P_{out}/Z (W/mm), $f \simeq 10$ GHz	1.0	0.5	2	4
I_{DC}/Z (mA/mm)	100	70	150[b]	700[b]

[a]Z = gate length or emitter length.
[b]From Ref. 3.11.

$$Pf^2 = 1.5 \times 10^{10} \times 10^{10} = 1.5 \times 10^{20} \text{ W/s}^2$$

with a theoretical limit of 5×10^{20} W/s^2, an order of magnitude lower than GaAs. The measured Pf^2 product improves by about a factor of 2 under pulsed conditions [3.12]. The advantages of silicon are lower cost, higher thermal conductivity, and lower $1/f$ noise. The device limitations on frequency response are the transit time of electrical charge and the rate of change of electrical charge. These limitations are discussed in this chapter. For completeness, the heterojunction transistors have been included in the tables. These transistors will become the microwave transistors of the future.

3.1 MICROWAVE BIPOLAR TRANSISTOR

3.1.1 Silicon Bipolar Small-Signal Model

The silicon bipolar transistor is shown in cross section in Fig. 3.1. The bipolar transistor is a current-controlled device, where base current modulates the collector current of the transistor. Normally, common-emitter or common-base configuration is chosen for power gain. For this case, the emitter-base junction is forward biased and the collector-base junction is reverse biased. The base current flows through a distributed RC line before reaching the active portion of the transistor, the emitter edges.

The small-signal equivalent circuit of the silicon bipolar transistor can be derived from the physical cross section of the device given in Fig. 3.1. For this device structure, the distributed T-equivalent circuit of Fig. 3.2 has been found to be an effective small-signal model at fixed-bias conditions. The parameter values for this equivalent circuit are given in Table 3.4 for three modern microwave transistors at the bias for a low noise figure and at the bias for high gain. The emitter pitch and emitter periphery for these transistors will determine the optimum frequency range of operation. These parameters are also given in Table 3.4.

Other bipolar transistor equivalent circuits are given in Fig. 3.3, including the distributed hybrid Π, the simplified hybrid Π, and the simplified T equivalent circuit. The simplified circuits are less accurate in broadband device simulations. The hybrid-Π is popular because of its similarity to the GaAs MESFET equivalent

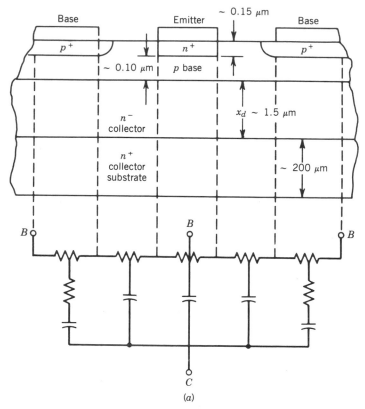

Figure 3.1 (*a*) Bipolar transistor cross section (NPIN); (*b*) bipolar T-equivalent circuit.

circuit described later. The bonding inductances to the base and emitter must also be included in the RF design, usually about 0.5 nH for the base and 0.2 nH for the emitter. Typical circuit values are given for a modern silicon bipolar transistor (Avantek AT-41400) at $V_{CE} = 8$ V, $I_{CE} = 25$ mA.

Although the distributed nature of the bipolar requires an effective value of r_b and C_c, the figure of merit for the bipolar is

$$f_{max}^2 = f_t / 8\pi r_b' C_c \qquad (3.2)$$

where f_{max} is the frequency at which unilateral gain becomes unity and f_t represents the delay time from emitter to collector (i.e., the transit time). The transit time is given by

$$\tau_{ec} = \tau_e + \tau_{eb} + \tau_{bc} + \tau_b + \tau_d + \tau_c \qquad (3.3)$$

where τ_e = emitter delay due to excess holes in emitter

τ_{eb} = emitter–base capacitance charging time through emitter = $r_e' C_{Te}$ = $(kT/qI_E) C_{Te}$

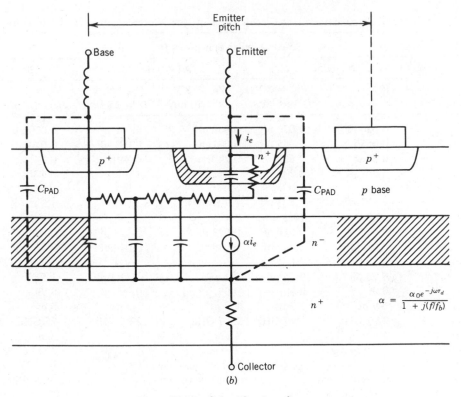

Figure 3.1 (*Continued*)

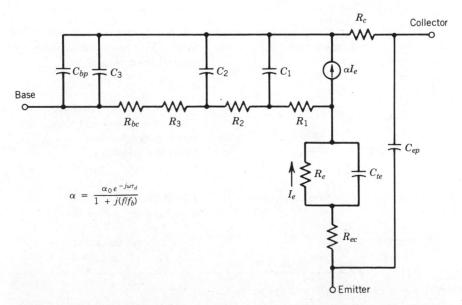

Figure 3.2 Small signal equivalent circuit of microwave bipolar transistor chip excluding bond wire inductances and package parasitics.

TABLE 3.4   Small-Signal Equivalent-Circuit Elements for Microwave Bipolar Transistors

	Silicon Bipolar Transistors from Avantek:				
	AT-60500		AT-41400		AT-22000
Parameter	$V_{CE} = 8$ V $I_C = 2$ mA	$V_{CE} = 8$ V $I_C = 10$ mA	$V_{CE} = 8$ V $I_C = 10$ mA	$V_{CE} = 8$ V $I_C = 25$ mA	$V_{CE} = 8$ V $I_C = 18$ mA
C_{ep}	0.026 pF	0.026 pF	0.032 pF	0.032 pF	0.020 pF
$C_{bp} + C_3$	0.055 pF	0.055 pF	0.091 pF	0.091 pF	0.040 pF
R_{ec}	0.66 Ω	0.66 Ω	0.24 Ω	0.24 Ω	0.2 Ω
$R_{bc} + R_3$	4.2 Ω	4.2 Ω	1.0 Ω	1.0 Ω	0.4 Ω
R_c	5.0 Ω	5.0 Ω	5.0 Ω	5.0 Ω	5.0 Ω
R_1	7.5 Ω	7.5 Ω	2.7 Ω	2.7 Ω	1.8 Ω
R_2	10.3 Ω	10.3 Ω	3.1 Ω	3.1 Ω	2.0 Ω
C_1	0.010 pF	0.010 pF	0.023 pF	0.023 pF	0.020 pF
C_2	0.039 pF	0.039 pF	0.048 pF	0.048 pF	0.015 pF
R_e	12.9 Ω	2.6 Ω	2.6 Ω	1.1 Ω	1.6 Ω
C_{te}	0.75 pF	0.75 pF	2.1 pF	2.1 pF	1.5 pF
α_0	0.99	0.99	0.99	0.99	0.99
τ_d	6.9 ps	7.3 ps	6.9 ps	7.3 ps	8 ps
f_b	22.7 GHz	22.7 GHz	22.7 GHz	22.7 GHz	25 GHz
Emitter pitch	6 μm		4 μm		2μm
Emitter length (Z)	125 μm		350 μm		300 μm
Die size	0.3 mm $\times$ 0.3 mm $\times$ 0.1 mm		0.3 mm $\times$ 0.3 mm $\times$ 0.1 mm		0.3 mm $\times$ 0.3 mm $\times$ 0.1 mm

Symbol	Definition	Symbol	Definition
C_{ep}	Emitter bond pad capacitance	C_{te}	Emitter–base junction capacitance
C_{bp}	Base bond pad capacitance	α	Common-base current gain
R_{ec}	Emitter contact resistance		$\alpha = \dfrac{\alpha_0 e^{-j\omega\tau_d}}{1 + jf/f_b}$
R_{bc}	Base contact resistance		
		α_0	Low-frequency common-base current gain
R_c	Collector resistance	τ_d	Collector depletion-region delay time
R_1		τ_b	Base-region delay time
R_2 }	Distributed base resistance	f_b	Base cutoff frequency
R_3			$f_b = \dfrac{1}{2\pi\tau_b}$
C_1		r_b'	Base resistance
C_2 }	Distributed collector–base capacitance		$r_b' = R_{bc} + R_1 + R_2 + R_3$
C_3			
R_e	Emitter resistance $R_e = kT/qI_e = r_e'$	C_c	Collector–base capacitance $C_c = C_{bp} + C_1 + C_2 + C_3$

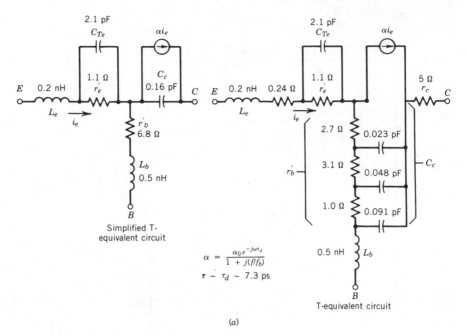

(a)

Figure 3.3   Equivalent circuits for AT-41400 at $V_{CE} = 8$ V, $I_{CE} = 25$ mA: (a) T-equivalent circuit.

τ_{bc} = base–collector capacitance charging time through emitter = $r'_e C_c$
τ_b = base transit time
τ_d = collector depletion-layer delay time = $X_d/2v_s$
τ_c = base–collector capacitance charging time through collector

The frequency at which the common-emitter current gain ($|h_{21e}|$) reduces to unity is defined by f_t and is determined by the delay time from emitter to collector τ_{ec} according to the equation

$$f_t = 1/2\pi\tau_{ec} \tag{3.4}$$

The calculation for the transit time of the Avantek AT-41400 transistor chip at $V_{CE} = 8$ V, $I_{CE} = 25$ mA follows:

$$\tau_e = \frac{X_{jeb}^2}{2D_{pe}\beta_0} = \frac{(0.15)^2 (10^{-8})}{(2)(2)(100)} = 0.56 \text{ ps}$$

where β_0 = 100 = low-frequency or dc value of current gain, h_{FE}
D_{pe} = 2 cm²/s = hole diffusion coefficient in emitter
X_{jeb} = 0.15 μm = depth of emitter–base junction

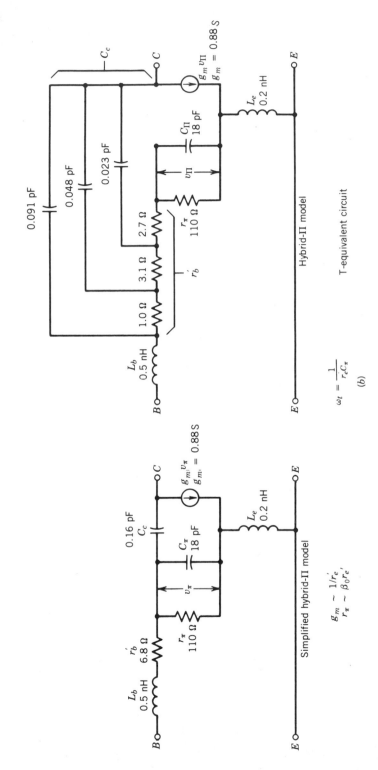

Figure 3.3 (b) hybrid-π equivalent circuit.

$$\tau_{eb} = r'_e C_{Te} = (1.1)(2.1) = 2.31 \text{ ps}$$

$$\tau_{bc} = r'_e C_c = (1.1)(0.16) = 0.18 \text{ ps}$$

$$\tau_b = \frac{W^2}{n D_{nb}} = \frac{10^{-5} \times 10^{-5}}{(2.2)(8)} = 5.68 \text{ ps}$$

where W = base length = 0.10 μm
D_{nb} = 8 cm^2/s = electron diffusion coefficient in base
n = empirical factor to account for built-in aiding electric field due to base impurity gradient = 2.2

$$\tau_d = \frac{X_d}{2v_s} = \frac{1.1 \times 10^{-4}}{1.6 \times 10^7} = 6.88 \text{ ps}$$

where X_d = depletion width of collector = 1.1 μm
v_s = saturated drift velocity for silicon = 0.8 $\times$ 10^7 cm/s

$$\tau_c = r_c C_c = (5)(0.16) = 0.80 \text{ ps}$$

$$\tau_{ec} = 16.4 \text{ ps}$$

$$f_t = \frac{1}{2\pi(16.4) \times 10^{-12}} = 9.7 \text{ GHz}$$

$$r'_b C_c \simeq \frac{(6.8)(0.16)}{2} = 0.54 \text{ ps}$$

$$f_{max} = \sqrt{\frac{9.7 \times 10^9}{8\pi(0.54)(10^{-12})}} = 26.7 \text{ GHz}$$

A fair approximation for the bipolar is

$$U \simeq G_{ma} \simeq (f_{max}/f)^2 \tag{3.5}$$

which gives an estimate of the transistor gain. The G_{ma} is usually 2 to 5 dB lower than U in practice. The S parameters of the transistor should be used to give a more accurate calculation of G_{ma}, using the equations in Chapter 1.

For optimum design of silicon bipolar transistors, the parasitic resistances and capacitances must be minimized. In addition, (3.2)–(3.4) show that the minimum values of r'_b, C_c, and τ_{ec} will give the maximum frequency of operation and therefore the maximum gain.

An important observation for the bipolar transistor is the large transconductance, which can be shown to follow from the forward-biased emitter–base junction. Since the emitter current is given by

$$I_E = I_s \left[\exp\left(\frac{qV_{\text{in}}}{kT}\right) - 1 \right] \tag{3.6}$$

the transconductance is

$$g_m = \frac{\partial I_c}{\partial V_{\text{in}}} = \frac{\alpha_0 I_E q}{kT} = \frac{I_E(\text{mA})}{26} \tag{3.7}$$

Since this will scale with size, the transistor gain parameter at high gain bias is

$$\frac{g_m}{Z} = \frac{I_E(\text{mA})}{26\, Z} \simeq \frac{1\ \text{S}}{0.35\ \text{mm}} \simeq 3\ \text{S/mm}$$

$$\frac{I_E}{Z} = \frac{25\ \text{mA}}{0.35\ \text{mm}} \simeq 70\ \text{mA/mm}$$

where Z is the emitter length or periphery. These are the values reported in Table 3.3 and achieved from the microwave silicon bipolar structure used at Avantek.

The superior microwave performance of the AT-220 bipolar transistor is a result of the reduction of the emitter pitch to 2 μm. The proportional increases in the ratio of the emitter periphery to base area lead to increase in f_{max} and reduction in noise figure. The curves of gain versus frequency for the modeled transistor are given in Fig. 3.4. The f_{max} is extrapolated to 50 GHz, and further improvements will be obtained when another reduction in emitter pitch can be achieved [3.13].

3.1.2 Silicon Bipolar Noise Model

The high-frequency or microwave noise of a silicon bipolar transistor can be modeled by the three noise sources in Fig. 3.5: the base thermal noise e_b due to r_b', the shot noise of the forward-biased emitter–base junction e_e, and the collector partition noise i_{cp}, which is strongly correlated to the emitter–base shot noise. These noise terms and the generator thermal noise are given by ($\Delta f = 1$ Hz)

$$\overline{e_g^2} = 4kTR_g \tag{3.8}$$

$$\overline{e_b^2} = 4kTr_b' \tag{3.9}$$

$$\overline{e_e^2} = 2kTr_e \tag{3.10}$$

$$\overline{i_{cp}^2} = \frac{2kT(\alpha_0 - |\alpha|^2)}{r_e} \tag{3.11}$$

$$r_e = kT/qI_e \tag{3.12}$$

following the development by Hawkins [3.14]. We use the definition of noise figure as the ratio of the output noise power to that from a noiseless but otherwise identical device:

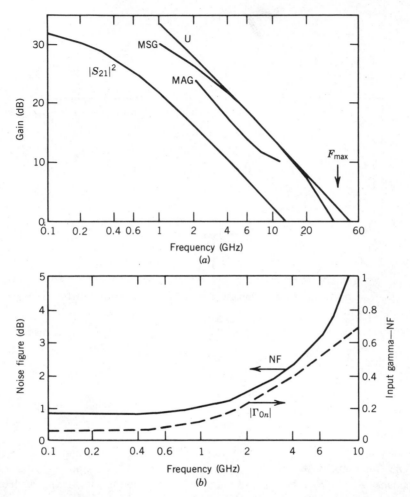

Figure 3.4 Common-emitter performance of 2-μm-pitch silicon bipolar transistor chip (AT-22000): (*a*) S_{21} gain, maximum available gain G_{ma}, maximum stable gain G_{ms}, and unilateral gain U versus frequency. (*b*) Minimum noise figure and associated input reflection coefficient versus frequency. (From Ref. 3.13.)

$$F = \overline{i_L^2}/\overline{i_{LO}^2} \tag{3.13}$$

where i_{LO} is the value of i_L due to the generator e_g alone. Using the definition of α,

$$\alpha = \frac{\alpha_0}{1 + jf/f_b} \tag{3.14}$$

we find that the following relation describes the noise figure:

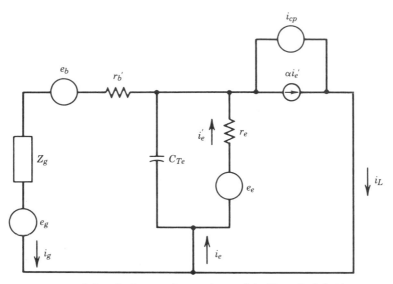

Figure 3.5   Bipolar transistor noise model. (From Ref. 3.14.)

$$F = 1 + \frac{r_b'}{R_g} + \frac{r_e}{2R_g} + \left(\frac{\alpha_0}{|\alpha|^2} - 1\right) \frac{(R_g + r_b' + r_e)^2 + X_g}{2r_e R_g}$$

$$+ \frac{\alpha_0}{|\alpha|^2} \frac{r_e}{2R_g} \left[\omega^2 C_{Te}^2 X_g^2 - 2\omega C_{Te} X_g + \omega^2 C_{Te}^2 (R_g + r_b')^2\right] \quad (3.15)$$

Earlier versions of this equation were reported by Nielsen [3.15] and Cooke [3.16].

The minimum noise figure F_{min} and the corresponding optimum generator impedance $Z_{opt} = R_{opt} + jX_{opt}$ are found by differentiating (3.15) with respect to X_g and then R_g. This equation is of the form

$$F = A + BX_g + CX_g^2 \quad (3.16)$$

where, by introducing the factor

$$a \equiv \left(1 - \frac{|\alpha|^2}{\alpha_0} + \omega^2 C_{Te}^2 r_e^2\right) \frac{\alpha_0}{|\alpha|^2} \quad (3.17)$$

the coefficients can be written

$$A = a \frac{(R_g + r_b')^2}{2r_e R_g} + \frac{\alpha_0}{|\alpha|^2} \left(1 + \frac{r_b'}{R_g} + \frac{r_e}{2R_g}\right) \quad (3.18)$$

$$B = -\frac{\alpha_0}{|\alpha|^2}\frac{\omega C_{Te}r_e}{R_g} \qquad (3.19)$$

$$C = \frac{a}{2r_eR_g} \qquad (3.20)$$

Differentiating with respect to X_g and setting dF/dX_g to zero for the optimum generator reactance gives

$$\left.\frac{dF}{dX_g}\right|_{X_{\text{opt}}} = 0 = B + 2CX_{\text{opt}} \qquad (3.21)$$

Thus

$$X_{\text{opt}} = \frac{-B}{2C} = \frac{\alpha_0}{|\alpha|^2}\frac{\omega C_{Te}r_e^2}{a} \qquad (3.22)$$

and the corresponding noise figure is

$$F_x = A - CX_{\text{opt}}^2 = a\frac{(R_g + r_b')^2}{2r_eR_g} + \frac{\alpha_0}{|\alpha|^2}\left(1 + \frac{r_b'}{R_g} + \frac{r_e}{2R_g}\right) - a\frac{X_{\text{opt}}^2}{2r_eR_g}$$

$$(3.23)$$

This must be optimized with respect to source resistance to give F_{min}. Separating powers of R_g, (3.23) can be written in the form

$$F_x = A + \frac{B}{R_g} + CR_g \qquad (3.24)$$

where

$$A = a\frac{r_b'}{r_e} + \frac{\alpha_0}{|\alpha|^2} \qquad (3.25)$$

$$B = a\frac{r_b'^2 - X_{\text{opt}}^2}{2r_e} + \frac{\alpha_0}{|\alpha|^2}\left(r_b' + \frac{r_e}{2}\right) \qquad (3.26)$$

$$C = \frac{a}{2r_e} \qquad (3.27)$$

Differentiating with respect to R_g, at the minimum:

$$\left.\frac{dF}{dR_g}\right|_{R_{\text{opt}}} = 0 = \frac{-B}{R_{\text{opt}}^2} + C \qquad (3.28)$$

Thus

$$R_{\text{opt}}^2 = \frac{B}{C} = r_b'^2 - X_{\text{opt}}^2 + \frac{\alpha_0}{|\alpha|^2} \frac{r_e(2r_b' + r_e)}{a} \tag{3.29}$$

$$F_{\text{min}} = A + 2CR_{\text{opt}} = a\frac{r_b' + R_{\text{opt}}}{r_e} + \frac{\alpha_0}{|\alpha|^2} \tag{3.30}$$

The factor a is a simple symmetrical function of f_e and f_b, as found by substituting for $|\alpha|$ and $C_{Te}r_e$ in (3.17):

$$a = \left[1 + \frac{f^2}{f_b^2} - \alpha_0 + \left(1 + \frac{f^2}{f_b^2}\right)\frac{f^2}{f_e^2}\right]\frac{1}{\alpha_0}$$

$$= \left[\left(1 + \frac{f^2}{f_b^2}\right)\left(1 + \frac{f^2}{f_e^2}\right) - \alpha_0\right]\frac{1}{\alpha_0} \tag{3.31}$$

The fourth parameter, the equivalent noise resistor R_n, is calculated by another derivation [3.17] to be

$$R_n \simeq \frac{r_b'}{\alpha} + \frac{r_e}{2}\left[1 + \left(\frac{r_b'}{r_e}\right)^2\left(\frac{f}{f_e}\right)^2\right] \tag{3.32}$$

Thus the four noise parameters of a bipolar transistor are determined as follows: F_{min} by (3.30), R_{opt} by (3.29), X_{opt} by (3.22), and R_n by (3.32). There are two limiting cases to consider for F_{min}, the base-limited case when the dominant time constant is τ_b and the emitter-limited case when the dominant time constant is τ_{eb}. For modern microwave transistors biased at a low emitter current for minimum noise figure, the second case seems to better fit the measured data [3.14].

For the base-limited case considered by Nielsen, C_{Te} and X_{opt} are zero and (3.31), (3.29), and (3.30) become

$$a = \left(1 + \frac{f^2}{f_b^2} - \alpha_0\right)\frac{1}{\alpha_0} \tag{3.33}$$

$$R_{\text{opt}}^2 = r_b'^2 + \frac{1 + f^2/f_b^2}{1 + (f^2/f_b^2) - \alpha_0}r_e(2r_b' + r_e)$$

$$= (r_b' + r_e)^2 + \frac{\alpha_0 r_e(2r_b' + r_e)}{1 + (f^2/f_b^2) - \alpha_0} \tag{3.34}$$

$$F_{\text{min}} = \left(1 + \frac{f^2}{f_b^2} - \alpha_0\right)\frac{r_b' + R_{\text{opt}}}{\alpha_0 r_e} + \left(1 + \frac{f^2}{f_b^2}\right)\frac{1}{\alpha_0} \tag{3.35}$$

Equations (3.34) and (3.35) can be shown to be the same as those given by Nielsen. In the emitter-limited case, $\alpha = \alpha_0$ and (3.31), (3.22), (3.29), and (3.30) become

$$a = \left(1 + \frac{f^2}{f_e^2} - \alpha_0\right)\frac{1}{\alpha_0} \tag{3.36}$$

$$X_{\text{opt}} = \frac{f}{f_e}\frac{r_e}{\alpha_0 a} \tag{3.37}$$

$$R_{\text{opt}}^2 = r_b'^2 + \frac{r_e(2r_b' + r_e)}{\alpha_0 a} - \frac{f^2}{f_e^2}\frac{r_e^2}{\alpha_0^2 a^2} \tag{3.38}$$

$$F_{\text{min}} = \left(1 + \frac{f^2}{f_e^2} - \alpha_0\right)\frac{r_b' + R_{\text{opt}}}{\alpha_0 r_e} + \frac{1}{\alpha_0} \tag{3.39}$$

Comparison of (3.35) and (3.39) shows that minimum noise figure increases more rapidly with frequency for the base-limited device, for the same values of R_{opt} and cutoff frequency f_e and f_b. In addition (3.34) and (3.38) show that R_{opt} is larger in the base-limited case, and this increases the difference between the values of F_{min} calculated from the two equations. Notice that F_{min} increases by f^2 above some corner frequency given by f_e, which is bias dependent.

3.1.3 Low-Frequency Noise in Transistors

The mechanisms causing low-frequency $1/f$ noise have been summarized by van der Ziel in a recent review [3.18]. An equivalent circuit for analyzing low-frequency noise in the bipolar transistor is shown in Fig. 3.6, where two noise sources are present in the input. The dominant source of flicker noise is the current generator. This is due primarily to minority carrier recombination in the emitter region [3.19]. The voltage noise source due to thermal noise of the resistance is usually a much lower contribution, so the noise power referred to the input is [3.20]

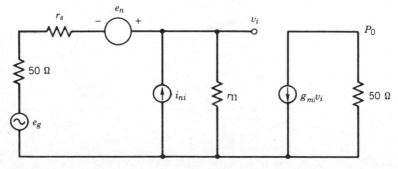

Figure 3.6 Low-frequency noise equivalent circuit of bipolar transistor.

$$P_{ni} = \frac{e_t^2}{200} + \frac{e_n^2}{200} + \frac{i_{ni}^2}{200}(r_s + 50)^2 + 2Ce_n i_{ni}(r_s + 50)$$

$$\simeq \frac{i_{ni}^2}{200}(r_s + 50)^2 = \frac{P_0}{G} \tag{3.40}$$

where $e_t = 4kTr_s\,\Delta f$, thermal noise voltage of r_s, assume zero

      e_n = equivalent input noise voltage, assume zero (determine by letting r_s = 0)

      C = correlation factor, assume zero

      P_0 = noise output power to 50-Ω load

      G = low-frequency gain

Since the noise is essentially current noise, the data are usually plotted in dBA/$\sqrt{\text{Hz}}$, as given in Fig. 3.7. For this measurement, a typical value of r_s is 1 kΩ.

It should be remembered that the finite base resistance, r_b', will allow a certain amount of the base current flicker noise, $i_b r_{bb}'$, to appear in the equivalent voltage source, e_n. Because the thermal noise due to r_{bb}' tends to dominate this flicker noise component, the e_n flicker noise corner frequency is well below that of the base current flicker noise. As the source impedance presented to the base approaches zero, the e_n noise source will begin to dominate i_{ni}.

A major advantage of silicon bipolar transistors is the low corner frequency for flicker noise. The data reported in Fig. 3.7 for the Avantek AT-22000 at a bias of

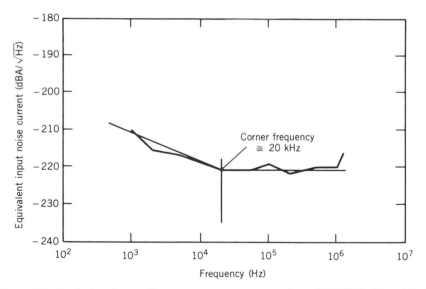

Figure 3.7   Equivalent input noise current versus frequency for an AT-22000 silicon bipolar transistor biased at $V_{CB} = 10$ V, $I_E = 10$ mA. (From Ref. 3.21.)

$V_{CB} = 10$ V, $I_E = 10$ mA gives a corner frequency of about 20 kHz at room temperature. Values below 100 kHz are typical of silicon bipolar transistors [3.21]. A comparison between bipolar and FET flicker noise is shown in a later figure (Fig. 3.21).

3.1.4 Heterojunction Bipolar Transistor

Because of the superior material properties of group III–V compounds such as GaAs, a bipolar transistor using this material has been a goal since 1957 [3.22]. The use of the heterojunction emitter–base has made the heterojunction bipolar transistor (HBT) a reality. Three primary advantages result from this structure (Fig. 3.8) [3.23]:

1. The forward-bias emitter injection efficiency is very high since the wider-bandgap Al GaAs emitter injects electrons into the GaAs base at a lower

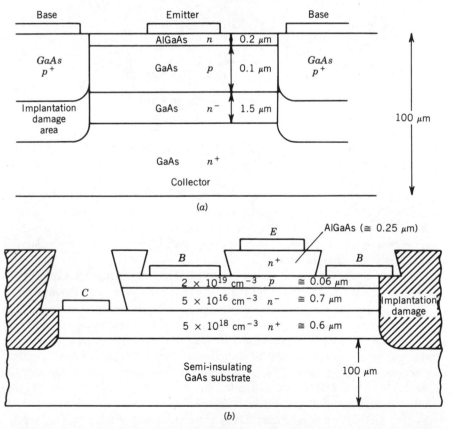

(a)

(b)

Figure 3.8 Heterojunction bipolar transistor structure (HBT): (*a*) single-chip structure; (*b*) HBT structure for GaAs monolithic circuits. [(*b*) from Ref. 3.23.] © 1987 IEEE.

energy level, but the holes are prevented from flowing into the emitter by an energy barrier.

2. The base can be doped heavily to reduce the base resistance.

3. Implant damage can be used to reduce the parasitic collector–base capacitance.

Other advantages for this bipolar transistor are high output current per device unit width or periphery, high current gain, and potentially low $1/f$ noise. Since the entire emitter-area cross section can carry current because of the lower base resistance, the power-handling capability of this structure will be very high. Output power of greater than 4.0 W/mm at 10 GHz has already been reported [3.10].

Although these transistors are not yet available commercially, excellent results have been reported [3.10, 3.11]:

$$f_t = 75 \text{ GHz}$$

$$f_{\max} = 175 \text{ GHz}$$

$$g_m/Z = 7 \text{ S/mm}$$

$$P/Z = 4.0 \text{ W/mm at 10 GHz}$$

$$1.5 \text{ W/mm at 36 GHz}$$

The high output power is a particularly useful feature of this transistor. If the maximum junction temperature can be made high, the realization of a high-power GaAs bipolar transistor may occur.

Another important feature of this transistor is the low $1/f$ noise, since the surface states of GaAs no longer contribute significant noise to the emitter current. A corner frequency below 1 MHz has been found for the HBT [3.24], which is becoming comparable to silicon bipolar transistors. This effect could be very significant for oscillator applications.

The small-signal equivalent circuit of a 1987 HBT from Texas Instruments is given in Fig. 3.9 and Table 3.5 for the *npn* transistor reported by Bayraktaroglu [3.25–3.27]. This is a millimeter-wave transistor with an emitter periphery of 60 μm and an emitter pitch of 4 μm. There are two emitter fingers each of length 15 μm. The total perimeter was calculated by including all of the emitter periphery, both sides. By a different method which includes only the length of emitter metal, one obtains 30 μm, a factor of 2 lower. This second method is the one used in Table 3.3. The nominal bias condition for this model is $V_{CE} = 4$ V, $I_c = 20$ mA.

The S parameters of this model are compared to the measured data on Fig. 3.10 up to 26.5 GHz. As low-noise oscillators these transistors have given the results summarized in Table 3.6.

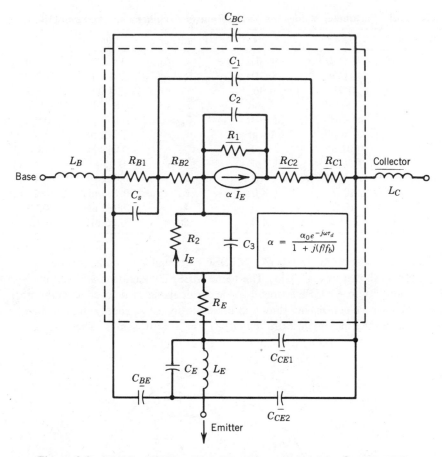

Figure 3.9   Model of HBT to 26.5 GHz. (From Ref. 3.26.) © 1988 IEEE.

3.2 MICROWAVE FET

3.2.1 Gallium Arsenide MESFET Small-Signal Model

The GaAs MESFET is more commonly used in microwave integrated-circuit designs because of higher gain, higher output power, and a lower noise figure in amplifiers. The higher gain is due to higher mobility of electrons (compared to silicon). The improvement in output power is due to the higher electric field and higher saturated drift velocity of the electrons [equation (3.1)]. The lower noise figure is partially due to the higher mobility of the electron carriers. Moreover, fewer noise sources are present in the FET (no shot noise) as compared to the bipolar transistor. A disadvantage of the GaAs MESFET is the higher $1/f$ noise compared to silicon bipolar transistors.

TABLE 3.5   Parameter Values for 60-μm Emitter Periphery *npn* and *pnp* HBTs

Parameter	*npn*	*pnp*	Parameter	*npn*	*pnp*
f_t	22 GHz	19 GHz	C_s	1.34 pF	0
f_{max}	40 GHz	25 GHz	R_{C1}	1 Ω	7.4 Ω
α_0	0.93	0.96	R_{C2}	4 Ω	3.3 Ω
τ	2 ps	4 ps	R_E	8.5 Ω	7.0 Ω
f_b	65 GHz	35 GHz	C_{BC}	0.012 pF	0.012 pF
C_1	0.06 pF	0.04 pF	C_{BE}	0.022 pF	0.022 pF
C_2	0.01 pF	0.1 pF	C_{CE1}	0.012 pF	0.012 pF
C_3	0.4 pF	0.3 pF	C_{CE2}	0.06 pF	0.08 pF
R_1	1.0×10^6	1.0×10^6	C_E	0.022 pF	0.03 pF
R_2	10 Ω	6.8 Ω	L_B	0.165 nH	0.26 nH
R_{B1}	17 Ω	3.0 Ω	L_E	0.032 nH	0.09 nH
R_{B2}	27.5 Ω	4.4 Ω	L_C	0.06 nH	0.134 nH

The cross section of the GaAs MESFET (metal-semiconductor field-effect transistor) is shown in Fig. 3.11. The name MESFET has been adopted because of the similarity to MOSFET (metal-oxide semiconductor field-effect transistor). In Fig. 3.11 the electrons are drawn to the drain by a V_{DS} supply that accelerates the carriers to the maximum drift velocity, $v_s \simeq 2 \times 10^7$ cm/s. The reverse bias

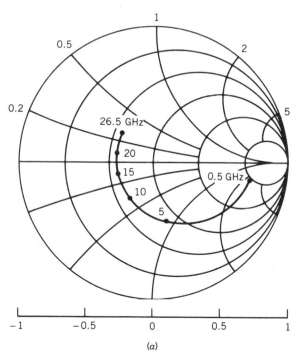

(a)

Figure 3.10   *S* parameters of HBT and model to 26.5 GHz: (a) S_{11}; (b) S_{22}; (c) S_{21}; (d) S_{12}. (From Ref. 3.27.) © 1988 IEEE.

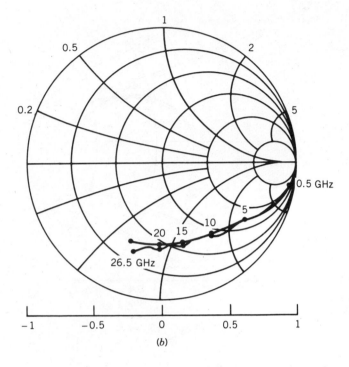

(b)

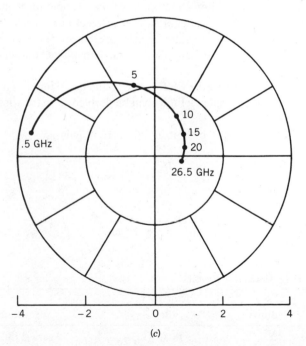

(c)

Figure 3.10   (*Continued*)

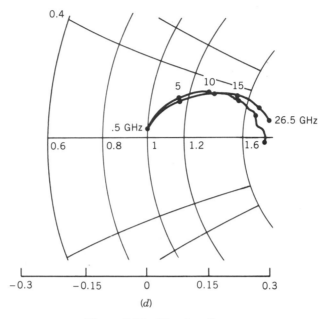

Figure 3.10 (*Continued*)

of the Schottky barrier gate allows the width of the channel to be modulated at a microwave frequency. Thus the majority carrier electrons are modulated by the input signal voltage applied across the input capacitance. There are several interesting contrasts between the FET and the bipolar, which are summarized in Table 3.7.

The frequency limitation of the FET is due to the gate length, which should be as short as possible. The frequency limits can be derived from the simplified hybrid π model in Fig. 3.12.

For the simplified model, the short-circuit current gain is

$$h_{21} = \frac{I_{\text{out}}}{I_{\text{in}}} = \frac{g_{m0} v_c}{I_{\text{in}}} \tag{3.41}$$

$$I_{\text{in}} = \frac{V_{\text{in}}}{R_c + 1/j\omega C_{gs}} \tag{3.42}$$

TABLE 3.6   HBT Oscillator Results

f (GHz)	P_0 (dBm)	$\mathcal{L}(f)$ (dBc/Hz)	Reference
4	10	-73 at 1 kHz	Agarwal [3.28] (Rockwell)
15.6	6.5	-60 at 10 kHz	Lesage et al. [3.29] (NEC)

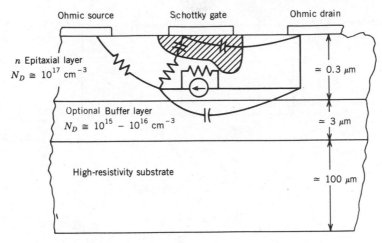

Figure 3.11   GaAs MESFET cross section.

At low frequencies,

$$I_{in} \simeq V_{in} j\omega C_{gs} \simeq v_c j\omega C_{gs} \qquad (3.43)$$

$$h_{21} \simeq \frac{g_{m0}}{j\omega C_{gs}} \qquad (3.44)$$

$$|h_{21}| = \frac{f_t}{f} = \frac{g_{m0}}{2\pi C_{gs}} \frac{1}{f} \qquad (3.45)$$

Thus the frequency where the short-circuit current gain becomes unity is

$$f_t = \frac{g_{m0}}{2\pi C_{gs}} \qquad (3.46)$$

which is an important figure of merit for the GaAs MESFET. The unilateral gain of the FET may be simply calculated from the y parameters in Fig. 3.12.

TABLE 3.7   Characteristics of Bipolar Transistor versus MESFET

Property	Common-Emitter Bipolar	Common-Source MESFET
Geometry	Vertical	Horizontal
Modulation	Base current	Gate voltage
Control signal	Current	Voltage
Frequency limitation	Base length	Gate length
Low-frequency transconductance	High	Low

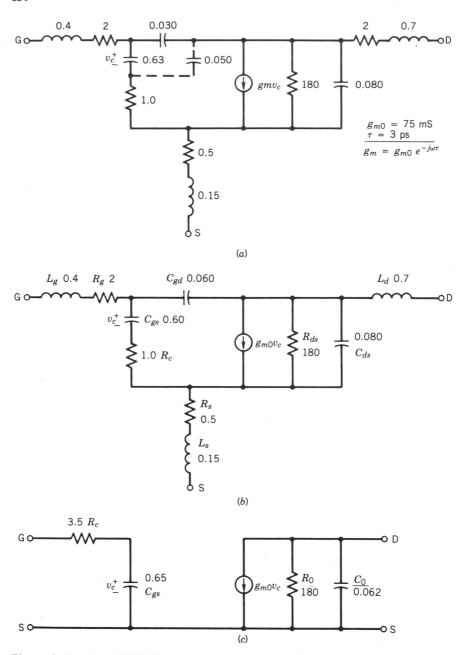

Figure 3.12 GaAs MESFET small-signal model (AT-8251); (*a*) complete model; (*b*) simpler model; (*c*) simplified model. $V_{DS} = 5$ V, $Z = 500$ μm; $I_{DS} = 50$ mA, $L_G = 0.3$ μm.

$$y_{11} = \frac{1}{R_c + 1/j\omega C_{gs}} \qquad (3.47)$$

$$y_{21} = g_{m0} \qquad (3.48)$$

$$y_{22} = 1/R_o + j\omega C_o \qquad (3.49)$$

$$y_{12} = 0 \qquad (3.50)$$

From (1.218),

$$U = \frac{|y_{21}|^2}{4 \, \mathrm{Re}(y_{11}) \, \mathrm{Re}(y_{22})} \qquad (3.51)$$

$$U = \frac{1}{4} \frac{1}{f^2} \left(\frac{g_{m0}}{2\pi C_{gs}} \right)^2 \frac{R_o}{R_c}$$

$$= (f_{max}/f)^2 \qquad (3.52)$$

Thus f_{max} is given by

$$f_{max} = \frac{f_t}{2} \sqrt{\frac{R_o}{R_c}} \qquad (3.53)$$

A high-gain FET requires a high f_t, a high output resistance, a low input resistance, and minimum parasitic elements. Under normal bias conditions the device is biased for maximum drift velocity of the electron carriers (about 3 kV/cm), so we have

$$f_t = \frac{g_{m0}}{2\pi C_{gs}} = \frac{1}{2\pi\tau} = \frac{v_s}{2\pi L_g} \qquad (3.54)$$

This equation shows the importance of short gate length L_g. Another interesting figure of merit for the FET is the g_{m0} per unit gate periphery (Z), given by

$$C_{gs} = \frac{\epsilon A}{d} = \epsilon \frac{L_g Z}{d} \qquad (3.55)$$

$$g_{m0} = \frac{v_s C_{gs}}{L_g} = \frac{v_s \epsilon Z}{d} \qquad (3.56)$$

$$\frac{g_{m0}}{Z} = \frac{v_s \epsilon}{d} \qquad (3.57)$$

For a typical ($Z = 500$-μm gate) FET, this parameter is

$$\frac{g_{m0}}{Z} \simeq \frac{2 \times 10^7 \text{ cm/s } 10^{-12} \text{ F/cm}}{0.13 \ \mu\text{m}}$$

$$= 150 \ \mu\text{S}/\mu\text{m} = 150 \text{ mS/mm}$$

$$g_{m0} \simeq 150 \times 0.50 = 75 \text{ mS}$$

which is in good agreement with measurements. Scaling the device larger in Z increases the transconductance, but the f_t and gain remain constant with scaling if parasitics are negligible. The g_{m0} is about a factor of 6 higher than a silicon MOSFET [3.30, 3.31] with an oxide thickness of 650 Å.

The velocity saturating effect of the GaAs MESFET has several interesting consequences. Referring to Fig. 3.13, we see that the channel can be considered to be two regions: a low-field region with a constant number of carriers and a high-field region with a "constant" velocity, which is discussed later. Since the current continuity is required,

$$I_{DS}/A = qn(x) v(x) \tag{3.58}$$

where

$$n(x)_{\text{max}} = N_D$$

$$v(x)_{\text{max}} = v_{\text{sat}} = v_s$$

the number of carriers must increase above N_D in region II. This causes an electron accumulation at the drain edge of the channel, followed by an electron depletion. In effect, a charge dipole occurs at the drain edge of the channel, which is a very small capacitive effect in the model ($\simeq 0.05$ pF in Fig. 3.12).

In GaAs, the electron carriers will slow down at an electric field greater than 3 kV/cm. The electrons move from a high-mobility state to a low-mobility state in about 1 ps, and thus the velocity of the carriers reaches a peak and slows down in the middle of the channel. The velocity versus electric field of GaAs and silicon is shown in Fig. 3.14. The change in velocity of the carriers in GaAs is the cause of the Gunn effect in Gunn diodes or TEOs (transfer electron oscillators).

In short-channel devices (less than 3 μm), a nonequilibrium velocity field characteristic must be considered. When the electrons enter the high-field region, they are accelerated to a higher velocity. This effect can cause peak velocities of about 4×10^7 cm/s, which relax to 1×10^7 cm/s after traveling about 0.5 μm. The overshoot in velocity reduces the transit time and shifts the dipole charge to the right of the channel.

An estimate of the drain current I_{DSS} can be made using (3.58). For a small-signal GaAs MESFET with

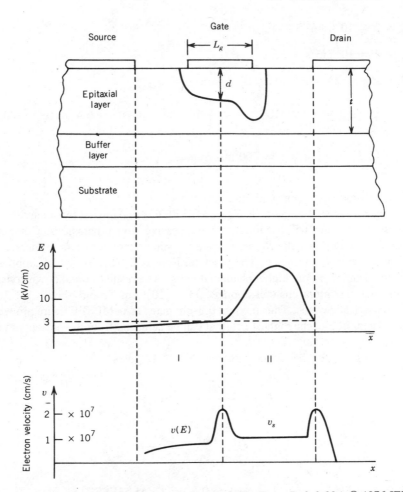

Figure 3.13   GaAs MESFET at high electric field. (From Ref. 3.32.) © 1976 IEEE.

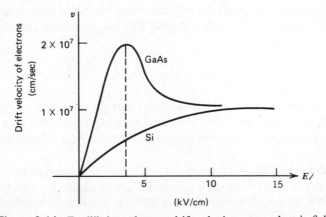

Figure 3.14 Equilibrium electron drift velocity versus electric field.

$$N_D \simeq 2 \times 10^{17} \text{ cm}^{-3}$$

$$v_{\text{sat}} = 2 \times 10^7 \text{ cm/s}$$

$$A = Z(t - d) = Z(0.03)(10^{-4}) \text{ cm}^2$$

$$\frac{I_{DSS}}{A} = qN_D v_{\text{sat}} = 1.6 \times 10^{-19} \times 2 \times 10^{17} \times 2 \times 10^7 = 6.4 \times 10^5 \text{ A/cm}^2$$

$$\frac{I_{DSS}}{Z} = (0.3)(6.4) \text{ A/cm} \simeq 200 \text{ mA/mm}$$

which is in good agreement with measurements.

The high-frequency gain of the GaAs MESFET is maximized by achieving the minimum gate length without introducing excessive device parasitics. Computer studies have shown that the R_G series gate resistance increases and R_o decreases as the gate length is shortened. The practical limit is $L_g/t > 1$, which implies a thin channel and therefore higher channel doping. As a result of breakdown considerations, the maximum channel doping is $4 \times 10^{17} \text{ cm}^{-3}$, and about 2×10^{17} cm^{-3} in practical devices. Thus modern 0.3-μm gate GaAs MESFETs are probably within a factor of 3 of the highest f_{max} that can be achieved from the present device structure.

Evaluating f_{max} for the device shown in Fig. 3.12 gives

$$f_t = \frac{0.075}{2\pi(0.60)} = 20 \text{ GHz}$$

$$f_{\text{max}} = \frac{20}{2}\sqrt{\frac{180}{3.5}} = 72 \text{ GHz}$$

These frequencies are typical of modern 0.3-μm GaAs MESFETS.

The small-signal models of several GaAs MESFET chips are given in Table 3.8. The perimeter of the transistor will determine the output power capability and the maximum frequency of broadband gain. The low-frequency broadband gain is given by

$$S_{21}(\text{LF}) = -2g_m Z_0 \tag{3.59}$$

which indicates that high gain requires a large gate perimeter for a large transconductance. The smaller perimeter devices provide more gain at higher frequency because the gate fingers are shorter (less phase shift) and the input capacitance C_{gs} is smaller.

The selection of a low-noise transistor is also based on the transistor perimeter. Above 12 GHz, a gate perimeter less than 250 μm is needed for a minimum noise figure [3.33] and high gain. At 4 GHz, a 500-μm perimeter is recommended and

TABLE 3.8 GaAs MESFET Chip Models from Avantek

Name and Bins	L_g (nH)	L_d (nH)	L_s (nH)	R_c (Ω)	R_g (Ω)	R_s (Ω)	R_{ds} (Ω)	C_{gs} (pF)	C_{gd} (pF)	C_{ds} (pF)	g_{m0} (mS)	Z (μm)
AT-10600, $Z = 250\ \mu$m												
3 V, 10 mA	0.6	0.7	0.15	2	5	5	275	0.16	0.03	0.06	27	250
5 V, 30 mA	0.6	0.7	0.15	2	5	5	275	0.26	0.015	0.06	42	250
AT-8251, $Z = 500\ \mu$m												
3 V, 20 mA	0.4	0.7	0.15	2	2.5	1	150	0.36	0.07	0.16	48	500
5 V, 50 mA	0.4	0.7	0.15	1	2	0.5	180	0.60	0.06	0.08	75	500
AT-8111, $Z = 750\ \mu$m												
3 V, 20 mA	0.4	0.5	0.15	1.5	1.5	1	180	0.70	0.10	0.14	68	750
5 V, 80 mA	0.4	0.5	0.15	1	1	0.5	180	1.2	0.08	0.15	115	750

at 2 GHz a 750-μm perimeter would usually give best noise figure and gain performance.

The dual-gate GaAs MESFET is simply two adjacent gates with a cascade connection normally used (CS followed by CG). The cross section in Fig. 3.15 is typical of the dual-gate transistor. The second gate can be used for AGC by varying the dc voltage at gate 2. As an amplifier, the dual-gate FET has higher gain with gate 2 RF grounded.

The dual-gate FET is also called a cascode [3.34] from the vacuum-tube prototype. From investigation of two device pairs with nine possibilities, the common-source (cathode) common-gate (grid) combination was found to have the lowest noise figure and was given the name cascode, which has continued to be used for the bipolar common-emitter and common-base pair, the FET common-source and common-gate pair, and the dual-gate GaAs MESFET, where the second gate is assumed to be at RF ground in the cascode connection. Since this device is simple

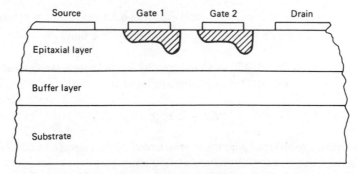

Figure 3.15 Dual-gate GaAs MESFET cross section.

to dc bias and has a low noise figure with high gain, it continues to be useful for many amplifier applications. In addition, the dual-gate FET can be used for mixers, multipliers, AGC amplifiers, and oscillators.

The dual-gate FET has an S_{11} similar to the common-source FET, an S_{22} similar to the common-gate FET, a very low S_{12} which is given by

$$S_{12} = \frac{(S_{12})_1 (S_{12})_2}{1 - (S_{22})_1 (S_{11})_2} \simeq 0 \tag{3.60}$$

and a high S_{21} given approximately by

$$S_{21} = \frac{(S_{21})_1 (S_{21})_2}{1 - (S_{22})_1 (S_{11})_2} \tag{3.61}$$

The effective transductance is given by

$$g_m = y_{21} = \frac{-(y_{21})_1 (y_{21})_2}{(y_{22})_1 + (y_{11})_2} \simeq g_{m1} \tag{3.62}$$

In dual-gate amplifier or oscillator applications, the two port S parameters can be used in the same manner as the single-gate GaAs FET; usually, the second gate is at RF ground for this two-port measurement.

3.2.2 GaAs MESFET Noise Model

The description of the noise performance of a MESFET begins with the junction field-effect transistor (JFET), originally proposed by Shockley [3.35], which treated the case of a gate length L_g which is at least three times the channel thickness a. This gave the well-known long-channel analysis where the channel is divided into two regions:

1. The gradual channel region from the source electrode to the pinch-off point
2. The depleted region from the pinch-off point to the drain electrode.

The thermal noise of a JFET was analyzed by Bruncke and van der Ziel [3.36] and shown to be represented by an equivalent input resistor.

$$R_n = 0.8/g_m \tag{3.63}$$

The high-frequency induced gate noise was added by van der Ziel [3.37, 3.38] to predict the increase in noise figure versus frequency.

The early analyses of GaAs MESFET noise performance were reported by W. Baechtold [3.39–3.41], but the effects of velocity saturation were not included.

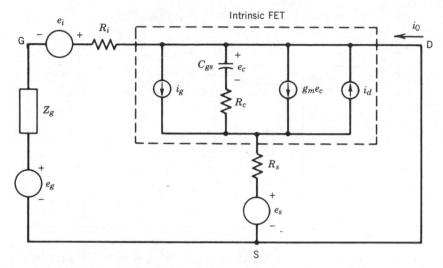

Figure 3.16 GaAs MESFET simplified noise model, including intrinsic and extrinsic noise sources. (From Refs. 3.45 and 3.46.) © 1976 IEEE.

The JFET model has been modified by many authors (see Ref. 3.42 for a complete discussion), until finally Turner and Wilson [3.43] first analyzed the velocity saturation effects in GaAs for short-channel transistors. This was later modified by Grebene and Ghandi [3.44] to allow the velocity saturation to occur in the channel (instead of at the drain electrode), which gave the present two-section model of the JFET given in Fig. 3.13. An analysis of this model was reported by Pucel et al. [3.45, 3.46], which used the equivalent circuit in Fig. 3.16.

The noise sources are the thermal noise of two resistors, R_i and R_s, the high-frequency induced gate noise i_g, and the channel thermal noise i_d. These noise terms and the generator thermal noise are given by ($\Delta f = 1$ Hz)

$$\overline{e_g^2} = 4kTR_g \tag{3.64}$$

$$\overline{e_i^2} = 4kTR_i \tag{3.65}$$

$$\overline{e_s^2} = 4kTR_s \tag{3.66}$$

$$\overline{i_g^2} = 4kTg_{gn} \tag{3.67}$$

$$\overline{i_d^2} = 4kTg_{dn} \tag{3.68}$$

$$\overline{i_g^* i_d} = jC\sqrt{\overline{i_g^2}\,\overline{i_d^2}} \tag{3.69}$$

The intrinsic noise of a FET is thermal noise in the channel, which can be represented by two current noise generators, one in the drain circuit and one in the gate circuit [3.37, 3.38]. Since the gate noise generator represents the noise induced in

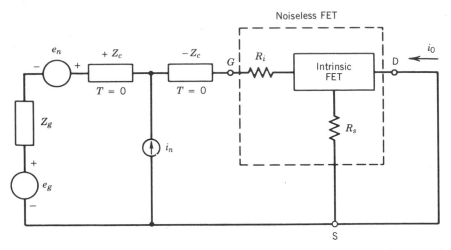

Figure 3.17 Noisy FET represented by a noiseless FET preceded by a noise network. (From Refs. 3.46 and 2.3.) © 1976 IEEE.

the gate by the thermal fluctuations in the drain current, these noise currents are partially correlated.

Because of velocity saturation effects, the channel has been divided into two regions which are separately analyzed for noise. This analysis will lead to three new parameters (K_r, K_g, K_c) [3.46], which describe the properties of the intrinsic noise generators i_g, i_d, and their correlation at the specified bias and temperature condition (usually $T_0 = 290$ K).

Following the noise descriptions given in Chapter 2, the noise model of Fig. 3.16 can be represented by a noise network and a noiseless FET, as shown in Fig. 3.17. The noise figure is given by

$$F = 1 + \left[1/R_g \left(r_n + g_n \left| Z_g + Z_c \right|^2 \right) \right] \qquad (3.70)$$

where R_g is the real part of the generator impedance at 290 K. The parameters r_n and g_n are the noise resistance and noise conductance, while z_c is the correlation impedance. The noise parameters are given by

$$r_n = (R_s + R_i) \frac{T_\alpha}{T_0} + K_r \frac{1 + \omega^2 C_{gs}^2 R_c^2}{g_m} \qquad (3.71)$$

$$g_n = K_g \frac{\omega^2 C_{gs}^2}{g_m} \qquad (3.72)$$

$$Z_c = R_s + R_i + \frac{K_c}{y_{11}} \qquad (3.73)$$

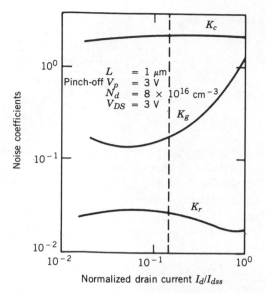

Figure 3.18 Drain current dependence of noise coefficients of a GaAs FET for specific set of design parameters and drain voltage. (From Ref. 3.46.) © 1976 IEEE.

$$\frac{1}{y_{11}} = R_c + \frac{1}{j\omega C_{gs}} \tag{3.74}$$

$$e_n = \sqrt{4kT_0 r_n} \tag{3.75}$$

$$i_n = \sqrt{4kT_0 g_n} \tag{3.76}$$

where T_α is the temperature of the FET.

The calculated noise coefficients are given in Fig. 3.18 as a function of drain bias current for a GaAs MESFET with a gate length of 1 μm. The dashed line at 0.15 I_d/I_{DSS} represents the bias for minimum noise figure [3.47, 3.48]. The minimum noise figure is achieved when the generator impedance is given by

$$R_g = R_{g0} = \sqrt{\text{Re}(Z_c^2) + r_n/g_n} \tag{3.77}$$

$$X_g = X_{g0} = -\text{Im}(Z_c) \tag{3.78}$$

and the corresponding noise figure is

$$F_{\min} = 1 + 2g_n\left[\text{Re}(Z_c) + R_{g0}\right] \tag{3.79}$$

which can also be expressed as

$$F_{\min} = 1 + 2\frac{\omega C_{gs}}{g_m} \sqrt{K_g[K_r + g_m(R_s + R_i)]}$$

$$+ 2\left(\frac{\omega C_{gs}}{g_m}\right)^2 [K_g g_m(R_i + R_s + K_c R_c)] \tag{3.80}$$

which shows the relative importance of the intrinsic and extrinsic noise sources. Notice that a high f_t [equation (3.46)] will lead to a low value for $F_{\min}$.

The analyses by Pucel et al. [3.42, 3.45, 3.46] have led to noise expressions based on theory. An alternative approach taken by Fukui leads to noise expressions based on measurements [3.49, 3.50], which are given by

$$F_{\min} = 1 + 0.016 f C_{gs} \sqrt{\frac{R_i + R_s}{g_m}} \tag{3.82}$$

$$R_n = \frac{0.8}{g_m} \tag{3.83}$$

$$R_{op} = 2.2\left(\frac{1}{4g_m} + R_i + R_s\right) \tag{3.84}$$

$$X_{op} = \frac{160}{f C_{gs}} \tag{3.85}$$

These parameters are easier to relate to transistor design parameters; as the processing improves, the parameters will need to be adjusted accordingly. Podell has also reported a simple empirical model [3.51] dependent on a parameter Q, which modifies the Bruncke–van der Ziel analysis to account for the equivalent noise resistance by

$$R_n = Q/g_m \tag{3.86}$$

where Q may vary from 0.6 to 1.2, depending on bias and processing details. This analysis has been applied by Gupta et al. [3.52] to show excellent agreement with measurements to 18 GHz. An up-to-date review paper by Cappy [3.53] summarizes noise modeling of both GaAs MESFETs and MODFETs.

For the FET, the low-frequency equivalent circuit is given in Fig. 3.19. The noise is caused by traps in the gate-channel depletion layer [3.54], traps in the substrate [3.55], and possibly surface states created by the passivation [3.56]. Much higher $1/f$ noise occurs in MOSFETs because of the traps in the oxide [3.57]. The noise power referred to the input becomes

$$P_{ni} \simeq e_n^2/200 = P_0/G \qquad (r_s = 0) \tag{3.87}$$

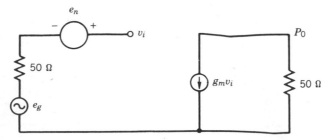

Figure 3.19 Low-frequency noise equivalent circuit of FET.

Since the noise is voltage noise, the data are usually plotted in $\mathrm{dBV}/\sqrt{\mathrm{Hz}}$ (Fig. 3.20).

Both sets of data have been plotted as noise power in Fig. 3.21 to demonstrate the superior, much lower, corner frequency for the silicon bipolar transistor. These data are representative of microwave transistor noise at low frequencies, but are very dependent on the process for making the transistors. As improvements are found, the corner frequencies should continue to decrease.

Note that straight-line approximations of device low-frequency noise are rarely accurate. Since most of the observed noise is due to discrete traps (which have a noise spectrum like a first-order low-pass filter) the device noise spectrum varies about the $1/f$ line. As MESFETs are cooled, the discrete trap frequencies are more apparent.

Finally, to summarize the 1988 noise performance of gallium arsenide MESFETs and silicon bipolar transistors, the minimum noise figure of these

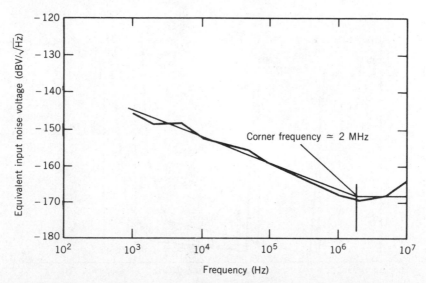

Figure 3.20 Equivalent input noise voltage versus frequency for an AT-10600 gallium arsenide MESFET at $V_{DS} = 3$ V, $I_D = 40$ mA. (From Ref. 3.21.)

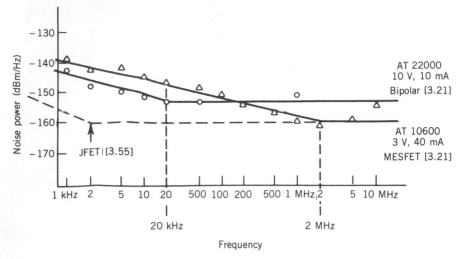

Figure 3.21 $1/f$ noise for microwave transistors. (From Refs. 3.21 and 3.55.)

transistors has been plotted in Fig. 3.22 for room temperature. The GaAs MESFETs will dominate the microwave region, but silicon bipolars will continue to find applications, especially for low-noise oscillators.

3.2.3   MODFET/HEMT

By using heterojunction semiconductor material, AlGaAs interfacing with GaAs, a new field-effect microwave semiconductor device can be manufactured with

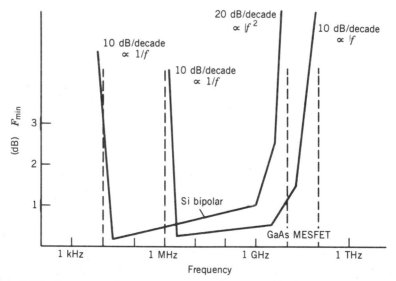

Figure 3.22 $F_{\min}$ versus frequency for low-noise silicon bipolar transistor and for low-noise gallium arsenide FET.

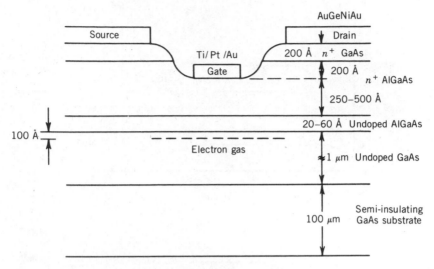

Figure 3.23 MODFET/HEMT structure. (From Refs. 3.1 and 3.58.)

superior microwave performance. This device is the MODFET (modulation-doped field-effect transistor), which is also called a HEMT (high electron mobility transistor), a SDHT (selectively doped heterostructure transistor), or a TEGFET (two-dimensional electron gas FET); the cross section of this transistor is given in Fig. 3.23 [3.58, 3.1].

The basic properties of the heterojunction can be understood from the difference in energy gap between the two materials, which causes band bending, resulting in an electron gas with high electron mobility in the undoped GaAs which is provided by the donors in the AlGaAs. The band bending results in a quantum well where a large population of electrons forms a two-dimensional gas which can easily be modulated by the gate voltage. This is analogous to an n-channel MOSFET, where the number of conduction electrons in the channel is controlled by the gate voltage. Since the band bending forces the electrons to be resident in the undoped GaAs layer, the electrons exhibit a very high mobility and high v_s even at room temperature, which accounts for the superior microwave performance.

The structure of the MODFET can be explained by treating the region beneath the gate metal and the GaAs buffer layer (Fig. 3.23). If the width of the n^+AlGaAs donor layer is very thin (ca. 250 Å), the depletion layer below the Schottky gate metal will extend into the undoped GaAs electron gas and interrupt the electron gas; this gives an enhancement-mode FET, since no channel flows if $V_{GS} = 0$ (a positive V_{GS} is needed). If the n^+ AlGaAs donor layer is thicker (ca. 500 Å), the depletion region only reaches the undoped AlGaAs layer, which is also depleted; this gives a depletion-mode FET. The voltage V_{GS} will modulate the population of electrons in the quantum well and therefore the I_{DS} of the FET. Since the electrons travel in an undoped GaAs region with few ionized donors, the mobility and v_S are larger for this structure compared to a normal GaAs FET with $N_D \simeq 10^{17}$ cm^{-3} (see Table 3.1).

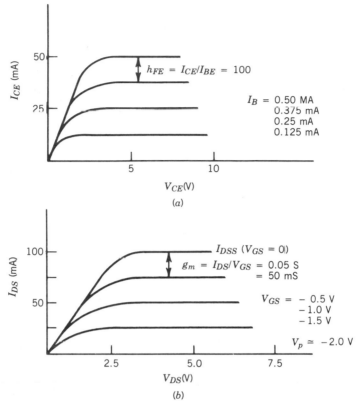

Figure 3.24 Dc characteristics of (*a*) a silicon bipolar transistor (AT-41400) and (*b*) a GaAs MESFET (AT-8251).

The output power of this structure is limited by the sheet carrier concentration of about 10^{12} electrons/cm^2, which limits the maximum output current. A technique for raising the sheet concentration is the use of multiple heterojunctions to form series superlattice structures. As an example, a four-layer MODFET gave about three times the output power of the comparable single-layer MODFET at 10 GHz (see Ref. 3.59, Fig. 4).

TABLE 3.9   Transistor Bias Points

Application	Si Bipolar (AT-41400)	GaAs MESFET (I_{DSS} = 100 mA) (AT-8251)
Low noise	V_{CE} = 8 V, I_{CE} = 10 mA	V_{DS} = 3 V, I_{DS} = 20 mA
High gain	V_{CE} = 8 V, I_{CE} = 50 mA	V_{DS} = 3–5 V, I_{DS} = 100 mA
High output power and low distortion	V_{CE} = 8 V, I_{CE} = 25 mA	V_{DS} = 5–7 V, I_{DS} = 50 mA
Class B	V_{CE} = 10 V, I_{CE} = 0 (with P_{in} = 0)	V_{DS} = 8 V, I_{DS} = 0 (with P_{in} = 0)

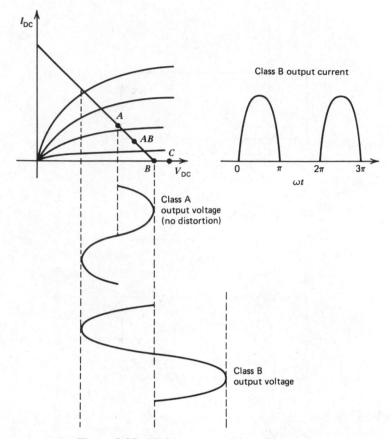

Figure 3.25 High-power operation of transistors.

3.3 DC BIASING

3.3.1 Simple Bias Circuitry

Before considering the RF design of an amplifier or oscillator, the dc biasing point should be established. Poor performance in a final design can often be traced to improper dc biasing.

The common-emitter or common-source biasing configuration is discussed here. Notice that the dc biasing is completely independent of the RF two-port configuration. For example, a transistor may operate as an amplifier in the common-gate mode, but the dc biasing circuit is in the common-source mode. The dc collector and drain characteristics of a typical bipolar and FET are given in Fig. 3.24.

The output current for both types of transistors is controlled by one biasing voltage. For the bipolar transistor, V_{BE}, the voltage across the forward-biased emitter–base junction determines the base drive current and therefore the collector output current. For the FET $- V_{GS}$, the voltage across the reverse-biased gate–source junction determines the channel width and therefore the drain output current.

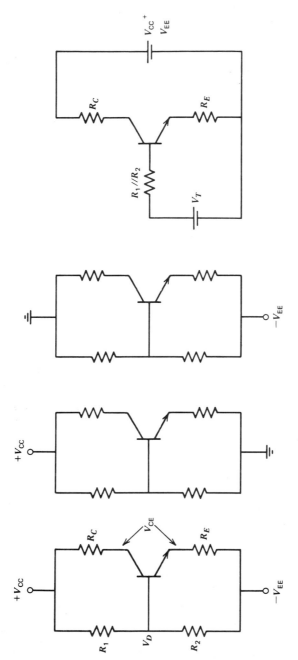

Figure 3.26 Bipolar transistor bias circuits in the simplest form.

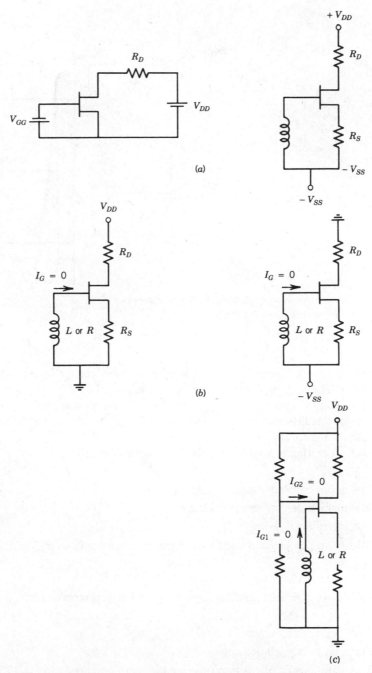

Figure 3.27 MESFET bias circuits: (*a*) two-supply bias; (*b*) single-supply bias; (*c*) dual-gate bias.

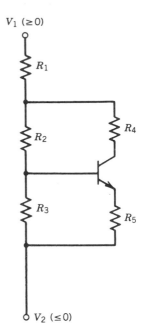

Figure 3.28 Passive bias circuit using two power supplies for either a bipolar transistor or a FET.

The biasing point depends on the application; some useful guidelines are listed in Table 3.9. The high-power biasing point is determined by the limitations of the particular transistor and its safe operating region. Always refer to the vendor's data sheet for this information. The definitions of class A, AB, B, and C are given graphically in Fig. 3.25.

The safe operating region for the bipolar is determined by

1. Maximum collector current
2. Maximum collector-emitter voltage
3. Secondary breakdown [3.2]
4. Maximum power dissipation (maximum junction temperature $\simeq 200°C$ for silicon)

For the GaAs MESFET, the safe operating region is determined by:

1. Maximum drain current
2. Maximum drain–source voltage
3. Maximum power dissipation (maximum junction temperature $\simeq 175°C$ for GaAs presently)
4. Maximum input power to gate

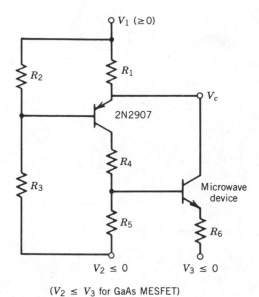

$V_1 \, (\geq 0)$

R_2

R_1

V_c

2N2907

R_4

R_3

Microwave
device

R_5

R_6

$V_2 \leq 0$

$V_3 \leq 0$

$(V_2 \leq V_3$ for GaAs MESFET)

Figure 3.29 Active bias circuit with PNP transistor and multiple power supplies.

For GaAs the junction temperature is limited by a possible chemical reaction or metal degradation [3.3, 3.4] to 175°C for high reliability. Moreover, lower thermal conductivity occurs for this material, so excellent heat sinking must be provided.

Some simple biasing circuits are tabulated in Fig. 3.26 for bipolars and in Fig. 3.27 for MESFETs. For silicon bipolars, the emitter–base is forward biased at about 0.7 V and the collector–base is reverse biased at a voltage dependent on device breakdown and desired P_{out}, typically, 8 to 12 V for small-signal microwave devices. For the FET, the gate to source is reverse biased and the drain to source is biased to the "saturation region," greater than 3 V, where the electrons travel at saturated drift velocity.

Bipolar circuits often use voltage-divider circuits. The design equations are

$$V_{CC} + V_{EE} = I_C R_C + V_{CE} + I_E R_E \tag{3.88}$$

$$V_D = 0.7 + I_E R_E \tag{3.89}$$

By making

$$R_E \gg \frac{R_B}{h_{\text{FE}}} = \frac{R_1 R_2}{R_1 + R_2} \frac{1}{h_{\text{FE}}} \tag{3.90}$$

the bias circuit becomes independent of h_{FE}.

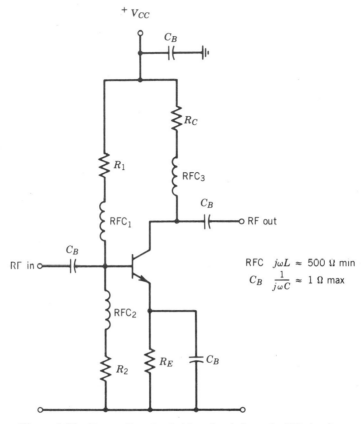

Figure 3.30 Decoupling the dc bias circuit from the RF circuit.

Normally, the single-supply circuits are preferred with the reverse-bias gate–source given by

$$V_{GS} = I_{DS}R_S \qquad (3.91)$$

The first two-supply circuit may be required when dc efficiency is important, as it has no dissipative losses in a source resistor. The single-supply dual-gate circuit is shown to indicate the simplicity of dual-gate bias.

To improve the dc efficiency of the voltage-divider bias circuits, it is possible to replace the dropping resistors (R_D or R_C) by RF transistors. This method of dc biasing is illustrated by examples in Chapter 4.

3.3.2 Temperature Stability

One of the most important considerations in circuit design is the temperature stability of the transistor bias point. The parameters that must be controlled are

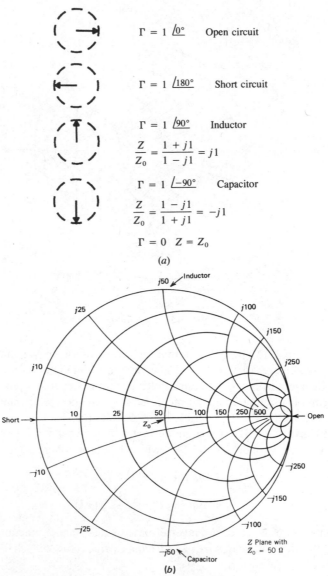

$\Gamma = 1 \; \underline{/0°}$ Open circuit

$\Gamma = 1 \; \underline{/180°}$ Short circuit

$\Gamma = 1 \; \underline{/90°}$ Inductor

$$\frac{Z}{Z_0} = \frac{1 + j1}{1 - j1} = j1$$

$\Gamma = 1 \; \underline{/-90°}$ Capacitor

$$\frac{Z}{Z_0} = \frac{1 - j1}{1 + j1} = -j1$$

$\Gamma = 0 \quad Z = Z_0$

(a)

(b)

Figure 3.31 Smith chart: (a) important points on the Smith chart; (b) coordinates in ohms, Z plane with $Z_0 = 50 \; \Omega$; (c) Y plane normalized to Y_0.

V_{BE} for bipolars and $-V_{GS}$ for FETs, which can be set with either a passive bias circuit or an active bias circuit.

For silicon bipolar transistors, the collector current will increase with temperature as h_{FE} increases due to the properties of the forward-biased emitter–base junction. A complete discussion of temperature and current effects on h_{FE} is given in Ref. 3.2 (pp. 134–147). Although the junction breakdown voltages increase

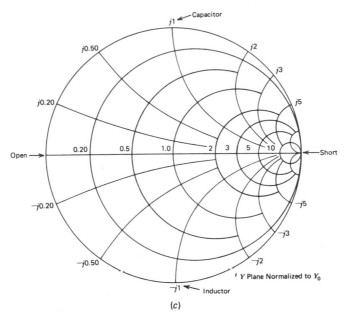

(c)

Figure 3.31 (*Continued*)

with temperature [3.2, p. 575], normal selection of dc bias point will never approach junction breakdown. Since h_{FE} increases with temperature, the S parameters will change, leading to variations in RF performance. Also, due to the higher collector current resulting from the increased h_{FE}, junction temperature will rise, lowering device reliability or possibly damaging the transistor.

For the GaAs MESFET, the drain current will decrease with temperature due to the temperature variations on (3.58), which is repeated here

$$I_{DS} = Aqn(x)\,v(x)$$

$$= Zqn(x)\,v(x)\big[t - d(x)\big] \qquad (3.58)$$

As the temperature increases the reverse-biased gate–channel junction will widen [3.2, p. 77], which increases $d(x)$ and reduces I_{DS}. This will change all the S parameters of the transistor.

TABLE 3.10   Shunt Microstrip-Line Matching Elements

Length, $2\beta l$ (deg)	Open Stub, Y/Y_0	Shorted Stub, Y/Y_0
45	$j0.42$	$-j2.42$
90	$j1.0$	$-j1.0$
180	∞	0
270	$-j1.0$	$j1.0$

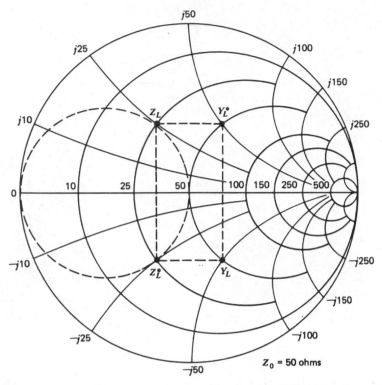

Figure 3.32 Smith chart with $Z_L/Z_0 = 0.5 + j0.5$. Coordinates in ohms ($Z_0 = 50\ \Omega$).

The most general passive bias circuit is given in Fig. 3.28. In this circuit some of the resistor values may be zero. The temperature stabilizing resistors are R_1 (collector or drain feedback) and R_5 (emitter or source feedback). If the collector current starts to increase, the increased voltage drop across the resistors will reduce the collector or drain current. A complete analysis of this problem is straightforward using SPICE or other nodal analysis programs.

TABLE 3.11 Impedance Matching Elements

Matching Element	Lumped	Distributed Microstrip Line
Inductance series	$Z = j\omega L$	Cascade microstrip line
Inductance shunt	$Y = \dfrac{1}{j\omega L}$	Parallel shorted stub
Capacitive series	$Z = \dfrac{1}{j\omega C}$	Not used
Capacitive shunt	$Y = j\omega C$	Parallel open stub

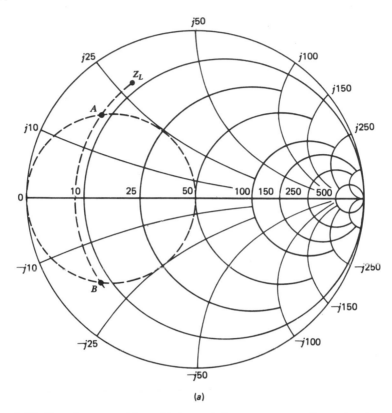

(a)

Figure 3.33 Impedance matching with lumped elements: (*a*) coordinates in ohms ($Z_0 = 50 \, \Omega$) ($Z_L/Z_0 = 0.15 + j0.60$); (*b*) four matching circuits.

The most common active bias circuit is shown in Fig. 3.29. This circuit can be used to bias both bipolar transistors and GaAs MESFETs. It is relatively insensitive to either h_{FE} or pinch-off voltage, and allows independent setting of bias voltage and bias current.

The bias voltage applied to the microwave device is V_{BE} of the PNP transistor above the voltage set by the divider formed by R_2 and R_3. The relatively constant V_{BE} of the silicon PNP transistor, which changes at $-2.5 \, \text{mV}/{}^\circ\text{C}$, produces a dc operating point with excellent temperature stability. The current flow through the divider is typically kept below 1 mA to minimize power dissipation. With the supply voltage and the bias voltage established, the microwave device bias current is set by the voltage drop across R_1.

R_4 isolates the PNP transistor from the input of the microwave device and optionally can be replaced by an RFC (RF choke). R_5 provides a path to ground for the current flowing through the PNP transistor, it is mandatory when biasing

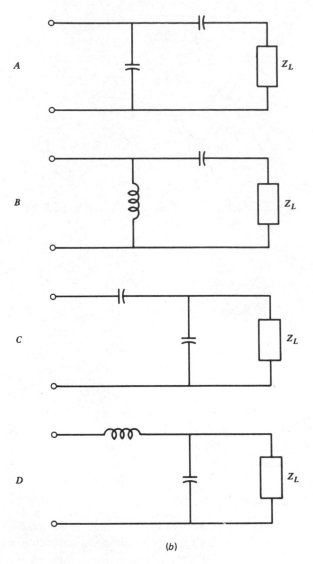

(b)

Figure 3.33   (*Continued*)

FETs that cannot tolerate gate current, but can be omitted when biasing bipolars that can accept base current. R_6 adds additional bias stability by acting as a series feedback stabilization element, and if used must be bypassed if full RF gain is to be maintained. A typical version of this circuit for bipolar transistors would ground V_2 and V_3, and (optionally) open circuit R_5. A typical version for GaAs MESFETs would set $R_6 = 0 \ \Omega$ and $V_2 = -5$ V, $V_3 = 0$.

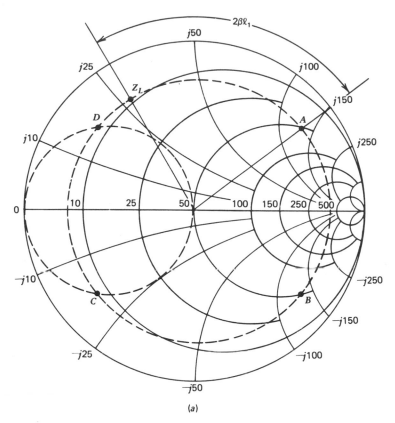

(a)

Figure 3.34 Impedance matching with a 50-Ω microstrip-line element: (a) coordinates in ohms ($Z_L/Z_0 = 0.15 + j0.60$); (b) six matching circuits.

3.3.3 Bias Decoupling

Bias decoupling networks prevent the dc bias circuit from presenting unwanted impedances to the microwave device. Improper impedance loading can reduce gain and output power capability, and often will result in oscillations. A typical decoupling network is shown schematically in Fig. 3.30.

Decoupling networks are most often designed to present a high impedance in the RF band, so the matching circuitry can establish the desired impedance match unperturbed. At low frequencies, where most microwave devices are potentially unstable and the RF matching circuitry usually reduces to an effective open circuit due to the dc blocking capacitors, decoupling networks can be used to provide an impedance termination (often 50 Ω) that is in the stable region of the microwave device. Since decoupling networks typically establish the low-frequency stability for an amplifier, they should be included in any stability analysis; they are usually omitted from simple RF matching analysis.

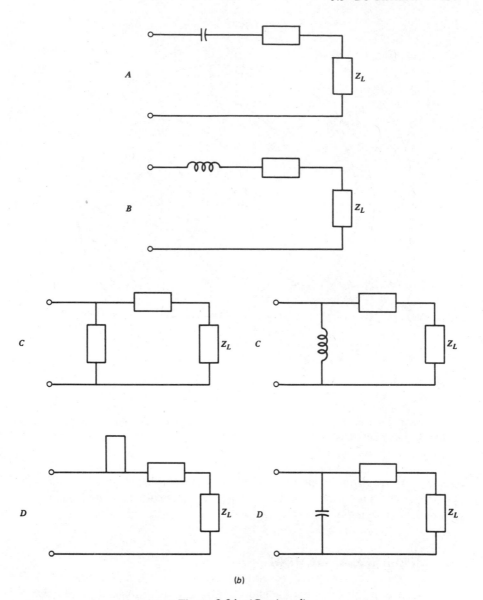

(b)

Figure 3.34 (*Continued*)

The RF chokes may be inductors, quarter-wave, high-impedance shorted stubs, or shorted stubs that are part of the RF matching circuit. Shorted stubs used as chokes are shorted through a bypass capacitor providing a low-impedance path to ground. If the resistors in the bias network are large (i.e., ×10) compared to the RF impedance at the point of attachment, RF chokes may not be necessary.

Extra decoupling is sometimes necessary when using PNP transistor active bias

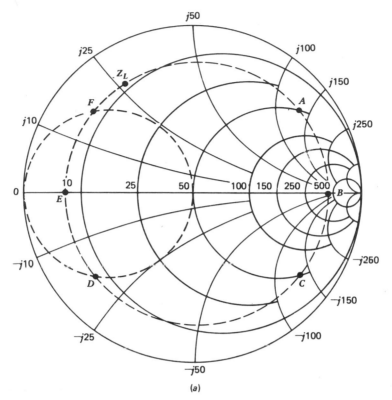

(a)

Figure 3.35 Impedance-matching problem: (*a*, *b*) Z-plane solutions; (*c*, *d*) Y plane solutions. [(*a*, *b*) coordinates in ohms ($Z_0 = 50 \ \Omega$) ($Z_L/Z_0 = 0.15 + j0.60$).]

circuits. Any or all terminals of the PNP may need to be ac grounded through bypass capacitors to ensure that there are no RF effects or oscillations due to the gain now present in the bias circuit.

3.4 IMPEDANCE MATCHING

An alternative title for this book could be *Applications of Impedance Matching*. The most important design tool in amplifier and oscillator design is the concept of impedance matching. All values of impedance can be plotted in the reflection coefficient plane, where $|\Gamma| < 1$ for positive resistance loads. Recalling (1.85) and Fig. 1.20, we see that the impedance at any point on a transmission line is

$$\frac{Z(x)}{Z_0} = \frac{1 + \Gamma_L e^{-j2\theta}}{1 - \Gamma_L e^{-j2\theta}} = \frac{1 + \Gamma(x)}{1 - \Gamma(x)} \qquad (3.92)$$

$$\Gamma(x) = \Gamma_L e^{-j2\theta} = \frac{Z(x) - Z_0}{Z(x) + Z_0} \qquad (3.93)$$

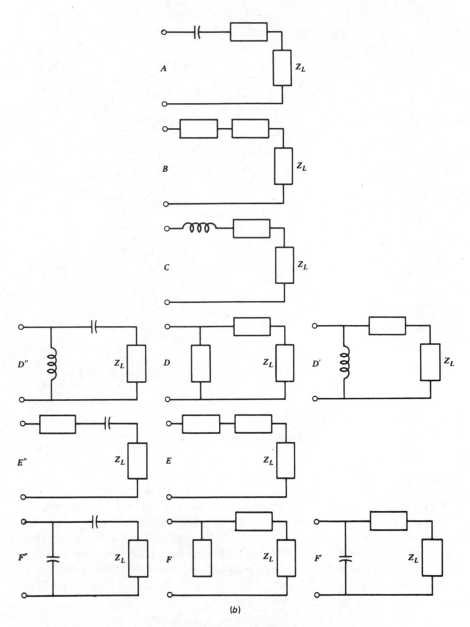

(b)

Figure 3.35   (*Continued*)

For any value of the reflection coefficient, a corresponding load impedance $Z(x)$ is found for the transmission line of characteristic impedance Z_0. For $|\Gamma_L| < 1$, only positive resistance values can occur. The polar plot of $|\Gamma_L| < 1$ is the Smith chart plane. This chart can be visualized as both a plot of $\Gamma_L e^{j2\theta}$ and a plot of $Z(x)$, the load impedance viewed at reference plane x. Notice that the phase shift

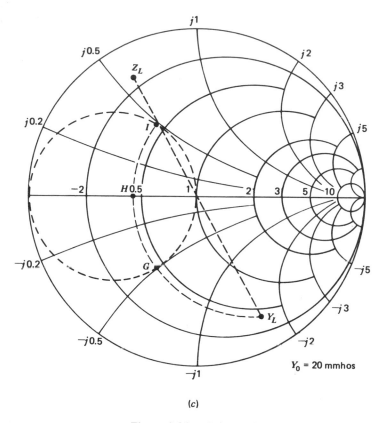

(c)

Figure 3.35   (*Continued*)

as one moves toward the generator is $e^{-j2\theta}$ ($\simeq e^{-j2\beta x}$ for lossless lines). Thus clockwise rotation on the Smith chart plane is movement toward the generator.

Several important points on the Smith chart [3.60] are given in Fig. 3.31. Both the impedance plane and the admittance plane are shown. For microstrip-line shunt elements, the admittance plane should be used, since these are shunt matching elements. Examples of lossless shunt matching elements are given in Table 3.10 for understanding the technique of impedance (admittance) matching. These elements should be considered lossless matching elements.

Impedance matching normally involves using lossless matching elements to move from a point on the impedance (or admittance) plane to the center of the chart where $Z/Z_0 = Y/Y_0 = 1$. This movement should be done with as few lossless elements as possible, usually two, which should have the minimum length if microstrip lines. The matching circuit could be a combination of lumped and distributed microstrip-line elements. Since an infinite number of solutions are possible, it is important to find several simple solutions so that practical considerations can be used to select the best design.

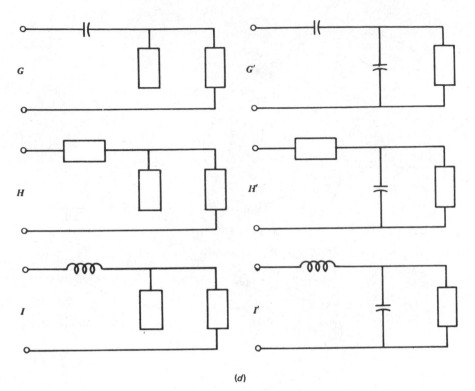

(d)

Figure 3.35 (*Continued*)

TABLE 3.12 Element Values for Solution of Fig. 3.35 ($f = 4$ GHz)

	First Element	Second Element
A	$Z_0 = 50\ \Omega,\ l = 0.112\lambda$	$C = 0.30$ pF
B	$Z_0 = 50\ \Omega,\ l = 0.163\lambda$	$Z_0 = 154\ \Omega,\ l = 0.250\lambda$
C	$Z_0 = 50\ \Omega,\ l = 0.214\lambda$	$L = 5.2$ nH
D	$Z_0 = 50\ \Omega,\ l = 0.362\lambda$	$Z_0 = 50\ \Omega,\ l = 0.058\lambda$
D'	$Z_0 = 50\ \Omega,\ l = 0.362\lambda$	$L = 0.70$ nH
D"	$C = 0.90$ pF	$L = 0.70$ nH
E	$Z_0 = 50\ \Omega,\ l = 0.413\lambda$	$Z_0 = 16.6\ \Omega,\ l = 0.250\lambda$
E"	$C = 1.3$ pF	$Z_0 = 19.4\ \Omega,\ l = 0.250\lambda$
F	$Z_0 = 50\ \Omega,\ l = 0.464\lambda$	$Z_0 = 50\ \Omega,\ l = 0.192\lambda$
F'	$Z_0 = 50\ \Omega,\ l = 0.464\lambda$	$C = 2.0$ pF
F"	$C = 3.0$ pF	$C = 2.0$ pF
G	$Z_0 = 50\ \Omega,\ l = 0.129\lambda$	$C = 0.60$ pF
G'	$C = 0.90$ pF	$C = 0.60$ pF
H	$Z_0 = 50\ \Omega,\ l = 0.161\lambda$	$Z_0 = 81\ \Omega,\ l = 0.250\lambda$
H'	$C = 1.3$ pF	$Z_0 = 81\ \Omega,\ l = 0.250\lambda$
I	$Z_0 = 50\ \Omega,\ l = 0.181\lambda$	$L = 2.6$ nH
I'	$C = 1.7$ pF	$L = 2.6$ nH

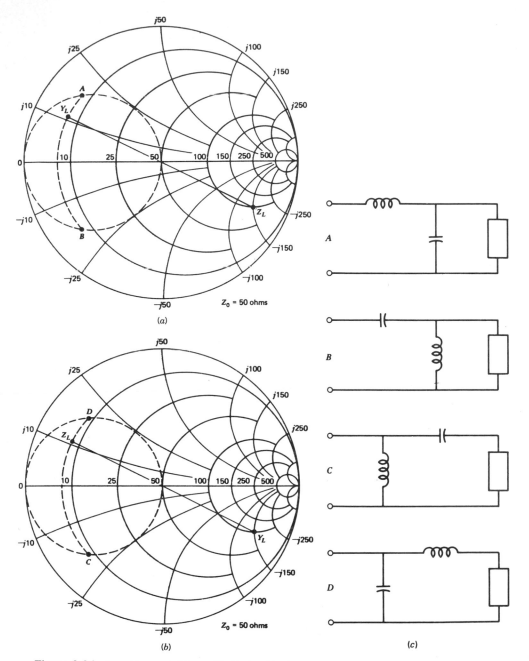

Figure 3.36   L-section matching with Z_L (or Y_L) inside unit resistance (or conductance) circle: (a, b) coordinates in ohms; (c) four matching circuits.

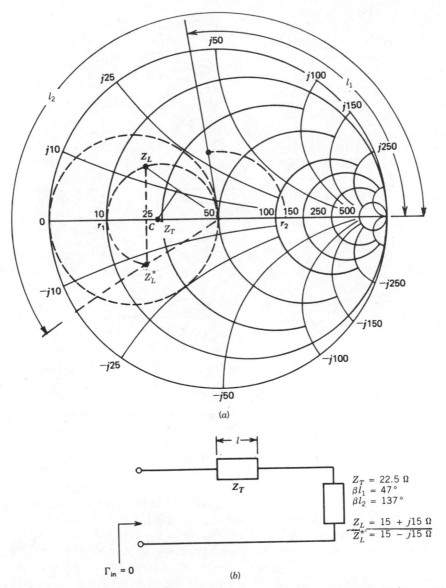

Figure 3.37 Single-section matching. (a) Coordinates in ohms. $Z_0 = 50\ \Omega$; $Z_L/Z_0 = 0.3 + j0.3$; $Z_T/Z_0 = \sqrt{0.2} = 0.45$; $Z_L'/Z_T = 0.67 + j0.67$; $r_2 = 2.22(50)\ \Omega$; $Z_{in}/Z_0 = 2.22(0.45) = 1.0$. Solution: $Z_T = 22.5\ \Omega$; $\beta l = 47°$. (b) Matching circuit.

The Smith chart is a useful tool for impedance matching. This is a circular chart that plots all values of Z_L or Y_L for positive values of R or G. Figure 3.32 shows the Smith chart and a value of $Z_L/Z_0 = 0.5 + j0.5$. Also plotted are Z_L^*/Z_0, Y_L/Y_0, and Y_L^*/Y_0, which gives a rectangle on the Smith chart. The usual imped-ance matching problem is to match Z_L to the center of the chart ($Z_0 = Z_G$), which

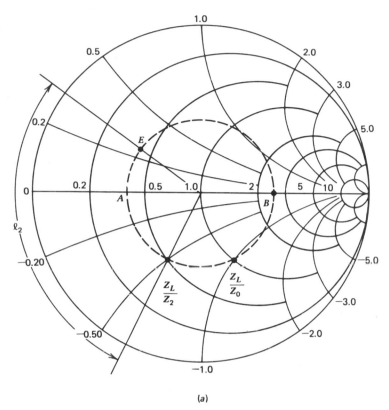

(a)

Figure 3.38 Two section matching. (*a*) Normalized coordinates. $A = 40 \ \Omega$; $B = 250 \ \Omega$; $Z_2 = 100 \ \Omega$; $E = 40 + j17.5 \ \Omega$; (*b*) Normalized coordinates. $A' = 40 \ \Omega$; $B' = 250 \ \Omega$; $Z_1 = 25 \ \Omega$; $D = Z_0/Z_1$; $E' = 40 + j17.5 \ \Omega$. (*c*) $Z_L/Z_0 = 1 - j1$; $Z_0 = 50 \ \Omega$. One solution: $l_2 = 0.119 \ \lambda$, $l_1 = 0.048\lambda$, $Z_2 = 100 \ \Omega$, $Z_1 = 25 \ \Omega$.

gives a zero reflection coefficient and maximum power transfer from the generator to the load. If Y_L is matched to the center of the chart, the same result has been achieved (i.e., zero reflection coefficient and maximum power transfer).

The lossless impedance matching elements are summarized in Table 3.11. The impedance matching can always be done with only lumped *LC*, or only microstrip line, or a combination of the two. A graphical solution on the Smith chart usually gives the best insight into the design.

The impedance-matching problem will be illustrated at $f = 4$ GHz by matching $Z_L/Z_0 = 0.15 + j0.60$ to the center of the chart (i.e., $Z_L/Z_0 = 1$). First, the problem will be solved by lossless L sections, which can give four solutions. Next, the problem will be solved using microstrip-line sections, which can give multiple solutions. Next, a complete solution in both impedance and admittance planes with seventeen designs will be given to illustrate the wide variety of exact solutions that is possible.

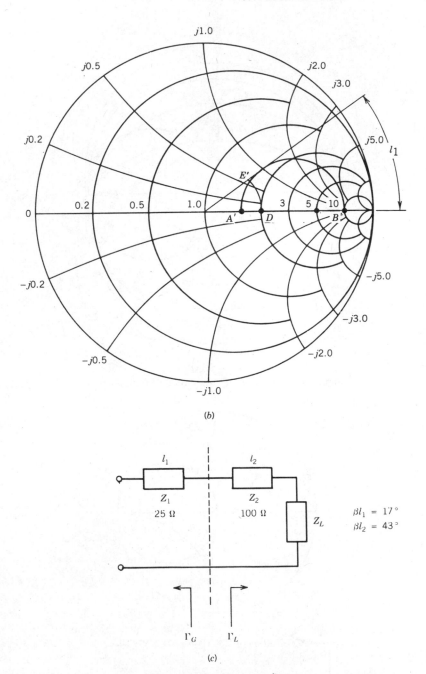

(b)

l_1 l_2

Z_1 Z_2 Z_L

25 Ω   100 Ω

$\beta l_1 = 17°$
$\beta l_2 = 43°$

Γ_G Γ_L

(c)

Figure 3.38 (*Continued*)

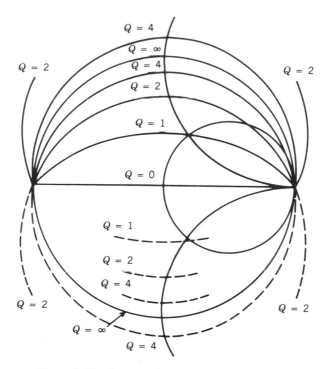

Figure 3.39 Constant Q contours on Smith chart.

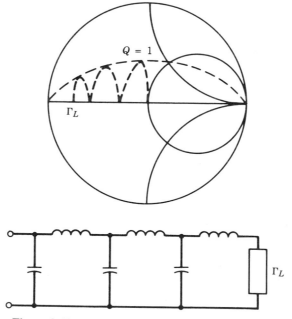

Figure 3.40 Graphical broadband matching example.

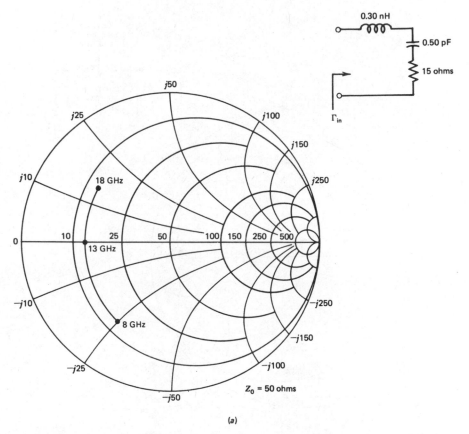

(a)

Figure 3.41 Broadband impedance matching. Coordinates in ohms. (a) Input impedance of GaAs MESFET; (b) Γ_1 plot; (c) Γ_2 plot; (d) final design.

Beginning with the lossless L sections in Fig. 3.33, we see that the first two solutions use a series capacitor followed by either a shunt capacitor (A) or a shunt indicator (B). The next solutions begin in the Y plane with a shunt capacitor followed by either a series capacitor (C) or a series inductor (D). For these solutions, part of the match is in the Z plane and part of the match is in the Y plane. The circle $R = 1$ and its mirror image are the primary curves required for the graphical solution.

Now consider the solutions in Fig. 3.34, drawing first the effect of a cascaded 50-Ω line. When the line intersects point A in the Z plane, a series capacitor completes the match. At point B a series inductor is required. For point C a shunt inductor or shorted microstrip line will give a match. At point D a shunt capacitor or open-circuited microstrip line completes the impedance match.

The complete matching problem with multiple solutions is given in Fig. 3.35.

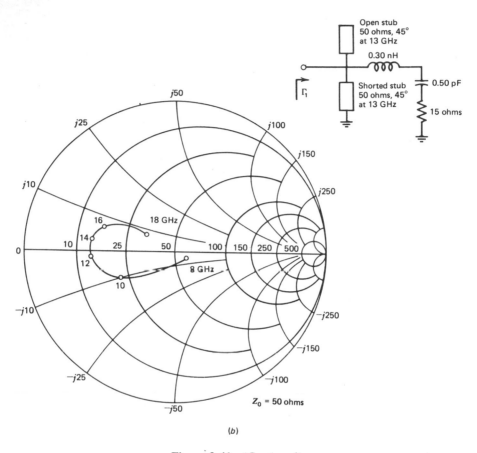

(b)

Figure 3.41 *(Continued)*

First, a circle of constant $|\Gamma_L|$ is drawn to illustrate the effect of a cascaded 50-Ω line. A total length of $\lambda_g/2$ will give an input impedance of Z_L/Z_0 again. This represents one rotation around the chart. The impedance matching problem is visualized as moving from Z_L/Z_0 to the center of the chart, which is the generator. Clockwise rotation around the Smith chart is required when transmission lines are cascaded. When the real axis is located (points *B* and *E*), a $\lambda/4$ transmission line will always match to the center of the chart with

$$Z_T^2 = Z_1 Z_2 \qquad (3.94)$$

with Z_1 = point *B* or *E*
$\quad$ Z_2 = center of chart = 50 Ω

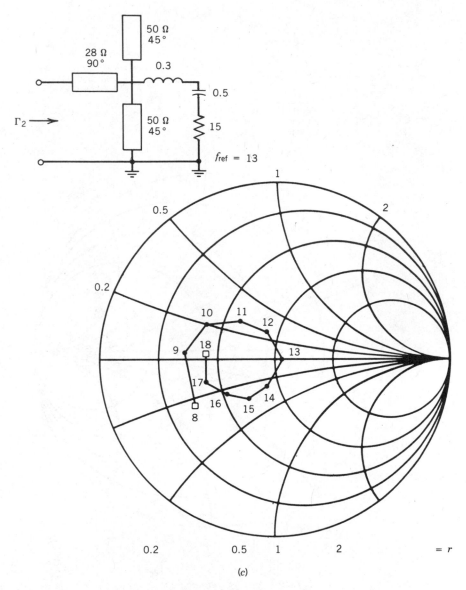

(c)

Figure 3.41 (*Continued*)

Notice that the mirror image of the circle $R = 1$ is useful for visualizing the L-section networks using shunt admittance elements (see Fig. 3.35a). Similarly, in Fig. 3.35c, the mirror image of the circle $G = 1$ is useful in plotting the shunt admittance values. Notice in Fig. 3.35c that the contour from Y_L is a constant admittance circle (not constant $|\Gamma_L|$), which is nearly the same contour for $|\Gamma_L|$

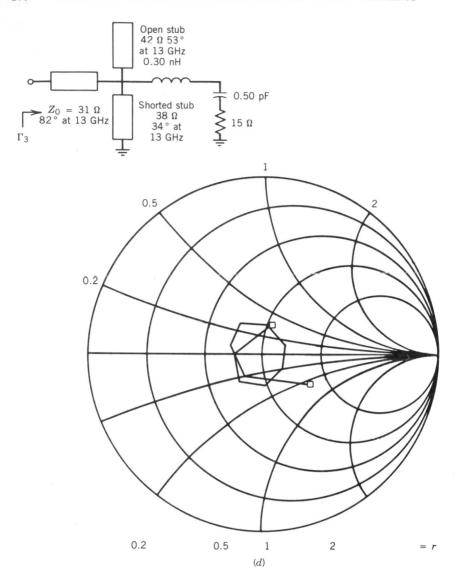

Figure 3.41 (*Continued*)

> 0.8. Similarly, for the solution D'', E'', F'' in Fig. 3.35b, a constant reactance circle is used. A total of 17 solutions has been found using only two lossless reactive elements. These solutions are given in Table 3.12 for the reader to verify the impedance-matching techniques.

Using only lumped elements, a total of four solutions have been found (D'', F'', G', I'). It can easily be shown that if the Z_L or Y_L point falls within the unit

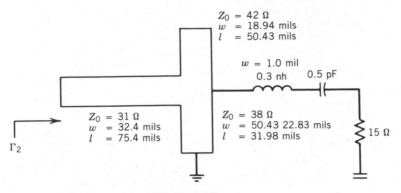

Figure 3.42 Microstrip cross.

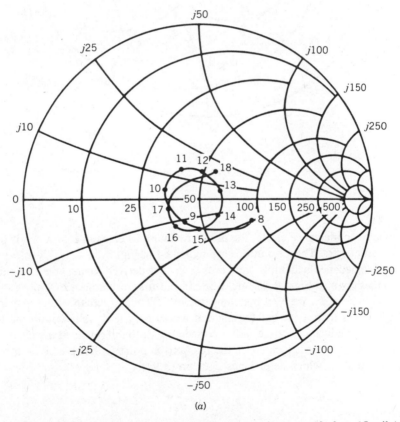

(a)

Figure 3.43 Broadband impedance matching in microstrip ($\epsilon_r = 10$, $h = 15$ mils): (a) without cross; (b) including cross; (c) final design, including cross (Table 3.13).

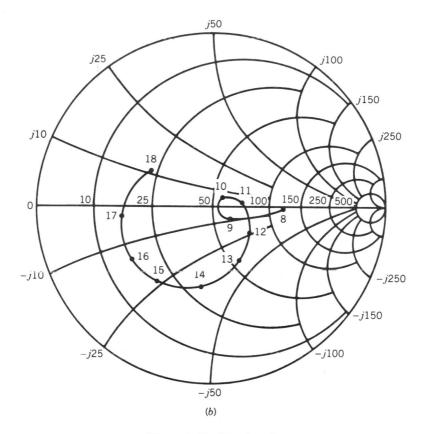

Figure 3.43   (Continued)

resistance or conductance circle, the number of lumped element L sections is only two instead of four. These solutions are indicated in Fig. 3.36.

Other impedance-matching networks can be derived using the graphical techniques shown in Figs. 3.37 and 3.38. In the first technique [3.61] Z_L can be considered a point on the $\lambda/4$ matching section. The total length of the matching structure is less than $\lambda/4$, but the method works only if Z_L falls within the unit resistance or conductance circle. After connecting Z_L to the center of the chart, the perpendicular bisector is located on the real axis at point C. Draw a circle around C, giving point r_1, which determines Z_T from (3.94).

$$Z_T/Z_0 = \sqrt{r_1} \tag{3.95}$$

Now renormalize Z_L to Z_T. Move toward the generator until the real axis is inter-

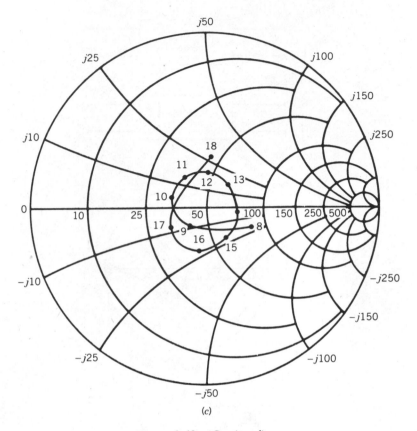

(c)

Figure 3.43 (*Continued*)

sected at r_2. The length of the matching section is given from the Smith chart construction in Fig. 3.37. Renormalizing from Z_T to Z_0 finds the load transferred to the center of the chart. The method also works for Z_L^*, giving a length of 137° for the matching section.

A variation on this technique is the use of two sections of cascaded transmission line, as shown in Fig. 3.38 [3.62]. This transformation is possible only if the

TABLE 3.13 Final Circuit Design of Fig. 3.43c

Element	Impedance, Z_0 (Ω)	Width (mils)	Length (mils)
Cascade line	29	32.6	66.7
Open stub	100	2.0	62.8
Shorted stub	100	2.0	18.5

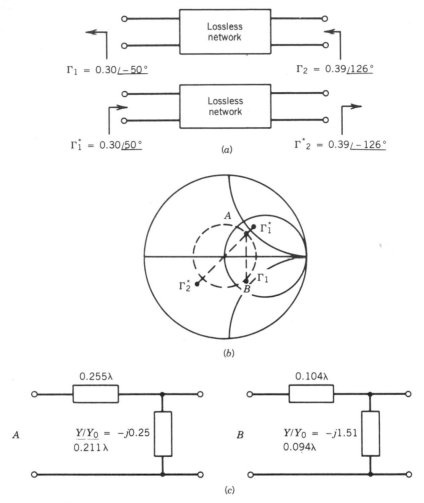

Figure 3.44 Graphical interstage match: (*a*) equivalent matching problems; (*b*) graphical solutions; (*c*) two solutions ($Z_0 = 50 \ \Omega$ for all elements).

values of Z_1 and Z_2 are chosen properly. First, Z_L/Z_2 is drawn as a circle moving toward the generator (clockwise). Next, circle AB is redrawn normalized to Z_1 to give circle $A'B'$. Now the generator is normalized to Z_1 (point D) and the generator point is moved toward the load to give point E' and l_1. Transforming E' to E gives length l_2. This technique shows that an infinite number of two-element solutions is always possible, since the range of Z_1 and Z_2 is very large.

The broadband impedance matching problem is considerably more difficult, since the load impedance is usually varying with frequency. The work of Bode

TABLE 3.14 S-Parameters of Low Noise Silicon Bipolar Transistor, AT-41400

Freq. (GHz)	S_{11} Mag.	S_{11} Angle	S_{21} dB	S_{21} Mag.	S_{21} Angle	S_{12} dB	S_{12} Mag.	S_{12} Angle	S_{22} Mag.	S_{22} Angle
				Noise Bias ($V_{CE} = 8$ V, $I_C = 10$ mA)						
0.1	0.73	−39	28.3	25.84	159	−39.2	0.011	75	0.94	−12
0.5	0.60	−121	22.2	12.91	113	−30.2	0.031	48	0.61	−28
1.0	0.57	−158	17.2	7.27	94	−28.0	0.040	51	0.50	−25
1.5	0.56	−172	13.7	4.84	84	−26.4	0.048	59	0.47	−25
2.0	0.57	176	11.4	3.71	77	−24.9	0.057	66	0.46	−24
2.5	0.57	170	9.5	2.97	71	−23.6	0.066	69	0.46	−26
3.0	0.60	164	8.0	2.52	64	−22.3	0.077	72	0.45	−28
3.5	0.60	157	6.8	2.18	61	−20.9	0.090	77	0.47	−29
4.0	0.61	152	5.5	1.89	55	−20.1	0.099	79	0.47	−30
4.5	0.63	147	4.7	1.72	51	−18.7	0.116	81	0.47	−36
5.0	0.63	144	3.7	1.53	46	−17.8	0.129	80	0.48	−40
5.5	0.65	139	3.1	1.42	42	−17.0	0.141	82	0.49	−44
6.0	0.66	136	2.1	1.28	38	−16.1	0.156	83	0.50	−47
				Gain Bias ($V_{CE} = 8$ V, $I_C = 25$ mA)						
0.1	0.56	−60	31.8	39.07	152	−40.9	0.009	69	0.87	−18
0.5	0.54	−145	23.5	15.00	104	−32.8	0.023	56	0.49	−28
1.0	0.54	−170	18.1	8.03	90	−29.6	0.033	65	0.42	−23
1.5	0.55	179	14.5	5.30	82	−26.9	0.045	72	0.41	−22
2.0	0.56	170	12.1	4.04	76	−24.7	0.058	75	0.41	−23
2.5	0.56	165	10.2	3.24	72	−23.1	0.070	78	0.40	−23
3.0	0.58	159	8.8	2.75	65	−21.6	0.083	79	0.40	−25
3.5	0.59	154	7.5	2.37	62	−20.4	0.096	82	0.41	−26
4.0	0.60	149	6.3	2.06	57	−19.3	0.108	83	0.42	−28
4.5	0.61	145	5.4	1.87	53	−18.1	0.124	84	0.42	−33
5.0	0.62	142	4.5	1.67	49	−17.3	0.136	83	0.43	−36
5.5	0.64	137	3.8	1.54	44	−16.5	0.150	85	0.42	−40
6.0	0.65	134	2.9	1.40	41	−15.7	0.165	84	0.44	−45

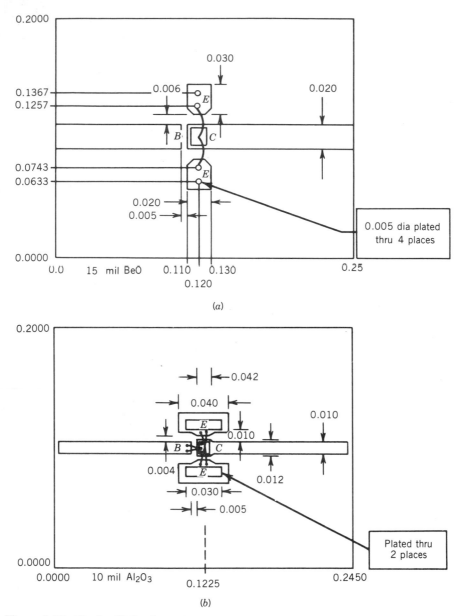

(a)

(b)

Figure 3.45 Bipolar 50-Ω microstrip-line test fixtures: (*a*) 15-mil BeO substrate; (*b*) 10-mil Al$_2$O$_3$ substrate. (Courtesy Avantek)

and Fano [3.63] has shown that there is a physical limitation on broadband matching when a reactive element is present. In the general case, Fano's limit gives a fundamental limitation on how well the matching can be achieved over a band of frequencies. For high Q load with a large reactance change with frequency, a relatively poor match will be achieved; a low Q load can be matched much better

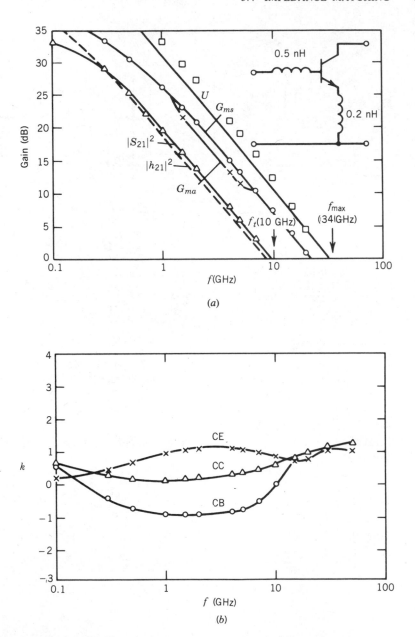

Figure 3.46 AT-41400 silicon bipolar transistor parameters versus frequency: (*a*) gain versus frequency (common-emitter); (*b*) stability factor versus frequency. V_{CE} = 8 V, I_C = 25 mA, L_E = 0.2 nH, L_B = 0.5 nH.

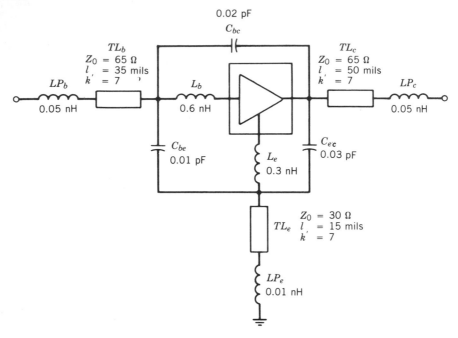

Figure 3.47 Equivalent circuit that can be used to model bipolar transistor packages; parameter values are given for 35-microstrip package from Avantek.

over a broad bandwidth. The best possible match for a parallel *RC* load of Q_1 over a bandwidth of Q_2 is given by

$$|\Gamma_{min}| = e^{-\pi Q_2 / Q_1} \qquad (3.96)$$

where

$$Q_1 = R_1 C_1 \omega_0 \qquad (3.97)$$

$$Q_2 = \frac{f_0}{\Delta f} = \frac{f_0}{BW} \qquad (3.98)$$

The design of wideband matching networks using the Smith chart requires avoiding the regions of high *Q*. These regions can be seen from the constant-*Q* contours of a reactive load in Fig. 3.39. By using multiple reactive elements which maintain

TABLE 3.15 S Parameters, Low Noise Silicon Bipolar Transistor, Common Emitter, AT-41435

Freq. (GHz)	S_{11} Mag.	S_{11} Angle	S_{21} dB	S_{21} Mag.	S_{21} Angle	S_{12} dB	S_{12} Mag.	S_{12} Angle	S_{22} Mag.	S_{22} Angle
				Noise Bias (V_{CE} = 8 V, I_C = 10 mA)						
0.1	0.80	-32	28.0	24.99	157	-39.2	0.011	82	0.93	-12
0.5	0.50	-110	21.8	12.30	108	-29.6	0.033	52	0.61	-28
1.0	0.40	-152	16.6	6.73	85	-26.2	0.049	56	0.51	-30
1.5	0.38	-176	13.3	4.63	71	-24.0	0.063	59	0.48	-32
2.0	0.39	166	11.0	3.54	60	-21.9	0.080	58	0.46	-37
2.5	0.41	156	9.3	2.91	53	-20.4	0.095	61	0.44	-40
3.0	0.44	145	7.9	2.47	43	-18.8	0.115	61	0.43	-48
3.5	0.46	137	6.7	2.15	33	-17.5	0.133	58	0.43	-58
4.0	0.46	127	5.6	1.91	23	-16.0	0.153	53	0.45	-68
4.5	0.47	116	4.7	1.72	13	-15.0	0.178	50	0.46	-75
5.0	0.49	104	4.0	1.58	3	-13.9	0.201	47	0.48	-82
5.5	0.52	91	3.3	1.45	-7	-13.0	0.224	40	0.47	-89
6.0	0.59	81	2.5	1.34	-17	-12.1	0.247	36	0.43	-101
				Gain Bias (V_{CE} = 8 V, I_C = 25 mA)						
0.1	0.63	-50	31.8	39.08	146	-40.0	0.010	83	0.84	-18
0.5	0.39	-137	22.9	13.97	99	-31.4	0.027	60	0.50	-26
1.0	0.36	-171	17.2	7.28	80	-27.1	0.044	67	0.45	-26
1.5	0.36	171	13.9	4.94	68	-23.5	0.067	66	0.43	-30
2.0	0.38	156	11.5	3.76	58	-21.6	0.083	63	0.41	-34
2.5	0.40	149	9.8	3.08	52	-19.6	0.105	63	0.39	-38
3.0	0.43	140	8.3	2.61	43	-18.3	0.122	64	0.38	-47
3.5	0.45	132	7.2	2.28	33	-16.8	0.144	59	0.39	-57
4.0	0.46	122	6.1	2.02	23	-15.6	0.165	55	0.40	-67
4.5	0.46	112	5.2	1.82	14	-14.6	0.185	50	0.42	-75
5.0	0.47	101	4.4	1.66	4	-13.7	0.207	45	0.43	-81
5.5	0.51	89	3.7	1.54	-5	-12.6	0.233	39	0.42	-89
6.0	0.58	79	3.0	1.41	-15	-11.8	0.257	33	0.37	-101

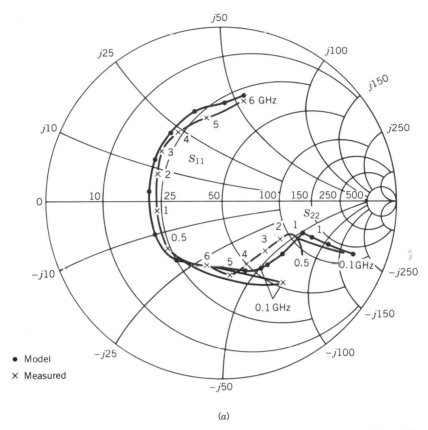

(a)

Figure 3.48 S parameter of model versus measured data for AT-41435 at bias of $V_C = 8.0$ V, $I_C = 25$ mA.

the reflection coefficient in a low-Q region, greater bandwidths can be achieved, approaching the limit discussed in Section 4.1.2 of the next chapter.

An application of this concept is given in Fig. 3.40, which matches a load of $\Gamma_L = 0.8/\underline{180°}$ to the center of the chart. This is a circuit often used in high-power bipolar designs.

Another technique for broadband matching consists of adding series–parallel–series resonant structures with the intent of approaching the $|\Gamma_{min}|$ given by (3.96). As an example of this procedure, consider the problem shown in Fig. 3.41. The *RLC* circuit of Fig. 3.41a represents a simple model of the input circuit of a high-Q, low-loss GaAs MESFET. The reflection coefficient of this circuit is plotted from 8 to 18 GHz on the Smith chart accompanying the circuit. To achieve a broadband match across this frequency range, the approach illustrated here is to series resonate the input impedance to the *RLC* network, then parallel resonate and

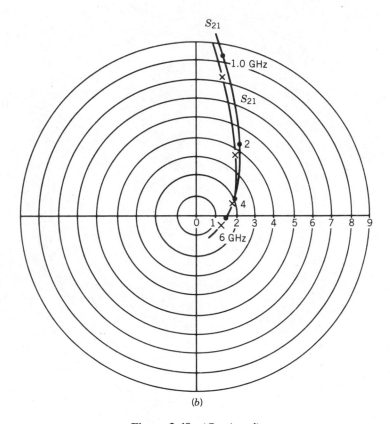

(b)

Figure 3.48 (*Continued*)

continue this process until a minimum reflection coefficient is realized. Figure 3.41b shows one step in this process where 50-Ω shunt open and shorted stubs have been used to improve the match. Notice from the accompanying Smith chart plot in Fig. 3.41b for the new reflection coefficient Γ_1 that the match has been improved considerably. Figure 3.41c shows the calculated circuit, which consists of the two shunt stubs and a series length of line. Next, the circuit is usually optimized using CAD tools to give the final circuit of Fig. 3.41d. The preceding discussion of distributed matching network design has focused on ideal lossless transmission-line circuits.

When physically realized in the form of a microstrip geometry, these circuits require consideration of the effects of the junctions between lines of varying widths. The physical and analytical microstrip models for these effects are considered in Appendix D, where equivalent-circuit models are given for the gap, the step junction, the bend, the tee junction, the cross junction, the slit, and the open-circuit stub fringing capacitance.

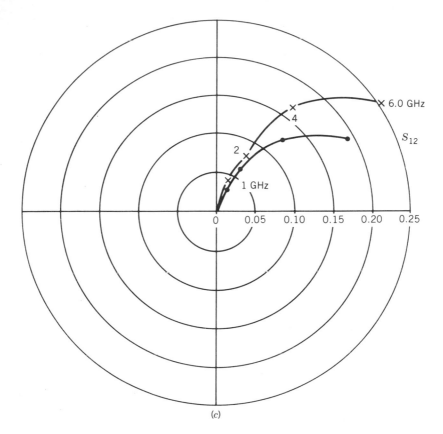

(c)

Figure 3.48   (*Continued*)

It is seen from the final matching network realization of Fig. 3.41*d* that the three microstrip-line elements are joined at a common junction along with the 0.3-nH inductor.

When two microstrip transmission lines are connected in shunt across a through line, as in the example of Fig. 3.41, the circuit configuration must be modeled

TABLE 3.16   Noise Parameters, Silicon Bipolar Transistor, AT-41400 (V_{CE} = 8 V, I_C = 10 mA)

f (GHz)	Γ_{min} (dB)	Γ_{on} Mag.	Γ_{on} Angle	R_n (Ω)
1.0	1.3	0.05	14	8
2.0	1.6	0.40	160	10
4.0	3.0	0.72	−156	16

Source: Ref. 3.64 (with changes in R_n).

TABLE 3.17 S and Noise Parameters, Silicon Bipolar Transistor, AT-41435 ($V_{CE} = 8$, $I_C = 10$ mA)

S Parameters

Freq. (GHz)	S_{11} Mag.	S_{11} Angle	S_{21} Mag.	S_{21} Angle	S_{12} Mag.	S_{12} Angle	S_{22} Mag.	S_{22} Angle
1.0	0.40	−152	6.73	85	0.049	56	0.51	−30
2.0	0.39	−166	3.54	60	0.080	58	0.46	−37
4.0	0.46	127	1.91	23	0.153	53	0.45	−68

Noise Parameters

Freq. (GHz)	F_{min} (dB)	Γ_{0n} Mag.	Γ_{0n} Angle	$\dfrac{R_n}{50} = r_n$
1.0	1.3	0.05	30	0.16
2.0	1.7	0.50	−163	0.20
4.0	3.0	0.68	−130	0.32

Source: Ref. 3.64 (with changes in R_n).

TABLE 3.18 S and Noise Parameters, Silicon Bipolar Transistor, AT-60535 ($V_{CE} = 8$, $I_C = 2$ mA)

S Parameters

Freq. (GHz)	S_{11} Mag.	S_{11} Angle	S_{21} Mag.	S_{21} Angle	S_{12} Mag.	S_{12} Angle	S_{22} Mag.	S_{22} Angle
1.0	0.64	−84	4.33	109	0.065	44	0.81	−27
2.0	0.42	−137	2.73	74	0.083	37	0.71	−38
4.0	0.36	148	1.57	27	0.110	42	0.67	−64

Noise Parameters

Freq. (GHz)	F_{min} (dB)	Γ_{0n} Mag.	Γ_{0n} Angle	$\dfrac{R_n}{50} = r_n$
1.0	1.4	0.50	55	0.32
2.0	1.8	0.40	−114	0.38
4.0	2.8	0.38	−153	0.40

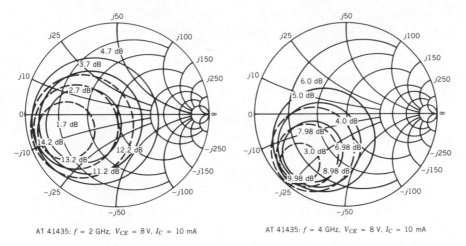

AT 41435: $f = 2$ GHz, $V_{CE} = 8$ V, $I_C = 10$ mA   AT 41435: $f = 4$ GHz, $V_{CE} = 8$ V, $I_C = 10$ mA

Figure 3.49 Noise figure and available gain circles in the Γ_g plane for the AT-41435 silicon bipolar transistor. (Courtesy Avantek)

physically by a cross junction. Figure 3.42 shows a microstrip realization for the current case employing the amplifier input circuit of Fig. 3.41. For purposes of analysis, this circuit assumes a relative dielectric constant of 10.0 (Al_2O_3) and a substrate thickness of 15 mils.

In the Smith chart plot of Fig. 3.43, we have shown the result of the effects of incorporating the microstrip cross junction model on the ideal response. From a comparison of the Smith chart plots of Fig. 3.43a and b, we see that a drastic alteration of the match has occurred for frequencies above 10 GHz. This illustrates the fact that great care must be exercised in modeling such a circuit for final realization. Figure 3.43c shows a plot of final circuit response when optimized to meet a $| \Gamma_{min} |$ value of less than 0.2. The final values for the transmission-line elements are given in Table 3.13.

A general impedance-matching problem that will occur in two-stage amplifier design is to match a reflection coefficient Γ_1 to a new reflection coefficient Γ_2 with lossless elements. An obvious solution is to match Γ_1 to the point 50 Ω and to match Γ_2 to the point 50 Ω, which requires four elements or less. For better bandwidth and a simpler design, a two-element solution should be considered.

Some graphical examples on the Smith chart will illustrate a useful technique. Consider the 4-GHz problem in Figure 3.44; match $\Gamma_1 = 0.30\underline{/-50°}$ to $\Gamma_2 = 0.39\underline{/126°}$. By considering both the Z and Y planes for Γ_1, multiple solutions can be generated using lumped or distributed elements. Keeping the matching philosophy of moving from the load impedance toward the center of the chart, the equivalent problem of matching the conjugates will be solved; match $\Gamma_2^* = 0.39\underline{/-126°}$ to $\Gamma_1^* = 0.30\underline{/50°}$. Move Γ_2^* to the Y plane and add an admittance which brings the reflection coefficient in, as shown in Fig. 3.44b. Then return to the impedance

TABLE 3.19 S Parameters, Silicon Bipolar Transistor, AT-41400 ($Z = 350$ μm, $V_{CE} = 8.0$ V, $I_C = 25$ mA, $L_B = 0.5$ nH, $L_E = 0.2$ nH)

Freq. (GHz)	S_{11} Mag.	S_{11} Angle	S_{21} Mag.	S_{21} Angle	S_{12} Mag.	S_{12} Angle	S_{22} Mag.	S_{22} Angle
				Common Emitter				
0.5	0.540	−145	15.0	104	0.023	56.0	0.490	−28.0
1.0	0.540	−170	8.03	90.0	0.033	65.0	0.420	−23.0
1.5	0.550	179	5.30	82.0	0.045	72.0	0.410	−22.0
2.0	0.560	170	4.04	76.0	0.058	75.0	0.410	−23.0
2.5	0.560	165	3.24	72.0	0.070	78.0	0.400	−23.0
3.0	0.580	159	2.75	65.0	0.033	79.0	0.400	−25.0
3.5	0.590	154	2.37	62.0	0.096	82.0	0.410	−26.0
4.0	0.600	149	2.06	57.0	0.108	83.0	0.420	−28.0
4.5	0.610	145	1.87	53.0	0.124	84.0	0.420	−33.0
5.0	0.620	142	1.67	49.0	0.136	83.0	0.430	−36.0
5.5	0.640	137	1.54	44.0	0.150	85.0	0.420	−40.0
6.0	0.650	134	1.40	41.0	0.165	84.0	0.440	−45.0
				Common Base				
0.5	0.939	177	1.92	−7.08	0.003	109	1.004	−4.48
1.0	9.956	174	1.93	−13.9	0.010	140	1.014	−8.73
1.5	0.977	170	1.94	−21.6	0.019	146	1.028	−13.4

2.0	1.02	167	1.96	−29.1	0.034	153	1.047	−17.9
2.5	1.04	163	1.95	−36.9	0.046	145	1.048	−22.9
3.0	1.11	159	2.02	−45.5	0.073	148	1.087	−28.3
3.5	1.15	155	1.99	−54.0	0.089	144	1.087	−33.0
4.0	1.21	149	2.01	−64.8	0.121	139	1.103	−39.2
4.5	1.27	144	2.02	−74.5	0.157	136	1.141	−45.6
5.0	1.30	138	1.98	−86.5	0.193	130	1.146	−52.6
5.5	1.38	130	2.01	−99.9	0.253	212	1.163	−62.0
6.0	1.37	123	1.89	−113	0.293	117	1.160	−69.2
6.5	1.33	116	1.72	−128	0.330	110	1.124	−76.8

Common Collector

0.5	0.978	−10.7	1.91	−5.84	0.113	71.8	0.912	170
1.0	0.962	−20.9	1.89	−13.3	0.218	67.9	0.902	161
1.5	0.934	−31.8	1.84	−20.0	0.329	60.6	0.877	151
2.0	0.895	−41.8	1.78	−25.8	0.422	53.0	0.850	143
2.5	0.844	−51.1	1.69	−31.2	0.500	45.8	0.811	135
3.0	0.823	−61.6	1.65	−37.1	0.588	40.4	0.797	126
3.5	0.756	−69.7	1.55	−41.0	0.647	33.0	0.750	119
4.0	0.703	−79.6	1.46	−45.7	0.715	26.3	0.712	111
4.5	0.674	−87.0	1.40	−49.4	0.753	20.7	0.675	105
5.0	0.628	−95.8	1.31	−52.8	0.799	14.9	0.625	98.2
5.5	0.605	−105	1.27	−57.1	0.839	9.76	0.602	92.5
6.0	0.564	−112	1.19	−59.0	0.864	4.05	0.545	86.6

TABLE 3.20 S and Noise Parameters, Field-Effect Transistor, AT-8251 ($V_{ds} = 3$ V, $I_d = 20$ mA)

S Parameters

Freq. (GHz)	S_{11} Mag.	S_{11} Angle	S_{21} Mag.	S_{21} Angle	S_{12} Mag.	S_{12} Angle	S_{22} Mag.	S_{22} Angle
0.5	0.98	−11	3.30	173	0.018	89	0.53	−3
1.0	0.98	−19	3.30	165	0.032	84	0.52	−6
2.0	0.96	−35	3.29	150	0.061	74	0.47	−12
3.0	0.94	−53	3.22	134	0.089	64	0.41	−23
4.0	0.87	−74	3.13	117	0.112	55	0.32	−33
5.0	0.79	−94	2.88	103	0.125	49	0.24	−43
6.0	0.74	−115	2.68	89	0.141	39	0.13	−60
7.0	0.70	−134	2.43	77	0.145	35	0.03	−167
8.0	0.66	−151	2.20	67	0.153	32	0.07	155
9.0	0.66	−163	2.09	57	0.160	25	0.13	138
10.0	0.67	−176	1.93	46	0.163	23	0.17	135
11.0	0.68	171	1.81	37	0.167	18	0.23	132
12.0	0.68	162	1.68	24	0.170	9	0.24	130

Noise Parameters

Freq. (GHz)	F_{min} (dB)	Γ_{0n} Mag.	Γ_{0n} Angle	R_n/Z_0
2.0	0.6	0.83	22	0.79
4.0	0.8	0.57	63	0.67
6.0	1.0	0.50	101	0.23
8.0	1.2	0.47	135	0.34

TABLE 3.21 S Parameters, Field-Effect Transistor, AT-8251 (V_{ds} = 5 V, I_d = 50 mA)

Freq. (GHz)	S_{11} Mag.	S_{11} Angle	S_{21} Mag.	S_{21} Angle	S_{12} Mag.	S_{12} Angle	S_{22} Mag.	S_{22} Angle
0.5	0.98	−14	5.40	173	0.011	88	0.69	−1
1.0	0.98	−25	5.32	163	0.020	82	0.67	−5
2.0	0.93	−47	5.16	144	0.038	71	0.60	−12
3.0	0.89	−70	4.82	127	0.056	65	0.53	−17
4.0	0.82	−94	4.48	110	0.068	52	0.46	−20
5.0	0.75	−117	3.88	97	0.075	50	0.40	−24
6.0	0.71	−137	3.51	83	0.080	46	0.31	−29
7.0	0.70	−159	3.07	71	0.083	42	0.22	−35
8.0	0.71	−169	2.73	63	0.086	41	0.16	−42
9.0	0.72	180	2.53	54	0.090	40	0.12	−46
10.0	0.73	172	2.31	44	0.093	39	0.09	−55
11.0	0.74	162	2.13	35	0.098	36	0.05	−131
12.0	0.73	150	2.08	21	0.104	34	0.08	−170

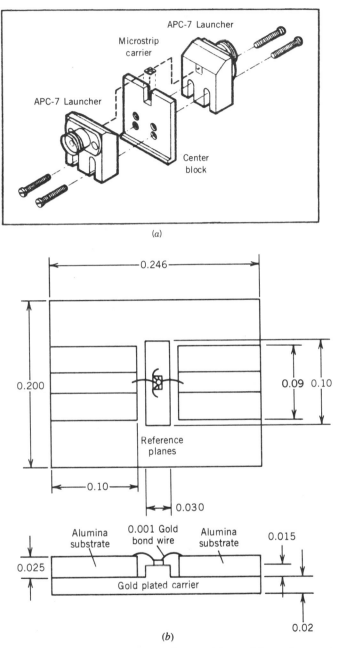

Figure 3.50 FET 50-Ω microstrip-line test fixture: (*a*) coaxial test fixture assembly; (*b*) details of microstrip carrier. (Courtesy Avantek)

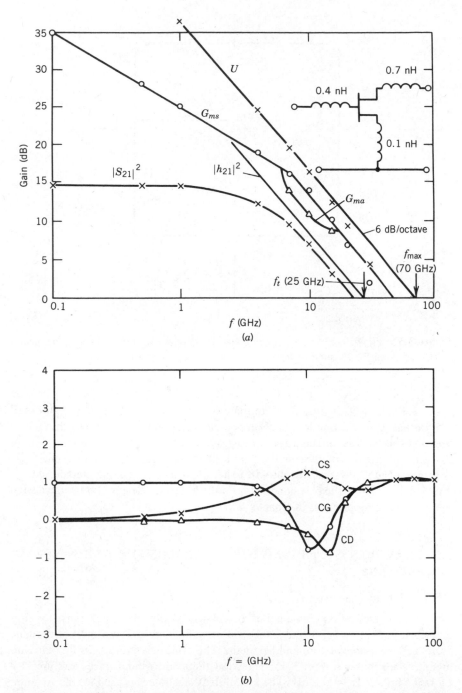

Figure 3.51 AT-8251 gallium arsenide FET parameters versus frequency: (*a*) gain versus frequency (common-source); (*b*) stability factor versus frequency. $V_{DS} = 5$ V, $I_D = 50$ mA, $Z = 500$ μm, $L_G = 0.40$ nH, $L_D = 0.70$ nH, $L_S = 0.10$ nH.

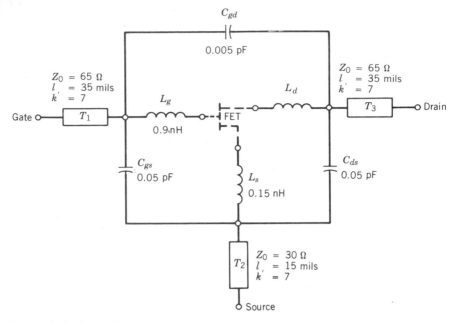

Figure 3.52 Equivalent circuit that can be used to model MESFET packages; parameter values are given for 35-microstrip package from Avantek. (Courtesy Avantek)

plane and add a 50-Ω line of appropriate length. Solution B in Fig. 3.44*c* will require less physical line length. Lumped-element solutions are more difficult to find and usually less practical for microwave designs.

   Although greater skill with the Smith chart is required to implement these designs, a better physical insight will result. A two- or even one-element design will usually result in the best bandwidth for the interstage design. Interstage design is discussed in more detail in Chapter 4.

3.5 EXAMPLES OF TRANSISTOR *S* PARAMETERS AND NOISE PARAMETERS

3.5.1 Bipolar Transistors

The common-emitter *S* parameters of the AT-41400 silicon bipolar transistor chip described in Section 3.1.1 and Table 3.4 are tabulated in Table 3.14 at the bias current for best noise figure and best gain. The 50-Ω test fixtures for bipolar chip mounting are shown in Fig. 3.45. The parasitic bonding inductances are typically $L_B = 0.5$ nH and $L_E = 0.2$ nH. The 15 mil BeO substrate was used for the Avantek catalog data; the 10-mil alumina substrate was used for the AT-22000 test data. The power gains and stability factors are plotted in Fig. 3.46 for the AT-41400 at 25 mA, giving a f_t of 10 GHz and a f_{max} of 34 GHz.

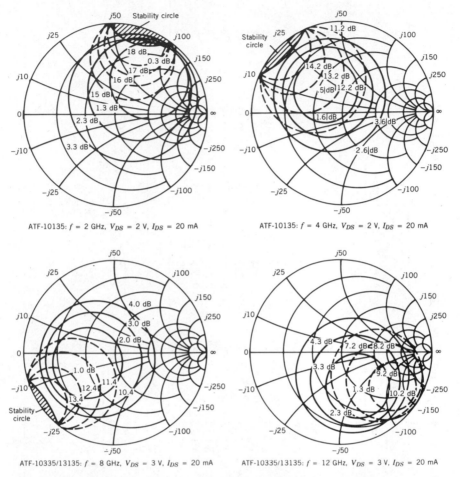

Figure 3.53 Noise figure and available gain circles in the Γ_g plane for the ATF-10135 and ATF-10335/13135 GaAs FETs. (Courtesy Avantek)

The microstrip-line packages simply rotate the S parameters clockwise on a Smith chart (i.e., the generator is moved farther from the load by nearly lossless transmission lines). The equivalent-circuit model for typical microstrip packages is given in Fig. 3.47. The new S parameters are given in Table 3.15 for the 35 package. The typical agreement between the modeled and measured S parameter data is shown in Fig. 3.48. The model was given in Table 3.4 for the chip and Fig. 3.47 for the package.

The noise parameters are summarized in Tables 3.16 to 3.18 for both chip and packaged transistors. Common-emitter configuration is nearly always used because of the higher available or associated gain $G_A(\Gamma_{0n})$. The noise and gain circles are given in Fig. 3.49 at 2 GHz and 4 GHz for the packaged AT-41435. The location of optimum noise measure can be shown to fall between Γ_{0n} and Γ_{0g} [3.65, 3.66].

TABLE 3.22 S and Noise Parameters, GaAs Field-Effect Transistor, ATF-10135 (V_{ds} = 2 V, I_d = 20 mA)

S Parameters

Freq. (GHz)	S_{11} Mag.	S_{11} Angle	S_{21} Mag.	S_{21} Angle	S_{12} Mag.	S_{12} Angle	S_{22} Mag.	S_{22} Angle
0.5	0.98	−18	5.32	163	0.020	78	0.35	−9
1.0	0.93	−33	5.19	147	0.038	67	0.36	−19
2.0	0.79	−66	4.64	113	0.074	59	0.30	−31
3.0	0.64	−94	4.07	87	0.110	44	0.27	−42
4.0	0.54	−120	3.60	61	0.137	31	0.22	−49
5.0	0.47	−155	3.20	37	0.167	13	0.16	−54
6.0	0.45	162	2.88	13	0.193	−2	0.08	−17
7.0	0.50	120	2.51	−10	0.203	−19	0.16	45
8.0	0.60	87	2.09	−32	0.210	−36	0.32	48
9.0	0.68	61	1.75	−51	0.209	−46	0.44	38
10.0	0.73	42	1.52	−66	0.207	−58	0.51	34
11.0	0.77	26	1.26	−82	0.205	−73	0.54	27
12.0	0.80	14	1.12	−97	0.200	−82	0.54	15

Noise Parameters

Freq. (GHz)	F_{min} (dB)	Γ_{on} Mag.	Γ_{on} Angle	R_n/Z_0
2.0	0.4	0.70	47	0.46
4.0	0.5	0.39	126	0.36
6.0	0.8	0.36	−170	0.12
8.0	1.1	0.45	−100	0.38
12.0	1.4	0.60	−41	1.10

TABLE 3.23 S and Noise Parameters, GaAs Field-Effect Transistor, ATF-10335 ($V_{DS} = 3$ V, $I_{DS} = 210$ mA)

S Parameters

Freq. (GHz)	S_{11} Mag.	S_{11} Angle	S_{21} Mag.	S_{21} Angle	S_{12} Mag.	S_{12} Angle	S_{22} Mag.	S_{22} Angle
1.0	0.96	−21	3.72	158	0.023	81	0.58	−19
2.0	0.95	−42	3.65	134	0.046	62	0.56	−37
3.0	0.87	−65	3.43	112	0.062	40	0.53	−47
4.0	0.84	−85	3.28	93	0.077	31	0.53	−54
5.0	0.78	−104	3.21	73	0.097	18	0.48	−62
6.0	0.69	−128	3.30	52	0.117	7	0.43	−76
7.0	0.59	−163	3.32	27	0.137	−12	0.35	−95
8.0	0.54	157	3.10	2	0.142	−27	0.26	−110
9.0	0.55	121	2.89	−19	0.153	−43	0.15	−119
10.0	0.54	93	2.71	−41	0.159	−58	0.07	−142
11.0	0.56	64	2.54	−61	0.155	−73	0.06	92
12.0	0.61	37	2.34	−84	0.144	−89	0.15	46
13.0	0.65	19	2.18	−102	0.132	−100	0.18	23
14.0	0.65	7	2.09	−120	0.125	−114	0.19	−2
15.0	0.67	−6	1.99	−139	0.122	−119	0.16	−27
16.0	0.68	−25	1.91	−170	0.116	−134	0.13	−30
17.0	0.59	−56	1.82	161	0.113	−148	0.08	−18
18.0	0.48	−87	1.74	140	0.130	−178	0.11	7

Noise Parameters

Freq. (GHz)	F_{min} (dB)	Γ_{On} Mag.	Γ_{On} Angle	$\frac{R_n}{50} = r_n$
4.0	0.70	0.63	85	0.32
6.0	0.90	0.47	130	0.06
8.0	1.0	0.37	−163	0.17
12.0	1.3	0.47	−65	0.74

TABLE 3.24 S Parameters, GaAs MESFET Chip, AT-8251 ($Z = 500$ μm, $V_{DS} = 5$ V, $I_D = 50$ mA, $L_G = 0.4$ nH, $L_D = 0.7$ nH, $L_S = 0.15$ nH) (Modeled Data)

Freq. (GHz)	S_{11} Mag.	Angle	S_{21} Mag.	Angle	S_{12} Mag.	Angle	S_{22} Mag.	Angle
				Common Source				
1.0	0.980	−25.0	5.32	163.0	0.020	82.0	0.670	−5.0
2.0	0.930	−47.0	5.16	144.0	0.038	71.0	0.600	−12.0
3.0	0.890	−70.0	4.82	127.0	0.056	65.0	0.530	−17.0
4.0	0.820	−94.0	4.48	110.0	0.068	52.0	0.460	−20.0
5.0	0.750	−117.0	3.88	97.0	0.075	50.0	0.400	−24.0
6.0	0.710	−137.0	3.51	83.0	0.080	46.0	0.310	−29.0
7.0	0.700	−159.0	3.07	71.0	0.083	42.0	0.220	−35.0
8.0	0.710	−169.0	2.73	63.0	0.086	41.0	0.160	−42.0
9.0	0.720	−180.0	2.53	54.0	0.090	40.0	0.120	−46.0
10.0	0.730	172.0	2.31	44.0	0.093	39.0	0.090	−55.0
11.0	0.740	162.0	2.13	35.0	0.098	36.0	0.050	−131.0
12.0	0.730	150.0	2.08	21.0	0.104	34.0	0.080	−170.0
				Common Gate				
1.0	0.489	180	1.49	−4.21	0.082	−2.71	0.670	−5.0
2.0	0.510	177	1.52	−8.9	0.088	1.62	0.916	−6.1

3.0	0.529	175	1.56	−13.4	0.090	−0.40	0.920	−8.1
4.0	0.573	173	1.63	−18.4	0.077	1.38	0.933	−11.6
5.0	0.618	172	1.68	−23.9	0.070	1.60	0.950	−16.0
6.0	0.685	170	1.80	−29.7	0.057	12.7	0.986	−22.0
7.0	0.788	169	1.93	−37.0	0.032	47.8	1.04	−30.3
8.0	0.846	167	2.05	−44.0	0.039	88.0	1.10	−38.3
9.0	0.962	164	2.29	−51.3	0.083	127	1.21	−46.4
10.0	1.10	160	2.70	−61.8	0.163	132	1.43	−57.3
11.0	1.37	152	3.32	−76.1	0.349	129	1.84	−75.1
12.0	2.38	134	6.18	−98.8	0.017	119	3.55	−98.3

Common Drain

1.0	0.997	−6.8	1.46	−3.57	0.078	83.9	0.550	173
2.0	1.00	−12.8	1.48	−8.03	0.141	82.7	0.579	166
3.0	1.01	−19.7	1.50	−12.4	0.209	81.1	0.607	159
4.0	1.03	−27.3	1.53	−17.2	0.289	79.7	0.652	153
5.0	1.02	−36.2	1.54	−22.1	0.368	71.6	0.670	147
6.0	1.05	−45.1	1.59	−28.2	0.441	69.1	0.726	141
7.0	1.07	−57.3	1.63	−35.1	0.526	62.9	0.768	135
8.0	1.11	−67.8	1.63	−41.8	0.598	58.6	0.782	130
9.0	1.18	−78.6	1.68	−49.2	0.690	55.2	0.829	124
10.0	1.32	−92.2	1.74	−60.0	0.826	50.1	0.870	114
11.0	1.51	−108	1.85	−73.0	0.932	44.6	0.919	106
12.0	2.09	−127	2.37	−94.7	1.25	37.2	1.15	88.1

The S parameters of the common-base and common-collector silicon bipolar transistors are given in Table 3.19. The bonding inductances must be specified.

3.5.2   Field-Effect Transistors

The common-source S parameters of the AT-8251 GaAs MESFET described in Section 3.2 (Table 3.8) are given in Tables 3.20 and 3.21. The 50-Ω test fixture for FET chip mounting is shown in Fig. 3.50, where the typical bonding inductances are $L_G = 0.4$ nH, $L_D = 0.7$ nH, and $L_S = 0.15$ nH.

Since the S parameters have been measured to higher frequencies, deembedding the test fixture is an important correction to the data. A through-delay method was used for these data [3.64]. The assumptions are:

1. $Z_0 = 50 \ \Omega$
2. Symmetry at both coaxial/microstrip transitions.

From measurements of a 50-Ω through line and a 50-Ω delay line, the S parameters of the test fixture can be subtracted from the data [3.67]. This correction is significant for frequencies above 6 GHz.

The power gains and stability factors of a 500-μm GaAs MESFET chip are given in Fig. 3.51, showing an f_t of 25 GHz and an f_{max} of 70 GHz. Notice the peak in the common-source stability factor, which has been predicted by the relations [3.68, 3.69]

$$f_0 = \frac{1}{2\pi \sqrt{L_s C_{eq}}} \tag{3.99}$$

$$C_{eq} = \frac{C_{gs} C_{ds}}{C_{gd}} \tag{3.100}$$

The microstrip-line packages shift the reference planes clockwise on the Smith chart as shown in Table 3.21. The package model is given in Fig. 3.52. The noise parameters are summarized in Tables 3.20, and 3.23 and in noise circles for the Γ_G plane shown in Fig. 3.53 [3.64]. The S parameters of the common-gate and common-drain GaAs MESFET chip are given in Table 3.24. The bonding inductances must be included with these data.

REFERENCES

3.1   T. Drummond, W. T. Masselink, and H. Morkoc, "Modulation-Doped GaAs/AlGaAs Heterojunction Field-Effect Transistors MODFET's," *Proceedings of the IEEE*, Vol. 74, No. 6, June 1986, pp. 773–822; correction, December 1986, p. 1803.

3.2 S. M. Sze, *Physics of Semiconductor Devices*, 2nd ed., Wiley, New York, 1981, pp. 19, 809, 170–175.

3.3 J. C. Irwin and A. Loya, "Failure Mechanism and Reliability of Low Noise GaAs FETs," *Bell System Technical Journal*, Vol. 57, 1978, pp. 2823–2846.

3.4 J. C. Irwin, "The Reliability of GaAs FET's," in *GaAs FET Principles and Technology*, J. V. De Lorenzo and D. K. Khandelwal, Eds., Artech House, Norwood, Mass., 1982, pp. 349–402.

3.5 E. O. Johnson, "Physical Limitations on Frequency and Power Parameters of Transistors," *RCA Review*, Vol. 26, June 1965, pp. 163–177.

3.6 C. Khandavalli, Avantek, to be published, 1989, Module MDL 140145-8.

3.7 H. T. Yuan, Y. S. Wu, and J. B. Kruger, "A 2-Watt X-Band Silicon Power Transistor," *IEEE Transactions on Electron Devices*, Vol. 25, No. 6, June 1978, pp. 731–736.

3.8 P. M. Smith, P. C. Chao, K. H. G. Duh, L. F. Lester, B. R. Lee, and J. Ballingall, "Advances in HEMT Technology and Applications," *IEEE MTT-S International Microwave Symposium Digest*, Vol. 11, 1987, pp. 749–752.

3.9 E. Sovero, A. K. Gupta, J. A. Higgins, and W. A. Hill, "35-GHz Performance of Single and Quadruple Power Heterojunction HEMT's," *IEEE Transactions on Electron Devices*, Vol. 33, No. 10, October 1986, pp. 1434–1438.

3.10 N. H. Sheng, M. F. Chang, P. M. Asbeck, K. C. Wang, G. J. Sullivan, D. L. Miller, J. A. Higgins, E. Sovero, and H. Basit, "High Power AlGaAs/GaAs HBTS for Microwave Applications," *Technical Digest of IEDM*, 1987, p. 619.

3.11 J. A. Higgins, Rockwell Science Center, private communication, 1988.

3.12 R. Allison, "Silicon Bipolar Microwave Power Transistors," *IEEE Transactions on Microwave Theory and Techniques*, Vol. MTT-27, No. 5, May 1979, pp. 415–422.

3.13 C. P. Snapp, "Microwave Silicon Bipolar Transistors and Monolithic Integrated Circuits," in *Handbook of Microwave and Optical Components*, Vol. 2, *Microwave Solid State Components and Tubes*, K. Chang, Ed., Wiley, New York, 1989.

3.14 R. J. Hawkins, "Limitations of Nielson's and Related Noise Equations Applied to Microwave Bipolar Transistors, and a New Expression for the Frequency and Current Dependent Noise Figure," *Solid-State Electronics*, Vol. 20, No. 3, March 1977, pp. 191–196.

3.15 E. G. Nielsen, "Behavior of Noise Figure in Junction Transistors," *Proceedings of the IRE*, Vol. 45, July 1957, pp. 957–963.

3.16 H. F. Cooke, "Transistor Noise Figure," *Solid-State Design*, February 1963, pp. 37–42.

3.17 H. Fukui, "The Noise Performance of Microwave Transistors," *IEEE Transactions on Electron Devices*, Vol. ED-13, March 1966, pp. 329–341.

3.18 A. van der Ziel, "Unified Presentation of 1/f Noise Sources," *Proceedings of the IEEE*, Vol. 76, No. 3, March 1988, pp. 233–258.

3.19 M. Conti, "Surface and Bulk Effects in Low Frequency Noise in NPN Planar Transistors," *Solid-State Electronics*, Vol. 13, No. 11, November 1970, pp. 1461–1469.

3.20 C. D. Motchenbacher and F. C. Fitchen, *Low-Noise Electronic Design*, Wiley, New York, 1973, p. 276.

3.21 V. Grande, Avantek, unpublished data, 1988.

3.22 H. Kroemer, "Theory of a Wide-Gap Emitter for Transistors," *Proceedings of the IRE*, Vol. 45, No. 11, November 1957, pp. 1535–1537.

3.23 P. M. Asbeck, M. F. Chang, K. C. Wang, D. L. Miller, G. J. Sullivan, N. H. Sheng, E. A. Sovero, and J. A. Higgins, "Heterojunction Bipolar Transistors for Microwave and Millimeter-Wave Integrated Circuits," *IEEE Microwave and Millimeter-Wave Monolithic Circuits Symposium Digest*, 1987, pp. 1–5.

3.24 P. M. Asbeck, A. K. Gupta, J. Ryan, D. L. Miller, R. J. Anderson, C. A. Liechti, and F. H. Eisen, "Microwave Performances of GaAs/(Ga, Al)As Heterojunction Bipolar Transistors," *Technical Digest IEDM*, 1984, pp. 844–845.

3.25 B. Bayraktaroglu, N. Camilleri, H. D. Shih, and H. Q. Tserng, "Al GaAs/GaAs Heterojunction Bipolar Transistors with 4 W/mm Power Density at X-Band," *IEEE MTT-S International Microwave Symposium Digest*, 1987, pp. 969–972.

3.26 B. Bayraktaroglu and N. Camilleri, "Microwave Performances of *npn* and *pnp* AlGaAs/GaAs Heterojunction Bipolar Transistors," *1988 IEEE MTT-S International Microwave Symposium Digest*, May 1988, pp. 529–532.

3.27 B. Bayraktaroglu, N. Camilleri, and S. A. Lambert, "Microwave Performances of *npn* and *pnp* AlGaAs/GaAs Heterojunction Bipolar Transistors," *IEEE Transactions on Microwave Theory and Techniques*, December 1988, pp. 1869–1873.

3.28 K. K. Agarwal, "Dielectric Resonator Oscillators Using GaAs/GaAlAs Heterojunction Bipolar Transistors," *IEEE MTT-S International Microwave Symposium Digest*, 1986, pp. 95–98.

3.29 S. R. Lesage, M. Madihian, N. Hayama, and K. Honjo, "15.6 GHz HBT Microstrip Oscillator," *Electronics Letters*, Vol. 29, No. 18, February 1988, pp. 230–232.

3.30 G. D. Vendelin, H. J. Sigg, T. P. Cauge, and J. Kocsis, "MOS Transistors for Microwave Receivers," *Proceedings of the 3rd Bicentennial Cornell Electrical Engineering Conference*, 1971, pp. 417–426.

3.31 H. J. Sigg, G. D. Vendelin, T. P. Cauge, and J. Kocsis, "D-MOS Transistor for Microwave Applications," *IEEE Transactions on Electron Devices*, Vol. ED-19, No. 1, January 1972, pp. 45–53.

3.32 C. A. Liechti, "Microwave Field-Effect Transistors-1976," *IEEE Transactions on Microwave Theory and Techniques*, June 1976, pp. 279–300.

3.33 R. W. Thill, W. Kennan, and N. K. Osbrink, "A Low Noise GaAs FET Preamplifier for 21 GHz Satellite Earth Terminals," *Microwave Journal*, March 1983, pp. 75–84.

3.34 H. Wallman, A. B. Mauer, and C. N. Gadsdeu, "A Low-Noise Amplifier," *Proceedings of the IRE*, June 1948, pp. 700–708.

3.35 W. Shockley, *Proceedings of the IRE*, Vol. 40, 1952, pp. 1365–1376.

3.36 W. C. Bruncke and A. van der Ziel, "Thermal Noise in Junction-Gate Field-Effect Transistors," *IEEE Transactions on Electron Devices*, Vol. ED-13, No. 3, March 1966, pp. 323–329.

3.37 A. van der Ziel, "Thermal Noise in Field-Effect Transistors," *Proceedings of the IRE*, Vol. 50, August 1962, pp. 1808–1812.

3.38 A. van der Ziel, "Gate Noise in Field-Effect Transistors at Moderately High Frequencies," *Proceedings of the IRE*, Vol. 51, March 1963, pp. 461–467.

3.39 W. Baechtold, "Noise Behavior of Schottky Barrier Gate Field-Effect Transistors at Microwave Frequencies," *IEEE Transactions on Electron Devices*, Vol. ED-18, No. 2, February 1971, pp. 97–106.

3.40 W. Baechtold, "Noise Behavior of GaAs Field-Effect Transistors with Short Gate Lengths," *IEEE Transactions on Electron Devices*, Vol. ED-19, No. 5, May 1972, pp. 674–680.

3.41   W. Baechtold, K. Daetwyler, T. Forster, T. O. Mohr, W. Walter, and P. Wolf, "Si and GaAs 0.5 μm-Gate Schottky-Barrier Field-Effect Transistors," *Electronics Letters*, Vol. 9, No. 10, May 1973, pp. 232–234.

3.42   R. A. Pucel, H. A. Haus, and H. Statz, "Signal and Noise Properties of Gallium Arsenide Microwave Field-Effect Transistors," *Advances in Electronics and Electron Physics*, Vol. 38, Academic Press, 1975, pp. 145–265.

3.43   J. A. Turner and B. L. H. Wilson, "Implications of Carrier Velocity Saturation in a Gallium Arsenide Field-Effect Transistor," *Proceedings of the International Symposium on Gallium Arsenide*, 1968, pp. 195–204, Paper 30.

3.44   A. B. Grebene and S. K. Ghandi, "General Theory for Pinched Operation of the Junction-Gate FET," *Solid-State Electronics*, Vol. 12, No. 7, July 1969, pp. 573–589.

3.45   H. Statz, H. A. Haus, and R. A. Pucel, "Noise Characteristics of Gallium-Arsenide Field-Effect Transistors," *IEEE Transactions on Electron Devices*, Vol. ED-21, No. 9, September 1974, pp. 549–562.

3.46   R. A. Pucel, D. F. Masse, and C. F. Krumm, "Noise Performance of Gallium Arsenide Field-Effect Transistors," *IEEE Journal of Solid-State Circuits*, Vol. SC-11, No. 4, April 1976, pp. 243–255.

3.47   G. E. Brehm, "Variation of Microwave Gain and Noise Figure with Bias for GaAs FET's," *Proceedings of the 4th Bicentennial Cornell Electrical Engineering Conference*, 1973, pp. 77–85.

3.48   G. Brehm and G. D. Vendelin, "Biasing FET's for Optimum Performance," *Microwaves*, February 1974, pp. 38–44.

3.49   H. Fukui, "Design of Microwave GaAs MESFET's for Broad-Band Low Noise Amplifiers," *IEEE Transactions on Microwave Theory and Techniques*, Vol. MTT-27, No. 7, July 1979, pp. 643–650.

3.50   H. Fukui, "Addendum to Design of Microwave GaAs MESFET's for Broad-Band Low-Noise Amplifiers," *IEEE Transactions on Microwave Theory and Techniques*, Vol. MTT-29, No. 10, October 1981, p. 1119.

3.51   A. F. Podell, "A Functional GaAs FET Noise Model," *IEEE Transactions on Electron Devices*, Vol. ED-28, No. 5, May 1981, pp. 511–517.

3.52   M. S. Gupta, O. Pitzalias, Jr., S. E. Rosenbaum, and P. T. Greiling, "Microwave Noise Characterization of GaAs MESFET's: Evaluation by On-Wafer Low Frequency Output Noise Current Measurement," *IEEE Transactions on Microwave Theory and Techniques*, Vol. MTT-35, No. 12, 1987, pp. 1208–1218.

3.53   A. Cappy, "Noise Modeling and Measurement Techniques," *IEEE Transactions on Microwave Theory and Techniques*, Vol. MTT-36, No. 1, January 1988, pp. 1–10.

3.54   C. Su, H. Rohdin, and C. Stolte, "1/f Noise in GaAs MESFET's," *Technical Digest of IEDM*, 1983, pp. 601–604.

3.55   H. F. Cooke, private communication, 1988.

3.56   R. A. Pucel and J. Curtis, "Near-Carrier Noise in FET Oscillators," *IEEE MTT-S International Microwave Symposium Digest*, 1983, pp. 282–284.

3.57   S. T. Hsu, "Surface Related 1/f Noise in MOS Transistors," *Solid-State Electronics*, November 1970, pp. 1461–1469.

3.58   T. Mimura, K. Joshin, and S. Kuroda, "Device Modeling of HEMTs," *Fujitsu Scientific and Technical Journal*, September 1983, pp. 243–277.

3.59   C. A. Liechti, "Heterostructure Transistor Technology—A New Frontier in Microwave Electronics," *15th European Microwave Conference*, 1985, pp. 21–29.

3.60 P. H. Smith, *Electronic Applications of the Smith Chart*, McGraw-Hill, New York, 1969.

3.61 G. N. French and E. H. Fooks, "The Design of Stepped Transmission-Line Transformers," *IEEE Transactions on Microwave Theory and Techniques*, Vol. MTT-16, No. 10, October 1968, p. 885.

3.62 G. N. French and E. H. Fooks, "Double Section Matching Transformers," *IEEE Transactions on Microwave Theory and Techniques*, Vol. MTT-17, No. 9, September 1969, p. 719.

3.63 R. M. Fano, "Theoretical Limitations on the Broad Band Matching of Arbitrary Impedances," *Journal of the Franklin Institute*, Vol. 249, January 1960, pp. 57–83, and February 1960, pp. 139–155.

3.64 G. D. Vendelin and W. C. Mueller, "Noise Parameters of Microwave Transistors," *Microwave Journal*, November 1987, pp. 177–186.

3.65 W. Baechtold, "Noise Behaviour of Schottky Barrier Gate Field-Effect Transistors at Microwave Frequencies," *IEEE Transactions on Electron Devices*, Vol. ED-18, February 1971, pp. 97–106.

3.66 G. D. Vendelin, W. Hooper, and G. Bechtel, "Designs and Application of Low-Noise GaAs FET Amplifiers," *Digest of International Conference on Communications*, June 1973.

3.67 W. Kennan, "Measurement and Modeling of GaAs FET Chips," *Avantek Applications Note ATP-1054*, October 1983.

3.68 G. D. Vendelin and M. Omori, "A Computer Model of the GaAs MESFET Valid to 12 GHz," *Electronics Letters*, Vol. 11, February 1975, pp. 60–61.

3.69 G. D. Vendelin and M. Omori, "Try CAD for Accurate GaAs MESFET Models," *Microwaves*, June 1975, pp. 58–70.

BIBLIOGRAPHY

Anderson, K. J., and A. M. Pavio, "FET Oscillators Still Require Modeling, but Computer Techniques Simplify the Task," *Microwave System News*, September 1983, pp. 60–72.

Archer, J. A., "Design and Performance of Small-Signal Microwave Transistors," *Solid State Electronics*, Vol. 15, March 1972, pp. 249–258.

Archer, J. A., "Low-Noise Implanted Base Microwave Transistors," *Solid State Electronics*, Vol. 17, April 1974, pp. 387–393.

Banerjee, I., P. W. Chye, and P. E. Gregory, "Unusual C-V Profiles of Si Implanted (211) GaAs Substrates and Unusually Low Noise MESFETs Fabricated on Them," *IEEE Electron Device Letters*, January 1988, pp. 10–12.

Berenz, J. J., K. Nakimo, and K. P. Weller, "Low Noise High Electron Mobility Transistors," *IEEE Microwave and Millimeter-Wave Monolithic Circuits Symposium Digest*, 1984, pp. 83–85.

Carlin, H. J., and P. Amstutz, "On Optimum Broad-Band Matching," *IEEE Transactions on Circuits and Systems*, Vol. CAS-28, May 1981, pp. 401–405.

Chen, J. T. C. and C. P. Snapp, "Bipolar Microwave Linear Power Transistor Design," *IEEE Transactions on Microwave Theory and Techniques*, Vol. MTT-27, May 1979, pp. 423–430.

Cooke, H. F., "Microwave Transistors: Theory and Design," *Proceedings of the IEEE*, Vol. 59, August 1971, pp. 1163–1181.

Dawson, R., "Equivalent Circuit of the Schottky-Barrier Field-Effect Transistor at Microwave Frequencies, *IEEE Transactions on Microwave Theory and Techniques*, Vol. MTT-23, No. 6, June 1975, pp. 499–501.

Day, P. E., "Transmission Line Transformation Between Arbitrary Impedance Using the Smith Chart," *IEEE Transactions on Microwave Theory and Techniques*, Vol. MTT-23, September 1975, p. 772.

Dilorenzo, J. and D. Khandelwal, *GaAs FET Principles and Technology*, Artech House, Norwood, Mass., 1982.

Duh, K. H. G., S. M. J. Liu, L. F. Lester, P. C. Chao, P. M. Smith, M. B. Das, B. R. Lee, and J. Ballingall, "Ultra-Low-Noise-Characteristics of Millimeter-Wave High Electron Mobility Transistors," *IEEE Electron Device Letters*, Vol. 9, October 1988, pp. 521–523.

Ferry, D. K. *Gallium Arsenide Technology*, Howard W. Sams, Indianapolis, Ind., 1985.

Fukui, H., Ed., *Low-Noise Microwave Transistors and Amplifiers*, IEEE Press and Wiley, New York, 1981.

Graham, E. D., and C. W. Gwyn, Eds., *Microwave Transistors*, Artech House, Norwood, Mass., 1975.

Hikosaka, K., et al., "A 30 GHz 1-W Power HEMT," *IEEE Electron Device Letters*, Vol. EDL-8, No. 11, 1987, pp. 521–523.

Hower, P. L., and G. N. Bechtel, "Current Saturation and Small-Signal Characteristics of GaAs Field-Effect Transistors," *IEEE Transactions on Electron Devices*, Vol. ED-20, March 1973, pp. 213–220.

Hsu, T. H., and C. P. Snapp, "Low-Noise Microwave Bipolar Transistor with Sub-Half-Micrometer Emitter Width," *IEEE Transactions on Electron Devices*, Vol. ED-25, June 1978, pp. 723–730.

Kirk, C. T., "A Theory of Transistor Cut-Off Frequency Falloff at High Current Density," *IEEE Transactions on Electron Devices*, Vol. ED-9, 1962, pp. 162–174.

Lamming, J. S., "Microwave Transistors," in M. J. Howes and D. V. Morgan, Eds., *Microwave Devices*, Wiley, New York, 1976.

Liechti C. A., E. Gowen, and J. Cohen, "GaAs Microwave Schottky-Gate FET," *Digest of Technical Papers, International Solid-State Circuits Conference*, 1972, pp. 158–159.

Mao, S., S. Jones, and G. D. Vendelin, "Millimeter-Wave Integrated Circuits," *IEEE Transactions on Electron Devices*, July 1968, pp. 517–523.

Matthaei, G. I., "Short-Step Chebyshev Impedance Transformers," *IEEE Transactions on Microwave Theory and Techniques*, Vol. MTT-14, August 1966, p. 372.

Mead, C., "Schottky-Barrier Gate Field-Effect Transistors," *Proceedings of the IEEE*, Vol. 14, No. 2, February 1966, pp. 307–308.

Milligan, T. A., "Transmission-Line Transformation Between Arbitrary Impedances," *IEEE Transactions on Microwave Theory and Techniques*, Vol. MTT-24, March 1976, p. 159.

Rohde, U. L., "Modeling Noise in Microwave Bipolar Transistors," *Microwave Systems News and Communications Technology*, February 1987, pp. 91–93.

Sigg, H., T. Cauge, J. Kocsis, and G. D. Vendelin, "Double-Diffused MOS Transistor Achieves Microwave Gain," *Electronics*, February 15, 1971, pp. 99–104.

Sigg, H., and G. D. Vendelin, "Low Noise Single and Dual Gate D-MOS Transistor," *Electron Devices Meeting*, Washington, D.C., December 1972.

Snapp, C. P., "Bipolars Quietly Dominate," *Microwave System News*, November 1979, pp. 45–67.

Somlo, P. I., "A Logarithmic Transmission Line Chart," *IEEE Transactions on Microwave Theory and Techniques*, Vol. MTT-10, July 1960, p. 463.

Vendelin, G. D., *Design of Amplifiers and Oscillators by the S-Parameter Method*, Wiley, New York, 1982.

Vendelin, G. D., "Feedback Effects in the GaAs MESFET Model," *IEEE Transactions on Microwave Theory and Techniques*, June 1976, pp. 383–385.

Vendelin, G. D., "Feedback Effects on GaAs MESFET Noise Performance," *1975 International Microwave Symposium*, May 1975, pp. 324–326.

Vendelin, G. D., "Five Basic Bias Designs for GaAs FET Amplifiers," *Microwaves*, February 1978, pp. 40–42.

Vendelin, G. D., W. Alexander, and D. Mock, "Computer Analyzes RF Circuits with Generalized Smith Charts," *Electronics*, 1974, pp. 102–109.

Vendelin, G. D., G. Brehm, and G. Bechtel, "Gallium Arsenide Field-Effect Transistor Applications," *Fairchild Application Note*, April 1974.

Wolf, P., "Microwave Properties of Schottky-Barrier Field-Effect Transistors," *IBM Journal of Research and Development*, Vol. 14, No. 2, March 1970, pp. 125–141.

Youla, D. C., "A New Theory of Broad-Band Matching," *IEEE Transactions on Circuit Theory*, Vol. CT-11, March 1964, pp. 30–50.

PROBLEMS

3.1. Using a bipolar low-frequency hybrid-Π model, show that the 50-Ω gain is

$$S_{21}(\text{LF}) = \frac{-2\, Z_0\, g_m\, r_\pi}{r_b + r_\pi + Z_0}$$

Estimate this gain for the AT-41400 model in decibels.

3.2. From a low-frequency model of the GaAs FET, show that the 50-Ω gain is

$$S_{21}(\text{LF}) = -2g_m Z_0$$

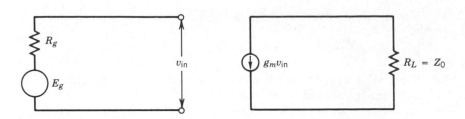

(*Note:* For an open-circuit line, $|v_{in}| = 2|E_g|$.) Estimate this gain for the transistors given in Table 3.4 at the high gain bias in decibels.

3.3. Show that the simplified hybrid Π can be derived from the simplified T-equivalent circuit. Give an estimate of C_π, r_π, and so on.

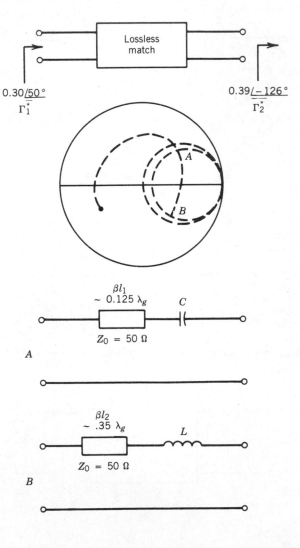

3.4. Find the element values for the interstage-matching networks given in the figure.

3.5. Using the concept of single-element matching, derive the three-element interstage-matching network shown.

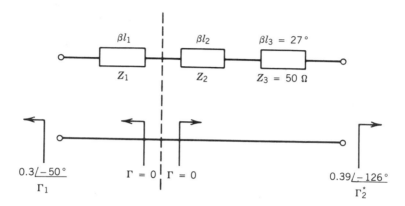

3.6. Find a single-element transmission line that performs the interstage match shown.

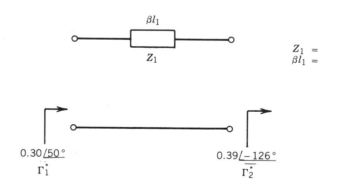

3.7. Complete or verify the following six interstage designs of $f = 4$ GHz. Show the six solutions on a Smith chart.

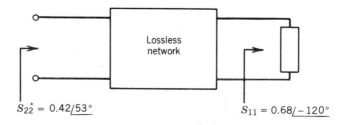

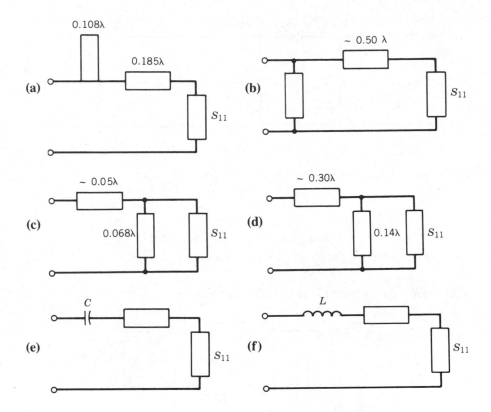

3.8. Design a three-element lossless lumped network M that matches the generator to the load at 10 GHz. What is the maximum VSWR for 20% bandwidth? Recommend modifications of M to increase bandwidth.

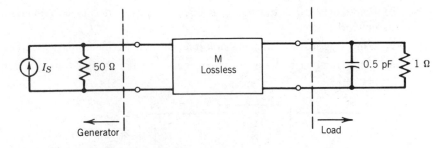

3.9. Repeat Problem 3.8 with a single-element $\lambda/4$ line and two elements of $\lambda/4$ line (i.e., match 50 Ω to 7.07 Ω, then match 7.07 Ω to 1 Ω). Which circuit gives the widest bandwidth?

3.10. Derive the S parameters of the following FET model (omiting L_g, L_s, L_d) for $\tau_d = 0$ at $f = 4$ GHz.

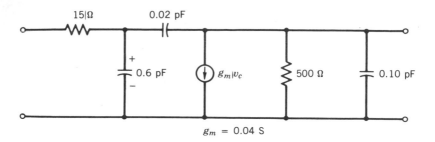

$g_m = 0.04$ S

3.11. Given $Z_L/Z_0 = 1 - j1.2$ with $Z_0 = 50$ Ω:

    (a) Find Γ_L, Γ_L^*, $1/\Gamma_L$, and $1/\Gamma_L^*$.

    (b) Find Y_L/Y_0 and Y_L^*/Y_0.

    (c) Find the admittance of $1/\Gamma_L$ and the admittance of $1/\Gamma_L^*$.

    (d) Plot these eight points on a compressed Smith chart.

3.12. Match the following load to a 50-Ω generator using lumped elements.

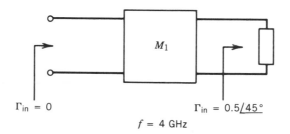

$\Gamma_{in} = 0$ $\Gamma_{in} = 0.5\underline{/45°}$

$f = 4$ GHz

3.13. **(a)** Design four lossless matching networks using lumped elements.

    (b) Using only 50-Ω transmission lines (lossless), design two matching networks.

    (c) Using any lossless transmission line, design two matching networks.

    (d) Using lumped and transmission-line elements, design four matching networks.

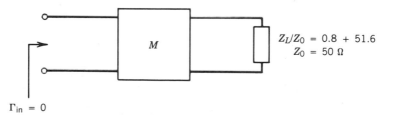

$Z_L/Z_0 = 0.8 + 51.6$
$Z_0 = 50$ Ω

$\Gamma_{in} = 0$

3.14. Find a single-section element that matches $Z_L/Z_0 = 1.4 - j0.6$ to a 50-Ω generator ($Z_0 = 50$ Ω).

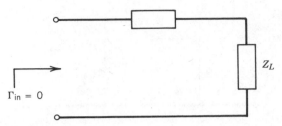

$\Gamma_{in} = 0$

3.15. A low-noise amplifier is shown at $f = 8$ GHz for a 50-Ω generator and load. Find Γ_{0n} for this transistor.

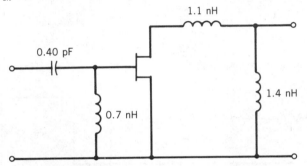

3.16. For the following distributed circuit, find the input impedance at $f = 2$ GHz and $f = 4$ GHz. Also calculate Γ_{in} at 2 and 4 GHz.

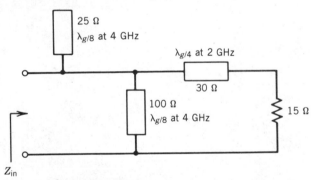

3.17. Calculate Z_{in} for the following two microstrip circuits. $f = 10$ GHz, $\epsilon_r = 10$, and $h = 25$ mils. Use Figs. 1.24 and 1.25 to calculate Z_0 and k'.

(a)

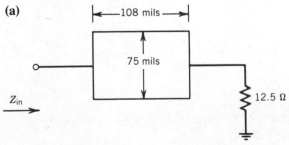

(b)

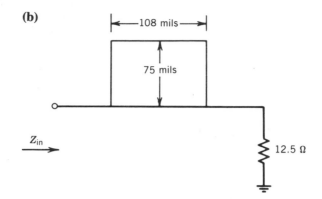

3.18. Given a LNA design, find Γ_{0n} for $Z_0 = 50\ \Omega$. $f = 6$ GHz.

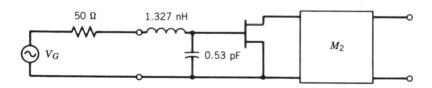

4 Amplifier Design

4.0 INTRODUCTION

In an earlier version of this book, an HFET-1001 GaAs MESFET chip from Hewlett-Packard with gate length 1 μm and gate width 500 μm was selected to illustrate the design of a 3- to 4-GHz single-stage amplifier using COMPACT to achieve a flat gain of 15.8 dB. In this update the AT-8251 GaAs MESFET chip from Avantek with a 0.3-μm gate length and a 500-μm gate width will be used to illustrate a one-stage design.

Before launching into the design procedures, which are usually begun on the Smith chart and finished on the computer, some important considerations must be reviewed. Stability is the most important issue in amplifier design. The amplifier must be stable at all frequencies: below the band, in the band, and above the band; moreover, it must be stable for all terminations that will be connected (i.e., not just 50-Ω terminations). Even for single-stage designs, this may not be an easy problem if you are restricted to lossless matching circuitry (Fig. 1.1). Therefore, there may be a need to add resistors to the amplifier for stability. Since this will always lower gain, the engineering challenge is to find a solution that does not require resistors. Nevertheless, several rigorous solutions using "device modification" will be discussed.

Another consideration is the reverse gain parameter S_{12}. A useful approximation for common-source FETs and common-emitter bipolar transistors is to assume that this parameter is zero and design the lossless matching networks. This is only a beginning point in the design, which will give a useful amplifier circuit topology. The effect of this assumption has been reported in the literature [4.1], and a design approach including the effects of S_{12} is presented in this chapter.

There are many different types of single-stage amplifiers to consider: low noise, high power, wide bandwidth, feedback designs, and balanced designs. All of these amplifier design procedures are discussed in this chapter. For multistage amplifiers, the design of a two-stage amplifier will first be considered a simple extension of lossless one-stage design. Once again, this is a simplification which assumes that S_{12} is zero for both transistors, only a beginning point in order to evaluate different amplifier circuit topologies. The special considerations for stability of a multistage amplifier are also treated in this chapter.

Another addition to this chapter in this edition is the distributed or traveling-wave amplifier, which is becoming very useful for wideband FET applications. This amplifier is discussed in Chapters 4 and 5.

4.1 ONE-STAGE DESIGN

4.1.1 Unilateral Design

As we recall from Fig. 1.1. the design of a one-stage amplifier consists of finding

1. An input lossless matching network M_1
2. An output lossless matching network M_2

so that the maximum or desired transistor gain is achieved over the operating bandwidth of the amplifier. Usually, the common-emitter or common-source configuration is chosen for highest gain per stage. If the stability factor k is greater than unity, these two networks can be found to give the maximum available gain G_{ma} given by (1.217). If the stability factor is less than or equal to unity, the amplifier could be terminated in a matching structure which causes oscillation; that is, G_{ma} is infinite. This should be avoided by locating the regions of instability in the Γ_G and Γ_L planes. The input and output terminations (Γ_G and Γ_L) must be designed to avoid the instability regions. Usually, these unstable regions are near the conjugate match for S_{11} and S_{22}. Thus a stable amplifier will require some input and or output mismatch if k is less than or equal to unity.

There are at least two alternative approaches for potentially unstable amplifiers:

1. Add resistive matching elements to make $k \geq 1$ and $G_{ma} \simeq G_{ms}$.
2. Add feedback to make $k \geq 1$ and $G_{ma} \simeq G_{ms}$.

The techniques are described in detail in Section 4.1.2, modifying the characteristics of a transistor.

For narrowband amplifiers it is usually recommended to accept a transistor with $k < 1$, design the amplifier for a gain approaching G_{ms}, and to ensure that the Γ_G and Γ_L terminations provide stability at all frequencies, both inside and outside the amplifier passband. The regions of instability for the output plane (the Γ_L plane) are given in Fig. 1.29, which was plotted from the transistor S parameters and from (1.135) and (1.136). Similar regions can be found for the input plane (Γ_G plane).

The design of an amplifier would usually have the following specifications:

Gain and gain flatness
Bandwidth and center frequency ($f_2 - f_1, f_0$)
Noise figure
Linear output power
Input reflection coefficient ($VSWR_{in}$)
Output reflection coefficient ($VSWR_{out}$).
Bias voltage and current.

For small-signal amplifiers, the small-signal S parameters are sufficient to complete the design. After selecting an appropriate transistor based on these specifications, a one-stage amplifier design should be considered if sufficient gain can be achieved; otherwise, a two-stage amplifier should be designed.

The circuit topology should be chosen to allow dc bias for the transistor. Usually, RF short-circuited stubs are placed near the transistor to allow dc biasing. If the topology does not allow dc bias, a broadband bias choke or high-resistance bias circuit that does not affect the amplifier performance must be used.

The following steps can be tabulated for the design of a one-stage amplifier (see Fig. 4.1):

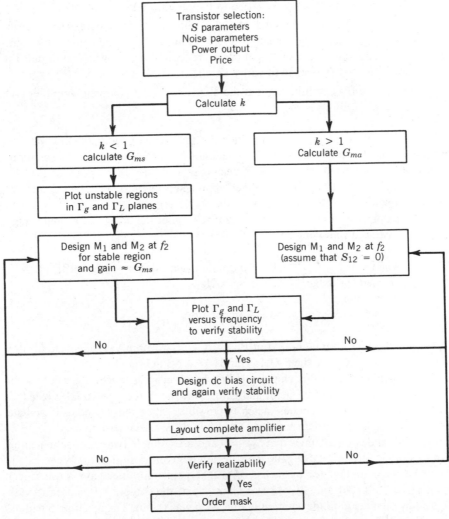

Figure 4.1 Simplified amplifier design procedure.

1. Select a transistor based on data sheet description of the S parameters, noise figure and linear output power.

2. Calculate k and G_{ma} or G_{ms} versus frequency.

3. For $k > 1$, select the topologies that match the input and output (and allow dc biasing) at the upper band edge f_2. Ideally, this will give G_{ma} and $S'_{11} = S'_{22} = 0$. Usually, $S_{12} = 0$ is assumed for the initial design. Next, the topology may be varied to flatten gain versus frequency at the expense of S'_{11} and S'_{22}.

4. For $k < 1$, plot the regions of instability on the Γ_G and Γ_L planes and select topologies that partially match the input and output at the upper band edge and avoid the unstable regions. The gain will approach G_{ms} as an upper limit. Next, the topology may be varied to flatten gain versus frequency.

5. After finding initial M_1 and M_2, plot the amplifier S parameters versus frequency; make adjustments in topology until the specifications for gain, input reflection coefficient, and output reflection are satisfied. Also plot Γ_G and Γ_L versus frequency to verify amplifier stability.

6. Design dc bias circuit. Lay out the elements of the complete amplifier and check realizability.

As a beginning point in an amplifier design, the circuit topologies may be designed using the impedance-matching techniques of Chapter 3, with the assumption that S_{12} is zero. Later, an exact design procedure will be described which includes the effects of S_{12}. The simplified approach is suggested to investigate two-element topologies with terminations in the stable region.

An example will illustrate the procedure. Consider the design of a single-stage amplifier with the following specifications:

Gain:	10 dB minimum (gain flatness $\pm$ 0.5 dB)
Frequency range	4.0 to 8.0 GHz
Bias voltage:	15 V
Bias current:	50 mA maximum

A GaAs MESFET chip (AT-8251 at $V_{DS} = 5$ V, $I_d = 50$ mA) is adequate for this single-stage design since the 8-GHz gain is 13.4 dB (G_{ma}).

The S parameters of this chip (including bonding inductance) are tabulated in Table 4.1 with the stability factor and gains over the range 1 to 10 GHz. Since $k > 1$ at 8 GHz, a simultaneous match is possible. Since the device can oscillate over 1 to 7 GHz, the circuit topology must avoid the unstable region.

The initial design assumes that $S_{12} = 0$ and uses simple two-element matching circuits based on the 8-GHz gain. This design is given in Fig. 4.2a. When calculated on the computer [or by means of (1.217)] this amplifier gave a transducer gain of 12.2 dB, but only 2.7 dB at 4 GHz. Next, the computer was used to vary Z_0 and the lengths until the solution in Fig. 4.2b was found. The resulting S parameters for the broadband design are given in Table 4.2. Notice the large value of

TABLE 4.1 Transistor S Parameters, AT-8251 (V_{ds} = 5 V, I_d = 50 mA)

Freq. (GHz)	S_{11} Mag.	S_{11} Angle	S_{21} Mag.	S_{21} Angle	S_{12} Mag.	S_{12} Angle	S_{22} Mag.	S_{22} Angle	G_{ma}/G_{ms} (dB)	k
1.0	0.98	−25	5.32	163	0.020	82	0.67	−5	24.249	0.099
2.0	0.93	−47	5.16	144	0.038	71	0.60	−12	21.329	0.280
3.0	0.89	−70	4.82	127	0.056	65	0.53	−17	19.349	0.338
4.0	0.82	−94	4.48	110	0.068	52	0.46	−20	18.188	0.537
5.0	0.75	−117	3.88	97	0.075	50	0.40	−24	17.138	0.684
6.0	0.71	−137	3.51	83	0.080	46	0.31	−29	16.422	0.846
7.0	0.70	−159	3.07	71	0.083	42	0.22	−35	15.681	0.987
8.0	0.71	−169	2.73	63	0.086	41	0.16	−42	13.446	1.066
9.0	0.72	180	2.53	54	0.090	40	0.12	−46	12.660	1.090
10.0	0.73	172	2.31	44	0.093	39	0.09	−55	11.750	1.131

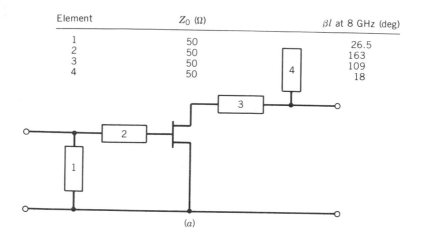

Element	Z_0 (Ω)	βl at 8 GHz (deg)
1	50	26.5
2	50	163
3	50	109
4	50	18

(a)

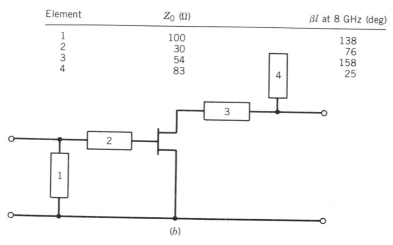

Element	Z_0 (Ω)	βl at 8 GHz (deg)
1	100	138
2	30	76
3	54	158
4	83	25

(b)

Figure 4.2   Amplifier design example: (a) initial design; (b) final design using CAD.

S_{11} at 4 GHz, but the gain is flat. Methods for realizing low VSWR at the input and output ports are discussed in a later section. Notice that it is inherently impossible to achieve flat gain, low input reflection coefficient, and low output reflection coefficient simultaneously from lossless matching.

The amplifier stability is checked in Fig. 4.3 for the final design over the range 1 to 10 GHz. The dc bias network uses a drain dropping resistor and a bypassed source resistor. The schematic for the entire amplifier is given in Fig. 4.4, where all elements are realizable. The biasing parasitics, especially the drain choke (RFC), should be included in the RF calculations before this amplifier is realized. A simple realization of the drain bias choke is a high-impedance $\lambda_g/4$ shorted stub.

TABLE 4.2 Amplifier Performance

Freq. (GHz)	S_{11} Mag.	S_{11} Angle	S_{21} Mag.	S_{21} Angle	S_{12} Mag.	S_{12} Angle	S_{22} Mag.	S_{22} Angle	S_{21} (dB)	k
1.0	0.990	95.4	3.82	−158	0.014	121	0.717	−42.7	11.7	0.099
2.0	0.912	−41.7	6.05	106	0.045	32.9	0.616	−94.5	15.6	0.280
3.0	0.920	−119	4.52	42.3	0.053	−19.7	0.495	−132	13.1	0.338
4.0	0.895	−156	3.81	−2.97	0.058	−61.0	0.437	−169	11.6	0.537
5.0	0.834	177	3.48	−42.5	0.067	−89.5	0.373	143	10.9	0.684
6.0	0.715	149	3.61	−87.4	0.082	−124	0.317	77.6	11.2	0.846
7.0	0.450	102	3.78	−139	0.102	−168	0.359	1.27	11.6	0.987
8.0	0.250	7.91	3.53	168	0.111	146	0.426	−60.1	10.9	1.07
9.0	0.596	−96.1	2.71	105	0.096	91.5	0.331	−111	8.55	1.09
10.0	0.947	−153	1.03	37.3	0.041	32.3	0.139	−91.9	−0.377	1.13

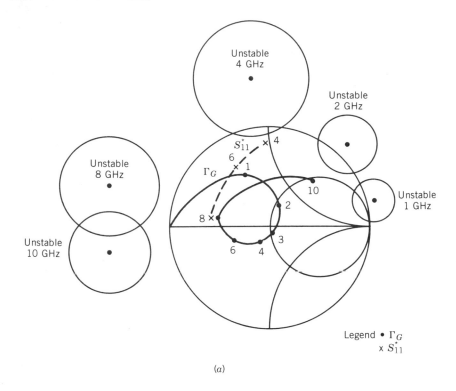

(a)

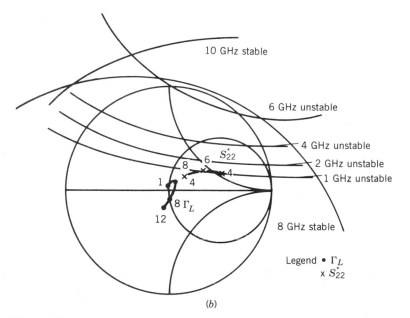

(b)

Figure 4.3 (a) Unstable regions in Γ_G plane; (b) unstable regions in Γ_L plane.

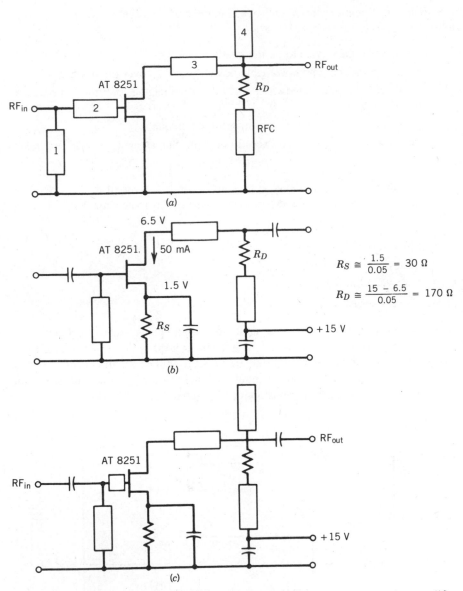

Figure 4.4 Amplifier schematic: (*a*) RF schematic; (*b*) dc schematic; (*c*) complete amplifier schematic.

4.1.2 Nonunilateral Design

Device Modification. The nonunilateralness ($S_{12} = 0$) of microwave transistors is an important factor in designing an amplifier. It has a direct influence on the amplifier sensitivity, the ease with which the different networks can be tuned, and the design itself. Nonunilateralness is taken into account from the outset of the design in the following procedure.

Device modifications at microwave frequencies are often required for the following reasons:

1. To stabilize the device to be inherently stable
2. To flatten the maximum available power gain G_{ma} over the passband
3. To reduce the gain–bandwidth constraints associated with the impedances to be matched
4. To improve the input or output match over frequency
5. To increase S_{11} and S_{22} to exceed unity by increasing S_{12} for an oscillator application

The same techniques can be used to achieve the first three of these goals. The basic requirement is to load the device with resistance in some way. The following options exist:

1. Shunt resistive loading
2. Series resistive loading
3. Voltage-shunt feedback
4. Current-series feedback

In wideband designs it is advisable to use frequency-selective loading or feedback in order to compensate for the inherent roll-off in the gain available from the device. This usually implies using an inductor or shorted stub in series with a resistor (shunt loading and voltage-shunt feedback) or a capacitor in parallel with a resistor (series loading). Various options are illustrated in Fig. 4-5.

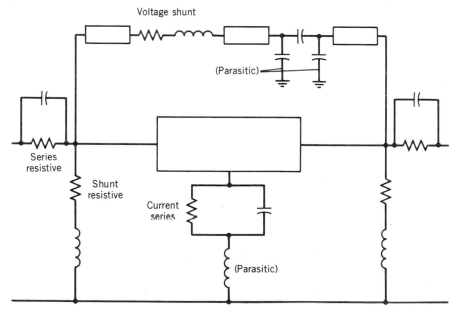

Figure 4.5 Different options available to modify the characteristics of an active device.

Current-series feedback usually degrades stability, but it can be a useful technique for modifying the input impedance (usually increasing it) or to change the optimum noise impedance or in some cases to improve the minimum noise figure obtainable with a device that is matched for a low-input VSWR.

The influence of current series feedback on the input reflection parameter S_{11} is mapped in Fig. 4.6 for a typical GaAs MESFET chip at 2 GHz to give an idea of the effect of this type of feedback. The overall effect on the other S parameters and the gain must also be considered before using feedback to improve S_{11}.

Since inherent stability is of interest, the question of how to stabilize the device, with minimum reduction in gain available from the device in the passband, arises. This usually implies using frequency-selective loading or feedback. The optimum values can be found iteratively.

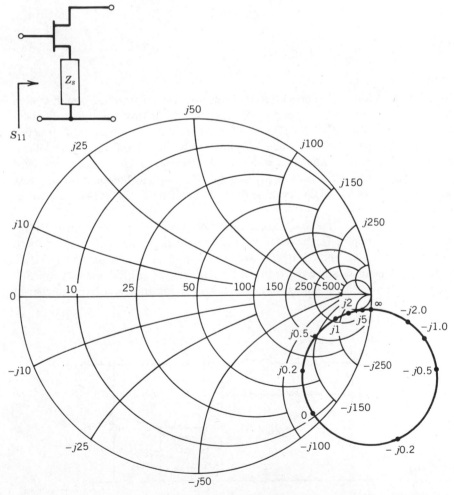

Figure 4.6 Series feedback for GaAs MESFET input reflection coefficient. Coordinates in ohms. AT-8251 chip data $V_{DS} = 5.0$ V, $I_{DS} = 50$ mA at $f = 2$ GHz.

From a realizability viewpoint, stabilizing a device with shunt resistive loading is often preferable. Stabilizing a device to be used in a wideband design with resistive loading at the input or with voltage-shunt feedback is also realistic. Shunt resistive loading and/or voltage-shunt feedback will give gain leveling. The values of the shunt loading resistor and series inductor or resistor and shorted stub required to equalize the gain at the passband edges can be calculated analytically [4.2] while those for voltage-shunt feedback can be found iteratively.

In broadband bipolar transistor amplifiers compound feedback, that is, both series and shunt feedback is commonly used. Reactive elements are usually used in the feedback paths to compensate for the inherent gain roll-off. This type of feedback is illustrated in Fig. 4.7, and this circuit is discussed in Section 4.1.5.

Initial values for R_f and R_e can be chosen by using the following low-frequency approximations:

$$Z_0^2 = R_f R_e \qquad (4.1)$$

which gives

$$Z_{\text{in}} = Z_{\text{out}} = Z_0 \qquad (4.2)$$

Notice that a series inductor with R_f removes this resistor from the circuit at high frequencies; also, a shunt capacitor with R_e removes this resistor at high frequencies, which could degrade S_{11} if the input impedance is low. Usually, R_1 is added for dc biasing and has a large value, typically greater than 300 Ω. By proper adjustment of R_f and R_e in combination with reactive elements, a flat S_{21} can be achieved over a wide bandwidth with low S_{11} and S_{22}. The same technique does not work well for the GaAs MESFET, because of the lower value of S_{21} at low frequencies (i.e., lower g_m).

The application of feedback to the dual-gate MESFET is shown in Fig. 4.8. For broadband amplifiers, the series–parallel combination is an interesting approach to reducing both S_{11} and S_{22} over a wide frequency range. The parallel feedback is not recommended, since the virtue of the dual-gate MESFET is a very low S_{12}. Since the termination at gate 2 may also be chosen arbitrarily, a large number of two-port circuits are possible.

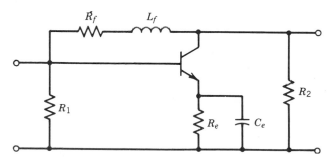

Figure 4.7   Compound feedback for wideband bipolar transistor amplifiers.

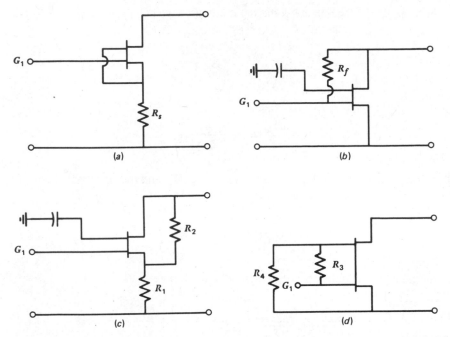

Figure 4.8 Dual-gate feedback networks: (*a*) series; (*b*) parallel; (*c*) series–parallel; (*d*) parallel–series.

Finding an equivalent set of parameters for the different circuits considered is a simple matter. The *y* parameters of networks in parallel (input currents adding and output currents adding, input voltages and output voltages the same; example voltage-shunt feedback circuit) add; the *z* parameters of networks in series (same input and output currents, input voltages and output voltages adding; example current-series feedback circuit) add; while the *h* parameters of series–parallel circuits (example Fig. 4.8*c*) add; and so on.

Tunability. An amplifier should be designed to have a low tunability factor where the tunability factor, δ, is defined as

$$\delta = \frac{\partial Y_{in}/Y_{in}}{\partial Y_L/Y_L} \tag{4.3}$$

A tunability factor of less than 0.3 is usually acceptable [4.3]. The tunability factor can be understood as follows. For δ = 0.3, if the load impedance or admittance changes by 10%, the input impedance or admittance changes by only 3%. The derivation of δ in terms of either *y* parameters [4.3] or *S* parameters [4.4] is presented below.

In terms of the y parameters, the input admittance is

$$Y_{in} = y_{11} - \frac{y_{21} \, y_{12}}{y_{22} + Y_L} \tag{4.4}$$

Then

$$\frac{dY_{in}}{dY_L} = \frac{y_{21} \, y_{12}}{(y_{22} + Y_L)^2} \tag{4.5}$$

The fractional change in input admittance is dY_{in}/Y_{in} and the fractional change in load admittance is dY_L/Y_L. Therefore, dividing (4.5) by

$$\frac{Y_{in}}{Y_L} = \frac{y_{11}}{Y_L} - \frac{y_{21} \, y_{12}}{Y_L(y_{22} + Y_L)} \tag{4.6}$$

gives the input tunability factor, which is defined to be

$$\delta = \frac{|dY_{in}/Y_{in}|}{|dY_L/Y_L|} = \frac{|y_{21} \, y_{12}| \, |Y_L|}{|y_{22} + Y_L| \, |y_{11} \, (y_{22} + Y_L) - y_{21} \, y_{12}|} \tag{4.7}$$

Tuning is easier as δ is made smaller. From (4.7), $\delta = 0$ when $Y_L = 0$ or when $Y_L \to \infty$, but such extreme admittances do not allow much power gain. Also, $\delta = 0$ when $y_{21} = 0$, but then there is no power gain. Making $y_{12} = 0$ for $\delta = 0$ is a consequence of unilateralization and the output circuit has no effect whatsoever on the input circuit.

For S-parameter analysis, the tuning factor is

$$\delta_{in} = \left| \frac{d\Gamma_{in}/\Gamma_{in}}{d\Gamma_L/\Gamma_L} \right| \tag{4.8}$$

which in terms of the two-port network S parameters can be written as

$$\delta_{in} = \frac{|S_{21}| \, |S_{12}| \, |\Gamma_L|}{|1 - S_{22} \, \Gamma_L| \, |S_{11} - D\Gamma_L|} \tag{4.9}$$

In practice, a value of $\delta_{in} < 0.3$ produces good tunability. Equation (4.9) shows that δ_{in} can be zero under some circumstances. That is, $\delta_{in} = 0$ when $S_{12} = 0$, which occurs when the unilateral assumption can be made. In this case, the output tuning does not affect the input. Also, $\delta_{in} = 0$ when $\Gamma_L = 0$, which is a very specific value of Γ_L that probably degrades the gain and noise performance of the amplifier. Of course, $\delta_{in} = 0$ when $S_{21} = 0$, that is, when there is no power gain.

Equation (4.9) can be solved for Γ_L in terms of δ_{in}, namely,

$$\cdot \quad |\Gamma_L| = \left| \alpha \pm \left| \alpha^2 - \frac{S_{11}}{S_{22}\,D} \right|^{1/2} \right| \qquad (4.10)$$

where

$$\alpha = \frac{D + S_{11}S_{22} + S_{12}S_{21}\delta_{in}^{-1}}{2\,S_{22}D} \qquad (4.11)$$

The value of $|\Gamma_L|$ obtained from (4.10) and (4.11) for a given δ_{in} is, in general, different from the value that produces maximum power gain or optimum noise performance. Therefore, this value of Γ_L produces good tunability but mismatches the amplifier.

The tunability factor, δ, is a strong function of the gain (and therefore load impedance) of an amplifier. If the gain is decreased enough and the optimum point on the constant-gain circle is chosen (tunability varies around any constant-gain circle), convergence will be quick when an amplifier is tuned. The highest tunable gain can be found iteratively and if necessary the optimum point on the relevant gain circle can be determined. The sensitivity of an amplifier to component changes is a function of the tunability of each device used in the amplifier.

Constant-Gain and Constant-Noise-Figure Circles and the Optimum Point on Each Circle for Small-Signal Amplifiers. It has been shown in Chapter 1 that the power gain of a two-port is constant for all load terminations lying on a constant-power gain circle on a Smith chart, while the available power gain is constant for all generator terminations lying on a constant-available-power-gain circle. The parameters of these circles are tabulated in Appendix C for convenience.

Although the gain on each constant-gain circle is constant, the stability factor and tunability is a function of the angle around the circle. The best tunable point is usually close to the most stable point on a gain circle and the optimum point on a gain circle can therefore be taken as the point with best tunability.

It is a simple matter to write a computer program to scan the performance on any gain circle and to choose the optimum point on it, which is the point of lowest tunability factor, δ_{min}. If the noise performance is also important, the noise figure at different positions on a gain circle must also be taken into account. The optimum point on an available power gain circle is then the point with lowest noise figure and acceptable tunability. The optimum point on a constant-noise-figure circle is usually the point with the highest tunable available power gain.

Defining the Equivalent Passive Impedance-Matching Problem. As long as a two-port is inherently stable or was compensated to be so, it is possible to transform the problem of finding a lossless network with input impedance on a constant-

power-gain circle to that of matching a given load impedance to a specific generator impedance to have a specific gain without introducing any approximations
[4.5]. The circle problem thus becomes a passive impedance-matching problem.

Similarly, it is possible to transform the problem of finding a lossless network
with output impedance on a constant-available-power-gain or constant-noise-figure
circle to that of matching a given generator impedance to a specific load impedance
to have a specific gain without introducing any approximations.

If the two-port is not inherently stable at some frequency in the passband, or
the performance is poor except close to the optimum point on a circle, the optimum
point on the relevant constant-gain or constant-noise-figure circle can be used to
complete the definition of the equivalent passive problem (a point match versus a
circle match). The relevant transformation equations are given in Appendix C for
convenience.

Computer-Aided Design of Impedance-Matching Networks. Having defined the
equivalent passive impedance matching problem to be solved in order to obtain
the specified operating or available power gain or noise figure versus frequency
response (narrowband and wideband), techniques are required to solve this
problem. In single-stage amplifiers the problem usually is that of matching a
complex load or source impedance to a resistive generator or load impedance to
have some prescribed transducer power gain versus frequency response, while it
can be to match a complex load impedance to a complex generator impedance with
prescribed gain performance when interstage matching networks for multistage
amplifiers are designed (Fig. 4.9).

Biasing the active devices imposes further constraints on the allowable topologies of the solutions to the equivalent passive impedance-matching problem. It is
possible to circumvent these problems (with the introduction of a possible nonop-
timality) by modifying the S parameters of the device to include the required shunt
inductors first (the requirements for dc blocking series capacitors are usually not a

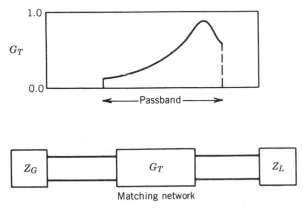

Figure 4.9   General amplifier impedance-matching problem.

problem). Impedance-matching problems can be solved analytically with subsequent optimization or iteratively with real-frequency techniques.

Solving the general impedance-matching problem with only analytical techniques is not feasible yet, primarily because the optimum gain function is not known beforehand and deriving the relevant constraints for complex terminations is not a simple task.

A problem with the analytical approach with subsequent optimization is that the solutions usually have been excessive number of components [4.6] which is usually not reduced in the subsequent optimization. The requirement for ideal transformers when fewer components are used is also a major stumbling block.

Despite these disadvantages, analytical techniques do provide insight into the matching problem. It is also useful to know that the input and output impedances of MESFETs and bipolar transistors can often be approximated to first order by using the equivalent circuits shown in Fig. 4.10. The analytical gain–bandwidth constraints associated with these impedances are well known from the work of Bode and Fano [4.7] both with positive slope as required for gain flattening by matching and unsloped as required for conjugate matching.

Fano's limit on the gain–bandwidth product of each of the two-element circuits are repeated in Figure 4.10. Right-hand plane zero's must obviously be avoided to obtain maximum bandwidth in these simple cases. It is also clear from these

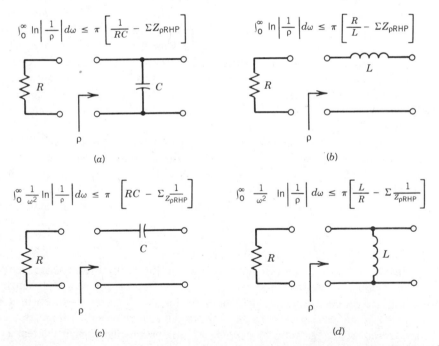

$$\int_0^\infty \ln\left|\frac{1}{\rho}\right| d\omega \le \pi\left[\frac{1}{RC} - \Sigma Z_{\rho RHP}\right]$$

$$\int_0^\infty \ln\left|\frac{1}{\rho}\right| d\omega \le \pi\left[\frac{R}{L} - \Sigma Z_{\rho RHP}\right]$$

(a)

(b)

$$\int_0^\infty \frac{1}{\omega^2} \ln\left|\frac{1}{\rho}\right| d\omega \le \pi\left[RC - \Sigma \frac{1}{Z_{\rho RHP}}\right]$$

$$\int_0^\infty \frac{1}{\omega^2} \ln\left|\frac{1}{\rho}\right| d\omega \le \pi\left[\frac{L}{R} - \Sigma \frac{1}{Z_{\rho RHP}}\right]$$

(c)

(d)

Figure 4.10 Approximate equivalent circuits for the input and output impedances of transistors and Fano's limit in each case: (a) MESFET output; (c) MESFET input.

equations that the bandwidth can be increased by reducing the gain of the network. The gain usually slopes upward with typically 3 to 9 dB/octave in the passband of interstage matching networks (i.e., if frequency selective feedback was not applied to the active device).

The first real-frequency techniques (specialized algorithms defined to solve impedance-matching problems iteratively in terms of real-frequency data, that is, using the actual generator and load terminations and gain specifications, instead of using equivalent circuits and an analytical expression for the required transducer power gain), were introduced by Carlin [4.8], Carlin and Komiak [4.9], Young and Scanlan [4.10], and Carlin and Yarman [4.11, 4.12]. Carlin and Amstutz [4.6] demonstrated the superiority of real-frequency techniques over solving impedance-matching problems analytically.

Although these real-frequency techniques are very powerful, there is no guarantee that they will find the optimum solution to a problem. The problem is mostly one of initialization (i.e., the initial direction the computer uses to iterate) and the high degree of nonlinearity that makes it susceptible to local minima. The requirement for ideal transformers when bandpass networks are synthesized can also be a major obstacle.

The algorithm of the first real-frequency technique was such that it could be initialized close to the optimum for low-pass problems when a conjugate match was required and the slope parameters used in it are fairly well decoupled. Therefore, it does perform very well when low-pass matching networks are synthesized.

The problem of initialization can be overcome to a large degree when six or fewer components are used (usually also a practical limit) by doing a systematic search with subsequent optimization. A parameter that lends itself very well to such an approach is the transformation Q's of a network at a specific frequency. This is simply the ratio of the input reactance and resistance or input susceptance and conductance calculated at any point in the network at a chosen frequency. It is this ratio that determines the amount by which the resistance will be transformed with the addition of each component to the network.

Each element of the network modifies the previous transformation Q. If the transformation Q at any point is too small, no transformation takes place, and if too high, the Q of the network will be high, which will result in narrowband performance. It is therefore possible to do a thorough search on the transformation Q's of a network in order to find initial solutions for a wideband problem, with a reasonable probability of finding the optimum in a realistic time (as defined from the viewpoint of using a personal computer), that is, if six or less elements are used and some additional constraints are imposed.

This particular technique also has the advantage that no transformers are required in the solution synthesized, and it is very easy to impose any of a wide range of topology constraints. Some of the details of this technique are outlined in Refs. 4.2 and 4.5. Notice that the Q of the network can also be observed from the Smith chart as shown in Fig. 3.39; by keeping the circuit impedance near the real axis, the bandwidth is maximized.

4.1.3 Low-Noise Amplifiers

For low-noise design, the transistor data must include S parameters at the low-noise bias and four noise parameters. One common noise parameter set is:

$$\begin{array}{ll} F_{\min} & \text{Minimum noise figure} \\ R_n = N/G_G & \text{Noise resistance} \\ Y_{0n} = G_{0n} + jB_{0n} & \text{Optimum noise admittance} \end{array}$$

The noise figure of the two-port is determined by the generator admittance (or impedance) presented to the input terminals and is calculated from (2.104), which is repeated here:

$$F = F_{\min} + \frac{R_n}{G_G}\left|Y_G - Y_{0n}\right|^2 \tag{4.12}$$

The output port is tuned for the maximum available gain if the amplifier is single-stage. In a two-stage, low-noise design the interstage circuit will probably be tuned for minimum second-stage noise figure. The N-parameter is invariant to lossless transformations [4.13].

The noise performance of a transistor is often visualized by plotting noise circles on the Γ_G plane as shown in Chapter 3 for bipolar transistors in Fig. 3.49 and for FETs in Fig. 3.53. Some additional effects on low-noise performance are 3-dB Lange coupler losses, microstrip-line losses, nonoptimum dc biasing, and resistive losses. Feedback can also be used in low-noise designs to vary $F_{\min}$ and Y_{0n}. When feedback is used, the noise figure may be reduced, but the available gain is also reduced. The minimum noise measure (defined in Section 4.2.2) is invariant to lossless feedback elements. It is not invariant to the common lead and is usually lowest for common-emitter or common-source transistors.

In the design of a single-stage amplifier, the specifications are usually clearly focused on either gain, noise figure, or output power. When the additional specification of input and/or output match is added by the system designer, a design trade-off will be necessary. For example, consider the one-stage design of a low-noise amplifier using the parameters at low noise bias. There are three basic designs to consider

1. Design M_1 for low noise,

$$\Gamma_G = \Gamma_{0n}$$

Design M_2 for

$$\Gamma_L = S_{22}'^* = \left(S_{22} + \frac{S_{12}S_{21}\Gamma_{0n}}{1 - S_{11}\Gamma_{0n}}\right)^*$$

This gives $F_{\min}$, high input VSWR, and low-output VSWR. This is the most common design technique, which is illustrated in Example 4.2.

2. Design M_1 for low noise,

$$\Gamma_G = \Gamma_{0n}$$

Design M_2 for

$$S'_{11} = \Gamma^*_{0n} = S_{11} + \frac{S_{12}S_{21}\Gamma_L}{1 - S_{22}\Gamma_L}$$

which gives

$$\Gamma_L = \frac{S_{11} + \Gamma^*_{0n}}{D - S_{22}\,\Gamma^*_{0n}}$$

This case gives $F_{\min}$, low-input VSWR, and high-output VSWR with a possibility of oscillation (if $k < 1$).

3. Trade-off of input and output VSWR, perhaps even at the sacrifice of increased noise figure. This case can be seen from the results shown in Fig. 4.12.

By considering a two-stage design, it may be possible to achieve $F_{\min}$, $S'_{11} = 0$, $S'_{22} = 0$, and stability at the expense of lower gain (see Section 4.2.2). Each case must be evaluated with the parameters provided; usually, design trade-offs are required to meet optimum system performance.

Matching circuits M_1 and M_2 (Fig. 1.1a) provide two degrees of freedom which can be used to match the input and output of the amplifier stage (for $k > 1$). If M_1 is used to provide a particular value of Γ_G, [e.g., Γ_{0n} (for optimum noise performance or whatever other consideration)], only one degree of freedom remains available and an option has to be made for input match only, output match only or a suitable compromise between the two.

On a Smith chart diagram at the output plane Γ_L (see Fig. 4.11) of the two-port transistor, the point (A) represents the impedance (or Γ_L) presented by M_2 for output match. Point B represents the input impedance (or Γ_L) of M_2, which leads to an input reflection coefficient of the transistor of Γ^*_{0n}, in other words, matching conditions at the input port ($\text{VSWR}_{\text{in}} = 1.0:1$).

The contours C_1, C_2, and so on, represent conditions of increasing levels of reflected signal (e.g., C_1 has as a parameter VSWR_{in}, $1.5:1$, C_2 has VSWR_{in} $2.0:1$, and so on). A random point C on C_2 is associated with $\text{VSWR}_{\text{in}} = 2.0:1$ and VSWR_{out} related to the distance between A and C. Obviously, the best matching conditions for a point on C_2 are realized at D, where this contour is tangent to a contour of constant $|\Gamma|$ surrounding A (leading to minimum output VSWR).

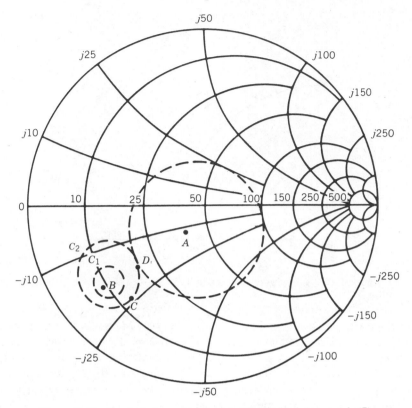

Figure 4.11 Γ_L plane for output match (A) and for input match (B).

Each pair of tangent contours of constant VSWR, centered on A and B, give a pair of values $VSWR_{in}$, $VSWR_{out}$ and determine a representative point in a plane of coordinates $VSWR_{in}$, $VSWR_{out}$. The totality of such points constitutes a curve (Fig. 4.12) whose intercepts are the points representing matching at the output (A) or input (B) and which should as well include intermediate situations that may offer a more acceptable compromise.

If some amount of noise figure increase is acceptable, point B can be replaced by a contour B_1 surrounding it and defined by the acceptable ΔF (Fig. 4.13). This contour reaches closer to A than point B was, so that the output mismatch accompanying an input match can be smaller. If a value $VSWR_{in} > 1.0:1$ is considered, each point situated on B_1 is replaced by a closed contour (two such contours are shown on Fig. 4.13) and all the contours generated in this way are included in a contour B_1', extending beyond B_1 and reaching even closer to A (better output match).

By repeating the operation for the same ΔF but different pairs of values $VSWR_{in}$, $VSWR_{out}$, a second curve can be drawn on Fig. 4.12 which comes closer to the ideal matching situation ($VSWR_{in} = VSWR_{out} = 1.0:1$). Various trade-offs are therefore possible between input match, output match, and noise figure.

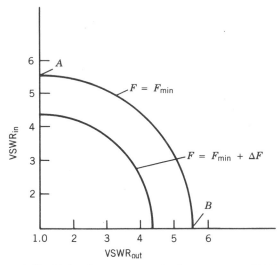

Figure 4.12 Noise-figure contours in the VSWR$_{in}$ and VSWR$_{out}$ plane.

All the above was based on the assumption that only M_1 is designed for purposes other than impedance matching (in the example discussed, the goal was low noise) and M_2 is dedicated only to achieving as good a matching condition as possible. Obviously, the same holds true if M_2 has a specific role (like maximum output power) and M_1 is available for matching. A family of curves like in Fig. 4.12 can be obtained having as a parameter output power or gain.

4.1.4 High-Power Amplifiers

Consider the design of a high-power amplifier to be the dual of that of a low-noise amplifier, where now special consideration for the output network is needed to give the desired output power. For high-power amplifier design, the large-signal parameters of the transistor should be obtained. For class A operation (see Fig. 3.25), the small-signal S parameters are a rough approximation of the large-signal performance. For class AB, B, or C the small-signal S parameters become progressively less useful.

The large-signal performance of the transistors should be measured under the maximum allowed dc bias. The optimum large-signal generator and load impedances (Γ_{GP} and Γ_{LP}) must be measured at each frequency of operation. These parameters will resemble S_{11}^* and S_{22}^*, but usually the output Γ_{LP} is significantly changed from S_{22}^*. The gain is reduced in comparison to G_{ma} or G_{ms}.

The contours of constant output power can be plotted in the Γ_L and Γ_G planes to determine the effect of nonoptimum match. This is analogous to noise circles in the Γ_G plane, but the contours are not necessarily circles. By assuming circles, we may write the effect of output load mismatch on large-signal gain, G_{LS}.

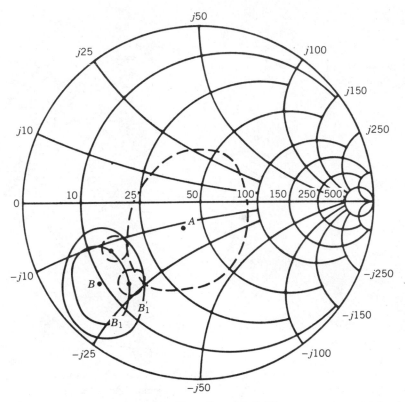

Figure 4.13 Γ_L plane for the condition $F = F_{\min} + \Delta F$. The new output match is given by B_1'.

$$G_{LS} = G_{\max} - \frac{R_p}{G_L} \left| Y_L - Y_{L\text{opt}} \right|^2 \qquad (4.13)$$

There are four large-signal parameters of importance in the output plane:

$G_{\max}$ = maximum large-signal gain

R_p = empirical factor in ohms (power resistance)

$Y_{L\text{opt}} = G_{L\text{opt}} + j\, B_{L\text{opt}}$, the optimum load admittance

The input matching Γ_{GP} and the operating bias point would complete the data needed for large-signal design. If a wideband high-power amplifier is required, the input match must be designed to flatten the gain while the output match is designed to give maximum power, Γ_{LP}.

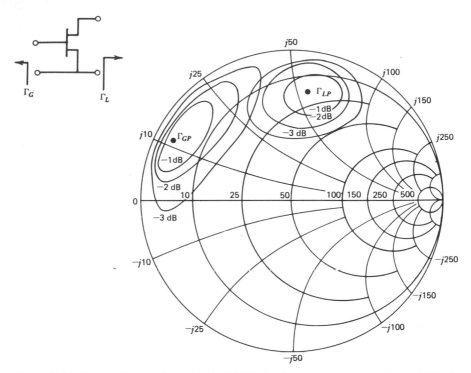

Figure 4.14 Large-signal tuning of GaAs MESFET. Coordinates in ohms. Dexcel 3501A chip. $V_{DS} = 8.0$ V; $I_{DS} = 0.5I_{DSS}$; $f = 12$ GHz. Optimum tuned for $P_{out} = +18.5$ dBm at 1 dB gain compression. Gain = 6.5 dB. (From Ref. 4.14.)

An example of large-signal detuning is plotted in Fig. 4.14 for the Dexcel 3501A chip at 12 GHz [4.14]. Contours of Γ_G detuning and Γ_L detuning are both plotted. The value of Γ_{GP} is very close to S_{11}^*, but Γ_{LP} has changed significantly from S_{22}^* at 12 GHz. A plausible explanation of the elliptical power contours by Cripps [4.15] shows that the optimum load resistance for maximum linear power is given by

$$R_{opt} = \frac{V_{DSS}}{0.5I_{DSS}} \qquad (4.14)$$

where V_{DSS} is the quiescent dc bias voltage point drain to source and I_{DSS} has the usual definition of I_{DS} with $V_{GS} = 0$. By showing that if $Z_L < R_{opt}$, the maximum value of X_L is given by

$$P_L = (0.5I_{DSS})^2 R_L \qquad (4.15)$$

$$P_L/P_{opt} = R_L/R_{opt} \qquad (4.16)$$

$$|X_L| \le R_{opt}^2 - R_L^2 \qquad (4.17)$$

and if $Z_L > R_{opt}$, the maximum value of B_L is given by

$$P_L = (0.5V_{DSS})^2 G_L \tag{4.18}$$

$$P_L/P_{opt} = G_L/G_{opt} \tag{4.19}$$

$$|B_L| \le (G_{opt}^2 - G_L^2) \tag{4.20}$$

Cripps achieves elliptic curves for the -1-dB and -2-dB power contours. Then by including the parasitic output capacitance and bonding inductance in the device impedance, the output device characteristic becomes the elliptical pattern given in Fig. 4.14.

Another important amplifier specification is the dynamic range, which is the range of input or output power with linear gain. At low powers this is limited by the noise figure or the minimum detectable signal, and at high powers it is limited by the power level where small-signal gain has been compressed by 1 dB. This is indicated schematically in Fig. 4.15. Defining the minimum detectable signal as 3 dB above thermal noise, we have

$$(\text{MDS})_{out} = kTB + 3 \text{ dB} + NF + \overset{.}{G} \tag{4.21}$$

$$(\text{MDS})_{in} = kTB + 3 \text{ dB} + NF \tag{4.22}$$

$$kTB = -114 \text{ dBm}/\text{MHz} \tag{4.23}$$

The dynamic range becomes

$$DR = P_{1 \text{dBc}} - (\text{MDS})_{out} \tag{4.24}$$

When two signals close in frequency are present in the amplifier, a third-order product will appear in the passband at $2f_2 - f_1$ or $2f_1 - f_2$ as a result of the nonlinear components of g_m. Since the intercept point of the third-order two-tone products is usually 10 dB above the 1-dB gain compression point, the spurious free dynamic range can be calculated graphically. The spurious free dynamic range is given by

$$DR_f = \tfrac{2}{3} \left[\text{TOI} - G - (\text{MDS})_{in} \right] \tag{4.25}$$

For an amplifier with a bandwidth of 30 MHz, transducer gain of 30 dB, and noise figure (NF) of 6 dB, the input minimum detectable signal is

$$(\text{MDS})_{in} = -114 \text{ dBm} + 15 \text{ dB} + 6 \text{ dB} + 3 \text{ dB}$$

$$= -90 \text{ dBm}$$

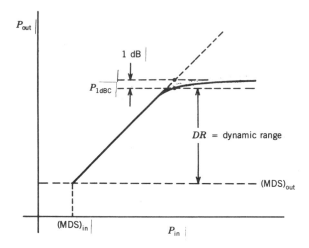

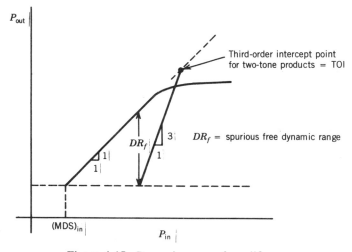

Figure 4.15   Dynamic range of amplifiers.

If the 1-dB compression point is +15 dBm (at the output), the linear dynamic range is 75 dB, and the spurious free dynamic range is about 57 dB.

High-power designs can also be optimized for any of the following:

1. High power ($\Gamma_L = \Gamma_{LP}$)
2. High gain ($\Gamma_L \simeq S_{22}^*$)
3. High efficiency (no source bias resistors)
4. Low third-order products
5. Low second harmonics
6. Low third harmonics

Each of these cases will usually require a different value of Γ_L, the load tuning parameter. There is a trade-off between all of these parameters.

The final consideration for high-power amplifiers is the thermal impedance of the transistor. The reliability or MTBF of the transistor is directly related to the maximum operating junction temperature or channel temperature of the transistor. At present the maximum recommended channel temperature of GaAs MESFETs is 175°C, and the maximum recommended junction temperature for the silicon bipolar transistor is 200°C. The higher number for Si is a result of the maturity of this technology. On the basis of the wider band gap of GaAs, a high operating channel temperature for GaAs should become possible (see Table 3.1), but at present it is limited to 175°C by a gate failure mechanism which is due to either a chemical reaction or metal migration [4.16–4.18] at higher temperatures. For a medium-power GaAs MESFET on a conventional 100-μm-thick chip, the thermal resistance of the chip (θ_{jc} or θ_{ch-c} is about 200°C/W for a 500 μm $\times$ 0.5 μm gate at a channel temperature of 150°C (e.g., the AT-8251 chip of Table 3.21). When operating at 0.5 I_{dss}, the dc power dissipation is

$$I_{DSS} = 0.10 \text{ A}$$

$$P_{\text{diss}} = P_{\text{dc}} + P_{\text{in}} - P_{\text{out}}$$

$$= 5(0.05) + 0.005 - 0.125 = 0.13 \text{ W}$$

$$\Delta T = P_{\text{diss}} \, \theta_{\text{ch-c}} = 0.13(200) = 26°C$$

$$T_{\text{ch}} = T_A + \Delta T = 25 + 26 = 51°C$$

This calculation ignores Θ_{CA}, the case to ambient thermal impedance, which is about 60°C/W for chips mounted on typical circuit carriers. This thermal impedance (Θ_{CA}) applies to the AT-8251 chip mounted on a 25-mil alumina substrate mounted on a 15-mil Kovar carrier in a 250-mil-thick aluminum case; each system must be evaluated for this additional thermal impedance. At an ambient temperature of 90°C, the channel temperature is $T_{CH} \simeq 90 + 34 = 124°C$, a reasonable temperature for high reliability. These temperatures should be verified by liquid crystal or IR scanning methods.

A similar set of design trade-offs apply to one-stage high-power designs (which assume small-signal S parameters), which are the dual of low-noise amplifiers. The three design techniques are:

1. Design M_2 for high power:

$$\Gamma_L = \Gamma_{0p}$$

Design M_1 for

$$\Gamma_G = S_{11}'^* = \left(S_{11} + \frac{S_{12}S_{21}\Gamma_{0p}}{1 - S_{22}\Gamma_{0p}} \right)^*$$

This gives maximum output power, low-input VSWR for maximum power transfer, and high-output VSWR. This is the most common design technique.

2. Design M_2 for high power:

$$\Gamma_L = \Gamma_{0p}$$

Design M_1 for

$$S_{22}' = \Gamma_{0p}^* = S_{22} + \frac{S_{12}S_{21}\Gamma_G}{1 - S_{11}\Gamma_G}$$

which gives

$$\Gamma_G = \frac{S_{22} + \Gamma_{0p}^*}{D - S_{11}\Gamma_{0p}^*}$$

This case gives high output power, low-output VSWR, and high-input VSWR with a possibility of oscillation (if $k < 1$).

3. Trade-off of input and output VSWR, perhaps even at the sacrifice of reduced output power. This case can be understood from the dual of Fig. 4.12.

A two-stage design may be used to give optimum power and minimum input and output VSWR (see two-stage low-noise design, Section 4.2).

This discussion has only treated the design at the fundamental frequency using linear small-signal S parameters. The large-signal harmonics and intermodulation products must also be included in high-power designs using SPICE, Microwave Harmonica, and so on (see Chapter 8).

4.1.5 Feedback

Amplifiers will have less sensitivity to transistor parameters when negative feedback is used. Negative feedback means that the feedback signal cancels the input signal or reduces the input signal to the transistor. For positive feedback the gain will increase, which is useful for peaking the gain at the upper band edge (or for making high-frequency oscillators).

Two types of feedback are considered here: shunt-series and transformer feedback. Four versions of shunt/series or voltage–current feedback can be used, as shown in Fig. 1.34 [4.19]. Each of these cases can be considered a modification to the z, y, h, or g parameters. When series or current feedback is used, terminal impedances increase; when parallel or voltage feedback is used, terminal impedances decrease. Current feedback can be visualized by an unbypassed emitter resistor R_E, while voltage feedback is R_F (Fig. 4.7). Usually, the noise figure is increased with this technique, although lossless elements can actually decrease the noise figure [4.20]. Five useful low-frequency transistor feedback circuits have been summarized in a book by Rohde and Bucher [4.21, p. 229].

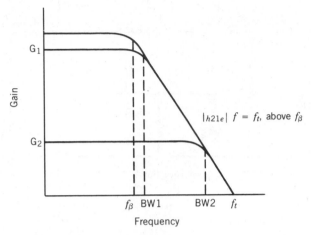

Figure 4.16 Current gain versus frequency.

Transformers are another form of lossless feedback that can be used for low-noise (or high-power) designs. Circuits that use this principle are also voltage or current feedback. Since transformers are lossless, this is also called noiseless feedback.

Feedback is also used to reduce the nonlinear distortion products in an amplifier. As the output power increases, the intermodulation products will increase in magnitude. Resistive feedback will reduce the magnitude of these products. The most important application of feedback is in broadband amplifiers. This can be understood by referring to Fig. 4.16, which shows the common-emitter (or common-source) current gain decreasing for $f > f_\beta$ at 6 dB/octave until it reaches unity at the frequency f_t, which is the gain–bandwidth figure of merit. Broadband amplifiers of increasing bandwidth can be realized simply by increasing the negative feedback, which of course lowers gain. When higher gain is required, more stages are cascaded in the final amplifier. This type of amplifier is described in detail in the next section, and several examples are included in Section 4.1.7.

Compound Feedback. If the noise figure is not a major concern, a compound series (R_1)/shunt (R_2) type of resistive feedback is often used. Referring to the simplified hybrid-π models of Chapter 3, at low frequencies both CE bipolars and CS FETs can be modeled by Figure 4.17, where R_1 is the series feedback and R_2 is the parallel feedback. The low-frequency analysis of this circuit will give an understanding of the circuit operation. This circuit works very well for bipolar transistors because of the large transconductance; for FETs it requires a large gate perimeter [equation (3.57)] which makes this circuit less popular for FET designs. The primary application of this amplifier is broadband gain with good input/output matches. The trade-off between gain and bandwidth is given in Fig. 4.16, where f_t is the well-known current gain–bandwidth product.

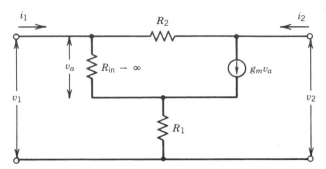

Figure 4.17   Compound feedback circuit at low frequency.

If $R_{in} = \infty$, the y parameters are

$$i_1 = G_2 v_1 - G_2 v_2 \tag{4.26}$$

$$i_2 = (g'_m - G_2) v_1 + G_2 v_2 \tag{4.27}$$

$$g'_m = \frac{g_m}{1 + g_m R_1} \tag{4.28}$$

Notice that the effective transconductance g_m is reduced by R_1; therefore, if the design uses a GaAs MESFET, $R_1 = 0$. Converting this two-port matrix to S parameters and applying the ideal matching condition $S_{11} = S_{22} = 0$ gives the values of feedback resistors:

$$R_1 = 1/g'_m - 1/g_m \tag{4.29}$$

$$R_2 = g'_m z_0^2 \tag{4.30}$$

For these values of feedback resistors, the S parameters are

$$S_{11} = S_{22} = 0 \tag{4.31}$$

$$S_{12} = \frac{1}{g'_m z_0 + 1} \tag{4.32}$$

$$S_{21} = -(g'_m z_0 - 1) \tag{4.33}$$

Device selection is determined by the required output power and bandwidth. Below 3 GHz, bipolar transistors are normally used because of the high g_m, which is set by the bias current:

$$g_m = I_E q/kT = I_E(\text{mA})/26 \tag{4.34}$$

Above about 3 GHz, a GaAs MESFET with high g_m will be used because of the higher f_t. Using (4.30) and (4.33), the gain can be expressed as

$$S_{21} = 1 - g'_m z_0 = 1 - R_2/z_0 \qquad (4.35)$$

Since

$$z_0^2 = R_2/g'_m = R_2(R_1 + 1/g_m) = R_2 R_{eq} \qquad (4.36)$$

the gain can also be written

$$S_{21} = 1 - \sqrt{R_2/R_{eq}} \qquad (4.37)$$

where R_{eq} is the equivalent emitter resistance.

It is useful to illustrate the last equations with a simple example, because of possible confusion in signs. Design for 20-dB low-frequency gain in a 50-ohm system:

$$-10 = 1 - R_2/50$$

$$R_2 = 550 \ \Omega$$

$$R_{eq} = z_0^2/R_2 = 4.55 \ \Omega = R_1 + 1/g_m$$

To complete the microwave design, reactive components are added as needed as shown in Fig. 4.18. Reactive components may be used to match the transistor at either the input or output or to modify the feedback at high frequencies. Notice that the parasitic emitter bonding inductance was included in order to predict the high-frequency gain. Example 4.4 will illustrate this design procedure.

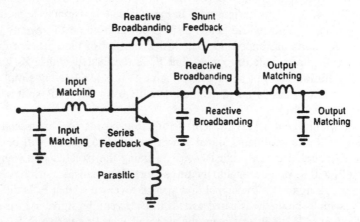

Figure 4.18 Elements of feedback amplifiers.

Transformer Feedback. Transformers or directional couplers can be used in both the RF and microwave ranges as feedback elements. Some examples of this feedback are described in this section. Because the feedback element is nearly lossless, this circuit type is also called noiseless feedback. Since the amplifiers in Figs. 4.17 and 4.18 usually have higher noise figure, the transformer circuits can give both low noise figure and wide bandwidth with high dynamic range.

There are many RF transistor amplifier design approaches using "broadband ferrite transformers." These designs include impedance-matching structures in and out of transistors, impedance transformations in conjunction with resistive feedback to reduce circuit losses, and various voltage or current feedback schemes as described in this section. Another design application uses ferrite directional coupler(s) as negative feedback element(s), where the ferrite directional coupler is a transformer arrangement having coupling ports exhibiting directional properties.

Before discussing directional feedback, a review of ferrite transformer attributes may be helpful. The type of transformer explained here is used in RF designs to transform impedances from the region tens to hundreds of ohms. These can have frequency bandwidths greater than two decades, depending on the transformation ratio, magnetic permeability, and physical size of the ferrite beads (cores). The upper-frequency limitation of ferrite transformers is around 1 GHz. The ferrite transformer core permeability decreases with increasing frequency. The construction usually has bifilar wound-wire combinations of a few turns around the ferrite core. This configuration displays minimal resistive circuit loss and wide frequency bandwidth.

Combinations of ferrite transformers are used to make power dividers displaying directional properties and minimal loss. These are commonly known as hybrid junctions and ferrite directional couplers. The hybrid junction is similar to the telephone-type isolation transformer. The directional tap (coupler) was developed and patented [4.22] in 1962 for use in the cable television industry. The ferrite directional coupler can divide power of other proportions than that of an even split.

A simple hybrid junction is a two-way in-phase power divider/combiner as shown in Fig. 4.19a. This design, used in most cable television systems, is called a "signal splitter." This hybrid junction splits power from port P_1 evenly between P_2 and P_3. All ports are impedance matched if the rest of the ports are matched. Ports P_2 and P_3 are isolated provided that P_1 is terminated with a Z_0 load. The theory of operation will not be discussed here, as it is a relatively simple hybrid junction. Insertion loss is no more then 1.0 dB over the 3-dB split within its operating bandwidth.

Also shown in Fig. 4.19b is a schematic of a four-port ferrite directional coupler using two ferrite transformers. Signal power supplied to port P_1 will be directed between the coupled port P_3 and P_2 with P_4 being the isolated port. The signal voltage at P_3 will be proportional to the turns ratio of the transformer, the remainder of the power going to P_2. The signal flow can be reversed so that P_4 is the coupled port to P_2, with P_3 being the isolated port. If P_1 is driven, the signal at P_3 is inverted 180° from that at P_2. If P_2 is driven, the signal at P_4 is 0° relative to P_1. There is added phase shift over the frequency band of these devices because of time delay that results from its nonzero physical dimensions.

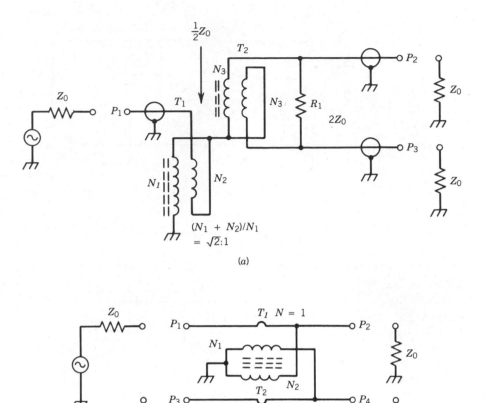

Figure 4.19 Ferrite transformers: (a) ferrite in-phase power splitter or hybrid junction; (b) four-port ferrite directional coupler.

Figure 4.20a shows a ferrite directional coupler incorporated with a negative-feedback transistor amplifier stage. This configuration was patented in 1971 [4.23]. Also in Fig. 4.20b is a transistor amplifier using two directional couplers, one in the input and the other in the output. This configuration was patented in 1977 [4.24]. In the circuit of Fig. 4.20a the coupler of Fig. 4.19 is connected to a transistor amplification stage. A signal source drives port P_3 of the coupler; the through output port P_4 is connected to the transistor base; the transistor collector output is connected to coupler port P_2; finally, the signal output of the circuit is at port P_1, connecting to the load. There is a feedback path between the base and collector of the transistor via the coupling between ports P_2 and P_4. Since the transistor is inverting to the signal, the amplified signal at P_2 adds in phase with the portion of the input signal coupled between ports P_3 and P_1. Thus the trade phrase "transformer lossless feedback" is used. Although there are some small

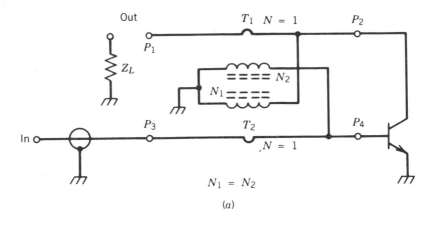

$$(a)$$

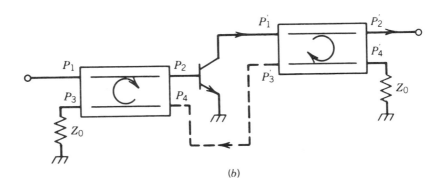

$$(b)$$

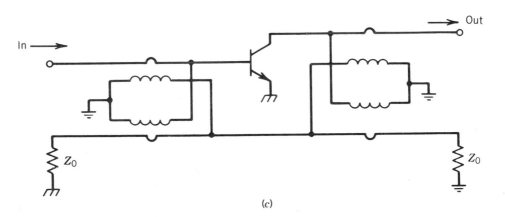

$$(c)$$

Figure 4.20 Feedback amplifier using (a) 4-port ferrite directional coupler [Ref. 4.21a]; (b) two directional couplers [Ref. 4.21b]; and (c) a physical realization of (b).

losses in the ferrite transformer(s), the circuit displays wide bandwidth and much lower noise figure than with a resistive feedback configuration. With fewer circuit losses than resistive feedback, the coupler feedback circuit usually displays a higher gain–bandwidth product. The word "usually" is necessary because the statement must be qualified with the phrase "within the operating bandwidth of the transformers used in the coupler circuit."

Noise figures as low as 1 dB can be realized with the first circuit of Fig. 4.20, and that is its major attribute. It has very poor output-to-input isolation, so care should be exercised in choosing the system application for that amplifier circuit. The single coupler circuit of Fig. 4.20 displays reverse isolation typically about 3 dB higher than forward gain. Resistive feedback circuits typically display 6 dB higher reverse isolation than their forward gain. The circuit of Fig. 4.20b, utilizing the two coupler circuit, exhibits much higher reverse isolation than does a resistive feedback circuit. At lower frequencies, reverse isolation can be 20 dB higher then forward gain. The noise figure of this two-coupler circuit is a few tenths of a dB higher because the coupled power in the input is directed to the Z_0 load at port P_3 of the input coupler. Some power gain is lost in the output coupler and the circuit is more complex, but this is a small price to pay for its attributes if the system needs high isolation.

The two-coupler circuit of Fig. 4.20 has the trade name of Power Feedback technology (Q-bit Corporation). "Power" is a scalar term, so mathematically it cannot be used to analyze the feedback loop. The term was adopted because the amplifier circuit has such a good input/output impedance match and high reverse isolation that the output signal power is proportional to the input power. The circuit shows a Z_0 resistive load at the coupled ports of the input and output coupler. This feature is what forces the circuit to normalize the input and output match to the Z_0 of the system. A signal applied to the amplifier output has two paths to follow to the input: No. 1 via the S_{12} of the transistor; and No. 2 through the two directional couplers. The coupler path is via the isolated ports of both the input and output coupler, so that path loss is very high. The isolation of the amplifier is basically the S_{12} of the transistor plus the through loss of the two couplers.

The amplifier designer will find that there are restrictions in certain design applications using ferrite transformers. Because the transformers have phase shift (i.e., time delay) over the frequency band, it is sometimes difficult to realize a flat gain over the same bandwidth. Another problem is that bias current must be supplied to the active transistor. This requires chokes and bypass capacitors around the ferrite transformers. Somewhere below the low end of the frequency band these elements can go into resonance and cause the amplifier to be unstable. Another design constraint is that a design with ferrite transformers can use much more real estate then a similar resistive-feedback configuration. If the system designer requires an amplifier with high reverse isolation, good impedance match, and high dynamic range, the attributes of Power Feedback make the design effort worth while.

The next example of transformer feedback is a three-winding transformer connected to a common-base transistor as shown in Fig. 4.21 [25]. At low frequen-

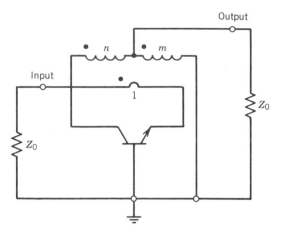

Figure 4.21 Feedback amplifier using the noiseless feedback technique. (From Ref. 4.25.)

cies for an ideal transistor with zero input impedance, infinite output impedance, and unity current gain, the amplifier will be matched at both ports when the transformer turns satisfy

$$\frac{R_G}{R_L} = \frac{n + m + 1}{m^2} \tag{4.38}$$

For $R_G/R_L = 1$, the turns ratio is

$$n = m^2 - m - 1 \tag{4.39}$$

With this choice of turns ratio, the power gain is m^2, the load impedance presented to the collector is $(n + m)Z_0$, and the source impedance presented to the emitter is $2Z_0$. Turns ratios for m equal to 2, 3, and 4 yield gains of 6, 9.5, and 12 dB and load impedances of 3, 8, and $15Z_0$, respectively. Since the turns ratio is high for reasonable gain, the bandwidth is limited.

The operation of the circuit depends on the completely mismatched conditions presented by the transistor to the circuit (i.e., the emitter presents a short circuit and the collector an open circuit). Hence there is no requirement to introduce resistive elements for impedance matching. Therefore, a noise-figure advantage is obtained with this circuit. Also, the source impedance of $2Z_0$ presented to the emitter tends to give optimum noise-figure performance with low collector currents, which also favors low noise figures. Finally, despite the small currents involved, relatively large output powers can be provided because of the high load impedance, which goes along with the higher-gain versions.

The main disadvantage of the circuit is that the high load impedance tends to limit the bandwidth. Also, low gain reduces the noise figure advantage, and stability is a concern since the isolation is not much higher than the gain. Nevertheless,

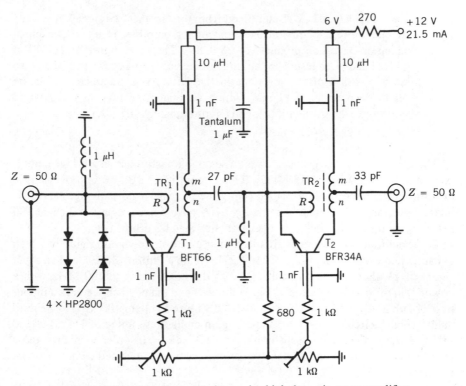

Figure 4.22 Two-stage ultralow-noise high dynamic range amplifier.

sufficient bandwidth can be achieved to provide broadband IF gain and noise figures competitive with those that could be obtained previously only in very narrowband units.

As this is a very convenient circuit, let us look at the actual working design shown in Fig. 4.22. Both transistors are made by Siemens. The antiparallel diodes at the input can be omitted. The core material used in the transformers depends on the frequency range. Siemens ferrite material type B62152-A8-X17, U17 for frequencies from 100 MHz up and K30 material for frequencies below, should be used; other ferrite suppliers are Krystinel and Fair-rite. This amplifier has the following characteristics:

1. Power gain: 19 dB
2. Noise figure: 1.3 dB
3. Third-order intercept point: 14 dBm at the input or 33 dBm at the output
4. 1-dB compression: +18 dBm
5. Input impedance: 50 ± 2 Ω
6. Bandwidth: 70 to 570 MHz (can be made to work at lower frequencies with the higher-permeability core)

7. Dynamic range: 102 dB (Determined from the fact that two signals separated in frequency of 3.17 mV at the input of the amplifier result in two intermodulation-distortion products of 25.4 nV. The noise figure of 1.3 dB at 2.4 kHz bandwidth results in the same noise voltage of 25.4 nV. If the noise floor of −138.8 dBm equivalent to the 25.4 nV is subtracted from the −36.96 dBm or the 3.17 mV, which generates the two intermodulation distortion products, the difference is approximately 102 dB.)

8. Power supply: 12 V/21 mA

The next transformer feedback circuit uses two transformers in a bridge configuration as shown in Fig. 4.23 [4.26]. This circuit offers a 30-dB improvement in the third-order intercept point, with a negligible increase in noise figure. An experimental version of this circuit using a 2N 5109 gave a noise figure of 2 dB at 100 MHz with a third-order intercept (TOI) greater than 50 dBm.

Some additional transformer circuits which give both low noise figure and high dynamic range are given in Fig. 4.24 [4.27]. Using a Phillips BFR94 and an input signal of 13 dBm, the first circuit gave 3.5-dB noise figure to 100 MHz with a power gain of 3.3 dB and excellent gain flatness and match. The second common-base circuit gives 4 dB power gain and 1.3 dB noise figure to 100 MHz. The third dual-feedback circuit gives 7.4 dB power gain and a noise figure of 1.4 dB to 100 MHz. The third-order intercept point was +37 dBm. In order to realize these circuits in the microwave range, the transformers could be realized with directional couplers.

Another example of a transformer feedback amplifier is shown in Fig. 4.25 [4.28] which is a base–emitter feedback amplifier. Depending on the grounding capacitors at the emitter and base, this may be considered a common-emitter or a common-base amplifier. Using a Siemens BFT66 bipolar transistor, a noise figure of 1 dB and a gain of 34 dB was achieved at 95 MHz. The performance is given

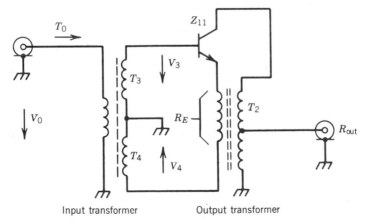

Input transformer   Output transformer

Figure 4.23 Schematic diagram of a bridge-type circuit which adds voltage feedback at input transformer and current feedback at the output transformer. (From Ref. 4.26.)

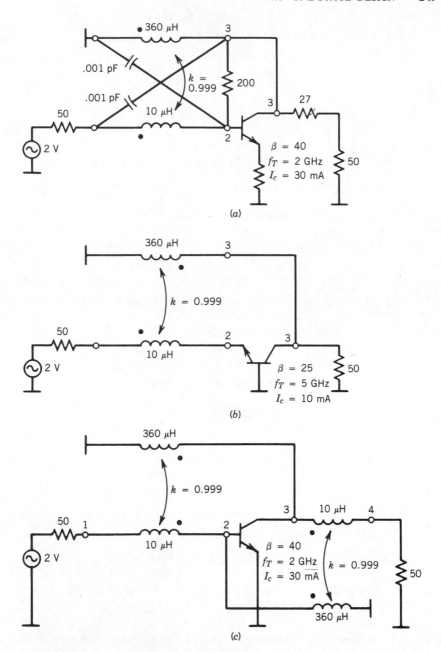

Figure 4.24 Transformer feedback circuits: (*a*) transformer coupled feedback amplifier with additional series and shunt feedback; (*b*) grounded-base low-noise feedback amplifier; (*c*) grounded-emitter dual feedback amplifier. (From Ref. 4.27.)

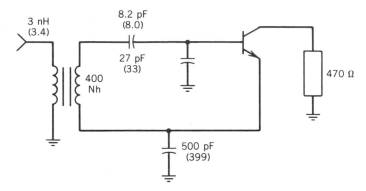

Figure 4.25 Base–emitter feedback amplifier optimized for 95 MHz operation. The values in parentheses are those obtained after optimization. (From Ref. 4.28.) (Reprinted with permission of Microwave Journal, © 1986 Horizon House-Microwave, Inc.)

in Fig. 4.26. A microwave version of the base–emitter feedback amplifier is shown in Fig. 4.27.

A final version of transformer feedback is given in Fig. 4.28, which is the UTO-221 from Avantek. This uses the AT41400 silicon bipolar transistors in a cascode arrangement. The advantages are the low current bias, high gain, low noise figure, and small size. The performance is typically as follows:

Frequency: 10–200 MHz

Gain: 28 dB

NF: 1.8 dB

P_{1dBc}: 16 dBm

VSWR: 1.5 : 1

TOI: 28 dBm

Bias: 15 V, 29 mA

The noise figure is slightly degraded by the resistive feedback, and the transformer turns ratio limits the bandwidth of the amplifier.

Positive Gain Slope. One additional feedback concept that is important in many practical applications is an amplifier with increasing gain versus frequency. Reactive feedback elements can be added to give this effect. The previous design equations set the low-frequency gain and the computer can be used to find appropriate reactive elements for low-input/output VSWR. An example is shown in Fig. 4.29 using the Avantek MSA-0885 amplifier gain block, with the calculated performance given in Fig. 4.30.

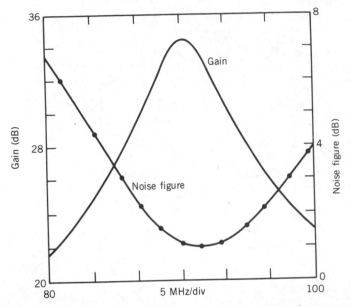

Figure 4.26 Gain and noise-figure predictions using software. ·'s refer to measured data using a Siemens BFT 66 microwave transistor. (From Ref. 4.28.) (Reprinted with permission of Microwave Journal, © 1986 Horizon House-Microwave, Inc.)

4.1.6 Balanced Amplifiers

There are at least three techniques for improving the input and output VSWR of the amplifier while simultaneously presenting the correct Γ_G and Γ_L to the transistor terminals. The most popular technique is that shown in Fig. 4.31a, the use of 3-dB Lange couplers [4.29] in a balanced configuration. In this technique the reflections from the identical amplifiers all appear at the termination port, so that the input port appears matched.

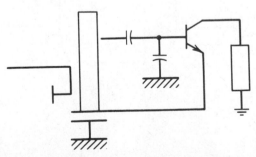

Figure 4.27 The microwave version of Fig. 4.25. A capacitor provides the necessary feedback at the "cool end" of the microstrip. (From Ref. 4.28.) (Reprinted with permission of Microwave Journal, © 1986 Horizon House-Microwave, Inc.)

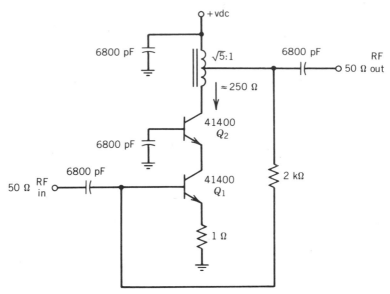

Figure 4.28 Cascade amplifier with transformer output.

One can show that

$$S_{11} = \tfrac{1}{2}(S_{11A} - S_{11B})$$ (4.40)

The proof is shown in Fig. 4.32, where the reflected voltages add out of phase at the input port and in phase at the 50-Ω terminated port. By similar arguments, the net gain of the balanced amplifier is

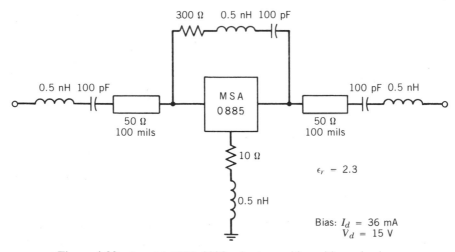

Figure 4.29 Avantek MSA 0885 gain stage with positive gain slope.

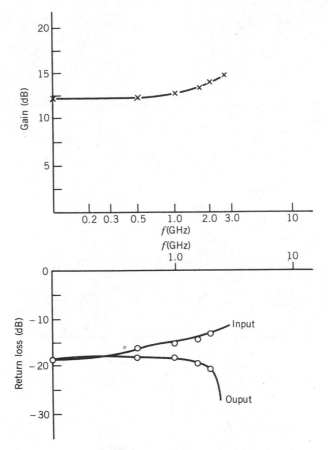

Figure 4.30 Gain and return loss versus frequency for amplifier in Fig. 4.29.

$$S_{21} = \tfrac{1}{2}(S_{21A} + S_{21B}) \tag{4.41}$$

and the overall noise figure is

$$NF = \tfrac{1}{2}(N_{FA} + N_{FB}) \tag{4.42}$$

For balanced amplifiers, if one stage opens, the gain will decrease about 6 dB and the noise figure will increase by about 3 dB; this is useful for troubleshooting amplifier modules.

The operating bandwidth of the amplifier is limited by the bandwidth of the coupler, which can be made very wide (more than two octaves). This technique of building one-stage balanced amplifiers has many advantages:

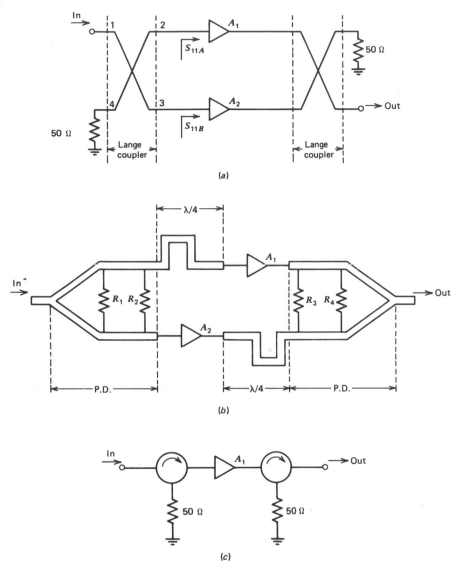

Figure 4.31   Low-VSWR amplifiers: (*a*) balanced amplifier using 3-dB Lange coupler; (*b*) balanced amplifier using 3-dB Wilkinson power divider (P.D.); (*c*) isolator amplifier.

1. The transistor can be intentionally mismatched for gain flatness, noise figure, output power, stability, and so on.
2. Each stage is isolated from the following stage by the Lange couplers. Very little interstage tuning is needed.
3. The reflections from A_1 and A_2 are terminated in 50 Ω, which usually guarantees stability.

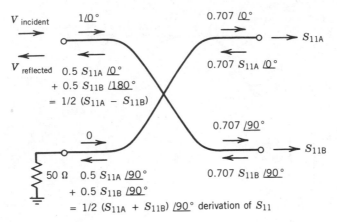

Figure 4.32 Analysis of 3-dB directional coupler.

4. If one transistor fails, the operating gain drops about 6 dB, which can be a useful feature in some applications.

5. The linear output power is 3 dB higher.

6. The labor in tuning these amplifiers is much less than in designing and tuning multistage unbalanced amplifiers.

The disadvantages of balanced amplifiers are the higher cost of more transistors and the higher dc power requirement.

Another popular choice for balanced amplifiers is the 3-dB Wilkinson power divider circuit shown in Fig. 4.31b. The power is split in phase, and a quarter-wave line is inserted in front of one amplifier and behind the opposite amplifier. If the input reflection coefficients S'_{11} of the amplifiers are identical, all reflected input signals appear 180° out of phase across R_2 and are dissipated. A low-frequency termination is also provided by R_2 for the sake of stability. Similarly, the output is terminated in R_3, and the signals are added in phase at the output. This technique will require more space than the Lange coupler and gives less bandwidth. The advantage is that a much simpler microstripline circuit can be used (no narrow coupling strips).

The relative bandwidths of these two approaches are shown in Fig. 4.33. At present the single-section Lange coupler is the most popular choice for double-octave wideband amplifiers. For wide bandwidths, the coupling is tighter than 3 dB (about 1.5 dB seems optimum). The third approach to low-VSWR amplifiers is the use of broadband isolators. This approach requires a thin-film ferrite circuit and a magnetic structure for realizing the isolators. Moreover, low-frequency broadband isolators are not available below 2 GHz. Unless a proven isolator capability exists, this approach is not commonly used.

In summary, the most common amplifier designs use the 3-dB Lange coupler to achieve low VSWR over broad bandwidths. These balanced amplifiers are easily cascaded to achieve high gain and high output powers.

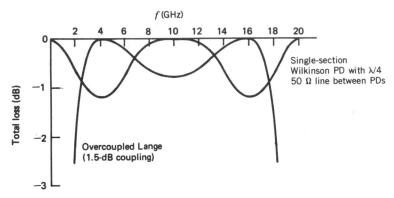

Figure 4.33 Comparison of 10-GHz cascaded Lange couplers versus cascaded single-section Wilkinson power dividers.

4.1.7 One-Stage Amplifier Design Examples

Example 4.1 Avantek AT-00572 Bipolar Transistor As an example of device modification, consider modifying the Avantek AT-00572 bipolar transistor to have a flat gain response with good input and output VSWRs over the CATV passband 100 to 400 MHz. The S parameters of the device over this frequency range are shown in Table 4.3. From a graphical display of these parameters it is clear that the impedances are higher than 50 Ω, and for this reason it may be that current-series feedback is not required.

The components required to equalize the maximum available power gain (G_{ma}) over the passband were determined with an iterative technique [4.2] and are listed as a function of the minimum and maximum G_{ma} in the passband in Table 4.4. It is clear on inspection of the table that the gain can be leveled very effectively with voltage-shunt feedback. By displaying the resulting S parameters on a Smith chart the VSWRs associated with the 389-Ω 2.2-nH solution were found acceptable and no further modification or impedance matching was required, (i.e., no series emitter resistor was needed for this design).

TABLE 4.3 S **Parameters of the AT-00572 Bipolar Transistor as Supplied by the Manufacturer ($V_{CE} = 8.0$ V, $I_C = 5$ mA)**

Freq. (MHz)	S_{11} Mag.	S_{11} Angle	S_{21} Mag.	S_{21} Angle	S_{12} Mag.	S_{12} Angle	S_{22} Mag.	S_{22} Angle
100	0.82	−22	8.99	150	0.027	75	0.87	−19
200	0.65	−40	8.57	129	0.044	72	0.72	−27
300	0.51	−54	7.55	116	0.061	69	0.64	−30
400	0.40	−61	6.29	103	0.074	66	0.57	−30

Source: Avantek 1987 Semiconductor Data Book.

TABLE 4.4 Voltage-Shunt Feedback Components Required to Flatten the Gain Performance of the Amplifier in Example 4.1 as a Function of the Minimum and Maximum Values of G_{ma} Obtained

G_{ma}		Components	
Min. (dB)	Max. (dB)	R_f (Ω)	L_f (nH)
16.9	17.4	689	142.8
16.3	16.8	612	88.9
15.8	16.2	545	54.5
15.2	15.6	486	30.9
14.7	15.1	434	14.2
14.1	14.6	389	2.1
13.6	14.1	349	0.010

The final design is shown in Fig. 4.34. The gain of the amplifier is 13.83 ± 0.48 dB, the input VSWR is lower than 1.32:1, and the output lower than 1.82:1 (Fig. 4.35). On calculating the tunability [see equation (4.3)] for this amplifier, it was found to vary between 0.64 and 0.65 in the passband. This is on the high side, but because of the simplicity of the design this is considered acceptable.

Example 4.2 ATF-13135 12-GHz Low-Noise Amplifier This example illustrates the design of a single-stage low-noise amplifier using the ATF-13135 GaAs MESFET from Avantek. The design specifications are:

Frequency	12 GHz
Noise figure	1.2 dB
Minimum gain	8.5 dB

The device data are tabulated below at 12 GHz:

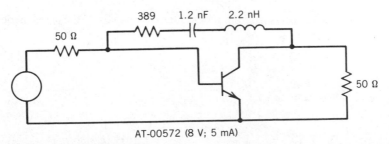

AT-00572 (8 V; 5 mA)

Figure 4.34 Wideband amplifier designed in Example 4.1 (gain: 13.83 ± 0.48 dB; VSWR-I ≤ 1.32; VSWR-0 ≤ 1.81).

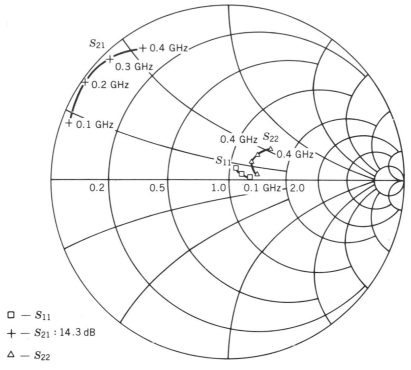

Figure 4.35 S parameters of the amplifier designed in Example 4.1 displayed on a Smith chart.

$$S = \begin{bmatrix} 0.61\ \underline{/37°} & 0.144\ \underline{/-89°} \\ 2.34\ \underline{/-84°} & 0.15\ \underline{/46°} \end{bmatrix}$$

$$F_{min} = 1.2\ \text{dB}$$

$$\Gamma_{0n} = 0.47\ \underline{/-65°}$$

$$R_n = 40\ \Omega$$

Using the impedance-matching techniques of Chapter 3, the input is designed with lossless elements for

$$\Gamma_g = \Gamma_{0n} = 0.47\ \underline{/-65°}$$

and the output is designed for

$$\Gamma_L = S_{22}'^* = \left[\left(S_{22} + \frac{S_{12}S_{21}\Gamma_{0n}}{1 - S_{11}\Gamma_{0n}} \right) \right]^* = 0.303\ \underline{/-85°}$$

Since the $|S'_{22}|$ is very low, representing a mismatch loss of

$$10 \log |1 - |S'_{22}|^2| = -0.42 \text{ dB}$$

the output matching circuit is optional. The transducer gain with no output match is given by (1.191):

$$
\begin{aligned}
G_T &= \frac{(1 - |\Gamma_G|^2) |S_{21}|^2 (1 - |\Gamma_L|^2)}{|(1 - S_{11}\Gamma_G)(1 - S_{22}\Gamma_2) - S_{12}S_{21}\Gamma_G\Gamma_L|^2} \\
&= \frac{(1 - 0.47^2)2.34^2}{\left|[1 - (0.61 \; \underline{/37°} \,)(0.47 \; \underline{/-65°} \,)](1) - 0\right|^2} \\
&= 7.406 \\
&= 8.70 \text{ dB}
\end{aligned}
$$

If the output is also matched, as shown in Fig. 4.36, the amplifier gain is

$$G_T = 9.12 \text{ dB}$$

The output port is conjugately matched and stable since $k > 1$. For lower-frequency designs when $k < 1$ a conjugate match at the output port may not be possible for a stable design. The overall performance of the amplifier in Fig. 4.36 is given below at 12 GHz:

$$F = 1.2 \text{ dB}$$
$$|S_{21}|^2 = 9.12 \text{ dB}$$
$$S_{11} = 0.38 \; \underline{/-124°}$$

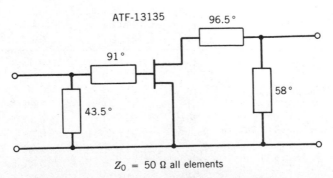

$Z_0 = 50 \; \Omega$ all elements

Figure 4.36   12-GHz low-noise amplifier design.

$$S_{22} = 0.002 \; \underline{/-104°}$$

$$k = 1.101$$

This low-noise design technique is extended to three stages in Example 4.10.

Example 4.3 Avantek M104H MODFET The design of a one-stage balanced low-noise module using Avantek M104H MODFET chips will illustrate the application of several practical design concepts. This FET has a gate length of 0.25 μm and a gate width of 150 μm. The specifications for this design at 25°C are as follows:

Frequency	7–17 GHz
Minimum gain	7.5 dB
Maximum gain flatness	±2.0 dB

TABLE 4.5 M104H MODFET Chip (V_{DS} = 3.5 V, I_{DS} = 13 mA, L_G = 0.25 μm, Z = 150 μm)

Freq. (GHz)	Reflection Coefficient In, S_{11}		Forward Gain, S_{21}		Isolation, S_{12}		Reflection Coefficient Out, S_{22}	
	Mag.	Angle	Mag.	Angle	Mag.	Angle	Mag.	Angle
12.0	0.61	−133.7	3.47	66.8	0.148	28.3	0.13	−25.2
13.0	0.59	−153.7	3.30	55.4	0.155	19.1	0.05	15.0
14.0	0.55	−174.2	3.20	45.4	0.161	10.0	0.07	72.8
15.0	0.55	162.9	3.10	33.4	0.170	1.1	0.11	96.2
16.0	0.57	139.7	2.90	21.0	0.176	−8.5	0.18	101.8
17.0	0.59	120.3	2.64	9.8	0.179	−17.2	0.24	97.1
18.0	0.64	106.6	2.42	0.8	0.183	−23.6	0.29	91.6
19.0	0.68	92.9	2.23	−10.1	0.187	−31.5	0.34	80.8
20.0	0.67	85.4	1.98	−17.1	0.185	−34.5	0.33	75.2

Derived Parameters

Freq. (GHz)	h_{21} (mag)	S_{21} (dB)	MSG (dB)	MAG. (dB)	k
12.0	3.81	10.79	13.71	0.00	0.88
13.0	3.46	10.37	13.29	0.00	0.92
14.0	3.26	10.10	12.98	0.00	0.96
15.0	3.12	9.81	12.61	0.00	0.96
16.0	2.99	9.26	12.17	0.00	0.96
17.0	2.82	8.43	11.68	0.00	0.98
18.0	2.67	7.68	11.22	0.00	0.98
19.0	2.54	6.95	10.76	0.00	0.96
20.0	2.43	5.93	10.29	8.97	1.05

Maximum noise figure	2.0 dB
Input/output VSWR	2 : 1
Minimum P_{out} (1 dBc)	+10 dBm

A balanced chip design is required to achieve the low input and output VSWR over this wide bandwidth. A MODFET was used to achieve this state-of-the-art noise performance, since a GaAs MESFET would give a typical noise figure of 2.5 dB over this frequency range with a minimum gain of 7.5 dB.

The S parameters and noise data for the M104H MODFET chip are given in Table 4.5 along with the stability factor. The GaAs MESFET made from the same mask set is given in Table 4.6 and Fig. 4.37 for a relative comparison in performance. Because of the coupler losses that are estimated at 0.2 dB maximum for 15-mil alumina, the first stage must have a noise figure of about 1.7 dB typical. The schematic for this design is given in Fig. 4.38. The input circuit has been

TABLE 4.6 M104L GaAs MESFET Chip (V_{DS} = 3.5 V, I_{DS} = 15 mA, L_G = 0.25 μm, Z = 150 μm)

Freq. (GHz)	Reflection Coefficient In, S_{11}		Forward Gain, S_{21}		Isolation, S_{12}		Reflection Coefficient Out, S_{22}	
	Mag.	Angle	Mag.	Angle	Mag.	Angle	Mag.	Angle
12.0	0.55	-117.5	3.85	70.7	0.123	44.7	0.33	-32.7
13.0	0.53	-134.1	3.72	60.8	0.132	37.5	0.26	-33.0
14.0	0.49	-151.2	3.66	51.8	0.140	31.0	0.23	-33.1
15.0	0.46	-171.1	3.55	41.3	0.151	22.5	0.17	-36.1
16.0	0.46	165.9	3.41	30.1	0.162	13.9	0.10	-33.8
17.0	0.49	144.1	3.17	18.8	0.171	5.4	0.05	3.2
18.0	0.53	127.4	2.93	9.1	0.179	-2.2	0.08	60.8
19.0	0.58	112.3	2.72	-1.5	0.187	-10.3	0.14	65.4
20.0	0.59	103.6	2.43	-9.6	0.190	-15.4	0.18	57.7

Derived Parameters

Freq. (GHz)	h_{21} (mag)	S_{21} (dB)	MSG (dB)	MAG. (dB)	k
12.0	4.52	11.71	14.95	0.00	0.91
13.0	4.10	11.40	14.50	0.00	0.94
14.0	3.86	11.28	14.17	0.00	0.96
15.0	3.66	11.00	13.71	0.00	0.98
16.0	3.47	10.66	13.23	0.00	0.98
17.0	3.25	10.03	12.70	0.00	0.99
18.0	3.01	9.34	12.15	0.00	0.98
19.0	2.81	8.69	11.62	0.00	0.97
20.0	2.65	7.71	11.06	10.38	1.01

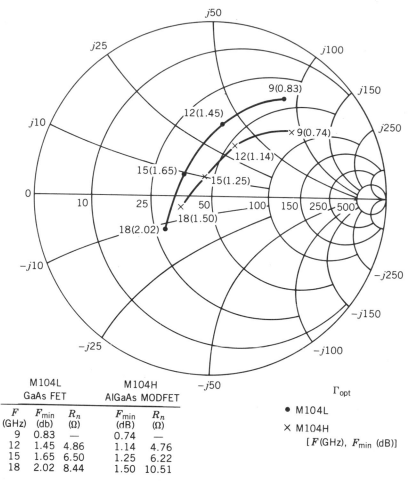

Figure 4.37 Noise parameters of two microwave FETs (M104L GaAs MESFET and M104H AlGaAs MODFET) at low noise bias ($Z = 150 \ \mu m$).

designed to present Γ_{on} to the transistor, and the output circuit flattens the gain. The input design is plotted in Fig. 4.39 to show the low noise match. The inductors are 1-mil gold bond wires. The circuit was realized on 15-mil alumina substrates with Lange couplers at input and output. The blocking capacitors were 2.0 pF. The dc bias for each transistor was

$$V_{DS} = 3.5 \text{ V}$$

$$I_D = 13 \text{ mA}$$

$$V_{GS} = -0.4 \text{ V}$$

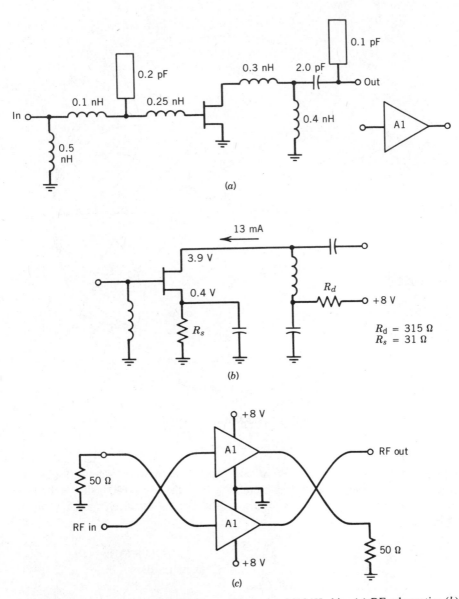

Figure 4.38 7- to 17-GHz MODFET amplifier using M104H chip: (*a*) RF schematic; (*b*) dc schematic; (*c*) balanced configuration.

and it was achieved with the schematic in Fig. 4.38. The noise figure of the final design was 1.8 to 2.0 dB and the gain was 9.5 ± 2.0 dB. Although the performance of this amplifier module, which is shown in Fig. 4.40, is considered state of the art for 1987, the design technique is very common for many of the amplifiers built in the microwave industry.

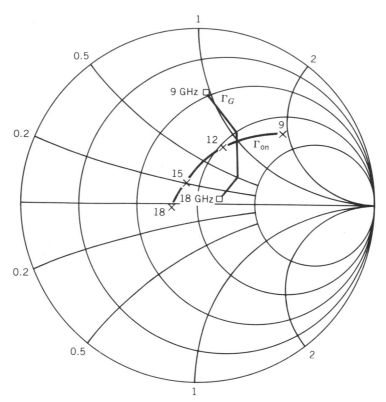

Figure 4.39 Γ_G and Γ_{ON} versus frequency for MODFET amplifier.

Example 4.4 Avantek AT-41400 Silicon Bipolar Transistor The next example is a feedback design on 25-mil alumina using the AT-41400 silicon bipolar transistor described in Chapter 3. The specifications for this design at 25°C are as follows:

Frequency	5–1000 MHz
Minimum gain	15 dB
Maximum gain flatness	±0.5 dB
Maximum noise figure	4.5 dB
Maximum input/output VSWR	2.0:1
Minimum P_{out} (1 dBc)	+10 dBm

The design uses the compound series/shunt resistive feedback (or voltage–current feedback) described in Section 4.5. The transistor is selected based on the output power requirement. Since the AT-41400 is designed for 8-V 25-mA operation, this device could deliver class A output power of

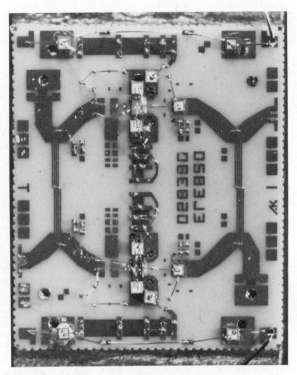

Figure 4.40 Photograph of 7- to 17-GHz MODFET amplifier. (Courtesy of Avantek.)

$$P_{\text{out}} \simeq 0.5(8)(.025) = 100 \text{ mW}$$

$$= 20 \text{ dBm}$$

Because of this high output power, a lower collector voltage can be used; the current must remain high to maintain g_m at 1.0 S. The bandwidth determines the gain from the gain–bandwidth product, which can be expressed as

$$f_t = |h_{21e}|f \simeq A_v f$$

Since the f_t is 10 GHz, the power gain is about 20 dB at 1 GHz. If additional bandwidth is needed, the lower gain should be used to maintain the low input and output VSWR.

The negative feedback resistors set the low-frequency match. Reactive elements are used to flatten the gain at the high end of the band, which is at a gain level within 10 dB of the G_{ma} of the device (i.e., about 0.4 GHz). The emitter resistor is given by the gain requirement

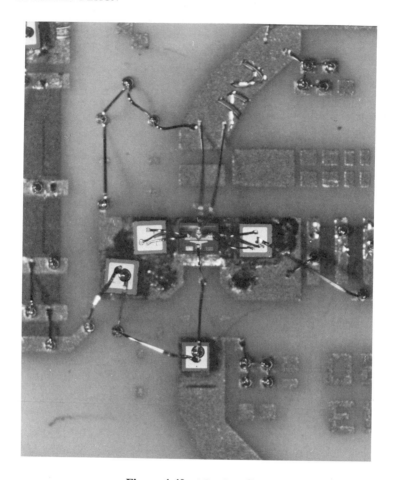

Figure 4.40 (*Continued*)

$$A_v = \frac{g_m}{1 + g_m R_E} R_L = 10 \qquad R_E = 4.0 \ \Omega$$

giving an effective transconductance of

$$g_m' = \frac{g_m}{1 + g_m R_E} = \frac{1}{1 + 4} = 0.20 \ \text{S}$$

which gives the RF feedback resistor

$$R_F = g_m' Z_0^2 = 0.20(2500) = 500 \ \Omega$$

The final RF and dc schematics are given in Figure 4.41, where the design has been optimized for flat gain and low VSWR. In some designs the emitter resistor R_E is bypassed for higher RF gain at high frequencies; however, this will also cause a degradation in input VSWR. The measured gain and return loss is plotted in Fig. 4.42. This amplifier is packaged in a TO-8U can for low-cost commercial applications; the product name is UTO-1013. Notice that active bias is used to maintain the gain constant over temperature.

Example 4.5 Darlington Feedback Amplifier   The next example is the design of a Darlington feedback amplifier, using the following specification:

Gain	12 dB minimum (at 100 MHz)
Frequency	0.1–4 GHz
Maximum input/output VSWR	2.0 : 1 (below 2 GHz)
Maximum noise figure	5.0 dB
Power supply	+5 V

The Darlington configuration is used because of the resulting increase in input impedance, which increases the S_{21} bandwidth compared to a single bipolar transistor [4.30]. The low-frequency feedback resistors are given by (4.36) and (4.37):

$$S_{21} = 1 - \sqrt{R_2/R_{EQ}}$$

$$Z_{in} = Z_{out} = \sqrt{R_2 R_{EQ}}$$

giving

$$R_{EQ} = 7.0 \ \Omega = 6.0 \pm 1.0 \ \Omega$$

$$R_2 = 360 \ \Omega$$

$$R_1 = 6.0 \ \Omega$$

which gives a conservative low-frequency gain of

$$S_{21} = 1 - 7.17 = -6.17 = 15.8 \ \text{dB}$$

When packaged in a microstrip package (MSA-0735) this amplifier gives a gain of 13.5 dB at 100 MHz and 6.2 dB at 4 GHz. The chip schematic is given in Fig. 4.43 and the packaged amplifier is shown in Fig. 4.44. The reduction in gain at 100 MHz (13.5 dB compared to the design value of 15.8 dB) is due to the other two resistors in Fig. 4.43, which are needed for dc bias.

This is a cascadable 50-Ω gain block with the same low cost and high reliability as silicon bipolar transisors. This example is the design of the Avantek MSA-0735

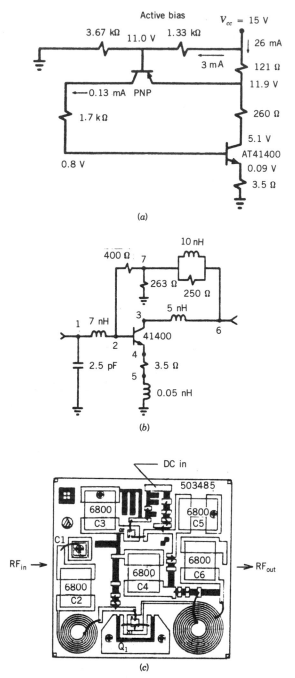

Figure 4.41 UTO-1013 feedback amplifier from Avantek: (a) dc schematic; (b) RF schematic; (c) complete mask layout. (Courtesy of Avantek.)

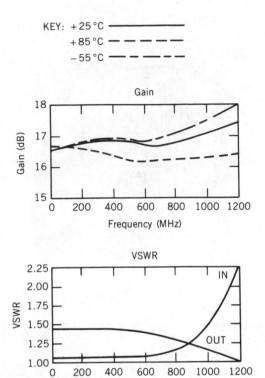

Figure 4.42   Gain and return loss for Example 4.4.

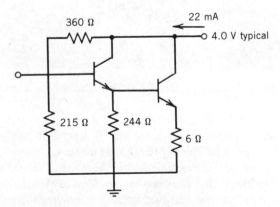

Figure 4.43   Circuit for Example 4.5.

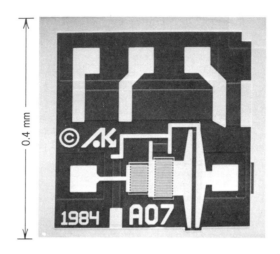

0.4 mm

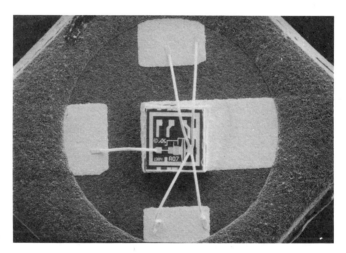

Figure 4.44 Photograph of MSA0735. (Courtesy of Avantek.)

(Fig. 4.45), which is specifically designed for 5-V operation and low noise figure. The trade name for this amplifier is MODAMP, meaning "monolithic Darlington amplifier." Other designs are available for high power to 1 W and wide bandwidth to 5 GHz. The primary disadvantage of the MODAMP structure is the low efficiency of resistive feedback networks and the common collector connection for all transistor stages. When this limitation has been overcome via an ISOSAT (isolated self-aligned transistor) process, microwave silicon MMIC applications will increase [4.31].

$T_A = 25\,°C,\ I_d = 22$ mA

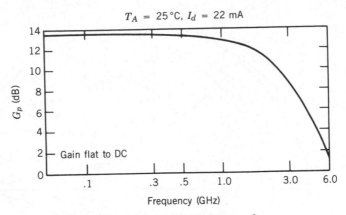

Figure 4.45 Gain of MSA0735 versus frequency.

Example 4.6 A Very Wideband Single-Stage Amplifier (2–18 GHz) In order to illustrate the reduction in the number of matching elements obtainable with real-frequency techniques compared to techniques based on analytical gain–bandwidth and network synthesis theory, the amplifier described by Medley in Ref. 4.32 was redesigned using Ref. 4.2. The device was modified in the same way to provide a basis for comparison. The designed amplifiers before and after optimization are shown in Fig. 4.46.

The gain obtained by using the S parameters supplied by the manufacturer (gain at 17 GHz adjusted) was 4.85 ± 0.44 dB, the input VSWR less than $1.59:1$, and

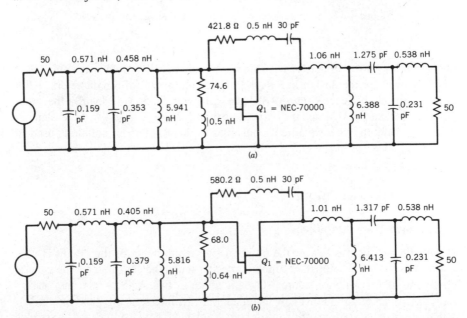

Figure 4.46 Amplifier designs for Example 4.6: (*a*) initial design; (*b*) optimized design.

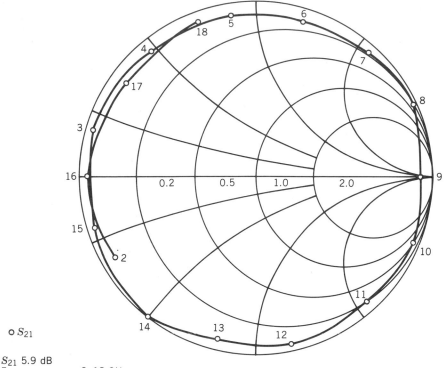

○ S_{21}

S_{21} 5.9 dB
Frequency range: 2–18 GHz

Figure 4.47 S_{21} parameter of the optimized design in Fig. 4.46.

the output VSWR less than 1.54:1. Similar results (4.91 ± 0.41 dB; 1.55; 1.42) were obtained for the design in Ref. 4.32 by using this set of S parameters. After optimization with Super-Compact PC the gain was 5.55 ± 0.35 dB, the input VSWR less than 1.73:1, and the output less than 1.76:1 (Fig. 4.47). Five-element matching networks were required, compared to the 10-element solutions used in Ref. 4.32.

4.2 MULTISTAGE DESIGN

4.2.1 High-Gain Amplifiers

If the gain per stage is low or the physical size requirement of the circuit is very small, a two-stage or multistage amplifier design should be considered. The general configuration for common-source stages is given in Fig. 4.48, where the passive network design requirements now include input M_1, interstage M_2, and output M_3. Other multistage designs to consider are cascaded balanced stages, cascaded feedback stages (see Example 4.9), or distributed amplifiers (see the next section).

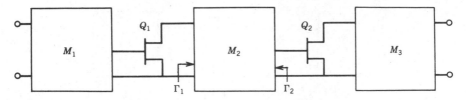

Figure 4.48 Two-stage amplifier design.

For a high-gain, two-stage amplifier, the lossless network M_2 is designed to give a conjugate power match between stages, which can be expressed by

$$(S_{11})_2 = \Gamma_2^* \tag{4.43}$$

$$(S_{22})_1 = \Gamma_1^* \tag{4.44}$$

where the assumption that $S_{12} = 0$ is used for simplification. Later this assumption will be dropped. Thus the design of a two-stage amplifier is basically the same as for the single-stage amplifier in Fig. 1.1 with the additional step of designing an interstage network M_2 that satisfies either (4.43) or (4.44). If M_2 is lossless, the conditions expressed by (4.43) and (4.44) are identical.

The two-port stability of the two-stage amplifier must be considered separately. Since two active devices are present, the stability factor of the two-stage amplifier does not guarantee overall stability. If the two-port k is greater than unity, there is no passive network connected to either M_1 or M_3 that can make the amplifier oscillate. However, the interstage network M_2 could cause the transistor to oscillate if either individual transistor stability factor is less than unity. At low frequencies, where the transistor stability factor is less than unity, the termination provided by the interstage network must be checked for stability of both Q_1 and Q_2. Since biasing stubs are required for Q_1 and Q_2, these shorted stubs usually guarantee stability in the interstage network.

Before turning to the two-stage design of low-noise and high-power amplifiers, it is helpful to consider the interstage match with the unilateral assumption for both stages, as shown in Fig. 4.49. The matching condition with this assumption is given for all three amplifier types. For a GaAs MESFET amplifier, the interstage will approximately match the terminations given by Fig. 4.50, where again the unilateral assumption was made for both stages. This step is helpful to investigate circuit topologies for the interstage network.

4.2.2 Low-Noise Amplifiers

For two-stage low noise design the most common design is for M_1,

$$(\Gamma_G)_1 = \Gamma_{On1} \tag{4.45}$$

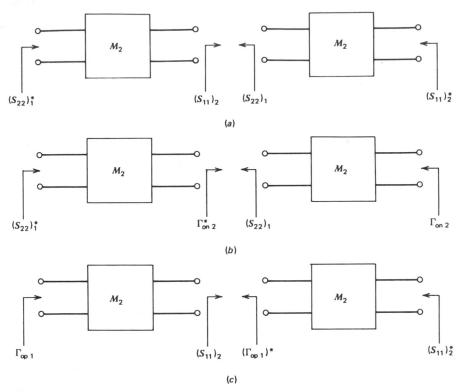

Figure 4.49 Interstage designs (assuming that $S_{12} = 0$): (a) high-gain interstage; (b) low-noise interstage; (c) high-power interstage.

and for M_2,

$$(\Gamma_G)_2 = \Gamma_{0n2} \tag{4.46}$$

The design of M_3 is a conjugate power match at the output. These conditions should be achieved from lossless matching networks, since resistors will add noise and reduce gain. Normally, a premium low-noise transistor is used in the first stage and a slightly higher noise figure in the second stage. The total noise figure of the

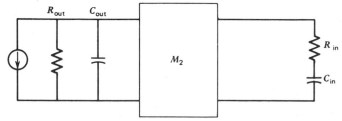

Figure 4.50 Interstage matching network for GaAs MESFET amplifier.

amplifier is [4.33; see (2.56)]

$$F = F_1 + (F_2 - 1)/G_{A_1} \qquad (4.47)$$

where F_1 is the first-stage noise figure and is a function of M_1, F_2 is the second-stage noise figure and is a function primarily of M_2 terminated by $(S_{22})'$, and G_{A_1} is the available gain of the first stage, which is a function of M_1. The noise figure of a multistage amplifier is given by the Friss noise equation [4.33] [also equation (2.57)]:

$$F_{\text{tot}} = F_1 + \frac{F_2 - 1}{G_{A_1}} + \frac{F_3 - 1}{G_{A_1}G_{A_2}} + \cdots \qquad (4.48)$$

where G_{A_1} = available gain first stage
$\quad\quad\;\; G_{A_2}$ = available gain second stage

If all stages are designed for minimum noise figure, we find that

$$(F_{\text{tot}})_{\text{min}} = (F_{\text{min}} - 1) + \frac{F_{\text{min}} - 1}{G_A} + \frac{F_{\text{min}} - 1}{G_A^2} + \cdots + 1 \qquad (4.49)$$

Using

$$\frac{1}{1 - x} = 1 + x + x^2 + \cdots \qquad (4.50)$$

we find a quantity $(F_{\text{tot}} - 1)$ which is defined as noise measure M. The minimum noise measure [also equation (2.60)]

$$(F_{\text{tot}})_{\text{min}} - 1 = \frac{F_{\text{min}} - 1}{1 - 1/G_A} = M_{\text{min}} \qquad (4.51)$$

refers to the noise of an infinite chain of optimum-tuned low-noise stages, so it represents a lower limit on the noise of the amplifier. Since it differs from $(F_{\text{tot}})_{\text{min}}$ by a factor of -1, the noise measure can be less than unity, but the noise figure cannot be less than unity.

The noise measure is usually lowest for common-emitter and common-source transistors; notice that this parameter will be minimum at an intermediate Γ_G between Γ_{0n} and Γ_{0G} (which is the generator reflection coefficient for maximum G_A) [4.34].

In the design of both M_1 and M_2, there is some trade-off between noise figure, input/output match, and gain; but normally for low-noise design a noise match is required. The other important considerations are overall stability at all frequencies and flat gain over the operating bandwidth.

Another design specification which is often encountered is low noise figure with a low S_{11}' (i.e., a good input match). This is normally accomplished with balanced stages, but this may be a rather expensive solution. With a two-stage design there are several options that might satisfy this production.

Since there are three matching circuits, we can satisfy any three design criteria. As an example, consider a two-stage low-noise design with the following requirements at a single frequency (Fig. 4.51):

$$F \simeq (F_{tot})_{min} = F_{min_1} + \frac{F_{min_2} - 1}{G_{A_1}} \simeq M_{min} + 1$$

$$S'_{11} = S'_{22} = 0$$

If the stability conditions are satisfied, the design can be achieved by properly designing the interstage network, which is an extension of the low-noise trade-offs in Section 4.1.3.

The schematic solution is given in Fig. 4.51. The design equations are

$$\Gamma_{L_1} = \frac{S_{11} + \Gamma^*_{0n1}}{D - S_{22}\Gamma^*_{0n1}} \qquad \Gamma_{L_2} = \frac{S_{11} + \Gamma^*_{0n2}}{D - S_{22}\Gamma^*_{0n2}}$$

The interstage must move $\Gamma^*_{L_1}$ to Γ_{0n2} using the techniques given in Chapter 3.

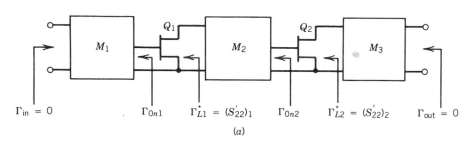

(a)

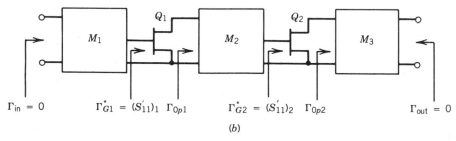

(b)

Figure 4.51 Two-stage amplifier design: (a) low noise; (b) high power.

4.2.3 High-Power Amplifiers

In addition to high output power, these amplifiers are designed for high third-order intercept point (TOI), as defined in Fig. 4.15. If the individual stages have a known value of TOI, the TOI of the combination can be estimated from the following formula, which assumes an in-phase addition of the third-order products generated by each stage [4.21, p. 104]:

$$\frac{1}{P_{\text{TOI}}} = \frac{1}{(P_{\text{TOI}})_2} + \frac{1}{G_2(P_{\text{TOI}})_1} \tag{4.52}$$

For the general multistage, case, the total TOI at the output is given for n stages by [4.35]

$$\frac{1}{P_{\text{TOI}}} = \frac{1}{P_{\text{TOI},n}} + \frac{1}{G_n P_{\text{TOI},n-1}} + \cdots + \frac{1}{G_n G_{n-1} \cdots G_k P_{\text{TOI},k-1}}$$
$$+ \cdots + \frac{1}{G_n G_{n-1} \cdots G_2 P_{\text{TOI},1}} \tag{4.53}$$

If all stages are identical, where $G_k = G$ and $P_{\text{TOI},k} = P$, this reduces to

$$\frac{1}{P_{\text{TOI}}} = \frac{1}{P} + \frac{1}{GP} + \frac{1}{G^2 P} + \cdots + \frac{1}{G^{n-1}P} \tag{4.54}$$

For an infinite chain of stages, this becomes

$$P_{\text{TOI}} = P(1 - 1/G) \tag{4.55}$$

which should be called the TOI measure, a transistor figure of merit analogous to noise measure.

As an example, consider a MSA-0435 amplifier at the recommended bias:

$$G = 8.3 \text{ dB}$$

$$P_{1\text{dBc}} = 12.5 \text{ dBm}$$

$$P_{\text{TOI}} \simeq 25.5 \text{ dBm}$$

$$V_D = 5.25 \text{ V}$$

$$I_D = 50 \text{ mA}$$

$$f = 1.0 \text{ GHz}$$

The calculated P_{TOI} is as follows:

n	P_{TOI} (dBm at output)
1	25.5
2	24.9
3	24.8
∞	24.8

This shows that usually the last two stages dominate the total P_{TOI}. In the practical cases where the stages are not identical, the calculation should be made with (4.53), which will show the stages affecting the total P_{TOI}.

Finally, the design of a multistage amplifier for wide dynamic range, as defined by Fig. 4.15, and (4.24) and (4.25), must be considered a combination of low noise design and high power design. The design equations are summarized below, where (4.56) is based on the approximation that P_{1dBc} is 10 dB below TOI:

$$\frac{1}{P_{1dBc}} = \frac{1}{P_{1dBc,n}} + \frac{1}{G_n P_{1dBc,n-1}} + \cdots + \frac{1}{G_n G_{n-1} \cdots G_2 P_{1dBc,1}} \quad (4.56)$$

where G is the power gain,

$$(MDS)_{out} = kTB + 3 \text{ dB} + F_{tot} + G_{tot} \quad (4.57)$$

where G_{tot} is the total transducer power gain, and

$$F_{tot} = F_1 + \frac{F_2 - 1}{G_{A_1}} + \cdots + \frac{F_n - 1}{G_{A_1} G_{A_2} \cdots G_{A(n-1)}} \quad (4.58)$$

where G_A is the available power gain.
If the amplifier is designed for wide dynamic range, we need

$$P_{1dBc} \leq P_{1dBc,n} \approx P_{1dBc,n}$$

$$(MDS)_{out} \geq (MDS)_{out,1} \approx (MDS)_{out,1}$$

which results in a best estimate of

$$DR \simeq P_{1dBc,n} - (MDS)_{out,1} \quad (4.59)$$

This equation should be used only as a guideline in the design. Notice that increasing the number of stages and gain will reduce the dynamic range. The optimum gain will depend on the receiver specifications.

4.2.4 Multistage Amplifier Design Examples

Example 4.7 Narrowband Two-Stage Amplifier with a Flat Gain Response and Low Input and Output VSWRs The narrowband two-stage (11.7 to 12.2 GHz) amplifier in Fig. 4.52*a* was designed to have a flat transducer power gain response and low input and output VSWRs [4.2]. This was achieved by designing the load-matching network for the maximum possible operating power gain at each frequency, concentrating the mismatch required to flatten the transducer power gain response in the interstage network and designing the input matching network

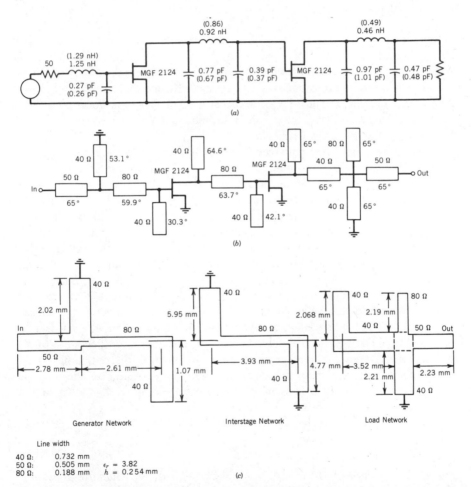

Figure 4.52 (*a*) Lumped-element two-stage amplifier of Example 4.7 ($G_T = 8.26 + 0.33$ dB; VSWR-I < 1.9:1; VSWR-0 < 1.39:1); (*b*) prototype commensurate distributed amplifier design ($G_T = 8.25 + 0.36$ dB; VSWR-I < 1.89:1; VSWR-0 < 1.51:1); (*c*) layout of the final amplifier in microstrip form ($G_T = 7.21 + 0.24$ dB; VSWR-I < 1.85:1; VSWR-0 < 1.63:1).

TABLE 4.7   S Parameters of the MGF2124 Mitsubishi FET Used in Example 4.7

Freq. (GHz)	S_{11}		S_{21}		S_{12}		S_{22}	
	Mag.	Angle	Mag.	Angle	Mag.	Angle	Mag.	Angle
11.7	0.741	71	0.865	−91	0.059	−74	0.532	151
11.8	0.748	68	0.883	−94	0.061	−76	0.530	149
11.9	0.751	65	0.886	−97	0.057	−79	0.528	146
12.0	0.755	62	0.895	−99	0.061	−82	0.524	144
12.1	0.755	58	0.910	−103	0.061	−84	0.522	142
12.2	0.758	55	0.905	−106	0.061	−86	0.520	139

to approximate a conjugate match. The overall response was then optimized with Super-Compact PC. The S parameters for the transistor are shown in Table 4.7. A design goal was to keep the element count as low as possible. The transducer power gain before optimization was 7.97 $\pm$ 0.42 dB; the input VSWR was lower than 1.95 : 1 and the output VSWR lower than 1.27 : 1. The transducer power gain of the optimized amplifier is 8.26 $\pm$ 0.33 dB; the input VSWR is lower than 1.9 : 1 and the output is lower than 1.39 : 1.

The commensurate distributed amplifier in Fig. 4.52b (line lengths specified at 12.2 GHz) was designed by transforming the input and interstage network of a lumped-element design to distributed equivalents and optimizing the networks thus obtained and by designing a commensurate distributed output network [4.2]. The gain of this amplifier varies between 7.91 and 8.64 dB (no overall optimization), the input VSWR is lower than 1.89 : 1, and the output is lower than 1.51 : 1.

This design was transformed to a microstrip equivalent. The substrate was assumed to have a dielectric constant of 3.82, the substrate thickness used was 0.254 mm, the cover height 5 mm, and the copper thickness 33 μm. Each microstrip matching network was then optimized with Super-Compact PC to approximate the original S parameters of the corresponding distributed network as closely as possible. The lengths of the stubs were then adjusted by taking the extension of the lengths of the open-ended stubs because of the open-end effects [4.36] and the shift in reference planes associated with T-junction discontinuities [4.37] into account (see Appendix D). The final step was to optimize the amplifier with Super-Compact PC. The resulting microstrip amplifier is shown in Fig. 4.52c. The expected gain is 7.21 $\pm$ 0.24 dB, the expected input VSWR smaller than 1.85 : 1, and that of the output smaller than 1.63 : 1. This gain is plotted in Fig. 4.53.

Example 4.8   Wideband Two-Stage Amplifier with a Flat Gain Response and Low-Input and Low-Output VSWRs   To design a wideband amplifier with a GaAs FET, it is usually necessary to reduce the gain–bandwidth constraints associated with the impedances to be matched by using resistive feedback or by loading the device with shunt or series resistance.

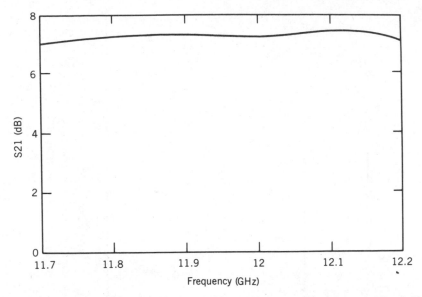

Figure 4.53 Expected gain of the distributed microstrip amplifier shown in Fig. 4.52*c*.

Designing a single-stage amplifier with the device first provides a good idea of the type and amount of the feedback or loading required. It is possible to design a single-stage amplifier with flat gain and low input and output VSWRs over the passband 2 to 6 GHz with the DEXCEL 1503A device if voltage-shunt feedback is used. A feedback resistor of 439 Ω in series with 0.5 nH flattens the G_{ma} to 8.31 $\pm$ 0.28 dB [4.2].

Matching networks for the amplifier were designed [4.2] and the resulting amplifier is shown in Fig. 4.54*a*. The gain obtained is 8.06 $\pm$ 0.13 dB, the input VSWR is less than 1.55, and the output less than 1.60. The *S* parameters for this transistor are shown in Table 4.8. It is clear from these results that a feedback resistor of the order of 439 Ω is required in order to reduce the gain–bandwidth constraints sufficiently for the single-stage amplifier. It was decided to use the same feedback resistor for the first stage and a 675-Ω resistor for the final stage of the two-stage amplifier (stage closest to the load). This would provide a total gain of approximately 18 dB, if obtainable.

The load-matching network was then designed for maximum operating power gain, the gain achieved was calculated, and the interstage network was used to flatten the gain response. The input matching network was designed to provide a low-input VSWR. The operating power gain obtained from the final stage and that assigned to the first stage are listed in Table 4.9. The transducer power gain obtained with the amplifier is 18.18 $\pm$ 0.16 dB. The input VSWR is lower than 1.72 : 1 and the output VSWR lower than 1.74 : 1. These results are plotted in Fig. 4.55.

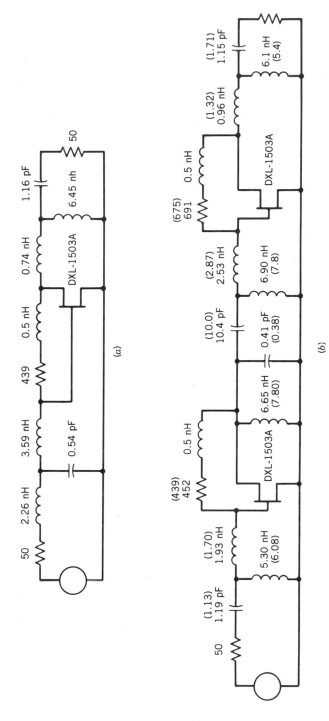

Figure 4.54 Amplifier designs for Example 4.8: (*a*) single-stage wideband amplifier (gain = 8.06 ± 0.13 dB; VSWR-I ≤ 1.55:1; VSWR-0 ≤ 1.60:1); (*b*) wideband two-stage amplifier (gain = 18.18 ± 0.16 dB; VSWR-I ≤ 1.72:1; VSWR-0 ≤ 1.74:1).

TABLE 4.8 Chip S Parameters of the DXL1503A (V_{ds} = 3.5 V, I_{dS} = 12 mA as Supplied by the Manufacturer)

Freq. (GHz)	S_{11} Mag.	S_{11} Angle	S_{21} Mag.	S_{21} Angle	S_{12} Mag.	S_{12} Angle	S_{22} Mag.	S_{22} Angle
2.0	0.97	−22	3.16	159	0.03	78	0.77	−10
3.0	0.93	−31	2.98	150	0.04	76	0.76	−13
4.0	0.89	−42	2.95	143	0.06	69	0.73	−16
5.0	0.85	−53	2.98	134	0.07	66	0.71	−19
6.0	0.80	−68	2.90	122	0.08	56	0.63	−22

After optimization with Super-Compact PC (optimization goal MS_{11} W = 1, MS_{22} W = 1, MS_{21} 18.2 dB W = 2) the gain is 18.18 ± 0.16 dB, the input VSWR is less than 1.72 : 1, and the output VSWR less than 1.74 : 1. The designed amplifier is shown in Fig. 4.54*b*.

Example 4.9   Two-Stage GaAs MMIC Feedback Amplifier   The next example is a low-cost two-stage GaAs MMIC feedback amplifier for the band at 2 to 6 GHz. The room-temperature specifications for this circuit are as follows: (MGA 64135)

Frequency	2 to 6 GHz
Minimum gain	9.0 dB (10.0 typical)
Maximum gain flatness	1 dB (±0.7 typical)
Maximum noise figure	8 dB (7 typical)
Minimum power output (1 dBc)	10 dBm (12 typical)

Using standard Avantek foundry design rules for the 0.3-μm gate, ion implantation process which were in effect in 1987, a two-stage feedback design was found to meet these specifications. The equivalent circuit for the FET is given in Fig. 4.56. The dc and RF schematics are given in Fig. 4.57 with a photograph of the final chip, the Avantek M641.

TABLE 4.9   Operating Power Gains Relevant to the Design of the Lumped-Element Amplifier of Example 4.8

Freq. (GHz)	Gain Obtained with Final Stage (dB)	Gain Required (dB)	Gain Specified for First Stage (dB)
2.0	10.54	18.20	7.66
3.0	10.17	18.20	8.03
4.0	10.06	18.20	8.13
5.0	10.12	18.20	8.08
6.0	9.94	18.20	8.26

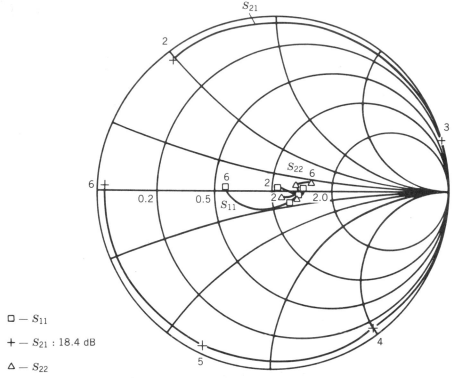

Figure 4.55 *S* parameters of Example 4.8 versus frequency.

This circuit is realized on a 100-μm-thick GaAs chip with a size of 24 mil by 35 mil. The typical performance is given in Fig. 4.58 when the chip is mounted in the 35 package shown in Fig. 4.57*d* and is tested in a 50-Ω microstrip system. This circuit has a minimum gain of 6 dB over 1.0 to 10 GHz.

The MGA-64135 is designed to be a high-volume low-cost amplifier gain block, essentially a microwave op amp with 50-Ω impedances. To accomplish this there are three major requirements. It must be easy to use, it must perform well, and it must be inexpensive. It turns out that these requirements are dependent on each other.

First, to make it easy to use, it must be in a simple, microstrip-compatible package. At this point, one can accomplish two objectives at once by choosing an inexpensive package. The package must be able to dissipate enough power to handle the heat generated in the device. One must have an adequate model of the package to be able to simulate the operation of the device inside the package. The package model becomes a part of the circuit model throughout the design.

One aspect of the cost of the IC is its size: the smaller the chip, the lower the cost; also, the device must fit comfortably inside the package. Another major cost driver is the yield of the circuit.

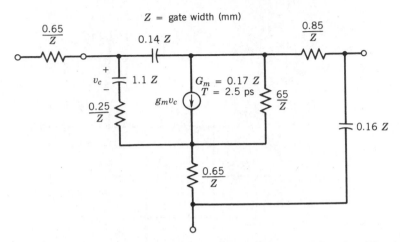

Figure 4.56 Normalized GaAs MESFET model using Avantek 1987 foundry. All resistors in ohms; all capacitors in picofarads.

The 641 is biased to share current (Fig. 4.57a). All the current through the input device (Q_1) comes from the output device (Q_2). One consequence of this scheme is the V_{gs} (gate-to-source voltage) of Q_1 controls both the current of Q_1 and the current of Q_2. While being a very effective scheme for biasing, the circuit is especially sensitive to variations in the pinch-off voltage of the FETs. Unfortunately, pinch-off is difficult to control, so the fabrication specification window is wide. A SPICE model was generated to take account of this difficulty. This model allows one to vary the pinch-off voltage and examine the current and voltage consequences of that change. The biasing resistors can thus be picked to ensure proper operation over a broad range of pinch-off voltage.

Example 4.10 Three-Stage Low-Noise Amplifier The next example is a three-stage low-noise amplifier using the Avantek ATF-10135, 10235, and 20135 packaged transistors. This example demonstrates the design of a typical 3.7–4.2 GHz TVRO (TV Receive Only) system. The RF schematic is given in Figure 4.59.

The device data are given in Tables 4.10 to 4.12, including both S parameters and noise parameters at the recommended dc bias for lowest noise. Using the data sheet values of noise and associated gain (G_A), the noise figure is estimated to be (at 4 GHz)

$$F_{\text{tot}} = F_1 + \frac{F_2 - 1}{G_{A_1}} + \frac{F_3 - 1}{G_{A_1} G_{A_2}}$$

$$= 1.122 + 0.202/20 + 0.318/20(20)$$

$$= 1.133$$

$$= 0.542 \text{ dB}$$

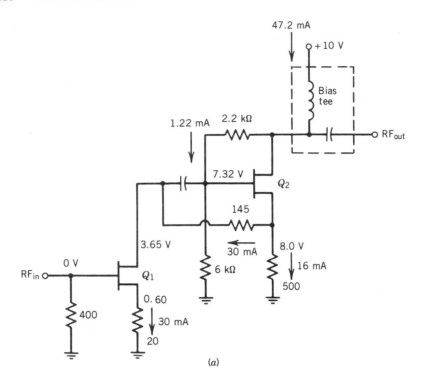

(a)

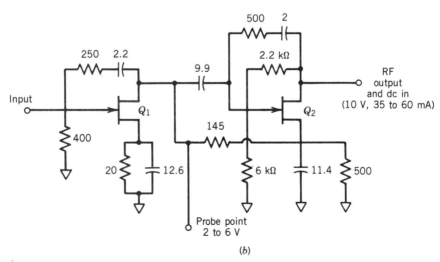

(b)

Figure 4.57 Circuit schematics for Example 4.9: (a) dc bias schematic; (b) RF schematic; (c) photograph of M641 GaAs MMIC (Courtesy of Avantek); (d) 35 package used for Avantek MGA 64135.

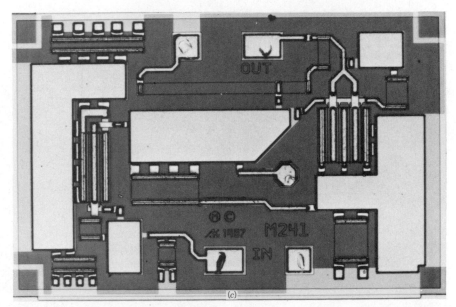

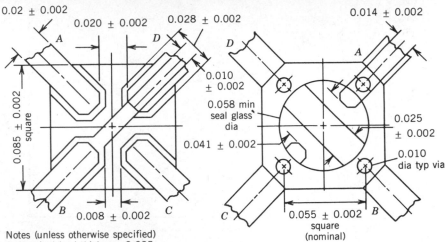

Notes (unless otherwise specified)
1. Nominal lead thickness 0.005
2. Nominal AL_2O_3 thickness 0.010 ± 0.002 − 0.001

Figure 4.57 (*Continued*)

$$T_e = 290(F - 1) = 38.6 \text{ K}$$

$$G_T = G_{A_1} + G_{A_2} + G_{A_3}$$

$$\simeq 13(3) = 39 \text{ dB}$$

The resistors are included for low-frequency stability, where $k < 1$. The calculated gain and noise figure is given in Table 4.13. This type of amplifier is in high-

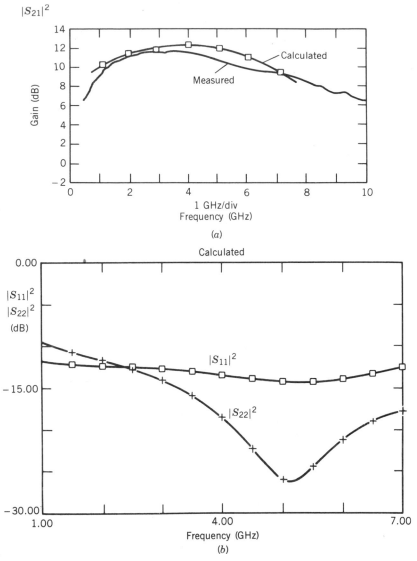

Figure 4.58 Measured and calculated performance of Example 4.9: (*a*) gain versus frequency; (*b*) return loss versus frequency; (*c*) reverse gain versus frequency.

volume production all over the world; the MODFET is expected to lower the noise figure even further.

4.3 SMALL-SIGNAL DISTRIBUTED AMPLIFIERS

Distributed amplifiers date back to 1937, when the concept was disclosed in a British patent [4.39]. By the 1940s, distributed or traveling-wave techniques had

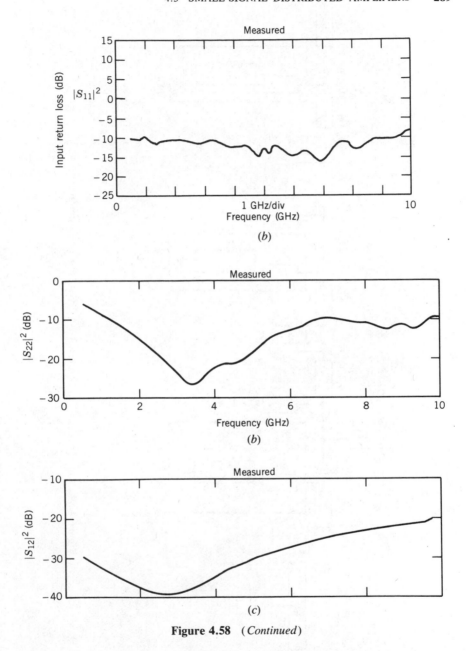

Figure 4.58 (*Continued*)

been used to design very broadband electron-tube amplifiers [4.39, 4.40]. However, it is only recently, with the availability of good-quality microwave GaAs FETs, that the distributed amplifier has again become popular [4.41–4.44]. The rebirth in popularity is in part due to the fact that excellent bandwidth performers are obtainable, because the input and output capacitances of the active devices are

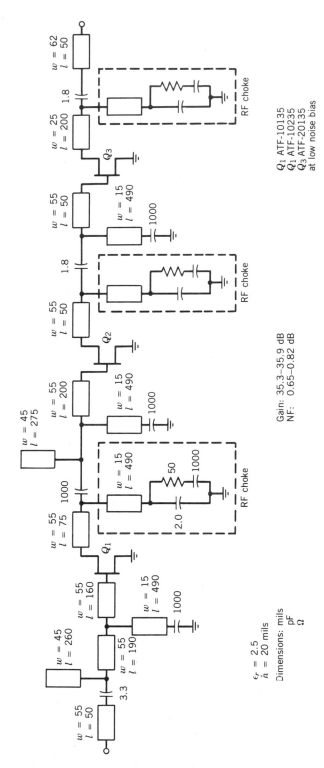

Figure 4.59 RF schematic of three-stage 3.7-4.2-GHz low-noise amplifier.

$\epsilon_r = 2.5$
$h = 20$ mils

Dimensions: mils
pF
Ω

Gain: 35.3–35.9 dB
NF: 0.65–0.82 dB

Q_1 ATF-10135
Q_1 ATF-10235
Q_3 ATF-20135
at low noise bias

TABLE 4.10   S and Noise Parameters, ATF-10135 ($V_{ds} = 2$ V, $I_d = 20$ mA)

S Parameters

Freq. (GHz)	S_{11} Mag.	S_{11} Angle	S_{21} Mag.	S_{21} Angle	S_{12} Mag.	S_{12} Angle	S_{22} Mag.	S_{22} Angle
0.5	0.98	−18	5.32	163	0.020	78	0.35	−9
1.0	0.93	−33	5.19	147	0.038	67	0.36	−19
2.0	0.79	−66	4.64	113	0.074	59	0.30	−31
3.0	0.64	−94	4.07	87	0.110	44	0.27	−42
4.0	0.54	−120	3.60	61	0.137	31	0.22	−49
5.0	0.47	−155	3.20	37	0.167	13	0.16	−54
6.0	0.45	162	2.88	13	0.193	−2	0.08	−17
7.0	0.50	120	2.51	−10	0.203	−19	0.16	45
8.0	0.60	87	2.09	−32	0.210	−36	0.32	48
9.0	0.68	61	1.75	−51	0.209	−46	0.44	38
10.0	0.73	42	1.52	−66	0.207	−58	0.51	34
11.0	0.77	26	1.26	−82	0.205	−73	0.54	27
12.0	0.80	14	1.12	−97	0.200	−82	0.54	15

Noise Parameters

Freq. (GHz)	F_{0n} (dB)	Γ_{0n} Mag.	Γ_{0n} Angle	R_n/Z_0
0.5	0.3	0.925	12	0.85
1.0	0.3	0.85	24	0.70
2.0	0.4	0.70	47	0.46
4.0	0.5	0.39	126	0.36
6.0	0.8	0.36	−170	0.12
8.0	1.1	0.45	−100	0.38
12.0	1.4	0.60	−41	1.10

absorbed in the distributed structures. The resulting amplifiers also exhibit very low sensitivities to process variations, can be designed as wideband low-noise amplifiers, and are relatively easy to simulate and fabricate. In addition, a great deal of design flexibility is possible since the number of devices, device size (gate periphery), transmission-line characteristic impedances, and amplifier upper cutoff frequency can all be varied to meet a particular component or chip design specification. The simplified amplifier structure shown in Fig. 4.60, can be used to help illustrate these design concepts.

As can be seen, the amplifier is composed of two main artificial transmission-line sections consisting of series inductances and shunt capacitances, which are usually supplied by the FET parasitics: hence two constant-k transmission lines which have different cutoff frequencies and attenuation characteristics result. Since

TABLE 4.11 S and Noise Parameters, ATF-10235 ($V_{ds} = 2$ V, $I_d = 20$ mA)

S Parameters

Freq.	S_{11}		S_{21}		S_{12}		S_{22}	
(GHz)	Mag.	Angle	Mag.	Angle	Mag.	Angle	Mag.	Angle
0.5	0.97	−20	5.68	162	0.023	76	0.47	−11
1.0	0.93	−41	5.58	143	0.050	71	0.45	−23
2.0	0.77	−81	4.76	107	0.086	51	0.36	−38
3.0	0.59	−114	4.06	80	0.120	35	0.30	−51
4.0	0.48	−148	3.51	52	0.149	18	0.23	−67
5.0	0.46	166	3.03	26	0.172	3	0.10	−67
6.0	0.53	125	2.65	1	0.189	−14	0.09	48
7.0	0.62	96	2.22	−20	0.191	−28	0.24	55
8.0	0.71	73	1.75	−39	0.189	−41	0.37	51
9.0	0.75	54	1.47	−55	0.184	−46	0.46	42
10.0	0.78	39	1.28	−72	0.180	−59	0.51	34
11.0	0.82	26	1.04	−86	0.179	−71	0.54	26
12.0	0.84	12	0.95	−101	0.177	−82	0.54	17

Noise Parameters

Freq.	F_{0n}	Γ_{0n}		R_n/Z_0
(GHz)	(dB)	Mag.	Angle	
2.0	0.6	0.73	74	0.33
4.0	0.8	0.45	148	0.15
6.0	1.0	0.42	−137	0.12
8.0	1.3	0.49	−80	0.45
12.0	1.6	0.65	−20	1.16

these lines are heavily loaded by the FET parasitic resistances R_i and R_{ds}, the number of active-device sections cannot be added indefinitely because the attenuation along the transmission lines will eventually exceed the gain obtained by adding an additional active device. The phase shift from section to section for both the drain and gate lines must also be made approximately equal so that the amplified signal from each FET will add along the output (drain) transmission line. Any wave traveling in the reverse direction on the output transmission line is absorbed by the termination.

One of the assumptions in the design of distributed amplifiers is that the artificial transmission lines of the gate and drain circuits be terminated with their characteristic impedances loads. This is somewhat difficult since the characteristic impedance of a constant-k transmission line is a function of frequency, and approaches infinity at cutoff. Hence there is no physical combination of elements that can provide a proper termination at all frequencies, but distributed structures can be

TABLE 4.12 S and Noise Parameters, ATF-20135 (V_{ds} = 2 V, I_d = 20 mA)

S Parameters

Freq. (GHz)	S_{11} Mag.	S_{11} Angle	S_{21} Mag.	S_{21} Angle	S_{12} Mag.	S_{12} Angle	S_{22} Mag.	S_{22} Angle
0.5	0.96	−20	5.90	162	0.024	77	0.50	−10
1.0	0.92	−40	5.77	144	0.046	66	0.48	−21
2.0	0.77	−76	4.92	109	0.086	52	0.39	−34
3.0	0.59	−107	4.2	83	0.111	40	0.33	−45
4.0	0.49	−136	3.64	57	0.137	24	0.26	−61
5.0	0.43	−179	3.15	32	0.167	9	0.14	−65
6.0	0.49	138	2.74	8	0.179	−5	0.05	22
7.0	0.57	106	2.32	−13	0.183	−18	0.19	60
8.0	0.68	81	1.92	−32	0.185	−33	0.33	57
9.0	0.73	62	1.62	−50	0.183	−40	0.42	46
10.0	0.77	47	1.41	−66	0.182	−52	0.46	38
11.0	0.82	36	1.12	−81	0.186	−67	0.50	27
12.0	0.85	22	0.98	−97	0.189	−75	0.51	15

Noise Parameters

Freq. (GHz)	F_{0n} (dB)	Γ_{0n} Mag.	Γ_{0n} Angle	R_n/Z_0
2.0	0.9	0.75	85	0.27
4.0	1.2	0.48	159	0.08
6.0	1.4	0.46	−122	0.08
8.0	1.7	0.53	−71	0.43
12.0	2.0	0.69	−14	1.04

designed for extremely wideband operation with the upper frequency limit approaching $0.8f_c$ of the distributed lines. However, ideal operation of these networks can occur only when lumped elements are used to realize the amplifier circuit. A complete design analysis of small- and large-signal distributed amplifiers is presented in Section 5.6, and only their general capabilities and performance considerations are described in this section.

The power gain of an n-section distributed amplifier can be shown to be [4.45]

$$G = \frac{g_m^2 R_{01} R_{02} \sinh^2[n/2(A_d - A_g)]\, e^{-n(A_d + A_g)}}{4[1 + (\omega/\omega_g)^2][1 - (\omega/\omega_c)^2]\sinh^2[1/2(A_d - A_g)]} \quad (4.60)$$

where $R_{01}[=(L_g/C_g)^{1/2}]$ and $R_{02}[=(L_d/C_d)^{1/2}]$ are the characteristic resistances of the gate and drain lines, respectively. The expressions for attenuation

TABLE 4.13   RF Performance of 3.7–4.2 GHz Three-Stage Low-Noise Amplifier

Freq. (GHz)	NF (dB)	S_{21} (dB)	S_{11} (dB)	S_{22} (dB)	k
0.5	14.36	−13.7	−1.93	−0.9	1.3×10^7
1.0	2.20	27.6	−0.69	−2.7	116.0
1.5	2.67	29.7	−0.85	−5.1	35.0
2.0	2.47	37.2	−1.35	−12.8	4.7
2.5	2.33	37.2	−1.22	−14.9	2.2
3.0	1.87	35.3	−1.90	−32.6	2.4
3.5	1.09	35.1	−1.91	−12.3	1.4
3.6	0.93	35.3	−2.04	−11.4	1.2
3.7	0.80	35.6	−2.28	−10.7	1.1
3.8	0.70	35.8	−2.66	−10.4	1.1
3.9	0.65	35.9	−3.20	−10.5	1.1
4.0	0.65	35.8	−3.89	−10.8	1.2
4.1	0.71	35.6	−4.51	−11.6	1.3
4.2	0.81	35.3	−5.17	−12.6	1.4
4.3	0.96	34.8	−5.93	−13.4	1.5
4.5	1.39	34.1	−8.52	−14.0	1.7
5.0	3.05	29.7	−3.35	−12.1	1.9
5.5	5.09	20.2	−0.49	−14.6	2.3
6.0	6.37	14.8	−0.40	−19.8	4.5

and phase velocity are well known for constant-k transmission lines and for the low-loss case. The attenuation factors for the gate and drain lines can be approximated as

$$A_g = \frac{(\omega_c/\omega_g)X_k^2}{\left\{1 - [1 - (\omega_c/\omega_g)^2]X_k^2\right\}^{1/2}} \tag{4.61}$$

$$A_d = \frac{\omega_d/\omega_c}{(1 - X_k^2)^{1/2}} \tag{4.62}$$

where $X_k = \omega/\omega_c$ is the normalized frequency, $\omega_g = 1/R_i C_{gs}$, $\omega_d = 1/R_{ds} C_{ds}$, and

$$\omega_c = \frac{2}{(L_g C_{gs})^{1/2}} = \frac{2}{(L_d C_{ds})^{1/2}} \tag{4.63}$$

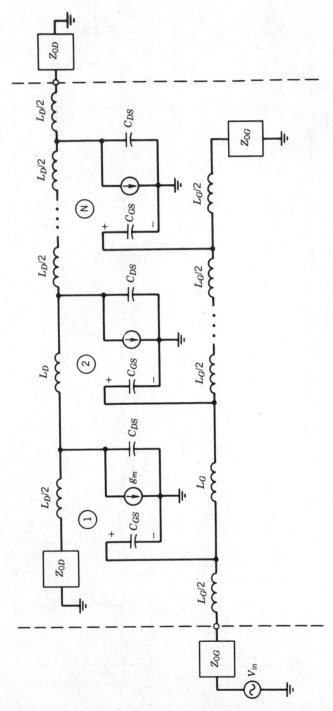

Figure 4.60 Simplified lumped-element distributed amplifier model illustrating constant-k transmission lines for gate-line and drain-line circuits.

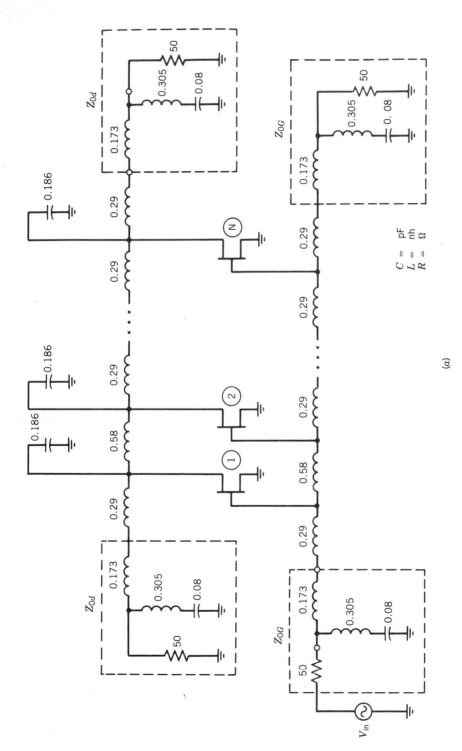

Figure 4.61 Typical lumped-element distributed amplifier: (*a*) circuit topology; (*b*) amplifier gain and frequency response as a function of the number of sections.

(*a*)

$C = $ pF
$L = $ nh
$R = $ Ω

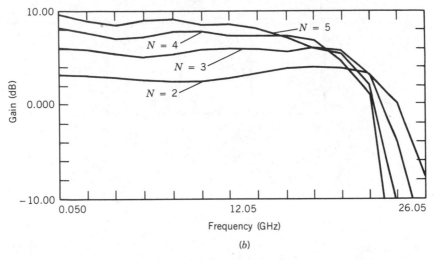

Figure 4.61 (*Continued*)

From the expressions above it can be seen that the frequency response of the amplifier is determined primarily by the gate line, since it exhibits a substantially larger attenuation factor than the drain line and a lower cutoff frequency. The gain is also limited by the gate-line attenuation factor because the gate-line attenuation will eventually exceed the added gain obtained by increasing the number of amplifier sections. The foregoing constraints usually limit the gain and bandwidth attainable with practical realizations.

The effects on gain and bandwidth as a function of the number of gain cells for a typical GaAs FET distributed amplifier are shown in Fig. 4.61. The FET used in the analysis is shown in Fig. 4.62. As can be seen, a reasonable improvement in gain is obtained as the number of amplifier sections (cells) is increased; but for n greater than 5, the gain increase begins to become negligible. It should also be noted that the flat bandwidth is also diminishing as n is increased. However, the amplifier flatness can be improved somewhat by optimizing the artificial transmission-line terminations. If dual-gate FETs were employed in the design, a slight gain improvement in overall gain (ca. 2 dB) could be obtained for the same number of sections.

A more subtle adverse effect of the gate-line frequency response is its effect on the amplifier's noise figure. Typically, the noise-figure performance of a distributed amplifier begins to degrade rapidly as the operating frequency approaches 60 to 70% of the gate-line cutoff frequency, while the amplitude response can be made flat beyond 80% of f_c. The amplifier's noise figure also degrades at very low frequencies, due to the effects of the gate-line termination resistor. Designing for low-noise operation is usually constrained, the only variable being the device gate

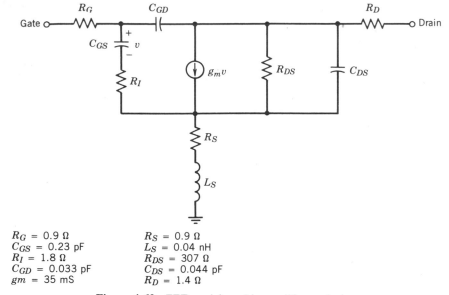

$R_G = 0.9 \; \Omega$
$C_{GS} = 0.23 \; \text{pF}$
$R_I = 1.8 \; \Omega$
$C_{GD} = 0.033 \; \text{pF}$
$gm = 35 \; \text{mS}$

$R_S = 0.9 \; \Omega$
$L_S = 0.04 \; \text{nH}$
$R_{DS} = 307 \; \Omega$
$C_{DS} = 0.044 \; \text{pF}$
$R_D = 1.4 \; \Omega$

Figure 4.62   FET model used in amplifier analysis.

periphery, which, unfortunately, also influences gain and frequency response. The calculated noise figure and gain performance for a nine-cell monolithic amplifier (Fig. 4.63) is shown in Fig. 4.64. Note the sharp increase in noise-figure performance as the operating frequency approaches the gate-line cutoff frequency.

The noise figure of a distributed amplifier has another important property of interest [4.46, 4.47]. Over the low-frequency portion of the operating frequency range, the noise figure will actually decrease as the number of stages increases, which clearly violates the Friis noise figure equation (4.48). For a simple distributed MESFET amplifier with n identical sections, Aitchinson [4.47] calculated the noise figure shown in Fig. 4.65 for a simple design operating at about 10 GHz. The parameters for this simple design are given in Table 4.14 for reference. The noise-figure equation for this calculation is [4.47]

$$F = 1 + \left(\frac{\sin n\beta}{n \sin \beta} \right)^2 + \frac{4}{n^2 g_m^2 Z_{0g} Z_{0d}}$$

$$+ \frac{Z_{0g} \omega^2 C_{gs}^2 R \, \Sigma_{r=1}^{n} f(r, \beta)}{n^2 g_m} + \frac{4P}{n g_m Z_{0g}} \qquad (4.64)$$

where $f(r, \beta) \simeq n^3/3$ for large n and R and P are noise parameters of the FET. The dominant terms for F are usually the last two terms of (4.64) for large n. If

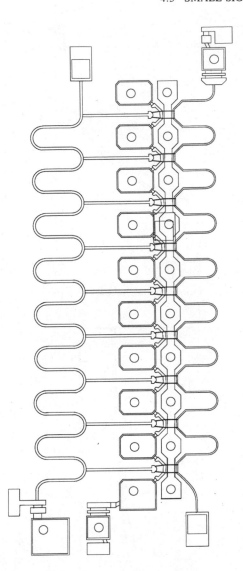

Figure 4.63 Low-noise monolithic dual-gate distributed amplifier topology.

the optimum value of n can be found when the last two terms of (4.64) are equal, we find that $n = 8$ for this simple analysis.

Because of its unique characteristics, such as low input/output VSWR, decade bandwidth performance, and low sensitivity to manufacturing tolerances, the distributed amplifier is rapidly making it the amplifier of choice for a variety of diverse microwave applications. Only in a few areas, such as very low noise and high-efficiency applications, has it been unable to gain a foothold. As the quality of the GaAs FET improves, the distributed amplifier is expected to encroach into

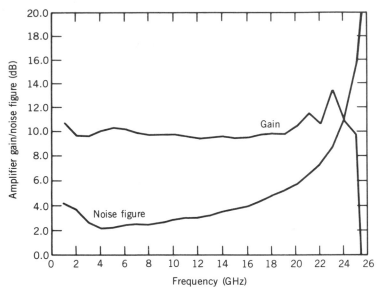

Figure 4.64 Noise figure and gain performance of a nine-section dual-gate FET monolithic amplifier.

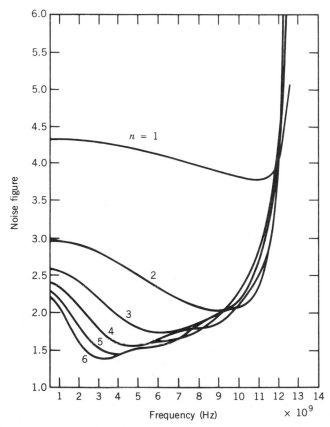

Figure 4.65 Noise figure of a simple distributed amplifier versus frequency and number of stages n. (From Ref. 4.47.) © 1985 IEEE

TABLE 4.14 Assumed MESFET Equivalent-Circuit Values Used to Compute Fig. 4.63

Z_{0g} (Ω)	50
C_{gs} (pF)	0.5
C_{ds} (pF)	0.2
g_m (S)	30×10^{-3}
R	0.2
P	0.6

the realm of high-power high-efficiency amplifiers currently dominated by reactively matched cascade designs.

REFERENCES

4.1 P. L. D. Abrie and P. Rademeyer, "A Method for Evaluating and the Evaluation of the Influence of the Reverse Transfer Gain on the Transducer Power Gain of Some Microwave Transistors," *IEEE Transactions on Microwave Theory and Techniques*, Vol. MTT-33, No. 8, August 1985, pp. 711–713.

4.2 *Complex Match User's Guide*, Compact Software, Paterson, N.J., 1988.

4.3 R. S. Carson, *High Frequency Amplifiers*, Wiley, New York, 1975.

4.4 G. Gonzales, *Microwave Transistor Amplifier Analysis and Design*, Prentice-Hall, Englewood Cliffs, N.J., 1984.

4.5 P. L. D. Abrie, *The Design of Impedance Matching Networks for Radio-Frequency and Microwave Amplifiers*, Artech House, Norwood, Mass. 1985.

4.6 H. J. Carlin and P. Amstutz, "On Optimum Broad-Band Matching," *IEEE Transactions on Circuits and Systems*, Vol. CAS-28, No. 5, May 1981, pp. 401–405.

4.7 R. M. Fano, "Theoretical Limitations on the Broadband Matching of Arbitrary Impedances," *Journal of the Franklin Institute*, Vol. 249, January 1960, pp. 57–83 and February 1960, pp. 139–155.

4.8 H. J. Carlin, "A New Approach to Gain–Bandwidth Problems," *IEEE Transactions on Circuits and Systems*, Vol. CAS-23, April 1977, pp. 170–175.

4.9 H. J. Carlin and J. J. Komiak, "A New Method of Broad-Band Equalization Applied to Microwave Amplifiers," *IEEE Transactions on Microwave Theory and Techniques*, Vol. MTT-27, February 1979, pp. 93–99.

4.10 G. P. Young and S. O. Scanlan, "Matching Network Design Studies for Microwave Transistor Amplifiers," *IEEE Transactions on Microwave Theory and Techniques*, Vol. MTT-29, No. 10, October 1981, pp. 1027–1035.

4.11 B. S. Yarman and H. J. Carlin, "A Simplified 'Real-Frequency' Technique Applied to Broad-Band Multistage Microwave Amplifiers," *IEEE Transactions on Microwave Theory and Techniques*, Vol. MTT-30, No. 12, December 1982, pp. 2216–2222.

4.12 H. J. Carlin and B. S. Yarman, "The Double Matching Problem: Analytic and Real-Frequency Solutions," *IEEE Transactions on Circuits and Systems*, Vol. CAS-30, No. 1, January 1983, pp. 15–28.

4.13   J. Lange, "Noise Characterization of Linear Two Ports in Terms of Invariant Parameters," *IEEE Journal of Solid-State Circuits*, Vol. SC-2, June 1967, pp. 37–40.

4.14   G. D. Vendelin, "Power GaAs FET Amplifier Design with Large Signal Tuning Parameters," *11th Asilomar Conference on Circuits, Systems, and Computers*, November 1977, pp. 139–141.

4.15   S. C. Cripps, "A Theory for the Prediction of GaAs FET Load-Pull Power Contours," *IEEE MTT-S International Microwave Symposium Digest*, 1983, pp. 221–223.

4.16   J. R. Black, "Electromigration Failure Model in Aluminum Metallization for Semiconductor Devices," *Proceedings of the IEEE*, Vol. 57, No. 9, September 1969, pp. 1587–1599.

4.17   J. C. Irvin and A. Loya, "Failure Mechanism and Reliability of Low Noise GaAs FET's," *Bell System Technical Journal*, Vol. 57, October 1978, pp. 2823–2846.

4.18   J. C. Irvin, "The Reliability of GaAs FET's," in *GaAs FET Principles and Technology*, J. V. DiLorenzo and D. Khandelwal, Eds., Artech House, Norwood, Mass., 1982, pp. 349–402.

4.19   P. R. Gray and R. G. Meyer, *Analysis and Design of Analog Integrated Circuits*, Wiley, New York, 1977, Chap. 8.

4.20   G. D. Vendelin, "Feedback Effects on GaAs MESFET Noise Performance," *IEEE MTT-S International Microwave Symposium Digest*, May 1975, pp. 324–326.

4.21   U. L. Rohde and T. T. N. Bucher, *Communication Receivers, Principle and Design*, McGraw-Hill, New York, 1988, pp. 104 and 229.

4.22   K. Simons, "Directional Coupler," U.S. Patent 3,048,798, August 1962, Jerrold Electronics.

4.23   D. Norton, "High Dynamic Range Amplifier," U.S. Patent 3,624,536, November 1971, Anzac Corp.

4.24   H. B. Mead and G. R. Callaway, "Broadband Amplifier," U.S. Patent 4,042,887, August 1977, Q-Bit Corp.

4.25   D. E. Norton and A. F. Podell, "Transistor Amplifier with Impedance Matching Transformer," U.S. Patent 3,891,934, June 1975, Anzac Corp.

4.26   U. L. Rohde, "Wide Band Amplifier Summary," *Ham Radio*, November 1979, pp. 34–35.

4.27   U. L. Rohde, "The Design of Wide-Band Amplifier with Large Dynamic Range and Low Noise Figure Using CAD Tools," *IEEE Long Island MTT Symposium Digest*, April 28, 1987, pp. 47–55.

4.28   U. L. Rohde, "Designing a Matched Low Noise Amplifier Using CAD Tools," *Microwave Journal*, Vol. 29, No. 10, October 1986, pp. 154–160.

4.29   J. Lange, "Interdigitated Stripline Quadrature Hybrid," *IEEE Transactions on Microwave Theory and Techniques*, Vol. MTT-17, December 1969, pp. 1150–1151.

4.30   J. F. Kukielka and C. P. Snapp, "Wideband Monolithic Cascadable Feedback Amplifiers Using Silicon Bipolar Technology," *IEEE Microwave and Millimeter-Wave Monolithic Circuits Symposium Digest*, June 1982.

4.31   C. P. Snapp, in *Handbook of Microwave and Optical Components*, Vol. 2, *Microwave Solid State Components and Tubes*, K. Chang, Ed., Wiley, New York, 1989.

4.32   M. W. Medley, "Network Synthesis CAD Program Applied to RF and Microwave Circuit Design," *Microwave Systems News and Communications Technology*, Vol. 17, Part I, January 1987, pp. 97–107; Part II, March 1987, pp. 88–93.

4.33   H. T. Friss, "Noise Figure of Radio Receivers," *Proceedings of the IRE*, Vol. 32, July 1944, pp. 419–422.

4.34 G. D. Vendelin, J. Archer, and G. Bechtel, "A Low-Noise Integrated S-Band Amplifier," *Microwave Journal*, February 1974, pp. 47–50; also "Digest of Technical Papers," *IEEE International Solid-State Circuits Conference*, February 1974, pp. 176–177.

4.35 D. Raicu, *Internal Report IR-11-84*, Eaton Corp., Electronic Instrumentation Division, Sunnyvale, Calif., November 1984; also in *Microwave Journal*, July 1988, pp. 136–142.

4.36 M. Kirschning, R. H. Jansen, and N. H. L. Koster, "Accurate Model for Open End Effect of Microstrip Lines," *Electronics Letters*, December 1980.

4.37 E. O. Hammerstad, "Equations for Microstrip Circuit Design," *5th European Microwave Conference Digest*, 1975.

4.38 W. S. Percival, "Thermionic Valve Circuits," British Patent 460,562, 1937.

4.39 W. H. Horton, J. H. Jasberg, and J. D. Noe, "Distributed Amplifier: Practical Considerations and Experimental Results," *Proceedings of the IRE*, July 1950, pp. 748–753.

4.40 E. L. Ginzton, W. R. Hewlett, J. H. Jasberg, and J. D. Noe, "Distributed Amplifications," *Proceedings of the IRE*, August 1948, pp. 956–969.

4.41 W. K. Chen, "Distributed Amplification Theory," in Proc. 5th Annual Allerton Conf. Circuit and System Theory, U. of Illinois, 1967, pp. 300–316.

4.42 A. S. Podgorski and L. Y. Wei, "Theory of Traveling-Wave Transistors," *IEEE Transactions on Electron Devices*, Vol. ED-29, No. 12, December 1982, 1845–1853.

4.43 Y. Ayasli, R. L. Mozzi, J. L. Vorhaus, L. D. Reynolds, and R. A. Pucel, "A Monolithic GaAs 1–13 GHz Traveling-Wave Amplifier," *IEEE Transactions on Microwave Theory and Techniques*, Vol. MTT-30, No. 7, July 1982, pp. 976–981.

4.44 Y. Ayasli, L. D. Reynolds, R. L. Mozzi, and L. K. Hanes, "2–20 GHz GaAs Traveling-Wave Power Amplifier," *IEEE Transactions on Microwave Theory and Techniques*, Vol. MTT-32, No. 3, March 1984, pp. 290–295.

4.45 J. B. Beyer, S. N. Prasad, R. C. Becker, J. E. Nordman, and G. K. Hohenwarter, "MESFET Distributed Amplifier Design Guidelines," *IEEE Transactions on Microwave Theory and Techniques*, Vol. MTT-32, No. 3, March 1984, pp. 268–275.

4.46 K. B. Niclas and B. A. Tucker, "On Noise in Distributed Amplifiers at Microwave Frequencies," *IEEE Transactions on Microwave Theory and Techniques*, Vol. MTT-31, August 1983, pp. 661–668.

4.47 C. S. Aitchinson, "The Intrinsic Noise Figure of the MESFET Distributed Amplifier," *IEEE Transactions on Microwave Theory and Techniques*, Vol. MTT-33, June 1985, pp. 460–466.

4.48 J. D. Sifri, "Matching Technique Yields Optimum LNA Performance," *Microwaves & RF*, February 1986, pp. 87–90.

BIBLIOGRAPHY

Archer, J. A. et al., "GaAs Monolithic Low-Noise Broad-Band Amplifier," *IEEE Journal of Solid State Circuits*, Vol. SC-15, December 1981, pp. 648–652.

Aylasi, Y., S. W. Miller, R. Mozzi, and L. Hanes, "Capacitively Coupled Traveling-Wave Power Amplifier," *IEEE Monolithic Circuits Symposium*, May 1984, pp. 52–54.

Basawapatna, G. R., "Design and Performance of a 2 to 18 GHz Medium Power GaAs FET Amplifier," *European Microwave Conference*, 1978, pp. 458–462.

Besser, L., "Synthesize Amplifiers Exactly," *Microwave Systems News*, October 1979, pp. 28–40.

Besser, L., and S. Swenson, "Take the Hassle Out of FET Amplifier Design," *Microwave Systems News*, September 1977, pp. 97–105.

Bingham, S. D., S. D. McCarter, and A. M. Pavio, "A 6.5 to 16 GHz Monolithic Power Amplifier Module," *IEEE Microwave and Millimeter-wave Monolithic Circuits Symposium Digest*, June 1985, pp. 38–41.

Bingham, S. D., S. D. McCarter, and A. M. Pavio, "A 6.5 to 16 GHz Monolithic Power Amplifier Module," *IEEE Transactions Microwave Theory Techniques*, Vol. MTT-33, No. 12, December 1985, pp. 1555–1559.

Black, H. S., "Stabilized Feedback Amplifiers," *Electrical Engineer*, Vol. 53, January 1934, pp. 114–120.

Camp, W. O., et al., "2-6 GHz Monolithic Microwave Amplifier," *IEEE Microwave and Millimeter-Wave Monolithic Circuits Symposium*, June 1983.

Camp, W. O., Jr., S. Tiwari, and D. Parsons, "2-6 GHz Monolithic Microwave Amplifier," *1983 Microwave Symposium Digest*, June 1983, pp. 76–80.

Chen, W. K., *Active Network and Feedback Amplifier Theory*, McGraw-Hill, New York, 1980.

Dueme, P., et al., "Miniaturization of an X-Band Monolithic GaAs Amplifier," *IEEE Microwave and Millimeter-Wave Monolithic Circuits Symposium*, June 1983.

Estreich, D. B., "A Monolithic Wide-Band GaAs IC Amplifier," *IEEE Journal of Solid State Circuits*, Vol. SC-17, No. 6, December 1982, pp. 1166–1173.

Estreich, D. B., "A Wideband Monolithic Gas IC Amplifier," *ISSCC Digest of Technical Papers*, February 1982, pp. 194–195.

Fornaciari, M. G., "Low Noise Amplifier Design for Satellite Receivers," *1979 WESCON Convention Digest*, August 1979, Session 25.

Honjo, K., et al., "Low Noise, Low Power Dissipation GaAs Monolithic Broadband Amplifiers," *IEEE GaAs IC Symposium Digest*, November 1982, pp. 87–90.

Honjo, K., T. Suguira, and H. Itoh, "Ultra-broadband GaAs Monolithic Amplifier," *IEEE Transactions on Microwave Theory and Techniques*, Vol. MTT-30, No. 7, July 1982, pp. 1027–1033.

Honjo, K., and Y. Takayama, "GaAs FET Ultra-Broadband Amplifiers for Gbit/s Data Rate Systems," *IEEE Transactions on Microwave Theory and Techniques*, Vol. MTT-29, No. 7, July 1981, pp. 629–636.

Hornbuckle, D., and R. L. Van Tuyl, "Monolithic GaAs Direct Coupled Amplifiers," *IEEE Transactions on Electron Devices*, Vol. ED-26, No. 2, February 1981, pp. 175–182.

Kennan, W., T. Andrade, and C. Huang, "A Miniature 2–18 GHz Monolithic GaAs Distributed Amplifier," *IEEE Monolithic Circuits Symposium*, May 1984, pp. 41–44.

Kim, B., H. Q. Tserng, and H. D. Shih, "44 GHz Monolithic GaAs FET Amplifier," *IEEE Electron Device Letters*, Vol. EDL-7, No. 2, February 1986, pp. 95–97.

Kriz, J. J., and A. M. Pavio, "A Monolithic Limiting Amplifier for Microwave FM Receiver Systems," *IEEE Gallium Arsenide Integrated Circuits Symposium*, October 1984, pp. 89–91.

Ku, W. H., and W. C. Petersen, "Optimum Gain–Bandwidth Limitations of Transistor Amplifiers as Reactively Constrained Active Two-Port Networks," *IEEE Transactions on Circuits and Systems*, Vol. CAS-22, June 1975, pp. 523–533.

Leung, C. C., T. C. Lo, M. Dutta, I. Kipnis, J. F. Kukielka, and C. P. Snapp, "Bipolar Process Produces Si MMIC Building Blocks," *Microwaves & RF*, May 1988, pp. 231–238.

Liechti, C. A., and R. L. Tillman, "Design and Performance of Microwave Amplifiers with GaAs Schottky-Gate Field-Effect Transistors," *IEEE Transactions on Microwave Theory and Techniques*, Vol. MTT-22, No. 5, May 1974, pp. 510–517.

Mamodaly, M., P. Quentin, P. Dueme, and J. Obregon, "100 MHz to 17 GHz Dual Gate Variable Gain Amplifier," *IEEE Transactions on Microwave Theory and Techniques*, Vol. MTT-30, June 1982, pp. 918–919.

Mellor, D. J., and J. G. Linvill, "Synthesis of Interstage Networks of Prescribed Gain Versus Frequency Slopes," *IEEE Transactions on Microwave Theory and Techniques*, Vol. MTT-23, No. 12, December 1975, pp. 1013–1020.

Meyer, F., "Wideband Pulse Amplifier," *IEEE Journal of Solid State Circuits*, Vol. SC-13, June 1978, pp. 409–411.

Meyer, R. G., and R. A. Blauschild, "A 4-Terminal Wideband Monolithic Amplifier," *IEEE Journal of Solid State Circuits*, Vol. SC 16, No. 6, December 1981, pp. 634–638.

Moghe, S., T. Andrade, H. Sun, and C. Huang, "A Manufacturable GaAs MMIC Amplifier with 10 GHz Bandwidth," *IEEE Monolithic Circuits Symposium*, May 1984, pp. 37–40.

Moghe, S. B., H. J. Sun, T. Andrade, C. C. Huang, and R. Goyal, "A Monolithic Direct-Coupled GaAs IC Amplifier with 12 GHz Bandwidth," *IEEE Transactions on Microwave Theory and Techniques*, Vol. MTT-32, No. 12, December 1984, pp. 1698–1703.

Niclas, K. B., "Active Matching with Common-Gate MESFET's," *IEEE Transactions on Microwave Theory and Techniques*, Vol. MTT-33, No. 6, June 1985, pp. 492–499.

Niclas, K. B., "The Exact Noise Figure of Amplifiers with Parallel Feedback and Lossy Matching Circuits," *IEEE Transactions on Microwave Theory and Techniques*, Vol. MTT-30, No. 5, 1982, pp. 832–835.

Niclas, K. B., "Multi-octave Performance of Single-Ended Microwave Solid State Amplifiers," *IEEE Transactions on Microwave Theory and Techniques*, Vol. MTT-32, August 1984, pp. 896–908.

Niclas, K. B., "On Design and Performance of Lossy Match GaAs MESFET Amplifiers," *IEEE Transactions on Microwave Theory and Techniques*, Vol. MTT-30, No. 11, November 1982, pp. 1900–1907.

Niclas, K. B., R. B. Gold, W. T. Wilser, and W. R. Hitchens, "A 12–18 GHz medium power GaAs MESFET Amplifier," *IEEE Journal of Solid-State Circuits*, Vol. SC-13, No. 4, August 1978, pp. 520–527.

Niclas, K. B., W. T. Wilser, R. B. Gold, and W. R. Hitchens, "The Matched Feedback Amplifier Ultrawide-Band Microwave Amplification with GaAs MESFET's," *IEEE Transactions on Microwave Theory and Techniques*, Vol. MTT-28, No. 4, April 1980, pp. 285–294.

Obregon, J., and R. Funk, "A 150 MHz–16 GHz FET Amplifier," 1981 *ISSCC Digest of Technical Papers*, February 1981.

Pauker, V., and M. Binet, "Wideband High Gain Small Size Monolithic GaAs FET Amplifiers," *1983 Microwave Symposium Digest*, June 1983, pp. 50–53.

Pavio, A. M., "A Network Modeling and Design Method for a 2–18 GHz Feedback Amplifier," *1982 International Microwave Symposium Digest*, June 1982, pp. 162–165.

Pavio, A. M., D. L. Allen, S. D. Thompson, and S. J. Goldman, "A Solid-State 2–10 GHz 1 Watt TWT Replacement Amplifier," *IEEE International Microwave Symposium Digest*, June 1984, pp. 221–223.

Pavio, A. M., and L. L. Cook, "A Comparison of Thin Film and Thick Film Ku-Band Amplifiers," *IEEE International Microwave Symposium Digest*, June 1986, pp. 431–432.

Pavio, A. M., and S. D. McCarter, "The Design of a Medium Power, 6–12 GHz GaAs FET Amplifier, Using High Dielectric Substrates," *IEEE International Microwave Symposium Digest*, June 1982, pp. 158–161.

Pavio, A. M., and S. D. McCarter, "Network Theory and Modeling Method Aids Design of a 6–18 GHz Monolithic Multi-stage Feedback Amplifier," *Microwaves Systems News*, December 1982, pp. 78–93.

Pavio, A. M., S. D. McCarter, and D. L. Peterson, "The Design of Broadband Power FET Amplifiers Employing Nonlinear Device Analysis and Matched Equalization Techniques," *IEEE International Microwave Symposium Workshop*, June 1983.

Pavio, A. M., S. D. McCarter, and P. Saunier, "A Monolithic Multi-stage 6–18 GHz Feedback Amplifier," *IEEE Microwave and Millimeter-Wave Monolithic Circuits Symposium Digest*, June 1984, pp. 45–48.

Pengelly, R. S., *Microwave Field-Effect Transistors Theory, Design and Applications*, 2nd ed., Research Studies Press, Division of Wiley, Chichester, West Sussex, England, 1987.

Peterson, W. C., D. R. Decker, A. K. Gupta, J. Dully, and D. R. Chen, "A Monolithic GaAs 0.1 to 10 GHz Amplifier," *1981 Microwave Symposium Digest*, June 1981, pp. 354–355.

Pucel, R. A. "Design Considerations for Monolithic Microwave Circuits," *IEEE Transactions on Microwave Theory and Techniques*, Vol. MTT-29, No. 6, June 1981, pp. 513–534.

Rigby, P. N., et al., "Broadband Monolithic Low Noise Feedback Amplifiers," *IEEE Microwave and Millimeter-Wave Monolithic Circuits Symposium*, June 1983.

Rigby, P. N., J. R. Suffolk, and R. S. Pengelly, "Broadband Monolithic Low-Noise Feedback Amplifiers," *1983 Microwave Symposium Digest*, June 1983, pp. 41–45.

Rohde, U. L., "Digital HF Radio: A Sampling of Techniques," *Ham Radio Magazine*, April 1985.

Rohde, U. L., *Digital PLL Frequency Synthesizers Theory and Design*, Prentice-Hall, Englewood Cliffs, N.J., 1983 (available from Compact Software, Paterson, N.J.).

Rohde, U. L., "Eight Ways to Better Radio Receiver Design," *Electronics*, February 10, 1975.

Rohde, U. L., "High Dynamic Range Receiver Input Stages," *Ham Radio Magazine*, October 1975.

Rohde, U. L., "I-F Amplifier Design," *Ham Radio Magazine*, March 1977.

Rohde, U. L., "Optimum Design for High-Frequency Communications Receivers," *Ham Radio Magazine*, October 1976.

Rohde, U. L., "Recent Developments in Communication Receiver Design to Increase the Dynamic Range," *ELECTRO/80*, Boston, May 1980.

Rohde, U. L., "Recent Developments in Shortwave Communication Receiver Circuits," *IEEE National Telecommunications Conference*, November 1979.

Rohde, U. L., "Required Dynamic Range and Design Guides for EMI/RFI Test Receivers," *ELECTRO/82*, Boston, May 1982.

Snapp, C. P., "Advanced Silicon Bipolar Technology Yields Usable Monolithic Microwave and High Speed Digital ICs," *Microwave Journal*, August 1983, pp. 93–103.

Snapp, C. P., J. K. Kulielka, and N. K. Osbrink, "Practical Silicon MMIC's Challenge Hybrids," *Microwaves & RF*, November 1982, pp. 93–99.

Terzian, P. A., D. B. Clark, and R. W. Waugh, "Broadband GaAs Monolithic Amplifier Using Negative Feedback," *IEEE Transactions on Microwave Theory and Techniques*, Vol. MTT-30, No. 11, November 1982, pp. 2017–2020.

Thompson, S. D., J. J. Kriz, and A. M. Pavio, "A Broadband Amplifier Utilizing Low Noise and Power MMICs," *Government Microcircuits Applications Conference*, November 1986.

Tserng, H. Q., S. R. Nelson, and H. M. Macksey, "2–18 GHz, High Efficiency, Medium-Power GaAs FET Amplifiers," *1981 Microwave Symposium Digest*, June 1981, pp. 31–33.

Tucker, R. S., "Gain–Bandwidth Limitations of Microwave Transistor Amplifiers," *IEEE Transactions on Microwave Theory and Techniques*, Vol. MTT-21, May 1973, pp. 322–327.

Ulrich, E., "Use Negative Feedback to Slash Wideband VSWR," *Microwaves*, October 1978, pp. 66–70.

Van Tuyl, R. L., "A Monolithic Integrated 4 GHz Amplifier," *1978 ISSCC Digest Technical Papers*, February 1978, pp. 72–73.

Wienreb, S., and M. Pospieszalski, "X Band Noise Parameters of HEMT Devices at 300°K and 12.5°K," *IEEE MTT-S International Microwave Symposium*, St. Louis, Mo., June 1985, pp. 539–542.

Wilkinson, E., "An N Way Power Divider," *IRE Transactions on Microwave Theory and Techniques*, January 1960, p. 916.

Vendelin, G. D., "Computer-Aided-Design of Broadband GaAs MESFET Microwave Integrated Circuits," *Conference Proceedings Military Electronics Defense Expo 79*, Wiesbaden, West Germany, 1979, pp. 149–157.

Vendelin, G. D., *Design of Amplifiers and Oscillators by the S-Parameter Method*, Wiley, New York, 1982.

Vendelin, G. D., "A Ku-Band Integrated Receiver," *International Solid State Circuits Conference*, February 1968; *IEEE Journal of Solid State Circuits*, September 1968, pp. 255–257.

Vendelin, G. D., "2–4 GHz FETs Complete with Bipolar for Low Noise Designs," *Microwave Systems News*, January 1977, pp. 71–75.

Vendelin, G. D., W. Hooper, and G. Bechtel, "Design and Application of Low-Noise GaAs FET Amplifiers," *Digest of Technical Papers, International Conference on Communications*, June 1973.

PROBLEMS

4.1. Given the S parameters of an AT-41400 chip at a bias of $V_{CE} = 8$ V, $I_c = 25$ mA, and $f = 4.0$ GHz,

$$S = \begin{bmatrix} 0.60 \ \underline{/149°} & 0.108 \ \underline{/83°} \\ 2.06 \ \underline{/57°} & 0.42 \ \underline{/-28°} \end{bmatrix}$$

(a) Design the RF schematic of a high-gain amplifier. What is the transducer gain, G_T? Use distributed elements.

(b) Design the dc bias circuit using only resistors; repeat using a PNP transistor for active bias. Assume two power supplies with $V_1 = +5$ V and $V_2 = -5$ V.

(c) Draw the complete RF and dc schematic for both of the dc circuits above.

4.2. (a) Using an ATF-13135 GaAs FET at $V_{GS} = -1.5$ V, $V_{DS} = 3$ V, and $I_{DS} = 20$ mA, design a two-stage low-noise amplifier at 12 GHz using distributed elements. What is the transducer gain and noise figure of this amplifier?

(b) With a power supply of -12 V, design a dc bias circuit.

(c) Give the complete RF and dc schematic diagram.

(d) Explain what you need to do to convert this design to a high-gain amplifier (do not redesign it); what is the new gain and the new noise figure?

$$S = \begin{bmatrix} 0.61 \ \underline{/37°} & 0.144 \ \underline{/-89°} \\ 2.34 \ \underline{/-84°} & 0.15 \ \underline{/46°} \end{bmatrix}$$

$$F_{min} = 1.2 \text{ dB}$$

$$\Gamma_{0n} = 0.47 \ \underline{/-65°}$$

$$R_n = 40 \ \Omega$$

4.3. Using the NEC-67383, design a single-stage LNA for 2.4 GHz using the following topology [4.48]:

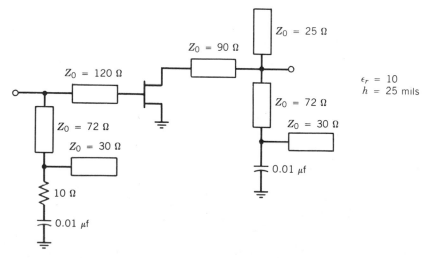

S parameters ($V_{DS} = 3$ V, $I_D = 10$ mA):

Freq. (GHz)	S_{11} Mag.	S_{11} Angle	S_{21} Mag.	S_{21} Angle	S_{12} Mag.	S_{12} Angle	S_{22} Mag.	S_{22} Angle
2.0	0.97	−46	3.30	136	0.020	70	0.63	−33
2.4	0.96	−53	3.30	132	0.029	64	0.62	−38
3.0	0.93	−60	3.0	118	0.045	50	0.62	−48
4.0	0.88	−79	2.64	107	0.06	35	0.61	−58

Noise parameters ($V_{DS} = 3$ V, $I_D = 10$ mA):

Freq. (GHz)	Γ_{On} Mag.	Γ_{On} Angle	R_n/Z_0	F_{min} (dB)
2.0	0.69	21	0.58	0.3
2.4	0.65	24	0.58	0.4
3.0	0.60	31	0.57	0.5
4.0	0.60	50	0.51	0.6

(a) Calculate the gain and noise figure of your design.

(b) Using a computer, calculate the gain and noise figure over the range 2 to 3 GHz. The goal is a noise figure of 0.8 dB maximum over the band.

4.4. Design an amplifier using the following data for the highest gain.

$$S = \begin{bmatrix} 0.65 \ \underline{/110^\circ} & 0.10 \ \underline{/23^\circ} \\ 1.96 \ \underline{/2^\circ} & 0.36 \ \underline{/-93^\circ} \end{bmatrix}$$

$$k = 1.149$$
$$f = 4 \text{ GHz}$$
$$G_{ma} = 10.59 \text{ dB}$$

$$\epsilon_r = 2.56 = k \text{ (relative dielectric constant)}$$

$$h = 0.031 \text{ in.}$$

$$V_{CE} = 8 \text{ V}$$

$$I_c = 10 \text{ mA}$$

$$V_{CC} = +15 \text{ V (power supply)}$$

(a) Give the RF design using lossless microstrip lines.

(b) Give the dc design.

(c) Draw the complete RF and dc schematic.

(d) From the microstrip-line design charts (Figs. 1.24 and 1.25), give the width and length of each matching element.

4.5. A feedback amplifier was designed with the following transistor at 8 GHz.

Using a y-parameter analysis, find the transducer power gain $|S_{21}|^2$ of the amplifier in decibels. What is S_{11} and S_{22} of the amplifier?

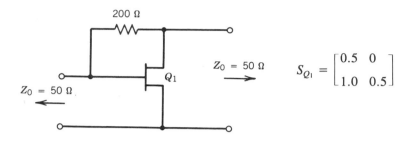

$$200\ \Omega$$

$$Z_0 = 50\ \Omega$$

$$Z_0 = 50\ \Omega$$

$$S_{Q_1} = \begin{bmatrix} 0.5 & 0 \\ 1.0 & 0.5 \end{bmatrix}$$

4.6. The S parameters of a GaAs MESFET are given below at $f = 8$ GHz. Find the stability factor k and G_{ma} and G_{ms} in decibels. Design an amplifier using distributed elements for $h = 25$ mils, $\epsilon_r = 10$. Give the dimensions of each matching element (length and width). Include a method of dc biasing the transistor. What is the gain of the amplifier? Draw a complete RF and dc schematic of the amplifier for $V_{DD} = 15$ V.

$$S = \begin{bmatrix} 0.8\ \underline{/180°} & 0.01\ \underline{/0°} \\ 1\ \underline{/90°} & 0.6\ \underline{/-90°} \end{bmatrix}$$

$$V_{DS} = 5\ \text{V}$$

$$V_{GS} = -1.5\ \text{V}$$

$$I_{DS} = 0.03\ \text{A}$$

4.7. (a) Design a 4-GHz LNA (low-noise amplifier) using the following packaged FET and lossless microstrip lines. Give the complete RF and dc schematic. Give the width and length of each matching element.

$$S = \begin{bmatrix} 0.65\ \underline{/-145°} & 0.125\ \underline{/7°} \\ 3.34\ \underline{/58°} & 0.37\ \underline{/-84°} \end{bmatrix}$$

$$\text{power supply} = -12\ \text{V}$$

$$F_{\min} = 0.80\ \text{dB} \qquad V_{DS} = 3\ \text{V}$$

$$\Gamma_{0n} = 0.46\ \underline{/143°} \qquad I_{DS} = 20\ \text{mA}$$

$$R_n = 12\ \Omega \qquad V_{GS} = -1.0\ \text{V}$$

Assume that a low-noise design is stable.

$$\epsilon_r = 2.56 = k$$

$$h = 0.031 \text{ in.}$$

(b) Redesign the input with lossless lumped elements.

(c) What is the gain of this amplifier (transducer power gain) in decibels?

4.8. Design a two-stage low-noise amplifier using the following GaAs FET at f = 8 GHz using microstrip-line matching elements:

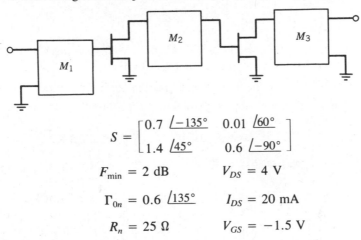

$$S = \begin{bmatrix} 0.7 \ \underline{/-135°} & 0.01 \ \underline{/60°} \\ 1.4 \ \underline{/45°} & 0.6 \ \underline{/-90°} \end{bmatrix}$$

$$F_{min} = 2 \text{ dB} \qquad V_{DS} = 4 \text{ V}$$

$$\Gamma_{0n} = 0.6 \ \underline{/135°} \qquad I_{DS} = 20 \text{ mA}$$

$$R_n = 25 \ \Omega \qquad V_{GS} = -1.5 \text{ V}$$

(a) Calculate k and G_{ma} or G_{ms} in decibels.

(b) Design M_1 for a low noise figure.

(c) Design M_2 for a low noise figure (assume that $S_{12} = 0$, if convenient).

(d) Design M_3 for gain (assume $S_{12} = 0$).

(e) Calculate the amplifier noise figure.

(f) Calculate the amplifier transducer gain.

(g) Design the bias circuit for $V_{DD} = 15$ V.

(h) Draw the complete amplifier schematic.

4.9. A two-stage low-noise amplifier is to be designed for the minimum possible noise figure. The noise parameters and S parameters are given below (at the low-noise bias point). Design M_1, M_2, and M_3 using transmission-line matching circuits. Calculate the total amplifier noise figure and the total amplifier transducer power gain.

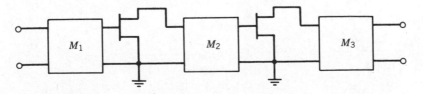

$$f = \text{GHz} \quad Z_0 = 50 \ \Omega$$

$$\text{Noise parameters: } F_{\min} = 3 \text{ dB}$$

$$R_n = \quad 30 \ \Omega$$

$$\frac{Y_{0n}}{Y_0} = \quad 0.23 - j0.55$$

$$S = \begin{bmatrix} 0.8 \ \underline{/-60^\circ} & 0 \\ 1.414 \ \underline{/60^\circ} & 0.9 \ \underline{/-50^\circ} \end{bmatrix}$$

4.10. Using the AT-10600 FET at 3 V, 10 mA, with low-noise bias, design a four-stage distributed amplifier for 1 to 18 GHz.

(a) Use C_{ADDED} in the drain to equalize the phase velocities.

(b) Use L_1 in series with the drain to equalize the phase velocities.

$$G = 6 \pm 0.7 \text{ dB} \qquad |S_{11}| < 8 \text{ dB R.L.}$$

$$\text{NF} < 8 \text{ dB} \qquad |S_{22}| < 8 \text{ dB R.L.}$$

4.11. Using the AT-8251 GaAs MESFET given in Table 3.21 at $V_{ds} = 5$ V, $I_d = 50$ mA, and $f = 4$ GHz, design a one-stage high-power amplifier for maximum dynamic range. Estimate the gain, the dynamic range, and the spurious free dynamic range.

$$R_p = 15 \ \Omega \qquad\qquad \text{Rn} = 35 \ \Omega$$

$$G_{\max} = 12 \text{ dB} \qquad F_{\min} = 1.0 \text{ dB}$$

$$\Gamma_{0p} = 0.3 \ \angle \ 18 \qquad \Gamma_{0n} = 0.5 \ \angle \ 60$$

$$P_{1\,\text{dBc}} = 21 \text{ dBm}$$

4.12. Using the AT-8251 GaAs MESFET given in Table 3.20 at $V_{ds} = 3$ V, $I_d = 20$ mA, and $f = 8$ GHz:

(a) Design a one-stage low-noise amplifier with $\Gamma_{\text{out}} = 0$; give the gain and noise figure and Γ_{in}.

(b) Design a one-stage low-noise amplifier with $\Gamma_{\text{in}} = 0$; give the gain and noise figure and Γ_{out}.

(c) Repeat design with Γ_{out} and Γ_{in} both low; give the gain and noise figure.

5 High-Power Amplifier Design

5.0 INTRODUCTION

During the last few years, state-of-the-art microwave transistor power amplifier design has shifted from the hybrid (discrete components) realm to the GaAs monolithic circuit arena. The change in design direction has occurred for two main reasons. First, the GaAs processing skills of the industry as a whole have markedly improved, to the point where cost, repeatability, and yield make it practical to employ monolithic designs in deliverable hardware. Second, considerable improvements in circuit modeling techniques have occurred, as well as the fact that CAD methods have become available to nearly all microwave designers. These new modeling and CAD methods will directly influence the final MMIC (monolithic microwave integrated circuit) cost by reducing and eventually eliminating the key component cost driver, the design iteration.

In the following sections, accurate model development and nonlinear CAD simulation will be stressed, so that the designer will be assured of the best possible chance of "first-time success" in developing some of the more demanding nonlinear two-port components. The determination of optimum load conditions, based on both modeling and measurements, will be discussed as it relates to the design of single- and multistage amplifiers. The chapter concludes with an illustration of several monolithic design/synthesis examples focusing on bandwidth, efficiency, and power output performance. Monolithic realization and design constraints are included in the examples as well as an analysis of the fundamental power output limitations in GaAs MESFET circuits.

5.1 DEVICE MODELING AND CHARACTERIZATION

The rapid growth in monolithic circuit technology has stressed the importance of accurate device modeling and characterization. The luxury of circuit tuning, which is common when microwave components are designed with discrete elements, is not available to the MMIC engineer. Unfortunately, the task of developing active-device models is tedious and usually requires extensive S-parameter measurements, and such engineering assignments are not associated with the glamor of MMIC design.

Although the development of a device model is heavily based on S-parameters, a discrete lumped-element model allows for circuit simulation flexibility. The elements in a FET model for example are strongly related to device physics and

313

can be used to predict performance variations due to temperature effects and process tolerances [5.1]. Such variations are difficult to accommodate in designs solely employing S-parameter device descriptions.

The engineer, whether designing a small-signal or nonlinear circuit, must first obtain a linear FET model such as the one shown in Fig. 5.1. The element values in the model are determined by various methods. The contact resistances R_d and R_g and the source resistance R_s are determined by dc measurements and are assumed to be frequency independent. The transconductance g_m and the drain-to-source resistance R_{ds} are easily determined from low-frequency RF measurements (100 MHz), which is much more accurate than determining these quantities at dc. The value of gate-to-source capacitance C_{gs}, an extremely important parameter, can be either measured or determined from optimization techniques with the remaining element values.

As mentioned above, the three resistances R_g, R_d, and R_s can be measured with the aid of a curve tracer (essentially dc). The configuration for measuring R_g and R_s is shown in Fig. 5.2. If a positive current step (ΔI_f) is applied to the gate, the FET drain can then be used as a probe to determine the voltage drop across the source resistance R_s. The value of R_s is then computed using the relation

$$R_s = \frac{\Delta V'_{ds}}{\Delta I_f} - R_{p2} = \frac{\Delta V_{ds}}{\Delta I_f} \tag{5.1}$$

where R_{p2} is the dc resistance of the probe. The resistance R_d can be determined similarly by reversing the roles of the FET source and drain. The configuration for measuring R_g is shown in Fig. 5.3. If the gate–source diode is forward biased, then R_g can be determined from the slope of the diode I–V characteristic. Thus

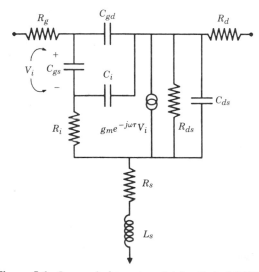

Figure 5.1 Lumped-element model for GaAs MESFET.

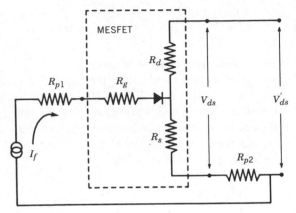

Figure 5.2 Configuration for measuring FET source resistance R_s.

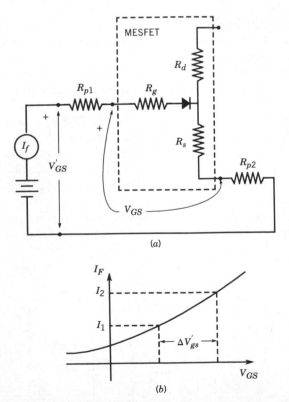

Figure 5.3 (a) Configuration for measuring FET gate resistance R_g; (b) gate diode I–V characteristic.

$$R_g = \frac{\Delta V'_{gs}}{I_2 - I_1} - R_s - R_{p1} - R_{p2} \qquad (5.2)$$

where R_{p1} and R_{p2} are the dc resistances of the probes.

At low RF frequencies, such as 100 MHz, the capacitive reactances of the elements C_{gs}, C_i, C_{dg}, and C_{ds} in a typical GaAs FET are much larger than their related resistances and can be ignored when measuring R_{ds} and g_m. Hence the values of g_m and R_{ds} can be deduced accurately from scalar S-parameter measurements. By assuming that the input impedance at these frequencies can be approximated by an open circuit, it is easy to show that

$$g_m = \left| \frac{R_s + R_{ds} + R_d + Z_0}{R_{ds}} \right| \frac{|S_{21}|}{2Z_0 - |S_{21}| R_{ds}} \qquad (5.3)$$

and

$$R_{ds} = \frac{1}{1 + g_m R_s} \left| Z_0 \frac{1 + S_{22}}{1 - S_{22}} - R_s - R_d \right| \qquad (5.4)$$

Although these equations are not independent, they are easily solved by iteration and converge rapidly. However, it should be remembered that at frequencies below 1 MHz the value obtained for R_{ds} will be much higher than the true RF value.

Obtaining the capacitive element values of the linear FET model is somewhat more difficult. The gate capacitance, which is the only capacitive element that can be measured indirectly, can be found using a method defined by DeLoach [5.2] to characterize microwave diodes. In this method, the FET gate, in conjunction with an external inductor (usually a bond wire), acts as a series resonant RLC network. The network is then placed across a through transmission line forming a single-pole band-reject filter. The DeLoach measurement circuit, with FET and gate inductor, is depicted in Fig. 5.4. The FET drain is RF bypassed to aid stability during measurement.

If the transmission loss of the circuit shown in Fig. 5.4 is measured as a function of frequency, the equivalent values of R, L, and C can be found. The analysis

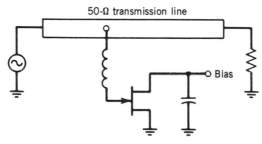

Figure 5.4 Measurement setup to determine C_{gs} by DeLoach method.

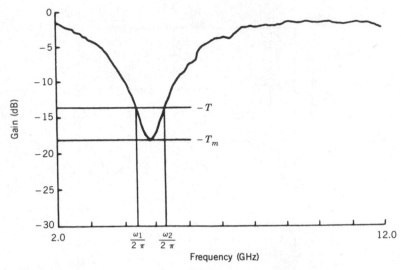

Figure 5.5 Band-stop filter response obtained from transmission loss measurement.

requires knowledge of the minimum transmission loss T_m, a second value of loss T, such that $T > T_m$, and the frequencies ω_1 and ω_2, where T occurs. A typical transmission-loss characteristic is shown in Fig. 5.5. Hence the values of R, L, and C can now be found from the following relations:

$$R = \frac{Z_0}{2[(T_m)^{1/2} - 1]} \tag{5.5}$$

$$L = \frac{Z_0}{2[(T_m)^{1/2} - 1](\omega_2 - \omega_1)} \left| \frac{T_m - 1}{T - 1} \right|^{1/2} \tag{5.6}$$

$$C = \frac{1}{\omega_1 \omega_2 L} \tag{5.7}$$

The gate–source capacitance is then computed from the relation

$$C_{gs} \approx (1 + g_m R_s)(C - C_{gd}) \tag{5.8}$$

where R_s and g_m have been determined from dc and RF measurements, respectively. Equation (5.8) is approximate and requires an estimated value for drain-to-source capacitance. This quantity is usually known and is typically 5 to 10% of the value of C. The dependence of C_{gs} on drain and gate bias can also be determined by observing the quantities ω_1 and ω_2, and by recomputing (5.6), (5.7), and (5.8).

The remaining model elements C_i, C_{dg}, C_{ds}, R_i, and τ are found by setting the computed model S-parameters equal to the measured device S-parameters and then

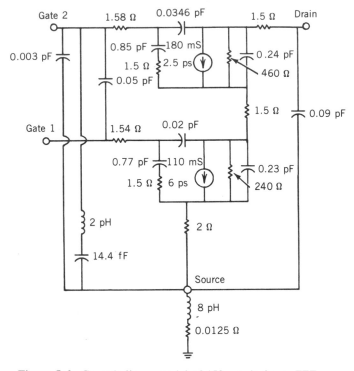

Figure 5.6 Cascode linear model of 450-μm dual-gate FET.

adjusting the values with the aid of an optimizer such as the one available in Super COMPACT. The dc measured element values should be fixed during optimization, while the elements C_{gs}, R_{ds}, and g_m should be tightly constrained. Excellent measured versus model performance can be obtained if the active device is carefully evaluated. A cascode dual-gate model (Fig. 5.6) and its computed versus actual performance (S-parameter) are shown in Fig. 5.7. As can be seen in the illustration, the model performance closely represents the measured S-parameter data. A dual-gate example was selected because it is more difficult to model than a single-gate device, thus illustrating the value of the approach.

The small-signal FET model formulation above is based very heavily on measured S-parameter data which are difficult and tedious to obtain accurately. However, there are some measurement techniques that yield good results provided that a proper test fixture is employed during device evaluation.

The measurement technique employs microstrip calibration standards for the automatic network analyzer (HP 8510) instead of coaxial shorts, opens, and sliding loads [5.3–5.5]. The microstrip standards help regain some of the accuracy typically lost if the system were calibrated in the coaxial line prior to the microstrip test fixture. The loss in accuracy would be due to the fact that the characteristics of the microstrip-to-coaxial transitions used on the fixture are imperfect. This calibration method, commonly called the TSD technique, requires the use of a

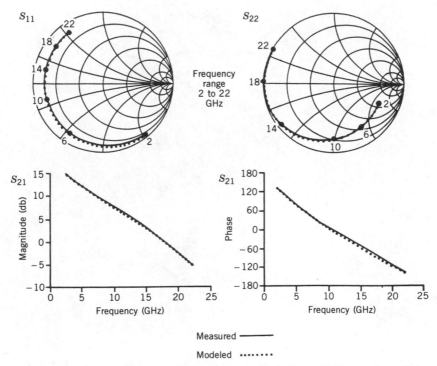

Figure 5.7 Measured versus computed response of dual-gate FET cascode model.

microstrip through line (T); microstrip short (S), whose reflection coefficient does not have to be known; and a delay line (D) that is approximately $3/8$ wavelength longer than the through line at the highest frequency of use. The TSD technique can provide heightened levels of accuracy over the conventional "short-open-load-through" calibration method. Also, the TSD technique can be further improved by using a microstrip open, consisting of an empty test fixture, which will allow the open-circuit fringing capacitances and leakage (gap) capacitance to be included in the calibration procedure. Although the exact characteristics of the standard need not be known, it is assumed that the characteristic impedance Z_0 is constant with frequency. Hence care must be taken in selecting microstrip dimensions at frequencies above 18 GHz. A typical test fixture that is used to evaluate discrete FETs and monolithic circuits is shown in Fig. 5.8.

Extending the foregoing small-signal model for nonlinear operation can be done in a variety of ways. The most obvious is to base a new circuit model on large-signal S parameters. This can easily be accomplished by measuring device S parameters at elevated power levels. The power level is usually chosen to correspond to the level encountered in the final circuit application. However, there are some negative aspects of using large-signal S parameters. First, as we have learned, S parameters are defined in a linear n-port system with a constant load impedance. Under large-signal conditions, the microwave n-port is not linear and large-signal

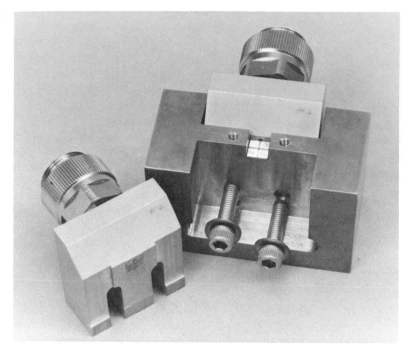

Figure 5.8 Typical *S*-parameter test fixture with APC-7 microstrip launchers. Courtesy of Texas Instruments.

S parameters cannot predict device performance for load impedances other than the one used during measurement (Z_0). Also, the value obtained for S_{22} is not the conjugate of the optimum load impedance.

To alleviate the limitations noted above, the active element must be characterized as a function of terminating impedance as well as a function of drive level. This technique, commonly called ''load pull,'' has the advantage that the FET is operated under conditions simulating actual circuit performance, but the method requires numerous measurements made at discrete frequencies and power levels [5.6, 5.7]. Although tedious, the concept is quite simple. The input impedance, power output, and gain of an active device are measured at a particular frequency and power level with a set load impedance. The load impedance can either be supplied passively with a preset tuner, or actively, by injecting a signal, which can vary in amplitude and phase, into the output of the device. The load-pull technique produces constant power contours, similar to constant-gain circles, on the Smith chart which are used in circuit synthesis. A set of constant power contours for a power GaAs FET is shown in Fig. 5.9, which is similar to Fig. 4.14.

The methods produce comparable results but are implemented quite differently. Elaborate mechanical tuners have been constructed that are controlled by the measurement system, which can step through a wide range of load impedances. These tuners are precalibrated at numerous frequencies, so that the data can be

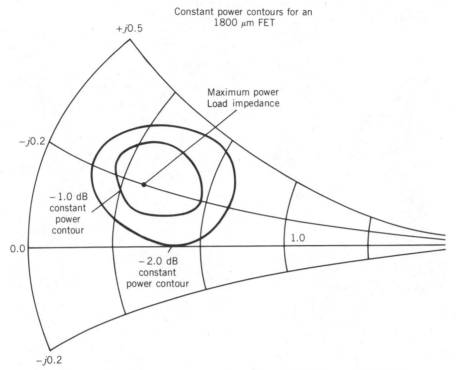

Figure 5.9 Constant-power contours for a typical power GaAs FET.

accumulated over a wide frequency range. The active load-pull method is typically implemented using a feed-forward technique so that the device under test (DUT) is terminated with a virtual load [5.8]. The impedance of the load is determined by injecting a backward traveling wave into the output of the DUT with the appropriate amplitude and phase. This signal must be coherent with the drive signal and is obtained by splitting the input signal and feeding it forward to the output port. A network analyzer is typically employed at the load to aid in system calibration. A typical measurement setup is shown in Fig. 5.10. The active load-pull method is somewhat easier to use than the passive technique and offers the advantage of being able to simulate any load impedance. Due to the loss inherent in mechanical tuners, very high or low impedances are difficult to simulate; hence the evaluation of very large power FETs, which require low load impedances, is not practical.

There is one key area in which load-pull data do not accurately predict nonlinear performance. In both the passive and active methods, the load impedance is synthesized at the fundamental frequency only. Harmonic terminations are typically ignored, although some effort has been made to synthesize fundamental and second harmonic loads. The problem is somewhat worse with mechanical tuner approaches, in that the harmonic terminations vary widely for small changes of the tuner elements. Active load-pull techniques usually have nearly constant termi-

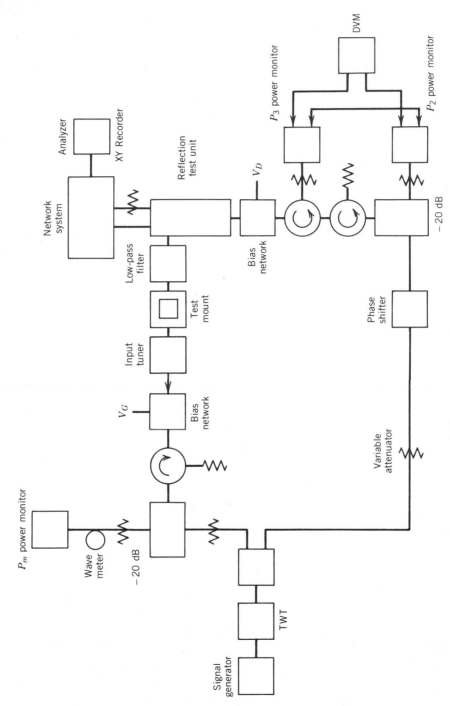

Figure 5.10 Active load-pull measurement setup.

nations, at least at low-order harmonics. Load-pull techniques have another drawback in that the data obtained cannot easily be integrated into a general nonlinear simulator.

A variety of numerical models, which have the potential of simulating in-circuit FET performance, have been proposed. These models range from the original low-frequency model presented by Shockley, which is valid for long-gate-length devices, to elaborate charge transport descriptions, which solve a set of partial differential equations in the time domain. Time-domain solutions are inefficient and are not practical for interactive (real-time) computing. A significant amount of time can be saved by approximating the solutions to the semiconductor differential equations; but adding the effects of the embedding circuit is difficult, since the FET input/output networks are usually characterized in the frequency domain. Another, more practical approach is to extend linear circuit theory to include nonlinear elements. The latter approach yields results similar to those obtained when designing with load-pull data [5.9].

It is evident from the discussion above that a numerical model, with the efficiency of a frequency-domain linear solution, is required for efficient interactive nonlinear design. Several such modeling approaches, commonly referred to as "harmonic balance" methods, are solved in the frequency domain but employ time-domain descriptions for the active-element nonlinearities. Thus nonlinear behavior of the total microwave circuit is obtained. Development of a CAD model of this type is relatively straightforward and begins with the linear FET model described previously (Fig. 5.1).

A nonlinear model may be formulated by taking the linear elements of the small-signal model and combining them with nonlinear descriptions of FET drain current, drain-to-source breakdown current, and gate–diode I–V characteristics [5.10]. Simulation is not complete unless the effects of input and output terminations as well as feedback are included. The resulting model is shown in Fig. 5.11. It should be noted that the elements R_{ds} is included in the nonlinear description of the drain current I_d and not as a separate model linear element; thus the effects on R_{ds} due to bias conditions are included.

The quantities Z_G and Z_L are the equivalent input and output impedances of the generator and load networks, which are usually determined from their two-port S-parameter descriptions, while the quantities Y_f and Z_{sf} are the two-port and one-port S-parameter descriptions of the shunt and series feedback networks, respectively. These S-parameter descriptions must include data at both fundamental and harmonic frequencies, and can be obtained from either laboratory measurements or computer simulation. The sources V_{GG} and V_{DD} contain both dc and RF components, and are selected to simulate actual circuit conditions. The nonlinear element values of the model are obtained solely from physical measurements.

The greatest contributor to the nonlinear characteristics of the FET is the drain current as a function of gate and drain voltage. The useful portion of the current at the FET drain terminal, which can be delivered to a load, is limited by the gate-to-drain breakdown current (I_b) and by the gate voltage swing, which is bounded by gate conduction and pinch-off. In this model the currents I_d, I_b, and I_g are

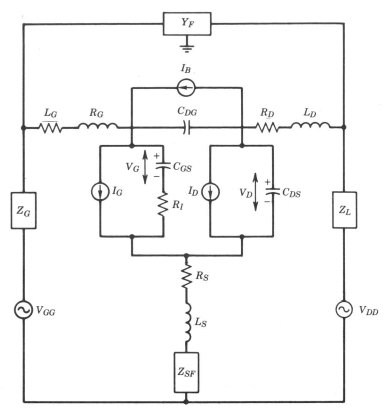

Figure 5.11   FET and circuit nonlinear model.

assumed to be functions of the internal gate voltage V_g and the drain voltage V_d. Thus low-frequency measurements can be used to determine the time-dependent characteristics of these quantities. Typical characteristics for I_d, I_b, and I_g for a 300-μm FET are shown in Fig. 5.12.

The drain current characteristics can be determined by measuring the current entering the FET as a function of gate and drain voltage. The burnout problems encountered during static measurements of this type can be eliminated by pulsing the drain voltage. Breakdown currents are also obtained if the drain voltage is made sufficiently large so that the FET operating point is in the drain breakdown region. However, pulsed I–V measurements of this type do not heat the FET to normal channel temperatures; hence an abnormally high value of g_m is obtained, and the pulses used are usually too short to yield an accurate value of R_{ds}. Although the lack of heating can be overcome by statically heating the FET itself, trapped charge at the surface of the FET still causes erroneous values of R_{ds} to be obtained with this method. The change in R_{ds} as a function of frequency is illustrated in Fig. 5.13.

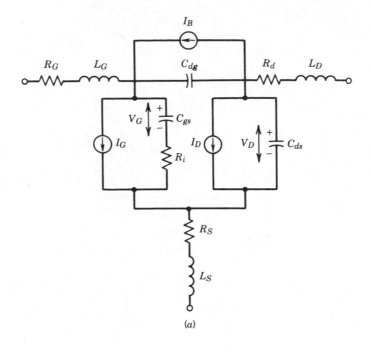

(a)

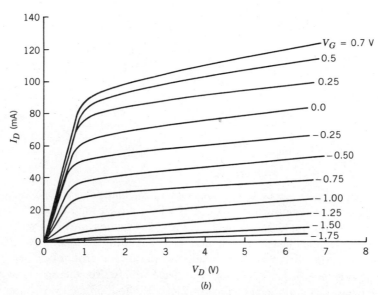

(b)

Figure 5.12 Typical characteristics for a 300-μm FET: (a) linear element values; (b) drain current characteristics; (c) drain-to-gate breakdown current as a function of terminal voltages; (d) forward gate diode I–V characteristic.

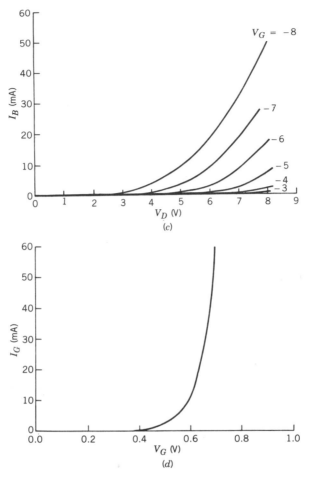

Figure 5.12   (*Continued*)

To obtain a more accurate representation of R_{ds} and g_m of an operating FET, a measuring system employing low-frequency RF signals can be utilized [5.11]. The 1-MHz measurement test set depicted in Fig. 5.14 applies large half-wave rectified RF voltages to both the gate and drain terminals. During evaluation, the FET drain current is continuously monitored while the phase relationship between gate and drain signals is varied. Hence, for every phase difference, the operating point of the FET traverses a different continuous closed contour on the FET's drain current *I–V* plane. A typical operating point path is shown in Fig. 5.15. The exact shape of the contour depends on the phase relationship between the gate and drain voltages. By measuring a sufficient number of contours, the entire drain current characteristic can be mapped as shown in Fig. 5.16. Computer algorithms are then used to generate FET drain current characteristic curves (Fig. 5.12), in which R_{ds}

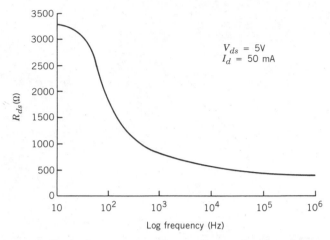

$$V_{ds} = 5V$$
$$I_d = 50 \text{ mA}$$

Figure 5.13 Drain-to-source resistance R_{ds} as a function of frequency.

and g_m can be obtained from the digitized data. Gate-to-drain breakdown current characteristics can also be constructed by monitoring gate current and utilizing sufficiently large RF voltages. A comparison between the FET characteristics obtained from conventional pulsed I–V measurements and rectified sine-wave measurements is shown in Fig. 5.17. A CAD FET model can use the data directly in look-up-table form, or analytical functions, describing FET characteristics, can be formulated.

Analysis of the nonlinear model of Fig. 5.11 begins by applying Kirchhoff's laws in the frequency domain, resulting in a coupled, complex simultaneous algebraic equation set with V_g and V_d as the independent variables. For each Fourier component (dc, signal, and harmonics), there are two complex equations [5.10]

$$AV_g + BV_d = C \tag{5.9}$$

and

$$DV_g + EV_d = F \tag{5.10}$$

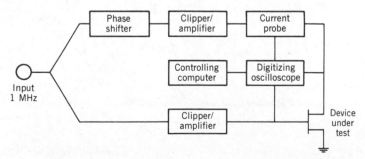

Figure 5.14 Measurement apparatus configuration for 1-MHz RF FET evaluation.

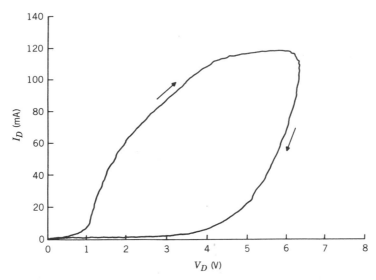

Figure 5.15 Typical FET operating path obtained during low-frequency RF evaluation.

The quantities A, B, D, and E are linear combinations of model element values and frequency, while the quantities C and F are functions of circuit values, voltage generators, and the nonlinear currents I_d, I_b, and I_g. These nonlinearities are also functions of V_g and V_d; thus (5.9) and (5.10) are best solved with a nonlinear iterative numerical solver.

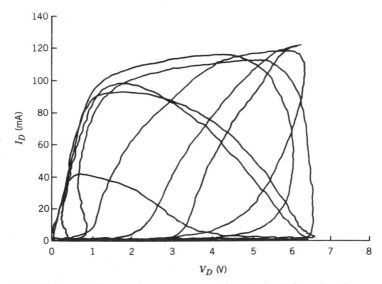

Figure 5.16 Entire FET operating region mapping obtained from low-frequency RF evaluation.

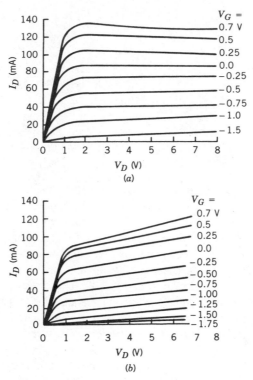

Figure 5.17 Comparison of FET drain current characteristics obtained from (*a*) pulsed *I–V* and (*b*) rectified sine-wave evaluation.

The solution of the harmonic balance problem above, illustrated in Fig. 5.18, begins by solving a linearized version of (5.9) and (5.10). Initially, the forward gate and breakdown currents are set to zero and the drain current is calculated by assuming a linear value of conductance. As each iteration is done, Fourier transforms produce time-domain functions for V_g and V_d which are used to generate the nonlinear current functions I_d, I_b, and I_g. Frequency-domain solutions for the nonlinear elements are obtained by using an inverse transform and the total circuit problem is again solved. The iteration process continues until a self-consistent set of equations are obtained form the nonlinear solver. This self-consistent equation set is checked by applying Kirchhoff's laws for each circuit loop (voltage) and at each circuit node (current). The consistency test must be applied at all frequencies; dc, signal, and harmonics—hence the name "harmonic balance."

With most practical circuit problems, sufficient accuracy is obtained when the dc component, signal, and five harmonics are used in the solution. Using fewer than three harmonics can lead to serious errors in the calculation of optimum load impedance, power output, and gain. Care must also be taken when describing the frequency-domain characteristics of the source and load impedances, since they are evaluated through the highest harmonic.

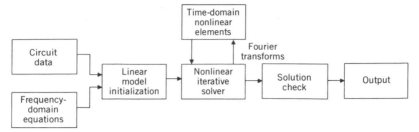

Figure 5.18 Large-signal model solution process.

The following circuit data illustrate the accuracy obtainable when using the characterization technique and harmonic balance methods described above. In Fig. 5.19 the performance of a 300-μm GaAs FET connected in a 50-Ω system is depicted. The saturation and harmonic characteristics of the amplifier are shown as the FET is driven at various power levels. A monolithic two-stage amplifier (Fig. 5.20a) was analyzed using a multi-FET model, and the measured versus modeled compression curve is shown in Fig. 5.20b. The model can be extended to accommodate multiple-drive signals at various frequencies [5.12]. A model of this type was used to calculate the performance of a single-gate FET mixer with the gate and source terminals driven. The mixer's measured versus computed performance is shown in Fig. 5.21.

Although the computer solution is done in both the time and frequency domains, all circuit quantities are available from the model. For example, because all the currents and voltages are known at each point in the circuit, voltage waveforms may be extracted or S-parameters may be calculated. Also, since the load network can be numerically varied, the effects of harmonic terminations can be studied or

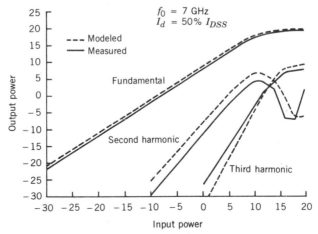

Figure 5.19 Modeled versus measured output harmonic content for a 300-μm FET in a 50-Ω system.

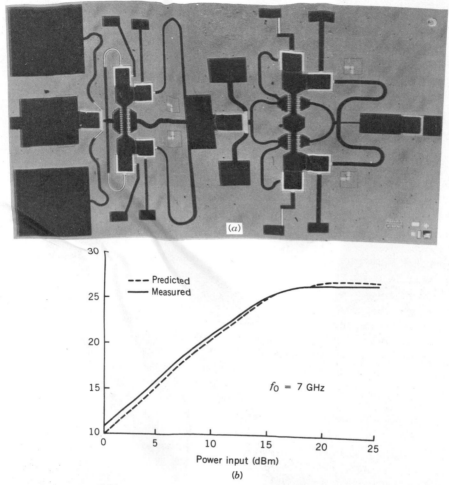

Figure 5.20 Monolithic two-stage power amplifier (400 mW) used in nonlinear analysis: (a) chip photograph (Courtesy of Texas Instruments); (b) comparison between measured and predicted output power performance.

optimum load contours, similar to the contours obtained from load-pull measurements, can easily be generated for any set of circuit conditions.

The nonlinear behavior of other active devices, such as Si and GaAs bipolar transistors, can be modeled using the same techniques. A simplified nonlinear bipolar model that includes both the forward and reverse transistor current sources is shown in Fig. 5.22a [5.13]. The diodes D1 and D2 are used to determine some of the device breakdown characteristics and forward base–emitter voltage. The charge storage effects are included in the elements C_{be} and C_{cb}.

The model can be solved by using either time-domain or harmonic balance techniques, provided that the characterisics of the current sources are known. All

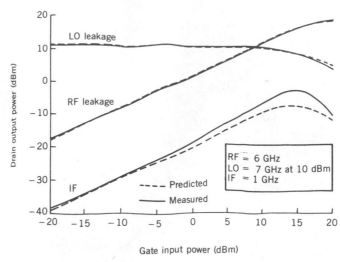

Figure 5.21 Measured versus predicted performance of 300-μm FET mixer in a 50-Ω system.

the current sources shown in the model are nonlinear and are functions of the internal circuit voltages as well as controlling currents, where applicable. The current sources representing D1 and D2 can be modeled in the forward direction by the diode equation controlled by the voltage across their associated capacitances and by polynomials or piecewise linear approximations in the reverse direction. They can also be modeled using look-up-table approaches based on physical measurements. The remaining current source characteristics can also be determined from physical measurements or analytically.

The small-signal model of Fig. 3.2, which applies to transistors operating at a fixed bias for a drive level remaining in the linear region, can be expanded and modified for nonlinear operation (Fig. 5.22b). This can be accomplished by using the time-domain capabilities of SPICE, which calculates all voltages and currents in a model versus time.

The regions of nonlinearity are best described by Fig. 5.23, which gives five limitations on linear operation for a common-emitter bipolar transistor. The turn-on nonlinearities occur at low voltages and low currents. The Kirk effect [5.14] limits high current operation when the collector current density becomes comparable to the doping level in the collector region, which causes the electric field to change in a direction that widens the base. Avalanche breakdown will limit the region of high collector–emitter voltage. The fifth limitation is nonlinear performance due to temperature effects such as thermal runaway and changes in $R_e = kT/QI_E$. Each of these five effects must be included in a large-signal SPICE model.

The large-signal SPICE equivalent circuit given in Fig. 5.22b should be compared to the small-signal distributed T-equivalent circuit of Fig. 3.2. The distributed collector–base capacitance has been replaced by the SPICE model of a

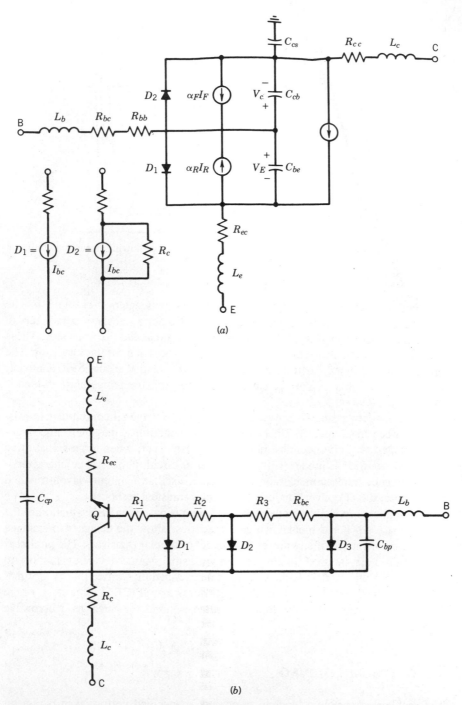

Figure 5.22 Nonlinear BJT models: (*a*) simple nonlinear model suitable for harmonic balance or SPICE analysis; (*b*) large-signal SPICE model.

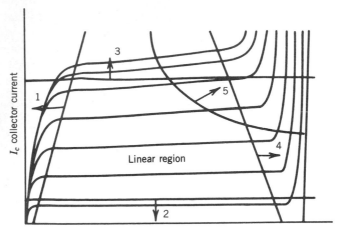

Figure 5.23 Bipolar transistor current–voltage characteristics showing linear and nonlinear regions.

unit diode, which is described by the seven diode parameters in Table 5.1. The intrinsic unit bipolar transistor is given by the 18 bipolar transistor parameters in the table. A unit of area is 1×1 μm for both the diode and the transistor. These parameters are determined by the semiconductor processing for the transistor. The temperature coefficients of the resistor should be included in the SPICE model. The device performance may be calculated versus ambient temperature, which is an important feature.

SPICE model parameters are given in Table 5.1 for the silicon bipolar transistors described in Chapter 3. Under small-signal conditions, the small- and large-signal models will give identical performance. However, when large-signal drive is applied, the SPICE model will yield nonlinear circuit performance characteristics such as compression, harmonic distortion, oscillator output waveforms and impedances, mixer conversion loss, and circuit transient response.

The nonlinear modeling method above, which is based on RF evaluation of the active device's $I–V$ and breakdown characteristics, allow the designer to evaluate fully the performance of any power field-effect or bipolar transistor. The nonlinear data obtained can be used, in conjunction with small-signal device evaluation, to formulate analytic models such as the model described by Curtice [5.15] for time-domain and harmonic balance simulation (Microwave SPICE, Libra, etc.), or the data can be loaded directly into a nonlinear simulator such as Microwave Harmonica.

5.2 OPTIMUM LOADING

The problem of optimum loading, to achieve a specified performance parameter such as maximum gain or power, becomes more difficult as circuit nonlinearities

TABLE 5.1   Large-Signal SPICE Model Parameters for Microwave Bipolar Transistors

Symbol	Definition	Units	Value for AT-41400 and AT-60500
	Transistors (Q)		
IS	Junction saturation current	A	1.65×10^{-18}
BF	Maximum forward beta		100
BR	Maximum reverse beta		5
NF	Current emission coefficient		1.03
VA	Early voltage	V	20
IK	Corner for high-current beta roll-off		0.10
ISE	Base–emitter leakage saturation current	A	5×10^{-15}
NE	Base–emitter leakage emission coefficient		2.5
CJE	Base–emitter zero-bias junction capacitance	F	1.8×10^{-15}
PE	Base–emitter built-in potential	V	1.01
ME	Base–emitter junction grading factor		0.60
FC	Forward bias depletion capacitance coefficient		0.50
TF	Ideal forward transit time	ps	12
XTF	TF bias dependence coefficient		4
VTF	TF dependency on V_{bc}	V	6
ITF	TF dependency on I_c	A	0.3
PTF	Excess phase at $1/(2\pi TF)$	Degree	35
XTB	Beta Temperature coefficient		1.818
	Diodes (D1, D2, D3)		
IS	Saturation Current	A	1×10^{-25}
CJO	Zero-bias junction capacitance	F	2.45×10^{-16}
VJ	Junction potential	V	0.76
M	Junction grading coefficient		0.53
FC	Forward-bias depletion capacitance coefficient		0.50
BV	Reverse breakdown voltage	V	45
IBV	Reverse breakdown current	A	1×10^{-9}

increase. In the preceding section, several modeling and simulation methods were described, which varied from simple linear FET approximations to elaborate CAD simulations. The circuit performance obtained when using the previous modeling methods to determine proper device loading conditions, as they relate to maximizing power output performance, will now be described.

The simplest method commonly used to determine the optimum load impedance (Z_{opt}) for a GaAs FET is the load-line approach. In this method the operating point

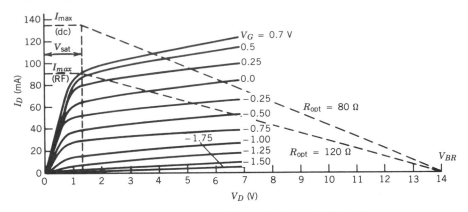

Figure 5.24 Calculation of R_{opt} using dc and RF load lines.

of the FET is approximated by a straight line extending from the maximum current point, on the drain current axis, to the maximum voltage (breakdown) point, on the drain voltage axis, on the FET characteristic curves. If dc measurements, such as a curve tracer, are used to obtain FET drain characteristics, the load resistance obtained will be quite different from the value determined from the RF drain characteristics. Two such load lines are illustrated in Fig. 5.24. The first line is drawn from I_{max}, obtained from pulsed I–V measurements (dc), to V_{br} and yields a load resistance value 80 Ω. The second line is drawn from the true value of I_{max} (RF) to the same value of V_{br}. When the slope of the RF load line is calculated, the value obtained for R_{opt} is 120 Ω. A load error of this magnitude can seriously degrade the power output performance of any amplifier.

The computation of R_{opt} using an RF load-line approach can be further refined by absorbing the FET and mounting parasitics into the output matching network. The parasitic absorption effectively places the value of R_{opt}, which was calculated from an RF load line, directly across the FET current source, thus simulating the results obtained for Z_{opt} from the load-pull method of Fig. 5.10. This technique is illustrated in Fig. 5.25.

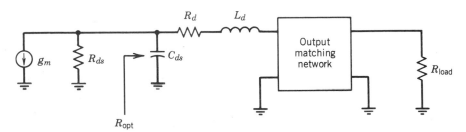

Figure 5.25 FET parasitic absorption into output-matching network.

Absorbing the device parasitics in the manner described above is a very effective method in designing broadband amplifiers when load-pull data or a nonlinear simulator are unavailable. In practice, about 1 dB of output power is sacrificed using this approach. Improvements in amplifier efficiency, accomplished by properly selecting harmonic terminations, also cannot be accomplished with the simple technique described above.

The most accurate approach in determining optimum load impedance is with the use of a nonlinear simulator, provided that the active device is properly modeled. Constant power contours can be generated by varying the output impedance terminating the FET (or active device) and observing the calculated circuit power output. The effect of other circuit parameters, such as dc bias, feedback, and input terminating impedance on power output should be included in the search for Z_{opt}. The maximum power output (P_{max}) and several constant power contours for a simple amplifier consisting of a FET with a 50-Ω input termination are shown in Fig. 5.26a. The FET used in the analysis is the device described in Fig. 5.12.

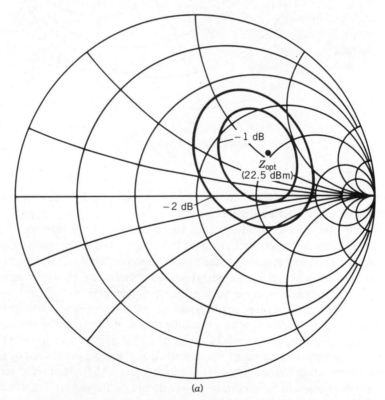

(a)

Figure 5.26 Nonlinear power output performance of typical FET: (a) constant power contours obtained from harmonic balance simulator; (b) FET amplifier power output for various load terminations.

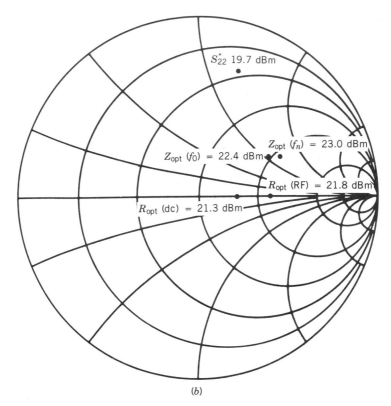

(b)

Figure 5.26 (*Continued*)

It should be noted that 50-Ω harmonic terminations were used in the analysis, thus simulating an active load-pull measurement.

Optimizing the load termination at the signal harmonics can enhance amplifier efficiency and power output. In the case of broadband design, the designer usually does not have flexibility in selecting the proper harmonic terminations since several harmonics can fall within the amplifier's operating bandwidth. However, harmonic termination optimization can be very effective for amplifiers designed to meet narrowband phased-array radar applications or in the efficiency enhancement of class B power designs. To illustrate the importance in selecting the correct FET terminating impedance, an analysis of the same FET amplifier used in the example above was conducted using all of the described R_{opt} and Z_{opt} calculations above, including terminating the amplifier with a load equal to S_{22}^{*}. Harmonic terminations were also adjusted to maximize power output performance. The results are shown in Fig. 5.26b. As can be seen, a good approximation to the true value of Z_{opt} can be had by using the RF load-line method and parasitic absorption techniques.

5.3 SINGLE-STAGE DESIGN

The synthesis techniques for power GaAs FET single-stage amplifiers closely parallels the method employed in small-signal design with the exception of several major areas. The problem of gain shaping to achieve a specified frequency response, which typically means gain flattening, is even more constrained when designing power amplifiers, because the output matching network must be designed for optimum power output performance. Terminating the FET with the optimum power load also implies that the small-signal output return loss of the finished amplifier will be poor. Since no gain shaping and impedance matching can be accomplished with the output-matching network, the entire burden of gain flatness falls on the input network. Unless lossy matching or feedback techniques are employed in the design of the input network, the input reflection coefficient of the resulting amplifier will also be poor. If the operating bandwidth is sufficiently narrow, a stand-alone amplifier stage can exhibit low input VSWR and flat gain performance. However, when broadband performance is desired, two identical amplifier stages can be combined with the use of 90° quadrature couplers (balancing) in order to achieve a low-input and low-output VSWR, as discussed in Section 4.1.6.

Drain and gate bias networks are usually ignored in small-signal design but can be very troublesome when realizing power gain stages. Special consideration must be given to the drain bias decoupling network, because drain currents of several amperes are not uncommon when using large power GaAs MESFETs. Amplifier stability requirements, especially at very low frequencies, further complicate bias circuit design.

It should also be remembered that the fusing current of 0.001-in.-diameter gold wire is less than 1 A; hence hybrid bias circuits can be comprised of several parallel bond wires. The fusing current problem becomes intensified in MMIC design since high-impedance microstrip decoupling lines are less than 12.5 μm wide. Similar to hybrid design, multiple-drain bias networks are often employed to supply sufficient current to high-power FETs.

The problems encountered in gate biasing are somewhat different. Although out-of-band stability requirements must be considered in gate bias network design, gate current requirements, which do not exist in small-signal design, cannot be forgotten. At low drive levels and with reverse bias applied to the FET gate, no gate current is present. However, at high drive levels and with the same reverse bias applied, the RF signal can easily forward bias the FET for a short portion of the RF cycle. During these forward-bias excursions, considerable gate currents can be developed, even though the average current might only be several milliamperes. Large current spikes present on the gate bias line can drastically alter the operating point of the FET. These shifts in operating point will degrade the amplifier's power output performance. Most bias problems, however, can be eliminated with a "stiff" or well-regulated negative voltage supply. Bias and matching network realization problems are easily illustrated with the following example.

A single-stage amplifier, designed to operate within the 9- to 10-GHz frequency band, can be synthesized using the 300-μm FET described in the preceding section. This FET, when properly biased and terminated, can develop approximately 22.5 dBm of output power when driven into saturation. However, excellent power output and efficiency performance can be obtained only if the amplifier has optimum harmonic terminations.

The design begins by synthesizing the output load network [5.16–5.18]. The network must transform the 50-Ω load impedance to the optimum load impedance (Z_{opt}) required by the FET. The optimum load impedance, which is shown in Table 5.2, was determined with the aid of a nonlinear solver for the fundamental frequency as well as for the second and third harmonics. Using these target values for the terminating impedance, a lumped-element matching network was synthesized using the method developed by Levy. The bandpass network, which employs a shunt inductor and series capacitor to facilitate biasing and decoupling, is shown in Fig. 5.27. The lumped-element realization was then converted to a distributed structure by employing high-impedance transmission lines for inductors, low-impedance transmission lines for the output parallel L-C resonator and a monolithic capacitor for the series-matching capacitor [5.19]. A quarter-wavelength open-circuit resonator (third harmonic) was used to simulate the series L-C network at the FET end of the network. Computer optimization was then used to trim the network performance. The resulting output-matching network is shown in Fig. 5.28. The load impedances presented to the FET are summarized in Table 5.3. As can be seen in the table, the load obtained after optimization using nonideal elements is very close to the desired values presented in Table 5.2. A Smith chart plot of the target versus actual values is also shown in Fig. 5.29.

The input-matching network was synthesized in a similar manner in lumped-element form. The source and load impedances used were 50 Ω and S_{11}^* of the FET/load network combination, respectively. The network was then converted to microstrip monolithic topology. Amplifier gain flatness was achieved by optimizing only the input network of the complete circuit. The resulting small-signal gain is shown in Fig. 5.30.

Although the design above employed harmonic terminations, a variety of output network topologies were analyzed to aid in illustrating the improvements in efficiency and power performance obtainable with this design approach. A summary of key performance variations as a function of output network design philosophy is presented in Table 5.4. It should be remembered that $Z_{opt}(f_0)$ was synthesized

TABLE 5.2 Optimum Load Impedance at Fundamental and Harmonic Frequencies

f_0 (GHz)	Z_{opt} (desired)		
	Fundamental	Second Harmonic	Third Harmonic
9.0	0.490 $\underline{/23°}$	0.999 $\underline{/80°}$	0.999 $\underline{/180°}$
9.5	0.520 $\underline{/28°}$	0.999 $\underline{/82.5°}$	0.999 $\underline{/180°}$
10.0	0.560 $\underline{/31°}$	0.999 $\underline{/85°}$	0.999 $\underline{/180°}$

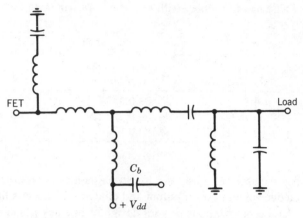

Figure 5.27 Lumped-element bandpass output network.

to simulate the optimum load impedance obtained from a load-pull measurement setup (harmonics terminated in 50 Ω).

The optimum network synthesized is substantially more complex than a simple two-element low-pass matching structure, but the added power and efficiency usually outweigh this drawback. It should also be noted that the networks presented, which were not designed to load the harmonics with optimum terminations, were also not allowed to degrade amplifier performance. Degradation in performance

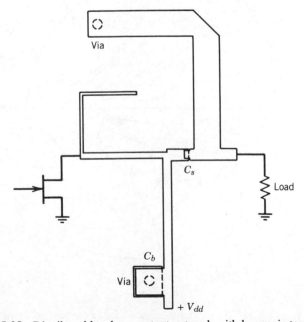

Figure 5.28 Distributed bandpass output network with harmonic terminations.

TABLE 5.3 FET Load Impedance with Monolithic Distributed Bandpass Matching Network

f_0 (GHz)	Z_{load} (actual)		
	Fundamental	Second Harmonic	Third Harmonic
9.0	$0.510 \underline{/25°}$	$0.749 \underline{/134°}$	$0.961 \underline{/-166°}$
9.5	$0.460 \underline{/28°}$	$0.752 \underline{/70°}$	$0.961 \underline{/177°}$
10.0	$0.480 \underline{/29°}$	$0.830 \underline{/0°}$	$0.954 \underline{/159°}$

can occur when the second, third, and fourth harmonics are terminated in such a manner as to reduce circuit power output and efficiency. This condition can easily occur with linear design methods or with design techniques which concentrate on the fundamental frequency of operation, such as most load-pull techniques.

A nonlinear solver [5.10] was also used to evaluate several popular forms of biasing, such as resistive dividers on the negative gate dc supply line and conventional self-biasing in the FET source path. In the self-biasing approach the source

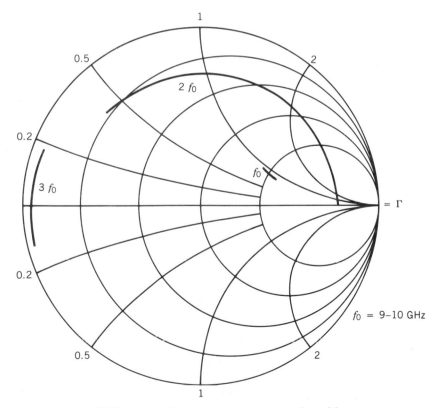

Figure 5.29 Desired load impedance as a function of frequency.

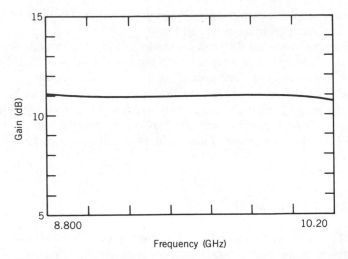

Figure 5.30 Calculated small-signal gain performance of single-stage amplifier.

resistor was adjusted so that the small-signal drain current was exactly the same as with an ideal dual-power-supply scheme. However, as the input power to the amplifier was increased, the drain current decreased and the saturated output power achieved was approximately 0.5 dB less than that obtained in the example above. The source resistor was then adjusted so that the FET drain current, under large signal conditions, was equivalent to the original small-signal case. Although the power output of the amplifier increased, it could never be restored completely. Amplifier efficiency was also markedly reduced.

The situation was similar when a high dc impedance (1 kΩ) gate bias supply was employed. The power output of the amplifier dropped several decibels when it was driven near saturation. However, no amount of bias voltage adjustment

TABLE 5.4 Key Performance Variations of a Single-Stage Amplifier as a Function of Output Matching Network Characteristics

f_0 (GHz)	Z_{load} (actual)	Z_{opt} (f_0)	R_{opt} (dc)	R_{opt} (RF)	Z_{opt} (ideal)	S_{22}^*
		Power Output (dBm)				
9.0	22.6	22.0	21.4	21.9	23.0	19.8
9.5	22.8	22.0	21.4	21.9	23.0	19.8
10.0	22.8	21.7	21.3	21.8	22.9	19.7
		Power Added Efficiency (%)				
9.0	43.3	37.0	27.2	34.0	48.0	19.8
9.5	44.0	37.3	27.9	34.9	47.7	17.5
10.0	45.1	36.5	26.3	31.1	46.3	15.6

could restore amplifier performance. When the power supply impedance reduced to 50 Ω, amplifier performance was still poor.

The example above has shown that considerable improvement in power performance and efficiency can be obtained when the output load network is designed to optimally terminate the fundamental and harmonic frequencies. Matching techniques, which are commonly used in broadband design, can be employed successfully in designing practical narrowband networks, particularly if monolithic realization is desired. However, careful bias design must be used in conjunction with the RF techniques above so that the final amplifier will perform to expectations.

5.4 MULTISTAGE DESIGN

Before the widespread availability of computer-aided analysis and synthesis, microwave multiple-stage power amplifiers were usually composed of cascaded balanced single-stage gain blocks. Although this approach allowed for ease of design, optimum performance and size could not be achieved. Presently, the design philosophy is shifting toward direct cascading of gain stages, which is due in part to the industry's demand for smaller, more efficient circuits, accurate modeling techniques, and to the availability of numerous CAD amplifier synthesis packages. These commercially available computer programs are excellent in designing small-signal amplifiers, but provide only starting values for multistage power amplifier design. The design problem is further compounded when broadband performance is required.

These complications are easily illustrated by examining the general design procedure of the small-signal two-stage amplifier depicted in Fig. 5.31. The amplifier consists of three basic networks: (1) input network, (2) interstage network, and (3) output network. Typically, the input and output networks are designed to provide an impedance transformation between the FET (or other active device) and measurement system, while the interstage network is used to provide the necessary gain shaping required to compensate for the 6-dB/octave roll-off of each FET. However, this technique cannot be applied directly to the design of power amplifiers.

As in the case of single-stage design, the output-matching network, for large-signal applications, must be designed to terminate the FET with the optimum load

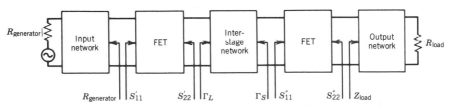

Figure 5.31 Representative two-stage amplifier. Courtesy of Texas Instruments.

impedance at both the fundamental and harmonic frequencies. Harmonic terminations unfortunately are not always possible in broadband designs because harmonics can easily fall within the operating bandwidth of the amplifier. Also, when the output network is designed for large-signal performance, the amplifier's output VSWR (small signal) will tend to be poor, although RF feedback may aid in reducing, but not eliminating some of the mismatch problems.

The design of the interstage network is similarly hampered because it must be designed for both gain shaping and power transfer performance between amplifier stages, with the latter condition taking precedence, especially at higher frequencies [5.20–5.22]. At the upper band edge, where the active-device gains are at their lowest value, sufficient drive power must be available at the output FET gate. Thus the size of the driver FET must be sufficiently large to supply this power. At the power band edge the output FET will exhibit more gain, hence requiring less drive power.

However, if the interstage network is designed to provide gain shaping by mismatching the output of the driver and the input of the output stage, the driver FET may be loaded in such a manner as to prevent it from delivering sufficient drive power. Even if the mismatch loss from the interstage network is small (less than 1 dB), the power transfer loss can easily be several decibels, due to the fact that the driver FET is not terminated with an impedance near Z_{opt}. Unfortunately, amplifiers designed using linear simulation methods usually exhibit excellent gain performance, but can easily have severe power output performance shortfalls, especially if the operating bandwidth is large. The input network of the two-stage power amplifier, if it is of the lossless form, must then provide the remainder of the frequency response contouring required by the total amplifier at the expense of input VSWR. Thus the completed amplifier can exhibit flat gain and power performance for broadband operation, but may exhibit unmatched terminal impedances.

The description above illustrates that the most difficult portion of designing multiple-stage amplifiers is the interstage network synthesis. A variety of network synthesis techniques have been described in the literature which were developed specifically to aid in the design of broadband amplifiers. These synthesis methods, which can be either low pass or bandpass approaches, provide gain sloped responses and generate matching networks for complex-to-real or complex-to-complex load impedances. The synthesized networks are usually of lumped-element form, which are later converted to distributed form. Excellent gain flatness performance for small-signal operation is obtained with these synthesis methods, such as the real frequency technique proposed by Carlin [5.23] or the bandpass synthesis technique described by Mellor [5.24]. Although synthesis methods produce networks with the required transfer function to achieve the amplifier's desired frequency response, the element values are often unrealizable. The terminal impedances are also unconstrained; thus the resulting amplifier may exhibit poor power output performance. However, broadband synthesis techniques do provide excellent starting values for matching network elements.

The problems in designing multistage amplifiers with lossless networks can be eliminated or at least significantly reduced by employing either (1) RF feedback techniques or (2) some form of lossy matching networks. At frequencies where substantial device gain is available, RF feedback can provide excellent gain shaping, stability, and impedance control while significantly reducing distortion. Transformer, resistive series and shunt, and bootstrapping are just a few common feedback methods that can be employed successfully at lower microwave frequencies.

In the upper end of the microwave spectrum, the benefits of RF feedback become diminished since there is little available excess gain. It then becomes necessary to employ other techniques, such as loading, to control impedance variations. Resistive loading techniques are quite effective in achieving input and interstage networks with upward-sloped gain performance characteristics that still exhibit well-controlled terminal impedances.

The synthesis of loaded (lossy) matching networks is considerably more empirical than conventional network synthesis and is usually accomplished through optimization methods. Several basic element topologies [5.19], which are composed of R's, L's, and C's, can be incorporated into input and interstage designs to dissipate power at the low-frequency end of the operating bandwidth while maintaining low loss at the higher frequencies. Typical topologies, with their respective frequency response characteristics, are shown in Fig. 5.32. The simple networks shown in Fig. 5.32a–c can easily be combined in an interstage network to bound the impedance presented to the driver stage. For example, when the operating bandwidth is large, high-order networks are required to provide the impedance transformation and gain contouring. However, the input impedance (driver end) of the network can traverse a path near the rim of the Smith chart as the operating frequency approaches the low end of the operating band. Hence these frequency-dependent series and shunt elements tend to confine the impedance seen by the driver FET to a region closer to the impedance at the high end of the operating band, while still providing the prescribed frequency response contour. The equalizer element in Fig. 5.32d is very useful in eliminating gain peaks that may arise through narrow frequency bands. The network depicted in Fig. 5.32e can be designed with a predescribed attenuation versus frequency performance for bandwidths exceeding several octaves and exhibits real terminal impedances at dc, thus aiding amplifier low-frequency stability. The foregoing networks, combined with bandpass-matching techniques, usually provide the best compromise between passband ripple performance, amplifier stability, and circuit complexity. Trade-offs of this type are illustrated in the following two-stage power amplifier design example. The design goals for a 6- to 18-GHz monolithic two-stage amplifier are shown in Table 5.5.

The first step in the design process is to determine the required output FET size so that the output power specification can be obtained. The FET can easily be sized by employing an RF load-line analysis. If we assume an RF breakdown voltage of 18 V and a pinch-off voltage of −5 V dc, the maximum RF voltage swing would be approximately 10.5 V. Hence the peak-to-peak RF current required to develop 1 W at the load must be a minimum of 0.76 A. But because of output circuit losses and the fact that the amplifier power output specification is at the

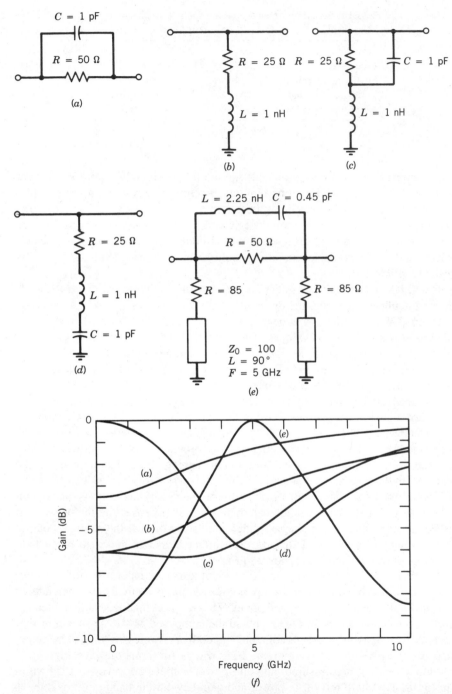

Figure 5.32 Typical amplitude equalization networks (a)–(e); respective frequency response characteristics (f).

TABLE 5.5   Performance Goals for a Two-Stage Monolithic 6- to 18-GHz Power Amplifier

Power output	1 W
Frequency response	6 to 18 GHz
Small-signal gain	7 dB
Temperature range	-55 to $+85°$C
Amplitude flatness	$< \pm 1.0$ dB

1-dB compression point and not at full saturation, considerably more RF current capacity in the output FET will be required. A current increase of approximately 50% should yield ample margin. With this in mind, an output FET size of 3000 μm was selected. The power output, neglecting circuit losses, with such a device should be approximately 32 dBm when fully saturated; thus a 30-dBm power specification at the 1-dB compression point should be feasible provided that the FET is correctly terminated.

The next step in the design is the output network synthesis. Based on RF load-line and nonlinear analysis, the optimum load impedance (Z_{opt}) for the FET was determined as a function of frequency, and the output network was then designed using bandpass synthesis techniques. The lumped-element network model was then converted to distributed form. The load impedance presented to the FET, after circuit optimization, is shown in Fig. 5.33 and the lumped-element and distributed networks are shown in Fig. 5.34.

The interstage design can then begin by synthesizing a matching network (complex source to complex load) using the S_{11} of the output FET and terminated output network as the load impedance, and Z_{opt}^* of the driver FET as the source impedance [5.23, 5.25]. The resulting network provides maximum power transfer between driver and output stage with no gain shaping. Gain shaping can now be provided by adding lossy elements to the interstage network and adjusting the reactive elements so that the driver is still terminated for maximum power output performance. Finally optimization can then occur after converting the network to distributed form and performing amplifier analysis at or near the saturated output power condition with proper dc bias. The lumped-element and distributed final interstage network models are shown in Fig. 5.35.

The final step in the design process is to synthesize the input network. This can be accomplished by employing bandpass filter synthesis methods and then adding loss, thus improving stability and input VSWR. The final distributed-element amplifier model is shown in Fig. 5.36 and the completed MMIC is shown in Fig. 5.37. As can be seen in the photograph, the networks were designed as symmetrical parallel pairs due to the extreme FET widths (in terms of wavelength) and low impedances. The measured gain and power output performance is shown in Fig. 5.38. The sharp roll-off in power and gain beyond the band edges is due, as expected, to the high-order networks that were employed in the design. These

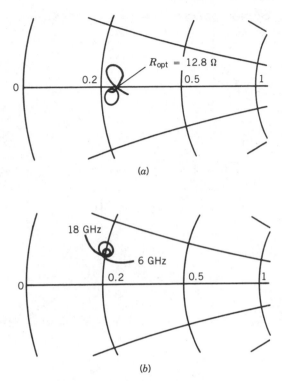

Figure 5.33 Optimum load impedance (actual and target) presented to output FET: (*a*) C_{ds} absorbed into output network ($\sim R_{opt}$); (*b*) actual load presented to FET (Z_{opt}).

high-order networks are also the reason why the in-band gain and power output performance is flat.

The design approach discussed above, although more empirical than small-signal design methods, does provide excellent results. However, it also requires some design experience and insight into network synthesis and topology limitations. Other nonlinear amplifier characteristics, such as third-order distortion, is best determined via nonlinear simulation. Common "rules of thumb" used to determine various distortion characteristics can easily give the design engineer erroneous results for broadband designs, especially when multiple stages are involved. As an example, the rule which states that the third-order intercept point is 10 dB greater than the amplifier's 1-dB compression point could yield a ±10-dB error for a single-stage single-gate power amplifier, depending on the harmonic terminations, device nonlinearity, and bias point. The error could be 20 or 30 dB for a dual-gate variable-gain design when the amplifier is biased for reduced gain operation. It should also be noted that the harmonic termination influences amplifier distortion products, a problem that is common in multistage designs with nonresistive stage coupling.

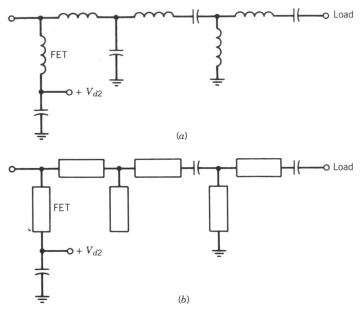

Figure 5.34 (a) Lumped-element and (b) distributed-element output network equivalent circuits.

5.5 POWER DISTRIBUTED AMPLIFIERS

Since the 1940s distributed or traveling-wave techniques have been used to design very broadband electron-tube amplifiers [5.26, 5.27]. However, it is only recently, with the availability of good-quality microwave GaAs FETs, that the distributed amplifier has again become popular [5.28–5.31]. The rebirth in popularity is in part due to the fact that excellent bandwidth performance is obtainable, because the input and output capacitances of the active devices are absorbed in the distributed structures. The resulting amplifiers also exhibit very low sensitivities to process variations and are relatively easy to design and simulate. In addition, a great deal of design flexibility is possible since the number of devices, device size (gate periphery), transmission-line characteristic impedances, and amplifier upper cutoff frequency can all be varied to meet a particular component or chip design specification. The simplified amplifier structure shown in Fig. 5.39 can be used to help illustrate these design concepts.

As can be seen, the amplifier is composed of two main artificial transmission-line sections consisting of series inductances and shunt capacitances, which are usually supplied by the FET parasitics. If the FET model of Fig. 5.40 is substituted into the amplifier above, two constant-k transmission lines, which have different cutoff frequencies and attenuation characteristics, result (Fig. 5.41). Since these lines are heavily loaded by the FET parasitic resistances R_i and R_{ds}, the number of active-device sections cannot be added indefinitely because the attenuation along

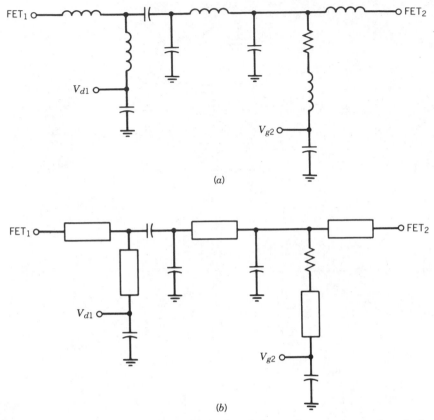

(a)

(b)

Figure 5.35   (a) Lumped-element and (b) distributed-element interstage network equivalent networks.

the transmission lines will eventually exceed the gain obtained by adding an additional active device. The phase shift from section to section for both the drain and gate lines must also be made approximately equal so that the amplified signal from each FET will add along the output (drain) transmission line. Any wave traveling in the reverse direction on the output transmission line is absorbed by the termination.

One of the assumptions in the design of distributed amplifiers is that the artificial transmission lines of the gate and drain circuits be terminated with their characteristic impedances loads. This is somewhat difficult since the characteristic impedance of a constant-k transmission line is a function of frequency and approaches infinity at cutoff. Hence there is no physical combination of elements that can provide a proper termination at all frequencies, but there is a class of circuit, known as the constant-k m-derived network, which can be used in conjunction with the termination resistor to properly terminate the artificial transmission lines through an operating frequency limit of $0.8f_c$. However, ideal operation of these

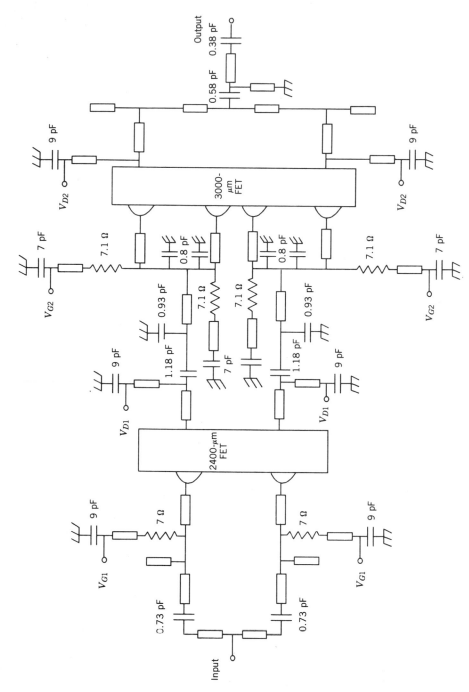

Figure 5.36 Final distributed-element amplifier circuit model.

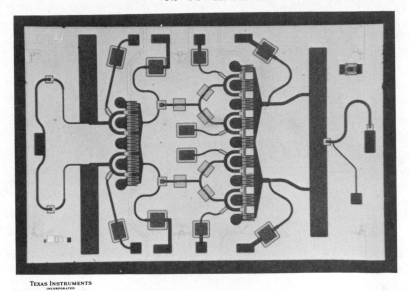

TEXAS INSTRUMENTS
INCORPORATED

Figure 5.37 Monolithic two-stage 6- to 18-GHz power amplifier.

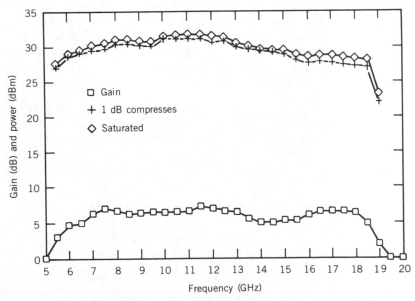

Figure 5.38 Measured gain and power output performance of monolithic two-stage amplifier.

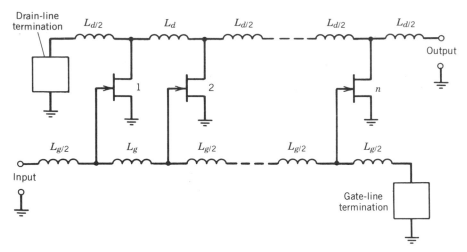

Figure 5.39 Schematic representation of a FET distributed amplifier [5.33]. © 1984 IEEE

networks can occur only when lumped elements are used to realize the amplifier circuit. A typical lumped-element distributed-amplifier circuit, which illustrates the use of optimized source and load terminations, is shown in Fig. 5.41c.

With these general concepts in mind, a more in-depth analysis can be formulated. If we define the gate circuit radian cutoff frequency as $\omega_g = 1/R_i C_{gs}$ and the drain circuit radian cutoff frequency as $\omega_d = 1/R_{ds} C_{ds}$, the propagation characteristics of the constant-k transmission-line sections are known. By requiring the phase shift between each gate-line section and drain-line section to be equal to assure proper amplifier performance, the cutoff radian frequency ω_c for both transmission lines must also be equal. With these constraints, the gain of the amplifier can be defined as [5.32, 5.33]

$$G = \frac{g_m^2 R_{01} R_{02} \sinh^2\left[n/2(A_d - A_g)\right] e^{-n(A_d + A_g)}}{4\left[1 + (\omega/\omega_g)^2\right]\left[1 - (\omega/\omega_c)^2\right] \sinh^2\left[1/2(A_d - A_g)\right]} \qquad (5.11)$$

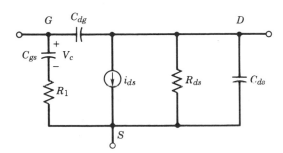

Figure 5.40 Simplified FET model.

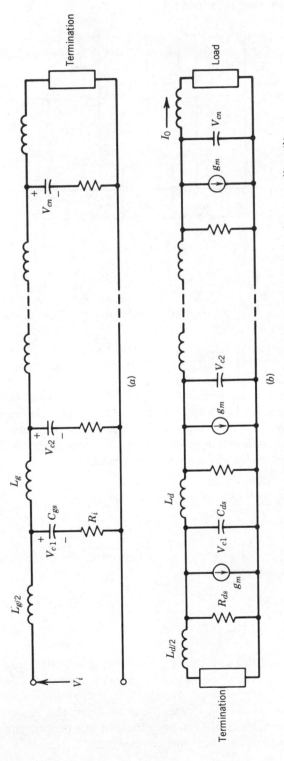

Figure 5.41 Constant-k lumped-element transmission lines: (a) input circuit or gate line; (b) output circuit or drain line; (c) simple lumped-element n-section amplifier showing m-derived terminations [5.33]. © 1984 IEEE

355

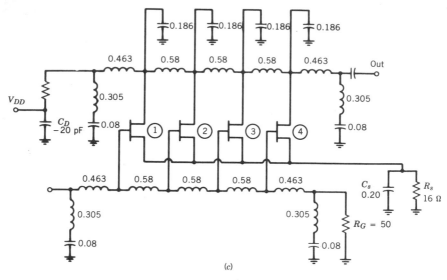

(c)

Figure 5.41 (*Continued*)

where R_{01} $[= (L_g/C_g)^{1/2}]$ and R_{02} $[= (L_d/C_d)^{1/2}]$ are the characteristic resistances of the gate and drain lines, respectively.

The magnitude of the amplifier's voltage gain for a single stage can be obtained from the power gain equation (5.11), provided the assumption is made that ideal impedance transformers are placed between cascade-connected amplifier stages which transform the drain-line impedance R_{02} to the succeeding gate-line impedance R_{01}. Thus the resulting voltage-gain expression is

$$A = \frac{g_m (R_{01}R_{02})^{1/2} \sinh\left[n/2(A_d - A_g)\right] e^{-n(A_d + A_g)/2}}{2\left[1 + (\omega/\omega_g)^2\right]^{1/2} \left[1 - (\omega/\omega_c)^2\right]^{1/2} \sinh\left[(A_d - A_g)/2\right]} \qquad (5.12)$$

By maximizing (5.12) for gain at a given frequency, the optimum number of devices n can be shown to be

$$N_{\text{opt}} = \frac{\log_e (A_d/A_g)}{A_d - A_g} \qquad (5.13)$$

It must be remembered that although N_{opt} can be large (> 10) if gate and drain circuit losses are low, in practice very little gain improvement is obtained for n greater than 8. Typically, with all but the best state-of-the-art devices, little improvement is obtained beyond 4. The choice of n is further constricted in the design of power-distributed amplifiers, since the total gate periphery, device drive level, and saturation characteristics must be considered.

The most critical factor in determining amplifier frequency response is transmission-line attenuation, with the gate line typically being the dominant contributor. The expressions for attenuation and phase velocity are well known for constant-k transmission lines, and for the low-loss case, the attenuation factors for the gate and drain lines can be approximated as

$$A_g = \frac{(\omega_c/\omega_g)X_k^2}{\left[1 - [1 - (\omega_c/\omega_g)^2]X_k^2\right]^{1/2}} \tag{5.14}$$

$$A_d = \frac{\omega_d/\omega_c}{(1 - X_k^2)^{1/2}} \tag{5.15}$$

where $X_k = \omega/\omega_c$ is the normalized frequency, $\omega_g = 1/R_iC_{gs}$, $\omega_d = 1/R_{ds}C_{ds}$, and

$$\omega_c = \frac{2}{(L_gC_{gs})^{1/2}} = \frac{2}{(L_dC_{ds})^{1/2}} \tag{5.16}$$

The attenuation characteristics of these lines as a function of frequency with ω_d/ω_c and ω_c/ω_g as parameters are shown in Fig. 5.42. It is evident from Fig. 5.42a that the frequency response is determined primarily by the gate line, while it is interesting to note that the attenuation on the drain line does not vanish at dc. Thus the low-frequency gain of the amplifier is determined primarily by the g_m of the FET, but is also a function of R_{ds}; hence dual-gate FETs should exhibit superior gain performance because their equivalent g_m is greater and R_{ds} is larger than a similar-sized single-gate FET.

The design challenge now becomes the minimization of the quantities ω_d/ω_c and ω_c/ω_g. For a given cutoff frequency ω_c, active devices with the smallest R_iC_{gs} (similarly, $R_{ds}C_{ds}$) product need to be selected, or a capacitor can be added in series with each FET's gate terminal [5.34]. However, a series capacitor acts as a voltage divider on the input of the FET, lowering the effective RF drive voltage, which in turn lowers the amplifier's gain. This is not always a poor performance trade, since it may allow the use of larger FETs for a given frequency response, resulting in an amplifier with greater power output. In a similar manner, drain-line losses can be lowered by padding C_{ds} with an external capacitor; but this will lower the drain-line cutoff frequency. In practice, the only padding added is a small series transmission line employed between the FET cells and the high-impedance transmission lines used to simulate the drain-line inductances. This is used to help equalize the phase shift from section to section between the gate and drain circuits.

A slightly different network, called the constant-R network [5.35, 5.36], can also be used to synthesize the amplifier's artificial transmission lines. This network, shown in Fig. 5.42c, if correctly synthesized, exhibits a cutoff frequency the square root of 2 times greater than a conventional constant-k network for the same value of shunt capacitance. However, in practice this performance improvement is rarely

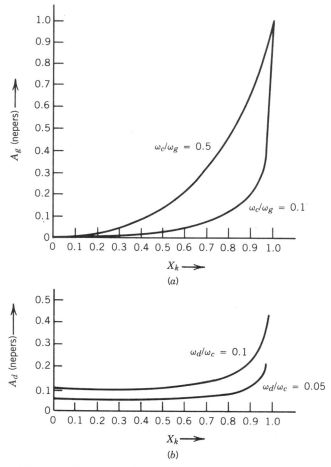

Figure 5.42 (*a*) Attenuation on gate-line versus normalized frequency; (*b*) attenuation on drain-line versus normalized frequency; (*c*) comparison between constant-*K* and constant-*R* filter sections [5.33]. © 1984 IEEE

achieved, due to the distributed nature of the network and the inability to achieve proper coupling between input and output inductors.

The attenuation constants A_g and A_d can also be expressed in terms of fractional bandwidth, number of devices, and circuit parameters. Equations (5.14) and (5.15) can then be written as

$$A_g = \frac{2aX_k^2}{n\left[1 + \left[(4a^2/n^2) - 1\right]X_k^2\right]^{1/2}} \tag{5.17}$$

$$A_d = \frac{2b}{n(1 - X_k^2)^{1/2}} \tag{5.18}$$

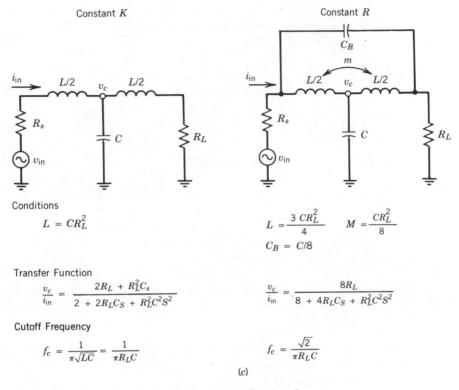

Figure 5.42 (*Continued*)

where

$$a = n\omega_c/2\omega_g \quad \text{and} \quad b = n\omega_d/2\omega_c \tag{5.19}$$

by using (5.17) and (5.18), an expression for normalized gain as a function of frequency can be derived from the voltage-gain expression (5.12). The expression

$$A_n = \frac{\sinh(b/n)e^b \sinh[(n/2)(A_d - A_g)]e^{-[(n/2)(A_g - A_d)]}}{\sinh(b)[1 + (4a^2/n^2)X_k^2]^{1/2}(1 - X_k^2)^{1/2} \sinh[(A_d - A_g)/2]} \tag{5.20}$$

where the gain (A_0) of the amplifier, at the low-frequency limit, is given by

$$A_0 = \frac{g_m(R_{01}R_{02})^{1/2} \sinh(b)e^{-b}}{2 \sinh(b/n)} \tag{5.21}$$

can be used to determine the 1-dB fractional bandwidth of any amplifier topology as a function of parameters a, b, and n. In Fig. 5.43, several frequency response

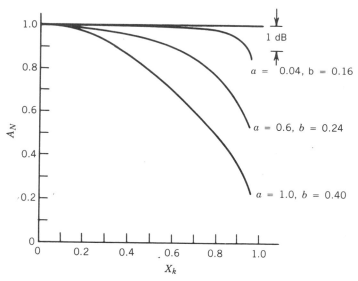

Figure 5.43 Normalized frequency response of $n = 4$ FET distributed amplifier for various values of a and b [5.33]. © 1984 IEEE

characteristics for an amplifier with four devices are shown for various values of a and b. By properly choosing a and b, the amplifier can be designed with a nearly flat frequency response throughout its operating bandwidth, with an upper-frequency limit close to the cutoff frequency of the transmission lines. Representative values for a and b which give the same fractional bandwidth are shown in Fig. 5.44. These values can be found by iteratively solving (5.20).

We have now developed an analytical approach to design distributed small-signal amplifiers given the desired bandwidth, low-frequency gain, FET characteristics, and number of stages, but there are other factors that must be considered before a power distributed amplifier can be designed. Also, the previous analysis assumes that ideal inductances and image terminations are employed in the design. However, in practice high-impedance lines are employed to simulate inductances, and image terminations are rarely used because of stability and gain peaking requirements. Regardless of the limitations with the previous analysis, it yields the most straightforward approach to the design of a microwave power amplifier.

The power limitation problems of distributed amplifiers are similar to the problems encountered in the design of conventional topologies such as cascaded reactively matched gain stages, except that distributed configurations are far less sensitive to device and circuit process variations. This process tolerance allows for a more efficient sizing of devices and sometimes leads, especially in broadband designs, to higher operating efficiency.

As with conventional approaches, the most basic dynamic range limitation mechanism is the maximum RF voltage swing that can be applied to the FET gate (Fig. 5.45). In a distributed amplifier, this usually translates to the largest signal

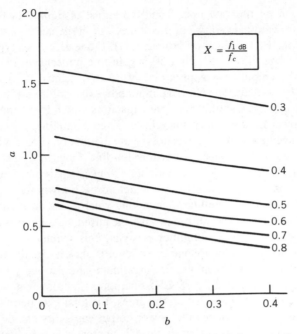

Figure 5.44 Representative values of a and b that yield the same fractional bandwidth [5.33]. © 1984 IEEE

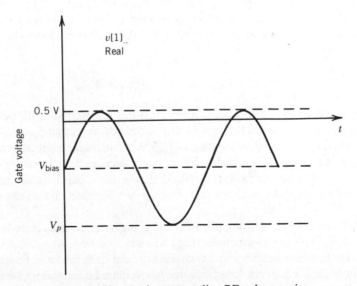

Figure 5.45 Maximum gate-line RF voltage swing.

that can be present on the gate line. Thus the signal is limited in the positive direction by the forward gate conduction voltage, and in the negative direction by the magnitude of the FET's pinch-off voltage (V_p). If one assumes a typical power FET pinch-off voltage of 5 V dc and a 50-Ω gate-line impedance, the maximum input power to the amplifier is approximately 75 mW. Thus the output power cannot exceed $P_{input} \times$ gain. The effective input power capabilities of the amplifier can be increased with the use of series gate capacitors which lower the gain.

The second power-limiting mechanism in distributed amplifiers, which is not a hard limit in conventional designs, is total gate periphery. For a given upper frequency of operation and a constant transmission-line impedance, there will exist a maximum total gate periphery. The periphery limit is easily understood if one considers that f_c is inversely related to C_{gs}, which forms a low-pass structure with L_g, and n_{opt} is proportional to gate-line attenuation, which also reduces the fractional bandwidth. The gate periphery limit can also be raised by adding a series gate capacitor. For practical power amplifier designs, this technique is a feasible compromise because the loss in voltage drive level is almost exactly offset by the increase in device g_m. As an example, the maximum gate periphery for a typical 0.5-μm FET amplifier designed for 18-GHz operation is approximately 1500 μm; but if series capacitors are employed on the gates, the total gate periphery can be increased beyond 3000 μm. Hence the power output capability can be raised but not necessarily doubled. The reason for this phenomenon will become apparent shortly.

Aside from the RF current limitation, which is proportional to gate periphery, the maximum RF voltage sustainable at the load is the other prime power-limiting mechanism. Typically, the limiting FET parameter is the drain-to-gate breakdown voltage.

In a traveling-wave structure, the last FET (nearest the load) will have the largest applied RF voltage present across the drain–source terminals, provided that there are no large standing waves on the drain line. The maximum approximate peak-to-peak voltage is defined as

$$V_{max} = V_{breakdown} + V_{pinch\text{-}off} - V_{knee} \qquad (5.22)$$

Using equation (5.22), a first-order approximation can be found for power at a specified load impedance. Thus in a 50-Ω system, a 1-W output power specification implies reverse breakdown voltages of 25 V, a parameter not readily obtained. These approximations are somewhat conservative because RF breakdown voltages are usually greater than dc values. It should also be noted that reducing the output load impedance may not be a feasible option since the amplifier's gain will also be reduced.

A more subtle power-limiting phenomenon is the optimum load impedance seen by individual FETs in the structure. If the transmission lines were ideal and there were no mismatched terminations, each FET would be terminated by an impedance equal to $Z_0/2$ (dc). An impedance this low is quite far away from the optimum load termination for all but the largest FETs (>1 mm). To approximate the true

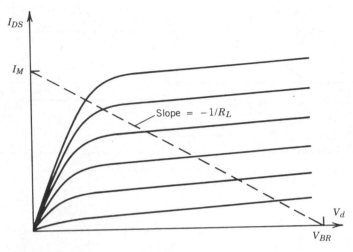

Figure 5.46 Optimum ac load impedance for class A operation.

optimum load for any FET, the ac load line can be drawn between V_{max} and I_{max} on the FET drain characteristic curves (Fig. 5.46). Unfortunately, the design engineer has very little control over the optimum terminating impedance in a distributed amplifier since the load line is usually predetermined by other circuit parameters.

Due to the above-mentioned constraints, maximum power output will be obtained from designs employing the largest possible FET cells. Nonoptimum loads also influence amplifier power added efficiency in a detrimental manner. However, it is possible to improve the efficiency and power output of a distributed amplifier by tapering the drain line impedance from section to section.

Without tapering or employing frequency-dependent terminations, a large fraction of the developed power propagates toward the termination end of the drain line. The intent of impedance tapering is to force the backward traveling current to zero at the termination, thus forcing all the developed current from each FET to travel in the forward direction only.

By operating the first FET into a section of line with impedance Z_0, all the current will flow into the next section. Then if the next section has a characteristic impedance of $Z_0/2$, one-third of the developed current from the second FET will cancel the reflected current from the first FET at the junction of the second FET. The remaining two-thirds of the current developed by the second FET and four-thirds of the current developed by the first FET now add and propagate toward the junction of the third FET as shown in Fig. 5.47. The transmission-line impedance after the third FET must now be equal to $Z_0/3$. This process continues where each successive transmission-line section has an impedance of Z_0/k, where k is the number of the stage of interest.

Several assumptions made in the analysis above should not be forgotten. First, it was assumed that each device delivers equal current. Due to unequal drive

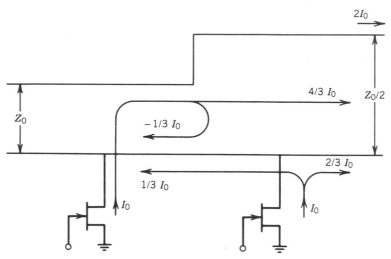

Figure 5.47   Current distribution in a correctly tapered drain circuit.

voltages on the gate line and FET process variations, current equalization is in practice difficult to achieve. The gate and drain lines are also dispersive, which makes the tapering accurate at only a single frequency. The second area of concern is in the area of fabrication. When microstrip realization is employed, there exists a small range of useful realizable impedances. Typically, these values range from 30 to 95 Ω; hence this $3:1$ ratio makes it difficult to realize the L/C transmission-line sections with the impedance values required for proper tapering.

There are also other practical considerations. Proper tapering requires that the load impedance presented to the amplifier be Z_0/k. A load impedance that low is probably unavailable in most instances. However, tapering may help keep the load impedance presented to the last several FETs low, minimizing the power loss due to low breakdown voltage. Also, microstrip-line width tapering eases the fusing current problems encountered in biasing power-distributed structures.

Now that the concepts and synthesis techniques have been developed, the design of a single- or dual-gate FET power amplifier can be illustrated. The design example chosen, which has been based on monolithic implementation, was selected specifically to highlight the problems of obtaining broadband frequency response and high power output. The performance goals for this design are summarized in Table 5.6.

TABLE 5.6   Performance Goals for Power Dual-Gate FET Distributed-Amplifier Design Example

Power output	1 W
Frequency response	2 to 18 GHz
Small-signal gain	8 dB
Input/output VSWR	$<2:1$
Amplitude flatness	$<\pm0.5$ dB

The design engineer must begin by selecting the general device type to be used. Typically, single-gate FET designs have suffered from low circuit gain, which is due to excessive drain-line losses, and poor device transconductance, which is associated with doping profiles selected for power output performance. Dual-gate designs, although difficult to stabilize, offer excellent gain performance and output voltage dynamic range improvements over a single-gate approach. They also have much higher values of R_{ds} for a given gate periphery, thus reducing drain-line loss problems. Hence a dual-gate design approach was selected.

Conventional power amplifiers, designed for broadband performance, usually require approximately 3 mm of gate periphery in the output stage to achieve 1 W performance. A similar amount of total gate periphery must then be employed in a distributed approach to achieve comparable performance. Since the number of stages in a typical amplifier must be at least four to achieve reasonable gain, and for complexity reasons should not exceed 10, it becomes evident that the device size must be between 300 and 750 μm.

Considering these constraints, a 450-μm intermediate-profile dual-gate FET was selected. A simplified single-gate equivalent model for this device is shown in Fig. 5.48. Using the FET element values given in Fig. 5.47, basic amplifier design parameters can be calculated.

If we assume 50-Ω input and output amplifier impedances, the gate- and drain-lines inductances can be calculated from the expression

$$Z_0 = (L_g/C_g)^{1/2} = (L_d/C_d)^{1/2} \qquad (5.23)$$

The gate- and drain-line attenuation constants, A_g and A_d, can be found by solving (5.14) and (5.15) using the expressions for the three cutoff frequencies ω_c, ω_g, and

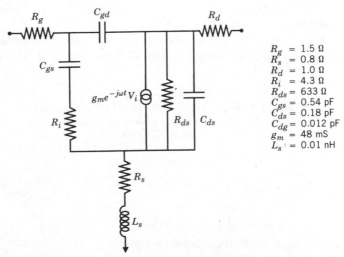

$$
\begin{aligned}
R_g &= 1.5\ \Omega \\
R_s &= 0.8\ \Omega \\
R_d &= 1.0\ \Omega \\
R_i &= 4.3\ \Omega \\
R_{ds} &= 633\ \Omega \\
C_{gs} &= 0.54\ \text{pF} \\
C_{ds} &= 0.18\ \text{pF} \\
C_{dg} &= 0.012\ \text{pF} \\
g_m &= 48\ \text{mS} \\
L_s &= 0.01\ \text{nH}
\end{aligned}
$$

Figure 5.48 Simplified single-gate equivalent model of dual-gate FET.

TABLE 5.7 Basic Amplifier Design Parameters
Obtained with FET Model of Fig. 5.48
(C_{gs} = 0.54 pF)

f_c	= 11.78 GHz
n_{opt}	= 7
L_g	= 1.35 nH
L_d	= 0.45 nH

By selecting $n = 6$ and $f = 10$ GHz:

A_0	= 16.0 dB
f_{1dB}	= 9.31 GHz

ω_d. By choosing an upper operating frequency limit such that $\omega < \omega_c$, the optimum number of stages (n_{opt}) can be determined from (5.13). Finally, the dc gain and 1-dB corner frequency for any value of n can be calculated from (5.20) and (5.21). The design parameters are shown in Table 5.7.

At first glance it appears that the selected design approach cannot achieve the required bandwidth performance. However, the cutoff frequency of the gate line can be raised by adding capacitors in series with each FET gate. The addition of a series capacitor approximately equal to the original value of C_{gs} (0.42 pF) will lower the gain by 7 dB, but will dramatically extend the amplifier's frequency response. The key amplifier parameters are shown in Table 5.8.

Before proceeding, the lumped-element circuit model must be replaced by a transmission-line equivalent. This is easily accomplished by substituting a high-impedance microstrip transmission line with a reactance equal to the reactance of L_g and L_d at the uppermost frequency of operation. Transmission-line models for capacitors, junctions, and terminations should be added at this time. Unfortunately, amplifier synthesis now becomes more empirical and will require the use of a linear circuit simulator such as Touchstone.

The gate transmission line can easily be tapered by adjusting the values of each gate capacitor so that the drive voltages at each FET gate are equal. This condition

TABLE 5.8 Basic Amplifier Design Parameters
Obtained by Adding a Series Gate Capacitor to
Extend Useful Operating Bandwidth (C'_{gs} = 0.24 pF)

f_c	= 26.58 GHz
n_{opt}	= 8
L_g	= 0.60 nH
L_d	= 0.45 nH

By selecting $n = 7$ and $f = 20$ GHz:

A_0	= 10.3 dB (corrected by 7 dB)
f_{1dB}	= 14.51 GHz

is most important at the high-frequency end of the operating range, assuring that each device contributes equal power to the drain line. Drive equalization offsets the normal losses encountered as the input signal propagates toward the termination end of the gate circuit. Hence the series capacitors are largest at the termination end of the line. A linear circuit optimizer is an excellent aid in determining the values of the series capacitors.

To maximize power output and to accommodate the total drain current, drain-line tapering was employed. The impedance range of the tapering was less than ideal, since the optimum impedance ratios for the lumped-element transmission-line sections could not be realized with microstrip inductors and capacitances supplied by the FETs. The final circuit transmission-line model is illustrated in Fig. 5.49. A photograph of the monolithic amplifier, which is 3.48×1.27 mm in size, is shown in Fig. 5.50 [5.37]. The complete dual-gate FET model (Fig. 5.6) was used in the final analysis to check key small-signal parameters such as stability, gain, and VSWR. The measured small-signal gain and power output performance are shown in Fig. 5.51. The nominal gain was 8.6 ± 1.2 dB and the saturated power output averaged 28.5 dBm throughout the frequency range 2 to 18 GHz.

Although linear simulators give excellent design results, they do not give the designer insight into selecting optimum loading and bias conditions and in understanding circuit power limitations. Fundamental design problems, such as voltage clipping levels, current saturation characteristics, and breakdown conditions can be analyzed only with the aid of a nonlinear solver.

As mentioned above, the gate line was tapered to provide equal drive levels, a condition that is imperative in obtaining optimum power output. To verify the drive equalization and other operating characteristics, a nodal nonlinear solver, Microwave SPICE, was employed. It should be noted that the nodal circuit consisted only of simple transmission-line elements since microstrip junction models were not available. An equivalent single-gate FET model was derived from the small-signal model shown previously and measured drain I–V (RF) and breakdown characteristics of a typical dual-gate FET at the appropriate bias conditions.

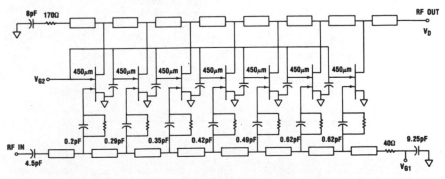

Figure 5.49 Transmission-line model of dual-gate FET distributed power amplifier.

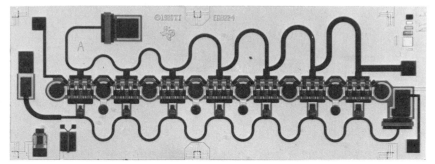

Figure 5.50 Monolithic dual-gate FET distributed power amplifier. Courtesy of Texas Instruments.

The cubic FET model parameters were then adjusted to simulate the measured 450-μm FET.

The measured versus predicted power output performance at the 1-dB compression point is shown in Fig. 5.52. As can be seen in the plot, the predicted performance agrees quite well with the SPICE calculation. Drive equalization is verified in Fig. 5.53, which shows that the gate voltage waveforms as a function of frequency are well matched. The drain voltage saturation characteristics can be seen in Fig. 5.54 as a function of FET position and frequency. The dramatic increase of RF voltage at the load end of the drain line can be seen clearly in the graphs. This analysis may explain the degradation in power output performance encountered at the upper end of the operating band, due to large swings in load impedance seen by each FET.

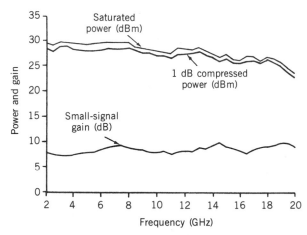

Figure 5.51 Measured gain and power output performance of monolithic distributed amplifier.

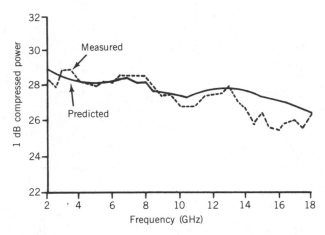

Figure 5.52 Measured versus predicted (Microwave SPICE) power output performance of monolithic distributed amplifier.

The design example above clearly illustrates the combination of analytical and empirical design techniques that must be employed in the synthesis of a microwave power distributed amplifier. It also emphasizes the importance of both linear and nonlinear simulators, in gaining a fundamental understanding of power output performance and amplifier limitations.

5.6 CLASS B AMPLIFIERS

There are probably very few applications where amplifier efficiency is as important as it is in the design of large active-element phased-array antennas. A variety of systems currently under development will require thousands of elements per array; thus an improvement in power amplifier efficiency of as little as 5% can greatly affect prime power requirements and thermal design. In fact, some systems will not be feasible unless total power amplifier chain efficiencies exceed 20%. As the bandwidth requirements for new systems expand beyond the 10% range to greater than an octave, high-efficiency design becomes even more difficult.

Operating RF amplifiers in other than the class A mode is quite common in frequency ranges between HF and UHF, but rarely used at microwave frequencies, and high-efficiency class C amplifiers are quite common for applications not requiring linear performance. When linear performance is required, either class B or class AB designs are employed.

The theoretical efficiency for true class B operation with sinusoidal signals is 78.5%, which is far greater than the 50% theoretical maximum obtainable with class A operation. Although efficiencies near the theoretical limits are obtainable at low frequencies, it is difficult to achieve better than 50% efficiency at microwave frequencies. Efficiency limitations are not just a function of device and circuit

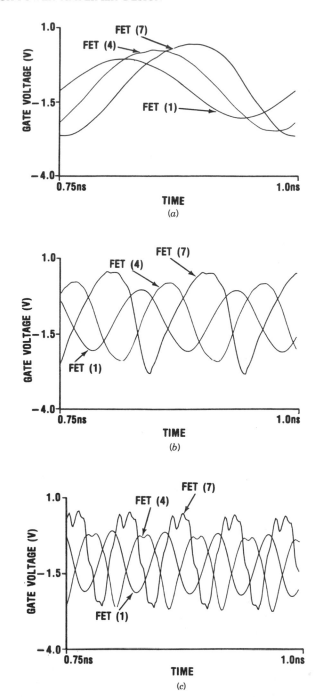

Figure 5.53 FET gate voltage waveforms as a function of frequency: (*a*) 4 GHz; (*b*) 10 GHz; (*c*) 16 GHz.

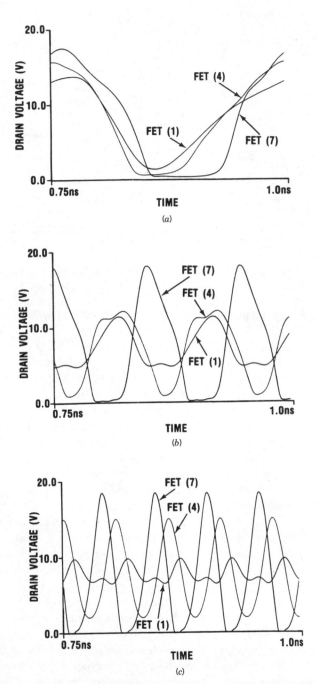

Figure 5.54 FET drain voltage waveforms as a function of frequency: (*a*) 4 GHz; (*b*) 10 GHz; (*c*) 16 GHz.

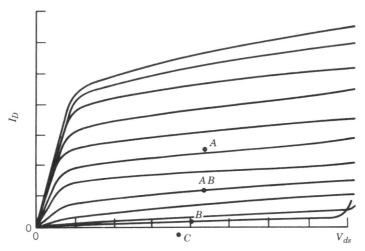

Figure 5.55 Typical operating point locations for various classes of amplifier operation.

losses, but are dependent on device $I-V$ characteristics as well as design trade-offs.

If we now consider the FET $I-V$ characteristics shown in Fig. 5.55 for a typical GaAs FET, it becomes apparent that biasing the device for true class B operation is impractical, since the g_m of the device at pinch-off approaches zero (small signal). Even biased at the zero gain point, the finite value of leakage current reduces the efficiency obtainable. There are also other bias conditions that must be considered.

With conventional class B design, the active device is biased to reproduce only one half of the sinusoidal input signal. Hence if the device is biased near cutoff and the drain (anode, collector, etc.) biased at breakdown, a positive input signal will cause the device to conduct. When the input signal reaches maximum, and the load and gain are sufficient, the voltage across the drain can approach zero. Thus a half-cycle sinusoid is reproduced at the device output. As the input signal swings negative the device is cut off, thus clipping the other half-cycle. If the device happens to be a GaAs FET with a finite gate-to-drain breakdown voltage, the drain should be biased back from the breakdown point by an amount equal to the magnitude of the input signal, since the gate will become more negative than V_p during negative half-cycles. The average current, neglecting leakage, under these conditions is $I_{max} \times 2/\pi$ during the conduction half-cycle.

The reduced conduction angle in class B operation causes another problem. The half-wave rectified sine wave at the output of single-ended designs is rich in harmonics or high in distortion. This problem is usually solved by employing push-pull topologies. Regardless of the push-pull approach (transformer coupled, complementary output, etc.), the output waveform is a composite of the waveforms generated by each half of the amplifier. Since each amplifier half is out of phase by 180°, the output waveform is a complete sine wave.

However, there are some practical issues with push-pull design at microwave frequencies that must be considered. First, complementary structures (transformerless topology) cannot be designed with GaAs FETs because p-channel FETs cannot be made to match the performance of n-channel devices, due to the vast difference between hole and electron mobilities. Second, although transformers or baluns can be used, they are difficult to design and implement and can have bandwidth limitations. Also, when either single-ended or push-pull approaches are employed, there is almost always sufficient Q in the output circuit to cause voltage ringing (flyback), which attempts to reconstruct the missing half-cycle from each device. This phenomenon can cause severe breakdown problems if the devices are biased at a drain voltage much above $V_{br}/2$. For class A operation the FET is biased at approximately the same drain voltage $V_{br}/2$. However, a considerably different load line results.

As shown previously, in the class A mode the operating point of the FET traverses a path from I_{max} ($V_{dd} = 0$) to V_{br} ($I_d = 0$), which is the typical dc load line. In the class B mode, the FET operating point traverses a path from I_{max} to approximately $V_{br}/2$. This difference in load line and bias point is illustrated in Fig. 5.56. The final practical consideration that we have aluded to is gain. Gain is also an efficiency driver, since we usually define power added efficiency as

$$\eta_{pa} = \frac{P_{out} - P_{in}}{V_{dd} \times I_{d(average)}} \qquad (5.24)$$

Most GaAs FETs fabricated today do not exhibit constant g_m as a function of drain current. Hence when the FET is biased near pinch-off, the small-signal gain

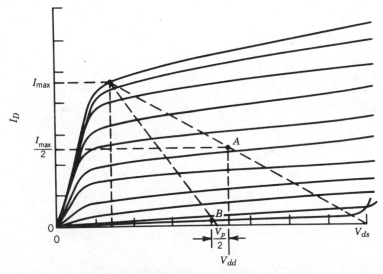

Figure 5.56 Class B versus class A load lines.

is essentially zero. The gain problem can be overcome partially by operating the FET at reduced rather than zero current, or class AB. With currents on the order of 20% of I_{dss} there can be sufficient gain, depending on the frequency of operation, to be practical. Typically about 6 dB of gain is lost by lowering the bias current from 50 to 60% of I_{dss} to the 20% level required for class AB operation. This gain-loss problem can also be reduced by adjusting the FET doping profile so that nearly linear g_m performance as a function of drain current is obtained.

Depending on how the FET channel is doped, the drain current versus gate voltage relationship can vary drastically. Several FET transfer characteristics versus doping profile are shown in Fig. 5.57. As can be seen, when power law doping is employed and n approaches infinity, a linear transfer characteristic results. A linear transfer characteristic is also possible when stepped doping is used. However, the doping used for typical GaAs FETs is essentially reversed from the cases above. For example, in the design of low-noise FETs, the doping is heaviest at the surface, giving low contact resistances, which is essential in obtaining low noise perfor-

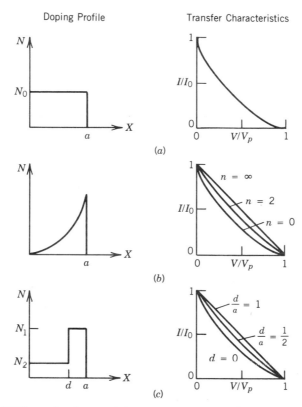

Figure 5.57 FET transfer characteristics versus doping profile: (*a*) uniform doping, $N = N_0$; (*b*) power law doping, $N = N_0 x^n$; (*c*) step doping, $N = N_1$ for $x < d$ or N_2 for $d \leq x \leq a$.

**TABLE 5.9   Performance of 2-Watt Push-Pull
Amplifier Reported by Westinghouse**

Frequency range	9.2 to 10.2 GHz
P_{out}	32.2 to 33.0 dBm
P_{in}	27.9 dBm (nominal)
η_{pa}	31 to 41%
Second harmonic level	40 dBc
Third-order distortion level	16 dBc

mance. It is also very difficult to fabricate FETs with stepped or spiked doping since it requires the growing of very thin, well-controlled layers. Unfortunately, these types of profiles are required to build FETs that exhibit excellent class B or AB performance.

Although most of the devices available currently are not optimized for class AB use, excellent efficiency can still be achieved with reasonable gain at frequencies through X-band. If the bandwidths are sufficiently limited as to allow for harmonic tuning, very impressive results can be achieved. Shown in Table 5.9 are the performance results of a class B (AB) push-pull amplifier that was reported by Westinghouse [5.38, 5.39]. Each amplifier half was designed using a 2400-μm FET with a retrograde doping profile to improve low bias current gain. Second harmonic tuning was also employed.

A dual push-pull amplifier, consisting of two 2-W push-pull amplifier stages combined with Wilkinson power dividers, was also reported. As can be seen in Table 5.10, this amplifier exhibited excellent performance throughout the same frequency band of interest with slightly lower efficiency. The lower efficiency was due to the added losses of the additional combining networks.

Only recently has there been any use of class B amplifiers in the microwave region. This was due the fact that when an FET is biased near pinch-off, its gain is substantially less than when it is operated at its maximum gain point. In the past, when FET gain performance was at a premium, the several decibels of gain reduction that resulted when the amplifier FETs were biased, even for AB operation, yielded a final amplifier with practically useless gain performance. Modern FETs exhibit substantially better performance; hence a 2- or 3-dB reduction in overall gain is quite tolerable and does not degrade the overall system power added efficiency, at least at X-band, appreciably.

**TABLE 5.10   Performance of 4-Watt Push-Pull
Amplifier Reported by Westinghouse**

Frequency range	9.2 to 10.2 GHz
P_{out}	36.0 to 36.3 dBm
P_{in}	30.9 dBm (nominal)
η_{pa}	35%
Gain (at 1-dB compression point)	5 dB

At lower frequencies, such as the HF and VHF regions, class B operation is well known. Class B or AB circuits can be implemented using either bipolar or field-effect transistors, and the design of such amplifiers proceeds as with any *S*-parameter synthesis approach. There are, however, some biasing and thermal constraints that differ from the GaAs FET, especially with the bipolar transistor that must be considered.

As with all semiconductors, the maximum power dissipation is closely related to the amount of heat that can be removed from the semiconductor die, provided that the heat generated by the device is distributed relatively uniformly throughout the structure. When bipolar transistors are used at high current levels, the emitter current of the device is concentrated at the emitter–base edge. Current crowding effects tend to forward bias the device at the edge of the emitter closest to the base contact. Hence the center of the emitter injects very little current compared to the area near the edges. Because of this effect, a high emitter periphery-to-area ratio is desired. In addition, most modern power BJTs employ emitter ballasting techniques that help equalize bias within the active device, thus eliminating thermal "hot-spotting." The ballasting is realized by controlling the resistance of the emitter contact area in such a way as to add a predetermined amount of series resistance in the emitter circuit, causing negative feedback, hence equalizing the device. This type of ballasting allows for a more efficient device and helps protect the final amplifier from high RF voltage breakdown problems caused by high VSWR. It also raises the saturation voltage and lowers the RF gain.

Another problem that is common with BJTs is thermal runaway. This condition is caused, particularly with class B or AB amplifiers, by the fact that the base-to-emitter voltage has a negative temperature coefficient. When the transistor is biased at low collector currents and no drive signal is present, the power dissipation and hence the device's operating temperature is low. As drive is applied and the amplifier approaches large-signal performance, the power dissipation and thus the operating temperature rise. If the bias voltage is fixed and the base-to-emitter voltage has dropped due to a temperature rise, the quiescent operating current could be substantially greater than during the initial no-drive state. With a large quiescent current, the device will operate even hotter, hence the problem worsens. However, these problems are usually avoided by proper bias supply design. These bias supplies, which are temperature compensated, typically employ a sensing diode with characteristics similar to those of the base–emitter junction, which is fabricated on the same transistor die or mounted adjacent to the power device. As the operating temperature of the device increases, the bias voltage also increases, thus maintaining a constant value of quiescent current and power dissipation.

Once the RF characteristics of the BJT are determined (Section 5.1), the design of a push-pull amplifier proceeds in the same manner as for the FET design described previously. The problem again becomes that of device matching and optimum load termination, with the additional task of balun synthesis (Chapter 7). A typical HF class AB push-pull amplifier is illustrated in Fig. 5.58 [5.40]. Transformer hybrids and matching transformers were employed in the intput and output networks to provide proper phasing and impedance functions. A temperature-

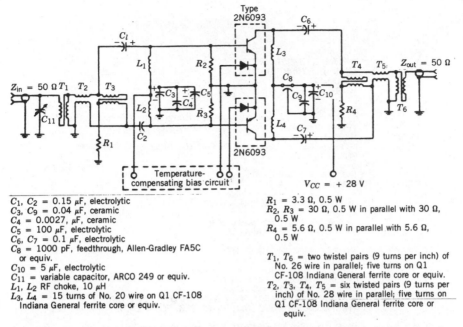

$C_1, C_2 = 0.15\ \mu F$, electrolytic
$C_3, C_9 = 0.04\ \mu F$, ceramic
$C_4 = 0.0027,\ \mu F$, ceramic
$C_5 = 100\ \mu F$, electrolytic
$C_6, C_7 = 0.1\ \mu F$, electrolytic
$C_8 = 1000$ pF, feedthrough, Allen-Gradley FA5C
  or equiv.
$C_{10} = 5\ \mu F$, electrolytic
$C_{11} =$ variable capacitor, ARCO 249 or equiv.
L_1, L_2 RF choke, 10 μH
$L_3, L_4 = 15$ turns of No. 20 wire on Q1 CF-108
  Indiana General ferrite core or equiv.

$R_1 = 3.3\ \Omega$, 0.5 W
$R_2, R_3 = 30\ \Omega$, 0.5 W in parallel with 30 Ω,
  0.5 W
$R_4 = 5.6\ \Omega$, 0.5 W in parallel with 5.6 Ω,
  0.5 W

$T_1, T_6 =$ two twistel pairs (9 turns per inch) of
  No. 26 wire in parallel; five turns on Q1
  CF-108 Indiana General ferrite core or equiv.
$T_2, T_3, T_4, T_5 =$ six twisted pairs (9 turns per
  inch) of No. 28 wire in parallel; five turns on
  Q1 CF-108 Indiana General ferrite core or
  equiv.

Figure 5.58 Typical class AB push-pull power amplifier employing bipolar junction transistors. Courtesy of RCA.

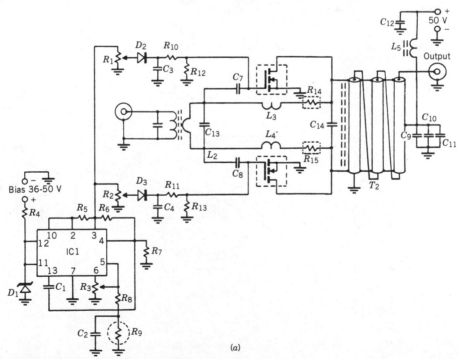

(a)

Figure 5.59 Push-pull TMOSFET 1-kW power amplifier; (a) circuit diagram; (b) parts list. Source: Motorola Application Note AN-758.

377

R1 - R2	1K SINGLE TURN TRIMPOTS
R3	10K SINGLE TURN TRIMPOT
R4	470 OHMS/2 WATT
R5	10 OHMS
R6, R12-R13	2K OHMS
R7	10K OHMS
R8	EXACT VALUE DEPENDS ON THERMISTOR R9 USED (TYPICALLY 5-10 K)
R9	THERMISTOR, KEYSTONE RL1009-5820-97-D1 OR EQUIVALENT
R10-R11	100 OHMS/1W CARBON
R14-R15	KD1 PYROFILM PPR 870-150-3 OR EMC TECHNOLOGY MODEL 5308 POWER RESISTORS, 25 OHMS

| D1 | 1N5357A OR EQUIVALENT |
| D2-D3 | 1N4148 OR EQUIVALENT |

| IC1 | MC1723 (723) VOLTAGE REGULATOR |

C1	1000 pF CERAMIC DISC CAPACITOR
C2-C4	0.1 uF CERAMIC DISC CAPACITOR
C5,C9	0.01 uF CERAMIC CHIP CAPACITOR
C6,C10	0.1 uF CERAMIC CHIP CAPACITOR
C7-C8	0.005 uF CERAMIC CHIP CAPACITOR
C11	1000 pF CERAMIC CHIP CAPACITOR
C12	0.47 uF CERAMIC CHIP CAPACITOR
C13	680 pF CERAMIC CHIP CAPACITOR
C14	UNENCAPSULATED MICA, 500 V. TWO 470 pF IN SERIES, CONNECTED IN PARALLEL WITH TWO 1000 pF IN SERIES. TOTAL VALUE APPROX. 700 pF

| L2-L3 | 15 nH, CONNECTING WIRES TO R14 AND R15, 2.5 cm EACH #20 AWG |
| L5 | 10 uH - 10 TURNS #12 AWG ENAMELED WIRE ON FAIR-RITE PRODUCT CORP. FERRITE TOROID #5961000401 OR EQUIVALENT |

| T1-T2 | 9:1 AND 1:9 IMPEDANCE RATIO RF TRANSFORMERS, TYPES RF800-3 AND RF2067-3 RESPECTIVELY (RF POWER SYSTEMS, 3038 E. CORRINE DR. PHOENIX, AZ 85032) |

UNLESS OTHERWISE NOTED, ALL RESISTORS ARE 1/2 WATT METAL FILM TYPE. ALL CHIP CAPACITORS ARE ATC TYPE 100/200B OR TANSITOR MPR 2/7, CASE 2

(b)

Figure 5.59 (*Continued*)

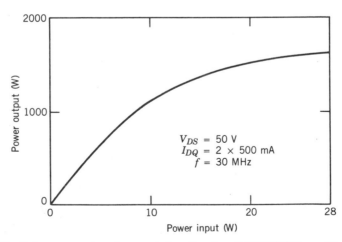

Figure 5.60 Gain compression characteristics of push-pull TMOSFET amplifier. Source: Motorola Application Note AN-758.

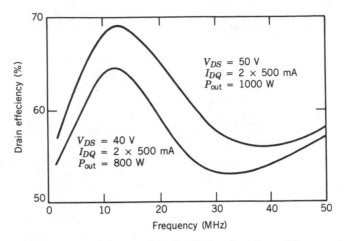

Figure 5.61 Power-added efficiency of TMOSFET push-pull amplifier as a function of drain voltage. Source: Motorola Application Note AN-758.

compensated bias network was also employed. The amplifier is capable of 120 W of power output in the frequency range 2 to 30 MHz, with an average power-added efficiency of 48%.

A typical high-power 2- to 50-MHz push-pull amplifier employing power TMOSFETs is shown in Fig. 5.59. The amplifier also employs broadband transformer techniques for matching and balancing. The amplifier's gain compression performance is shown in Fig. 5.60 and the drain efficiency is shown in Fig. 5.61. As can be seen, the amplifier exhibits excellent gain, efficiency, and power output performance.

Push-pull amplifiers of this type, employing BJTs or FETs, can be designed for a variety of frequency ranges extending to about 500 MHz [5.41]. As the operating frequency approaches 1 GHz, transformer hybrid techniques for power applications are replaced by classical microwave circuitry.

REFERENCES

5.1 J. E. Schutt-Aine, "Determination of a Small-Signal Model for Ion-Implanted Microwave Transistors," *IEEE Transactions on Electron Devices*, Vol. ED-30, No. 7, July 1983.

5.2 B. C. DeLoach, "A New Microwave Measurement Technique to Characterize Diodes and an 800-Gc Cutoff Frequency Varactor at Zero Volts Bias," *IEEE Transactions on Microwave Theory and Techniques*, January 1964, pp. 15–20.

5.3 D. Brubaker and J. Eisenberg, "Measure S-Parameters with the TSD Technique," *Microwaves & RF*, November 1985, pp. 97–102.

5.4 D. Brubaker, "The TSD Calibration Algorithm," *Microwaves & RF*, November 1985.

5.5 R. S. Tucker, "RF Characterization of Microwave Power FET's," *IEEE Transactions on Microwave Theory and Techniques*, Vol. MTT-29, No. 8, August 1981, pp. 776–781.

5.6 S. M. Perlow, "New Algorithms for the Automated Microwave Tuner Test System," *RCA Review*, Vol. 46, September 1985, pp. 341–355.

5.7 D. Poulin, "Load-Pull Measurements Help You Meet Your Match," *Microwaves*, November 1980, pp. 61–65.

5.8 C. Rauscher and H. A. Willing, "Simulation of Nonlinear Microwave FET Performance Using a Quasi-Static Model," *IEEE Transactions on Microwave Theory and Techniques*, Vol. MTT-27, No. 10, October 1979, pp. 834–840.

5.9 W. R. Curtice and M. Ettenberg, "A Nonlinear GaAs FET Model for Use in the Design of Output Circuits for Power Amplifiers," *IEEE Transactions on Microwave Theory and Techniques*, Vol. MTT-33, No. 12, December 1985, pp. 1383–1394.

5.10 D. L. Peterson, A. M. Pavio, and B. Kim, "A GaAs FET Model for Large-signal Applications," *IEEE Transactions on Microwave Theory and Techniques*, Vol. MTT-32, No. 3, March 1984, pp. 276–281.

5.11 M. A. Smith, T. S. Howard, K. J. Anderson, and A. M. Pavio, "RF Nonlinear Device Characterisation Yields Improved Modeling Accuracy," *1986 IEEE MTT-S International Microwave Sysmposium Digest*, pp. 381–384.

5.12 V. Rizzoli, A. Lipparini, and E. Marazzi, "A General Purpose Program for Nonlinear Microwave Circuit Design," *IEEE Transactions on Microwave Theory and Techniques*, Vol. MTT-31, February 1983, pp. 762–770.

5.13 J. J. Ebers and J. L. Moll, "Large-Signal Behavior of Junction Transistors," *Proceedings of the IRE*, Vol. 42, December 1954, pp. 1761–1772.

5.14 C. T. Kirk, "A Theory of Transistor Cutoff Frequency Roll-off at High Current Density," *IEEE Transactions on Electron Devices*, 1962, pp. 162–174.

5.15 W. R. Curtice, "GaAs MESFET Modeling and Nonlinear CAD," *IEEE Transactions on Microwave Theory and Techniques*, Vol. MTT-36, February 1988, pp. 220–230.

5.16 D. C. Youla, "A New Theory of Broadband Matching," *IEEE Transactions on Circuit Theory*, March 1964, pp. 30–50.

5.17 R. M. Fano, "Theoretical Limitations on the Broadband Matching of Arbitrary Impedances," *Journal of the Franklin Institute*, Vol. 249, January 1960, pp. 57–83.

5.18 R. M. Fano, "Theoretical Limitations on the Broadband Matching of Arbitrary Impedances," *Journal of the Franklin Institute*, Vol. 249, February 1960, pp. 139–155.

5.19 R. M. Cottee and W. T. Joines, "Synthesis of Lumped and Distributed Networks for Impedance Matching of Complex Loads," *IEEE Transactions on Circuits and Systems*, Vol. CAS-26, No. 5, May 1979, pp. 316–329.

5.20 B. S. Yarman, "Modern Approaches to Broadband Matching Problems," *IEE Proceedings*, Vol. 132, No. 2, April 1985, pp. 87–92.

5.21 B. S. Yarman, "Real Frequency Broadband Matching Using Linear Programming," *RCA Review*, Vol. 43, December 1982, pp. 626–654.

5.22 B. S. Yarman, "New Approaches to Broadband Matching Problems," *IMS Workshop*, June 1983.

5.23 H. J. Carlin and P. P. Civalleri, "On Flat Gain with Frequency-Dependent Terminations," *IEEE Transactions on Circuits and Systems*, Vol. CAS-32, No. 8, August 1985, pp. 827–839.

5.24 D. J. Mellor, "Improved Computer-Aided Synthesis Tools for the Design of Matching Networks for Wide-Band Microwave Amplifiers," *IEEE Transactions on Microwave Theory and Techniques*, Vol. MTT-34, December 1986, pp. 1276–1281.

5.25 R. Levy, "Explicit Formulas for Chebyshev Impedance-Matching Networks, Filters and Interstages," *Proceedings of the IRE*, Vol. 111, No. 6, June 1964, pp. 1063–1103.

5.26 W. H. Horton, J. H. Jasberg, and J. D. Noe, "Distributed Amplifier: Practical Considerations and Experimental Results," *Proceedings of the IRE*, July 1950, pp. 748–753.

5.27 E. L. Ginzton, William R. Hewlett, John H. Jasberg, and Jerre D. Noe, "Distributed Amplifications," *Proceedings of the IRE*, August 1948, pp. 956–969.

5.28 W.-K. Chen, "Distributed Amplification Theory," Department of Engineering, Ohio University, Athens, Ohio, 1967, pp. 300–316.

5.29 A. S. Podgorski and L. Y. Wei, "Theory of Traveling-Wave Transistors," *IEEE Transactions on Electron Devices*, Vol. ED-29, No. 12, December 1982, pp. 1845–1853.

5.30 Y. Ayasli, R. L. Mozzi, J. L. Vorhaus, L. D. Reynolds, and R. A. Pucel, "A Monolithic GaAs 1–13 GHz Traveling-Wave Amplifier," *IEEE Transactions on Microwave Theory and Techniques*, Vol. MTT-30, No. 7, July 1982, pp. 976–981.

5.31 Y. Ayasli, L. D. Reynolds, R. L. Mozzi, and L. K. Hanes, "2–20 GHz GaAs Traveling-Wave Power Amplifier," *IEEE Transactions on Microwave Theory and Techniques*, Vol. MTT-32, No. 3, March 1984, pp. 290–295.

5.32 J. B. Beyer, S. N. Prasad, R. C. Becker, J. E. Nordman, G. K. Hohenwarter, and Y. Chen, "Wideband Monolithic Microwave Amplifier Study," Department of Electrical and Computer Engineering, University of Wisconsin-Madison, Madison, WI, September 1983, pp. 1–90.

5.33 J. B. Beyer, S. N. Prasad, R. C. Becker, J. E. Nordman, and G. K. Hohenwarter, "MESFET Distributed Amplifier Design Guidelines," *IEEE Transactions on Microwave Theory and Techniques*, Vol. MTT-32, No. 3, March 1984, pp. 268–275.

5.34 R. Halladay, M. Jones, and S. Nelson, "A Producible 2 to 20 GHz Monolithic Power Amplifier," *IEEE 1987 Microwave and Millimeter Wave Monolithic Circuits Symposium*, pp. 19–21.

5.35 E. M. Chase and W. Kennan, "A Power Distributed Amplifier Using Constant-R Networks," *1986 IEEE MTT Digest*, pp. 811–815.

5.36 C. Hutchinson and W. Kennan, "A Low Noise Distributed Amplifier with Gain Control," *1987 IEEE MTT Digest*, pp. 165–168.

5.37 R. Halladay, A. M. Pavio, and C. Crabill, "A 1–20 GHz Dual-Gate Distributed Power Amplifier," *1987 IEEE GaAs IC Symposium*, pp. 219–222.

5.38 J. R. Lane, R. G. Freitag, H.-K. Hahn, J. E. Degenford, and M. Cohn, "High Efficiency 1-, 2-, and 4-Watt Class-B FET Power Amplifiers," *IEEE Transactions on Microwave Theory and Techniques*, Vol. MTT-34, December 1986, pp. 1318–1326.

5.39 R. G. Freitag, J. E. Degenford, and M. Cohn, "High Efficiency Single-Ended and Push-Pull Class B FET Power Amplifiers," *1985 GOMAC Conference*, November 1985.

5.40 RCA, *RF Power Transistor Manual*, RCA Solid State Division, Somerville, N.J., 1971.

5.41 H. O. Granberg, "Building, Push-Pull, Multioctave VHF Power Amplifiers," *Microwaves & RF*, November 1987.

BIBLIOGRAPHY

Evans, A. D., *Designing with Field-Effect Transistors*, McGraw-Hill, New York, 1981.

Gentili, C., *Microwave Amplifiers and Oscillators*, McGraw-Hill, New York, 1987.

Getreu, I. E., *Modeling the Bipolar Transistor*, Elsevier, Amsterdam, 1978.

Gilmore, R., "Nonlinear Circuit Design Using a Modified Harmonic Balance Algorithm," *IEEE Transactions on Microwave Theory and Techniques*, Vol. MTT-34, December 1986, pp. 1294–1307.

Kennan, W., and N. K. Osbrink, "Distributed Amplifiers: Their Time Comes Again," *Microwaves & RF*, November 1984, pp. 119–153.

Kennan, W., T. Andrade, and C. Huang, "A Miniature 2–18 GHz Monolithic GaAs Distributed Amplifier," *IEEE Microwave and Millimeter Wave Monolithic Symposium Digest*, May 1984, pp. 41–44.

Marshall, N., Optimizing Multi-stage Amplifiers for Low-Noise," *Microwaves*, April 1974, pp. 62–64.

Mellor, D. J., and J. G. Linvill, "Synthesis of Interstage Networks of Presescribed Gain Versus Frequency Slopes," *IEEE Transactions on Microwave Theory and Techniques*, Vol. MTT-23, December 1975, pp. 1013–1020.

Strauss, L., *Wave Generation and Shaping*, McGraw-Hill, New York, 1970.

PROBLEMS

5.1. Derive (5.3) and (5.4) by assuming that the input is an open circuit at low frequencies. Show (5.3) can be simplified to (3.59): $|S_{21}| \simeq -2 g_m Z_0$.

5.2. Derive the DeLoach relations given by (5.5) to (5.7).

5.3. Using the method outlined by DeLoach, determine the gate capacitance of a typical FET when $T = -10$ dB, $T_m = -15$ dB, $f_1 = 4.5$ GHz and $f_2 = 5.5$ GHz. Assume typical values of g_m, R_s, and C_{gd} for a 300-μm FET.

5.4. For the FET *I–V* characteristics illustrated in Fig. 5.24, calculate the RF and DC load lines when the breakdown voltage is 12 V.

5.5. Assuming that the output characteristics of an FET can be modeled using the following simple elements, design a four-element matching network for the 6- to 12-GHz frequency range that optimally terminates the FET for maximum power performance.

$$C_{ds} = 0.17 \text{ pF} \qquad R_d = 4 \text{ }\Omega$$

$$R_{ds} = 270 \text{ }\Omega \qquad R_{\text{opt}} = 77 \text{ }\Omega$$

$$L_d = 0.22 \text{ nH}$$

5.6. Determine the largest gate periphery that can be used in the design of a six-cell distributed amplifier for the 2- to 8-GHz frequency range. Use the single-gate equivalent dual-gate FET model illustrated in Fig. 5.48 and scale the element values accordingly.

5.7. Derive the power gain expression in (5.11) by first determining the current delivered to the output load on a FET-by-FET basis for the general amplifier case.

5.8. Determine the average value of drain current for a class B amplifier stage when driven for maximum output voltage swing and biased at a quiescent current of 15% I_{dss}.

5.9. Show that the ideal class A amplifier has an efficiency of 50% and the ideal class B amplifier has an efficiency of $\pi/4$ or 78.5%.

6 Oscillator Design

6.0 INTRODUCTION

Oscillator design is very similar to amplifier design. The same transistors, the same dc bias levels, and the same set of S parameters can be used for the oscillator design. The load does not know whether it is connected to an oscillator or an amplifier (see Fig. 1.1).

For the amplifier design, M_1 and M_2 can be designed with a normal Smith chart, since S'_{11} and S'_{22} are normally less than unity. For oscillators, S'_{11} and S'_{22} are both greater than unity for oscillation. Thus a compressed Smith chart that includes reflection coefficients greater than unity is a useful tool for oscillator design.

Oscillators can be designed from several points of view.

1. S-parameter design [6.1,6.2]
2. Small-signal negative resistance from a transistor model [6.3]
3. Series or parallel resonance
4. Low noise [6.4–6.6]
5. Large-signal analytic approach [6.7–6.9]
6. Nonlinear analysis [6.10,6.11]

Each of these viewpoints will give additional insights into the many challenges facing oscillator designers. Referring to the oscillator of Fig. 1.1, we can subdivide the problem into the low-loss resonator M_3, the active two-port with its S-parameter description, and the passive lossy load M_4.

In this chapter we emphasize the S-parameter design approach, which is the most useful to microwave designers. Before developing this concept, the compressed Smith chart or negative resistance Smith chart will be developed as a design tool. Next, the oscillator design is viewed as either series or parallel resonance, a one-port design. Then the various resonators available to microwave designers are reviewed. Then two-port (or n-port) oscillator design is presented. At this point some design examples are given to demonstrate some useful designs. These include bipolar designs using lumped elements, distributed elements, and dielectric resonators.

Negative resistance of an oscillator can also be derived from the transistor equivalent circuit. An example of this design method is given, leading to a wideband VCO design over 200 to 400 MHz. Some historically significant oscillator circuits are also given in this section.

The definition of oscillator Q is presented next. The two common techniques for measuring oscillator Q, load pulling and injection locking, are discussed.

384

The various descriptions of oscillator noise are developed leading to Leeson's noise model for the oscillator. Low-noise design examples are given, including both varactor tuned (VCO), YIG tuned, and DRO fixed tuned. In addition, noise degeneration is shown to produce lower oscillator noise.

The large-signal analytic approach to oscillator design will be presented with a design example at 5.3 GHz using a power GaAs MESFET. Then a nonlinear oscillator model is used to design an 8.8-GHz GaAs MESFET oscillator. The present state of the art for oscillators is summarized at the conclusion of this chapter.

The conditions for oscillation can be expressed as

$$k < 1 \tag{6.1}$$

$$\Gamma_G S'_{11} = 1 \tag{6.2}$$

$$\Gamma_L S'_{22} = 1 \tag{6.3}$$

The stability factor should be less than unity for any possibility of oscillation. If this condition is not satisfied, either the common terminal should be changed or positive feedback should be added. Next, the passive terminations Γ_G and Γ_L must be added to resonate the input and output ports at the frequency of oscillation. This is satisfied by either (6.2) or (6.3). It will be shown in Section 6.4 that if (6.2) is satisfied, (6.3) must be satisfied, and vice versa. In other words, if the oscillator is oscillating at one port, it must be simultaneously oscillating at the other port. Normally a major fraction of the power is delivered only to one port, since only one load is connected. Since $\left|\Gamma_g\right|$ and $\left|\Gamma_L\right|$ are less than unity, (6.2) and (6.3) imply that $\left|S'_{11}\right| > 1$ and $\left|S'_{22}\right| > 1$.

The conditions for oscillation can be seen from Fig. 6.1, where an input generator has been connected to a two-port. Using (1.180) for the representation of the generator, which is repeated here:

$$a_1 = b_G + \Gamma_1 \Gamma_G a_1 \tag{6.4}$$

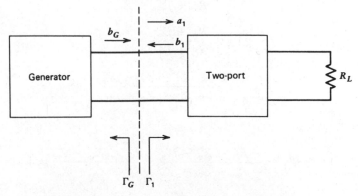

Figure 6.1   Two-port connected to a generator.

and defining

$$\Gamma_1 = S'_{11} \tag{6.5}$$

$$S'_{11} = \frac{b_1}{a_1} \tag{6.6}$$

gives

$$b_G = a_1(1 - \Gamma_1\Gamma_G)$$

$$= \frac{b_1}{S'_{11}}(1 - S'_{11}\Gamma_G) \tag{6.7}$$

$$\frac{b_1}{b_G} = \frac{S'_{11}}{1 - S'_{11}\Gamma_G} \tag{6.8}$$

Thus the wave reflected from the two-port is dependent on b_G, S'_{11}, and Γ_G. If (6.2) is satisfied, b_G must be zero, which implies that the two-port is oscillating. Since $|\Gamma_G|$ is normally less than or equal to unity, this requires that $|S'_{11}|$ be greater than or equal to unity.

The oscillator designer must simply guarantee a stability factor less than unity and resonate the input port by satisfying (6.2), which implies that (6.3) has also been satisfied. Another way of expressing the resonance condition of (6.2) is the following:

$$R_{in} + R_G = 0 \tag{6.9}$$

$$X_{in} + X_G = 0 \tag{6.10}$$

This follows from substituting

$$S'_{11} = \frac{R_{in} + jX_{in} - Z_0}{R_{in} + jX_{in} + Z_0} \tag{6.11}$$

$$\Gamma_G = \frac{R_G + jX_G - Z_0}{R_G + jX_G + Z_0}$$

$$= \frac{-R_{in} - Z_0 - jX_{in}}{-R_{in} + Z_0 - jX_{in}} \tag{6.12}$$

into (6.2), giving

$$\Gamma_G S'_{11} = \frac{-R_{in} - Z_0 - jX_{in}}{-R_{in} + Z_0 - jX_{in}} \cdot \frac{R_{in} + jX_{in} - Z_0}{R_{in} + Z_0 + jX_{in}} = 1$$

which proves the equivalence of (6.2) to (6.9) and (6.10).

TABLE 6.1   Typical Oscillator Specifications

Parameter	High-Q or Cavity-Tuned (e.g., YIG)	Low-Q or Varactor-Tuned VCO
Frequency	2 to 4 GHz	2 to 4 GHz
Power	+10 dBm	+10 dBm
Power variation versus f	±2 dB	±2 dB
Temperature stability versus f	±10 ppm/°C	±500 ppm/°C
Power versus temperature (−30 to 60°C)	±2 dB	±2 dB
Modulation sensitivity	10 to 20 MHz/mA	50 to 200 MHz/V
FM noise	−110 dBc/Hz at 100 kHz	−100 dBc/Hz at 100 kHz
AM noise	−140 dBc/Hz at 100 kHz	−140 dBc/Hz at 100 kHz
FM noise floor	−150 dBc/Hz at 100 MHz	−150 dBc/Hz at 100 MHz
All harmonics	−20 dBc	−20 dBc
Short-term post tuning drift	±2 MHz 1 μs	±2 MHz 1 to 100 μs
Long-term post tuning drift	±2 MHz 5 to 30 s	±2 MHz 5 to 30 s
Pulling of f all phases of 12-dB return loss	±1 MHz	±20 MHz
Pushing of f with change of bias voltage	5 MHz/V	5 MHz/V

Before proceeding with the oscillator design procedures, some typical oscillator specifications are given in Table 6.1 for the major types of oscillators. The high-Q or cavity-type oscillators usually have better spectral purity (see Section 6.7) than do the low-Q VCOs (voltage-controlled oscillators), which have faster tuning speeds. The resonators are described in Section 6.3. The FM noise is usually measured at about 100 kHz from the carrier in units of dBc, which means decibels below the carrier level, in a specified bandwidth of 1 Hz. If the measurement bandwidth is 1 kHz, the specification changes by 10^3, as discussed in Section 6.7.

In selecting a transistor to meet the specifications, the amplifier transistors with the same frequency and power performance are usually suitable. Lower close-in noise can be achieved from silicon bipolar transistors compared to GaAs MESFETs because of the $1/f$ noise difference described in Fig. 3.22.

6.1 THE COMPRESSED SMITH CHART

The normal Smith chart is a plot for the reflection coefficient of $|\Gamma| \leq 1$. The compressed Smith chart includes $|\Gamma| > 1$, and the chart is given in Fig. 6.2 for

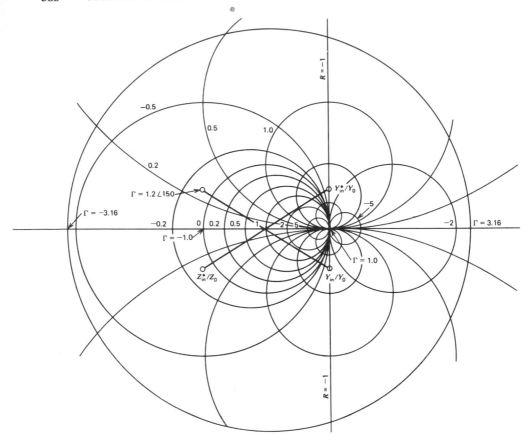

Figure 6.2 Compressed Smith chart.

$|\Gamma| \le 3.16$ (10 dB of return gain). This chart is useful for plotting the variation of S'_{11} and S'_{22} for oscillator design. The impedance and admittance properties of the Smith chart are retained for the compressed chart. For example, a Γ_{in} of 1.2 $\underline{/150°}$ gives the following values of Z and Y normalized to $Z_0 = 50\ \Omega$.

$$Z_{in}/Z_0 = -0.10 + j0.25$$

$$Z^*_{in}/Z_0 = -0.10 - j0.25$$

$$Y_{in}/Y_0 = -1.0 - j3.0$$

$$Y^*_{in}/Y_0 = -1.0 + j3.0$$

These values are plotted in Fig. 6.2 for illustration.

A frequency resonance condition simply requires the circuit imaginary term be zero. If the impedance resonance is on the left-hand real axis, this is a series resonance; that is, at frequencies above resonance the impedance is inductive and below resonance the impedance is capacitive. If the impedance resonance is on the right-hand real axis, the resonance is a parallel resonance; that is, at frequencies above resonance the impedance is capacitive and below resonance the impedance is inductive.

An oscillator resonance condition implies that both the circuit imaginary term and the circuit real term are zero, as given by (6.9) and (6.10). Impedances and admittances can be transformed on the compressed Smith chart by the methods discussed in Section 3.4; however, when $|\Gamma|$ is greater than unity, the goal of impedance transformation is usually to achieve either a series or a parallel resonance condition. Another method for visualizing negative resistance is to plot $1/S_{11}$ and multiply the result by -1. This allows the designer to use readily available Smith charts, with $|\Gamma| \leq 1$, to analyze circuits with $|\Gamma| \geq 1$. The proof of this concept can be shown by expressing the reflection coefficient of a one-port by

$$S_{11} = \frac{Z_s - Z_0}{Z_s + Z_0} \qquad (6.13)$$

$$\frac{1}{S_{11}} = \frac{Z_s + Z_0}{Z_s - Z_0} = \frac{Z_1 - Z_0}{Z_1 + Z_0} \qquad (6.14)$$

where $Z_1 = -Z_s$, which gives a negative resistance on Smith chart coordinates. For example, using the case in Fig. 6.2,

$$S_{11} = 1.2 \; \underline{/150^\circ}$$

$$\frac{1}{S_{11}} = 0.833 \; \underline{/-150^\circ}$$

$$\frac{Z_1}{Z_0} = 0.10 - j0.25$$

$$\frac{Z_s}{Z_0} = -0.10 + j0.25$$

The impedance of the one-port is plotted at Z_1 but understood to be Z_s.

6.2 SERIES OR PARALLEL RESONANCE

Oscillators can be classified into two types, series-resonant or parallel-resonant, as shown in Fig. 6.3. The equivalent circuit of the active device is chosen from the frequency response of the output port, that is the frequency response of Γ_G. For the series-resonant condition, the negative resistance of the active device must exceed the load resistance R_L at start-up of oscillation by about 20%. As the oscil-

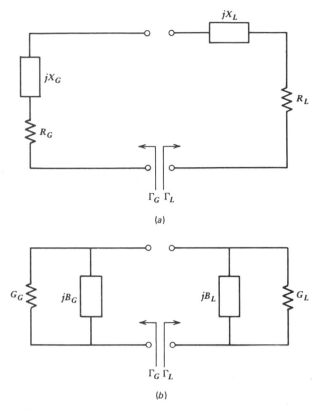

Figure 6.3 Oscillator equivalent circuits: (a) series-resonant; (b) parallel-resonant.

lation builds up to a steady-state value, the resonance condition will be reached as a result of limiting effects, which cause a reduction of R_G under large-signal drive.

For startup of oscillation

$$|R_G| > 1.2R_L \qquad (6.15)$$

for resonance

$$R_G + R_L = 0 \qquad (6.16)$$

$$X_G + X_L = 0 \qquad (6.17)$$

For the parallel resonant condition, the negative conductance of the active device must exceed the load conductance G_L at startup of oscillation by about 20%. The parallel resonant oscillator is simply the dual of the series resonant case. For start up of oscillation

$$|G_G| > 1.2G_L \qquad (6.18)$$

For resonance

$$G_G + G_L = 0 \qquad (6.19)$$

$$B_G + B_L = 0 \qquad (6.20)$$

To design the oscillator for series resonance, the reflection coefficient of the active transistor is moved to an angle of $180°$ (i.e., the left-hand real axis of the compressed Smith chart). Keeping in mind (6.2) for the input resonating port, we see that a nearly lossless reactance will resonate the transistor. For the example in Fig. 6.2,

$$\Gamma = 1.2 \; \underline{/150°} = S_{11}'$$
$$\Gamma_G = 0.83 \; \underline{/-150°} \simeq 1.0 \; \underline{/-150°}$$

The large-signal drive of the transistor will reduce S_{11}' to about $1.0 \underline{/150°}$. For parallel resonance oscillator design, the reflection coefficient of the active transistor is moved to an angle of $0°$ (i.e., the right-hand real axis of the compressed Smith chart). Alternatively, the reflection coefficient associated with impedance can be inverted to an admittance point, and the admittance can be moved to an angle of $180°$ (i.e., the left-hand real axis of the compressed Smith chart).

6.3 RESONATORS

Oscillators are often named by the type of resonator connected to the tuning port to give the desired Γ_G. The most common resonators are:

1. Lumped element
2. Distributed transmission line (microstripline or coaxial line)
3. Cavity
4. Dielectric resonator
5. YIG
6. Varactor

All of these structures can be made to have low losses and high Q. The first four types give a fixed-tuned or mechanically tuned oscillator. The YIG or varactor resonator will give a wideband tunable oscillator. For a high-Q resonator (>50), the reflection coefficient is simply the outer boundary of the Smith chart, with a phase depending on the transmission-line length between the resonator and the active two-port device.

The lumped-element resonators are high-Q capacitors and inductors, with associated parasitics. For example, a high-Q chip capacitor 0.050 in. in length has a typical parasitic inductance of 0.5 nH.

The distributed elements using microstripline are usually either open or shorted transmission lines of the correct length to give the proper angle to Γ_G. Other planar microstrip structures to consider are [6.12]:

1. Rectangular $\lambda/2$ microstripline
2. Circular disk
3. Circular microstrip ring
4. Triangular microstrip
5. Hexagonal microstrip
6. Elliptic microstrip

All of these resonators can be made high Q on low-loss dielectric substrates.

Cavity resonators can be made from low-loss coaxial line or waveguide. The simplest coaxial cavity is a $\lambda/4$ shorted stub, where the output is coupled by a shorted loop (magnetic coupling) or an open probe (electric coupling). A mechanical tuning screw near the open-circuit end can be used to shorten the line and therefore raise the resonant frequency.

The lowest-order rectangular waveguide cavity resonator is the TE_{101} mode, where the width and length of the cavity are $\lambda_g/2$ at the resonant frequency. Circular or elliptic waveguides can also be used as a cavity resonator.

6.3.1   Dielectric Resonators

A very popular low-cost resonator is the dielectric puck coupled to a microstrip structure, as shown in Fig. 6.4. The lowest-order resonant mode is the $TE_{01\delta}$ mode, which easily couples to the microstrip TEM mode. To use dielectric resonators effectively in microwave circuits, it is necessary to have an accurate knowledge of the coupling between the resonator and various transmission lines. Figure 6.4 shows the magnetic coupling between a dielectric resonator and a microstrip. A dielectric spacer may be added under the puck to improve the loaded Q by optimizing the

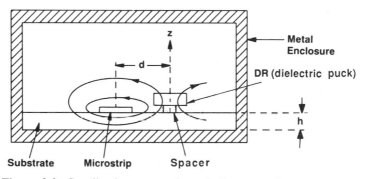

Figure 6.4   Coupling between a microstrip line and a dielectric resonator.

coupling. The resonator is placed on top of the microstrip substrate. The lateral distance between the resonator and the microstrip conductor primarily determines the amount of coupling between the resonator and the microstrip transmission line. Metallic shielding is required to minimize the radiation losses.

The $TE_{01\delta}$ mode in a dielectric resonator on top of a dielectric spacer can be approximated by a magnetic dipole of moment M. The coupling between the line and the resonator is accomplished by orienting the magnetic moment of the resonator perpendicular to the microstrip plane so that the magnetic lines of the resonator link with those of the microstrip line, as shown in Fig. 6.4. The dielectric resonator placed adjacent to the microstrip line operates like a reaction cavity that reflects the radio-frequency (RF) energy at the resonant frequency. It is similar to an open circuit with a voltage maximum at the reference plane at the resonant frequency. The equivalent circuit of the resonator coupled to a microstrip line is shown in Fig. 6.5, where L_r, C_r, and R_r are the equivalent parameters of the dielectric resonator, L_1, C_1, and R_1 are the equivalent parameters of the microstrip line, and L_m characterizes the magnetic coupling. The transformed resonator impedance Z in series with the transmission line is given by

$$Z = j\omega L_1 + \frac{\omega^2 L_m^2}{R_r + j\omega(L_r - 1/\omega^2 C_r)} \qquad (6.21)$$

Near resonance, ωL_1 can be neglected and Z becomes

$$Z = \omega Q_u \frac{L_m^2}{L_r} \frac{1}{1 + jX} \qquad (6.22)$$

where $X = 2Q_u (\Delta\omega/\omega)$, and unloaded Q and the resonant frequency of the resonator are given by

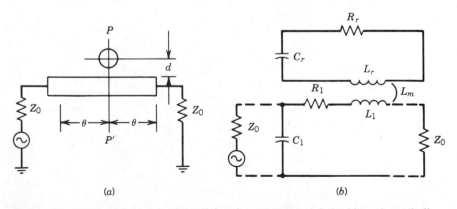

(a)   (b)

Figure 6.5 Equivalent circuit of the dielectric resonator coupled with a microstrip line.

$$Q_u = \frac{\omega_0 L_r}{R_r} \qquad (6.23)$$

$$\omega_0 = \frac{1}{\sqrt{L_r C_r}} \qquad (6.24)$$

At resonance, $X = 0$ and

$$Z = R = \omega_0 Q_u \frac{L_m^2}{L_r} \qquad (6.25)$$

Equation (6.25) indicates that the circuit shown in Fig. 6.5 can be represented by the parallel tuned circuit as shown in Fig. 6.6, where L, R, and C are given by

$$L = \frac{L_m^2}{L_r} \qquad (6.26)$$

$$C = \frac{L_r}{\omega_0^2 L_m^2} \qquad (6.27)$$

and

$$R = \omega_0 Q_u \frac{L_m^2}{L_r} \qquad (6.28)$$

The coupling coefficient β at the resonant frequency ω_0 is defined by

$$\beta = \frac{R}{R_{ext}} = \frac{R}{2Z_0} = \frac{\omega_0 Q_u}{2Z_0} \frac{L_m^2}{L_r} \qquad (6.29)$$

If S_{110} and S_{210} are defined as the reflection and transmission coefficients at the

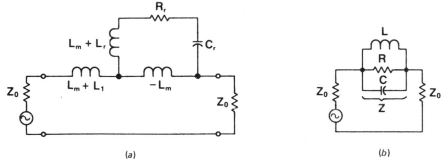

(a)   (b)

Figure 6.6 (a) Simplified equivalent circuit; (b) final equivalent circuit of a dielectric resonator coupled with a microstrip line.

resonant frequency of the resonator coupled to the microstrip, β can be shown to be given by [6.13]

$$\beta = \frac{S_{110}}{1 - S_{110}} = \frac{1 - S_{210}}{S_{210}} = \frac{S_{110}}{S_{210}} \tag{6.30}$$

This relation can be used to determine the coupling coefficient from the directly measurable reflection and transmission coefficients. The value of β can also be accurately calculated from a knowledge of the circuit configuration. The quantity L_m^2/L_r in (6.26) is a strong function of the distance between the resonator and the microstrip line for given shielding conditions and substrate thickness and dielectric constant. The relation between different quality factors is well known and is given by

$$Q_u = Q_L(1 + \beta) = Q_e\beta \tag{6.31}$$

The external quality factor Q_e is used to characterize the load coupling.

The S parameters of the dielectric resonator coupled to a microstrip with the lengths of transmission lines on input and output, as shown in Fig. 6.5, can be determined from the previous relations and are given by [6.14]

$$S = \begin{bmatrix} \dfrac{\beta}{\beta + 1 + jQ_u\Delta\omega/\omega_0} e^{-2j\theta} & \dfrac{1 + jQ_u\Delta\omega/\omega_0}{\beta + 1 + jQ_u\Delta\omega/\omega_0} e^{-2j\theta} \\[4mm] \dfrac{1 + jQ_u\Delta\omega/\omega_0}{\beta + 1 + jQ_u\Delta\omega/\omega_0} e^{-2j\theta} & \dfrac{\beta}{\beta + 1 + jQ_u\Delta\omega/\omega_0} e^{-2j\theta} \end{bmatrix} \tag{6.32}$$

where 2θ is the electrical line length between the input and output planes.

6.3.2 YIG Resonators

For wideband electrically tunable oscillators, we use either a YIG or a varactor resonator. The YIG resonator is a high-Q, ferrite sphere of yttrium iron garnet, $Y_2Fe_2(FeO_4)_3$, that can be tuned over a wide band by varying the biasing dc magnetic field. Its high performance and convenient size for applications in microwave integrated circuits make it an excellent choice in a large number of applications, such as filters, multipliers, discriminators, limiters, and oscillators. A YIG resonator makes use of the ferrimagnetic resonance, which, depending on the material composition, size, and applied field, can be achieved from 500 MHz to 50 GHz [6.15]. An unloaded Q greater than 1000 is usually achieved with typical YIG material.

6.3.3 Varactor Resonators

The dual of the current-tuned YIG resonator is the voltage-tuned varactor, which is a variable reactance achieved from a low-loss, reverse-biased semiconductor *pn* junction. These diodes are designed to have very low loss and therefore high Q. The silicon varactors have the fastest settling time in fast-tuning applications, but the gallium arsenide varactors have higher Q values. The cutoff frequency of the varactor is defined as the frequency where $Q_v = 1$. For a simple series RC equivalent circuit, we have

$$Q_v = \frac{1}{\omega RC_v} \tag{6.33}$$

$$f_{c0} = \frac{1}{2\pi RC_v} \tag{6.34}$$

The tuning range of the varactor will be determined by the capacitance ratio C_{max}/C_{min}, which can be 12 or higher for hyper-abrupt varactors. Since R is a function of bias, the maximum cutoff frequency occurs at a bias near breakdown, where both R and C_v have minimum values.

6.3.4   Resonator Measurements

Accurate characterization of microwave resonators is essential for their effective use. The important parameters that are required to describe fully a resonator for a given mode are the resonant frequency f_0, the coupling coefficient, and the quality factors Q_u (unloaded Q), Q_L (loaded Q), and Q_e (external Q due to resistive loading). The network analyzer displays the magnitude and phase of the reflection and transmission coefficients for the single-port resonator. Many methods for Q measurement are possible, but we will describe here only one simple technique using Q loci on the Smith chart.

The single-ended resonator is the most commonly used configuration for microwave resonant circuits. The parallel tuned circuit is known as the detuned short configuration, and the series-tuned circuit is known as the detuned open configuration. Either configuration can be converted to the other by displacing the reference plane by a quarter wavelength. The important parameters of these resonant circuits are defined in Table 6.2.

TABLE 6.2   Resonator Parameters

Parameter	Series Tuned	Parallel Tuned
f_0	$\dfrac{1}{\sqrt{LC}}$	$\dfrac{1}{\sqrt{LC}}$
Q_u	$\dfrac{\omega L}{R}$	$\dfrac{R}{\omega L}$
$\beta = \dfrac{Q_u}{Q_e}$	$\dfrac{Z_0}{R}$	$\dfrac{R}{Z_0}$
Q_L	$\dfrac{Q_u}{1+\beta}$	$\dfrac{Q_u}{1+\beta}$

Since analysis of the resonant circuits is similar for the two configurations, we restrict our discussion to the parallel tuned circuit. The input impedance of the parallel resonant structure can be written

$$\frac{1}{Z_{in}} = \frac{1}{R} + \frac{1}{j\omega L} + j\omega C \tag{6.35}$$

or

$$Z_{in} = \frac{R}{1 + 2jQ_u\delta} \tag{6.36}$$

where $\delta = (\omega - \omega_0)/\omega_0$ represents the frequency detuning parameter.

The locus of the impedance, using (6.36), can be drawn by varying frequency. As impedance is a linear function of frequency, a circular locus will be produced when plotted on the Smith chart as illustrated by circles A, B, and C in Fig. 6.7. Circle A, for which $R \simeq Z_0$ passes near the origin, is called the condition of critical coupling ($\beta = 1$ from Table 6.2) since it provides a perfect match to the transmission line at resonance. Circle C, with $R > Z_0$, is said to be overcoupled ($\beta > 1$), and circle B, with $R < Z_0$, is undercoupled. The coupled coefficient for any given impedance locus can be easily determined by measuring the reflection coefficient S_{110} at resonance.

For the undercoupled case,

$$\beta = \frac{1 - S_{110}}{1 + S_{110}} \tag{6.37}$$

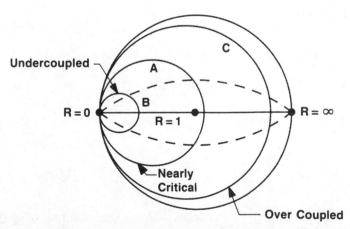

Figure 6.7 Input impedance of a resonant cavity referred to the detuned-short position plotted on the Smith chart for three degrees of coupling.

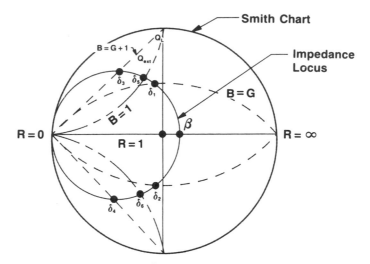

Figure 6.8 Identification of the half-power points from the Smith chart. Q_u locus given by $B = G$ $(X = R)$; Q_L by $B = G + 1$; Q_e by $B = 1$.

and for the overcoupled case,

$$\beta = \frac{1 + S_{110}}{1 - S_{110}} \qquad (6.38)$$

The evaluation of β locates the intersection of the impedance circle with the real axis, as shown in Fig. 6.8. To measure various quality factors, (6.36) can be written as

$$\bar{Z}_{\text{in}} = \frac{Z_{\text{in}}}{Z_0} = \frac{\beta}{1 + 2jQ_u\delta} = \frac{\beta}{1 + 2jQ_L(1 + \beta)\delta} = \frac{\beta}{1 + 2jQ_e\beta\delta} \qquad (6.39)$$

where Q_u, Q_L, and Q_e are interrelated by the well-known relation

$$Q_u = Q_L(1 + \beta) = Q_e\beta \qquad (6.40)$$

The normalized frequency deviations corresponding to various quality factors are given by

$$\delta_u = \pm\frac{1}{2Q_u} \qquad \delta_L = \pm\frac{1}{2Q_L} \qquad \delta_e = \pm\frac{1}{2Q_e} \qquad (6.41)$$

The impedance locus of Q_u, for example, can be determined by using (6.41) in (6.39) and is given by

$$(Z_{in})_u = \frac{\beta}{1 \pm j} \tag{6.42}$$

Equation (6.42) represents the points on the impedance locus where the real and imaginary parts of the impedance are the same. Figure 6.8 represents the locus of these points (corresponding to $B = G$) for all possible values of B. This locus is an arc whose center is at $Z = 0 \pm j$, and the radius is the distance to the point $0 \pm j0$. The intersection of this arc with the impedance locus determines the Q_u measurement points,

$$Q_u = \frac{f_0}{f_1 - f_2} \tag{6.43}$$

The frequencies f_1 and f_2 are called half-power points, because these points correspond to $R = X$ on the impedance locus, or $B = G$ on the admittance locus.

The loaded and external Q values can be determined in a similar way. The impedances corresponding to Q_e and Q_L are

$$(Z_{in})_e = \frac{\beta}{1 \pm j\beta} \tag{6.44}$$

and

$$(Z_{in})_L = \frac{\beta}{1 \pm j(1 + \beta)} \tag{6.45}$$

Using (6.44) and (6.45), the Q_e and Q_L loci can be easily determined. These loci are shown in Fig. 6.8.

6.4 TWO-PORT OSCILLATOR DESIGN

A common method for designing oscillators is to resonate the input port with a passive high-Q circuit at the desired frequency of resonance. It will be shown that if this is achieved with a load connected on the output port, the transistor is oscillating at both ports and is thus delivering power to the load port. The oscillator may be considered a two-port structure as shown in Fig. 1.1, where M_3 is the lossless resonating port and M_4 provides lossless matching such that all of the external RF power is delivered to the load. The resonating network has been described in Section 6.3. Normally, only parasitic resistance is present at the resonating port, since a high-Q resonance is desirable for minimizing oscillator noise. It is possible to have loads at both the input and the output ports if such an application occurs, since the oscillator is oscillating at both ports simultaneously.

The simultaneous oscillation condition is proved as follows. Assume that the oscillation condition is satisfied at port 1:

$$1/S'_{11} = \Gamma_G \qquad (6.46)$$

From (1.125),

$$S'_{11} = S_{11} + \frac{S_{12}S_{21}\Gamma_L}{1 - S_{22}\Gamma_L} = \frac{S_{11} - D\Gamma_L}{1 - S_{22}\Gamma_L} \qquad (6.47)$$

$$\frac{1}{S'_{11}} = \frac{1 - S_{22}\Gamma_L}{S_{11} - D\Gamma_L} = \Gamma_G \qquad (6.48)$$

By expanding (6.48), we get

$$\Gamma_G S_{11} - D\Gamma_L\Gamma_G = 1 - S_{22}\Gamma_L$$

$$\Gamma_L(S_{22} - D\Gamma_G) = 1 - S_{11}\Gamma_G$$

$$\Gamma_L = \frac{1 - S_{11}\Gamma_G}{S_{22} - D\Gamma_G} \qquad (6.49)$$

From (1.126),

$$S'_{22} = S_{22} + \frac{S_{12}S_{21}\Gamma_G}{1 - S_{11}\Gamma_G} = \frac{S_{22} - D\Gamma_G}{1 - S_{11}\Gamma_G} \qquad (6.50)$$

$$\frac{1}{S'_{22}} = \frac{1 - S_{11}\Gamma_G}{S_{22} - D\Gamma_G} \qquad (6.51)$$

Comparing (6.49) and (6.51) gives

$$1/S'_{22} = \Gamma_L \qquad (6.52)$$

which means that the oscillation condition is also satisfied at port 2; this completes the proof. Thus if either port is oscillating, the other port must be oscillating as well. A load may appear at either or both ports, but normally the load is in Γ_L, the output termination. This result can be generalized to an n-port oscillator by showing that the oscillator is simultaneously oscillating at each port [6.2, 6.12]:

$$\Gamma_1 S'_{11} = \Gamma_2 S'_{22} = \Gamma_3 S'_{33} = \cdots = \Gamma_n S'_{nn} \qquad (6.53)$$

Before concluding this section on two-port oscillator design, the buffered oscillator shown in Fig. 6.9 must be considered. This design approach is used to provide the following:

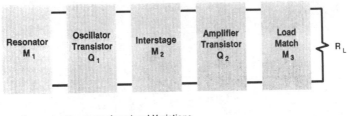

Decouples Resonator from Load Variations

Similar Devices for Q $_1$ and Q $_2$

P_{OUT} Q_2 > P_{OUT} Q_1

Figure 6.9 Buffered oscillator design.

1. A reduction in loading–pulling, which is the change in oscillator frequency when the load reflection coefficient changes.
2. A load impedance that is more suitable to wideband applications [6.1].
3. A higher output power from a working design, although the higher output power can also be achieved by using a larger oscillator transistor.

Buffered oscillator designs are quite common in wideband YIG applications, where changes in the load impedance must not change the generator frequency.

Two-port oscillator design may be summarized as follows:

1. Select transistor with sufficient gain and output power capability for the frequency of operation. This may be based on oscillator data sheets, amplifier performance, or S-parameter calculation.
2. Select a topology that gives $k < 1$ at the operating frequency. Add feedback if $k < 1$ has not been achieved.
3. Select an output load matching circuit that gives $|S'_{11}| > 1$ over the desired frequency range. In the simplest case this could be a 50-Ω load.
4. Resonate the input port with a lossless termination so that $\Gamma_G S'_{11} = 1$. The value of S'_{22} will be greater than unity with the input properly resonated.

In all cases the transistor delivers power to a load and the input of the transistor. Practical considerations of realizability and dc biasing will determine the best design.

For both bipolar and FET oscillators, a common topology is common-base or common-gate, since a common-lead inductance can be used to raise S_{22} to a large value, usually greater than unity even with a 50-Ω generator resistor. However, it is not necessary for the transistor S_{22} to be greater than unity, since the 50-Ω generator is not present in the oscillator design. The requirement for oscillation is $k < 1$; then resonating the input with a lossless termination will provide that $|S'_{22}| > 1$.

Table 6.3   HP2001 Bipolar Chip Common Base
(V_{CE} = 15 V, I_C = 25 mA)

L_B = 0	L_B = 0.5 nH
S_{11} = 0.94 $\underline{/174°}$	1.04 $\underline{/173°}$
S_{21} = 1.90 $\underline{/-28°}$	2.00 $\underline{/-30°}$
S_{12} = 0.013 $\underline{/98°}$	0.043 $\underline{/153°}$
S_{22} = 1.01 $\underline{/-17°}$	1.05 $\underline{/-18°}$
k = −0.09	−0.83

A simple example will clarify the design procedure. A common-base bipolar transistor (HP2001) was selected to design a fixed-tuned oscillator at 2 GHz. The common-base S parameters and stability factor are given in Table 6.3. Using the load circuit in Fig. 6.10, we see that the reflection coefficients are

$$\Gamma_L = 0.62 \ \underline{/30°}$$

$$S'_{11} = 1.18 \ \underline{/173°}$$

Thus a resonating capacitance of C = 20 pF resonates the input port. In a YIG-tuned oscillator, this reactive element could be provided by the high-Q YIG element. For a dielectric resonator oscillator (DRO), the puck would be placed to give $\Gamma_G \simeq 1.0 \underline{/-173°}$.

Another two-port design procedure is to resonate the Γ_G port and calculate S'_{22} until $|S'_{22}| > 1$; then design the load port to satisfy (6.3). This design procedure is summarized in Fig. 6.11.

An example using this procedure at 4 GHz is given in Fig. 6.12 using an AT-41400 silicon bipolar chip in the common-base configuration with a convenient value of base and emitter inductance 0.5 nH. The feedback parameter is the base

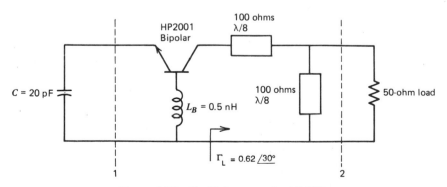

Figure 6.10   Oscillator example at 2 GHz.

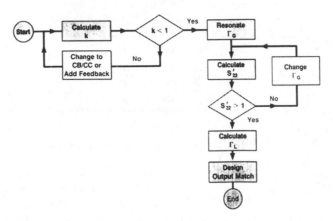

Figure 6.11 Oscillator design flowchart.

inductance, which can be varied if needed. The two-port common-base S parameters were used to give

$$k = -0.805$$

$$S'_{11} = 1.212 \underline{/137.7°}$$

Since a lossless capacitor at 4 GHz of 2.06 pF gives $\Gamma_G = 1.0 < -137.7°$, this input termination is used to calculate S'_{22} from (1.126) giving $S'_{22} = 0.637 < 44.5°$. This circuit will not oscillate into any passive load. Varying the emitter capacitor about 20° on the Smith chart to 1.28 pF gives $S'_{22} = 1.16 < -5.5°$, which will oscillate into a load of $\Gamma_L = 0.861 < 5.5°$. The completed lumped-element design is given in Fig. 6.13.

We now switch from the lumped design to a microstrip design that incorporates a dielectric resonator, which was described in Section 6.3. This oscillator circuit

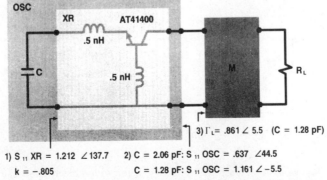

1) S_{11} XR $= 1.212 \angle 137.7$ 2) $C = 2.06$ pF: S_{11} OSC $= .637 \angle 44.5$
   $k = -.805$ $C = 1.28$ pF: S_{11} OSC $= 1.161 \angle -5.5$

3) $\Gamma_L = .861 \angle 5.5$ (C = 1.28 pF)

Figure 6.12 A 4-GHz lumped resonator oscillator using AT41400.

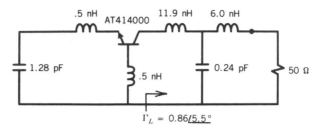

Figure 6.13   Completed lumped resonator oscillator (LRO).

is given in Fig. 6.14, where the dielectric resonator (DR) will serve the function of the emitter capacitor. This element is usually coupled to the 50-Ω microstripline to present about 1000 Ω of loading ($\beta \simeq 20$) at f_0, the lowest resonant frequency of the dielectric puck, at the correct position on the line. The load circuit will be simplified to 50 Ω ($\Gamma_L = 0$), so the oscillator must have an output reflection coefficient of greater than 100, thus presenting a negative resistance between -49 and -51 Ω. The computer file for analyzing this design is given in Table 6.4, where the variables are the puck resistance, the 50-Ω microstripline length, and the base feedback inductance. The final design is given in Fig. 6.15, where the 10-μH coils are present for the dc bias connections that need to be added to the design. It is important to check the stability of this circuit with the DR removed. The input 50-Ω termination will usually guarantee unconditional stability at all frequencies. The phase noise of this oscillator is very low at -117 dBc/Hz at 10 kHz frequency offset [6.16], which is discussed in Sections 6.7 and 6.10.

Another DRO example using the parallel feedback and the MSA-0835, a silicon MMIC described in Chapter 3, is shown in Fig. 6.16. In this oscillator the dielectric puck will load the input and output transmission lines with about 1000 Ω at the correct microstripline position to give an output reflection coefficient greater

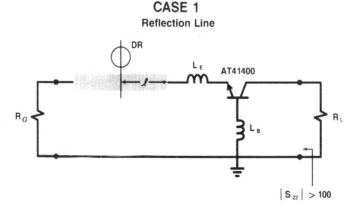

CASE 1

Reflection Line

Figure 6.14   Transmission-line oscillator with dielectric resonator.

TABLE 6.4 Super-Compact File for DRO Design in Fig. 6.15

```
*
-
* AT41400 AT 7.5V, 30mA IN DRO
* OSCILLATOR By Vendelin et al. Microwave Journal June
1986 pp. 151-152
BLK
        TRL         1    2            Z=50 P=250MIL K=6.6
        RES         2    3            R=?955.06?
        TRL         3    4            Z=50 P=?224.16MIL?
                                      K=6.6
        IND         4    0            L=1E4NH
        IND         4    5            L=.5NH
        TWO         6    7    5       Q1
        IND         6    0            L=?.33843NH?
        IND         7    0            L=1E4NH
        OSC:2POR    1    7
END
*
FREQ
        4GHZ
END
OUT
        PRI  OSC  S
END
OPT
        OSC
        MS22 = 100 GT
END
DATA
   Q1:S
    4    .8057 -176.14 2.5990 74.77 .0316 56.54 .4306
         -22.94
END
```

than 100. The completed circuit design is given in Fig. 6.17 for 4-GHz oscillation. This type of circuit can also be used as a self-oscillating mixer (SOM), where the signal is coupled to the input port and the IF is filtered from the output port with gain [6.17].

6.5 NEGATIVE RESISTANCE FROM TRANSISTOR MODEL

Two-port oscillator design requires a complete set of four complex parameters, (e.g., S parameters) to complete the design. An alternative approach is to use the

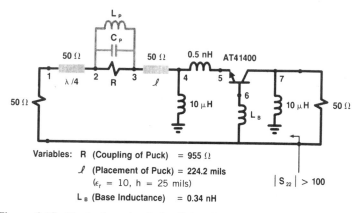

Variables: R (Coupling of Puck) $= 955\ \Omega$
ℓ (Placement of Puck) $= 224.2$ mils
$(\epsilon_r = 10,\ h = 25$ mils)
L_B (Base Inductance) $= 0.34$ nH

$|S_{22}| > 100$

Figure 6.15 Equivalent circuit for dielectric resonator oscillator (DRO).

transistor model for producing the negative resistance. An oscillator can be considered an amplifier with positive feedback, as shown in Fig. 6.18, where the gain is

$$A = \frac{\mu}{1 - \mu\beta} \tag{6.54}$$

and is infinite when the loop gain $\mu\beta$ is unity and the phase shift is 360°. This is the Barkhausen criterion for oscillation; a small portion of the output signal is fed back to the input in phase with the input signal. The initial input signal is generated by noise and the energy source is the dc bias supply.

Two networks that may be used to provide the feedback are given in Fig. 6.19, the Π and T networks. There are four degrees of freedom in the design, three

CASE 2

Feedback

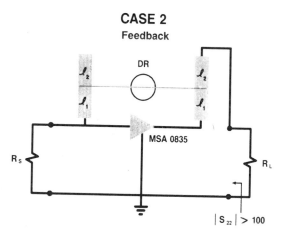

$|S_{22}| > 100$

Figure 6.16 Feedback oscillator using MSA 0835.

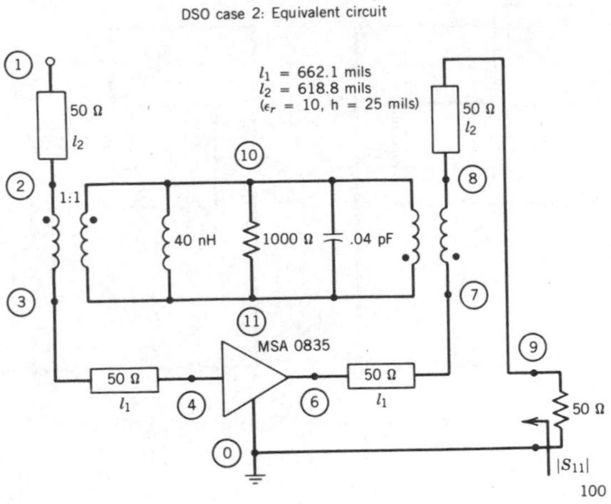

DSO case 2: Equivalent circuit

l_1 = 662.1 mils
l_2 = 618.8 mils
(ϵ_r = 10, h = 25 mils)

50 Ω l_2

1:1

40 nH   1000 Ω   .04 pF

MSA 0835

50 Ω l_1

50 Ω l_2

50 Ω l_1

50 Ω

$|S_{11}|$
100

Figure 6.17 Equivalent circuit for DRO using MSA 0835.

reactances and one resistive load. For the two-port design method, the four degrees of freedom are the magnitude and angle of Γ_G and Γ_L.

Several examples of amplifiers with feedback are given in Fig. 6.20 to 6.23, where the names are given to the original vacuum-tube prototypes. There is a large amount of energy stored in the resonant tank. Some of this energy is coupled to the load (not shown) and a small amount is coupled to the input in phase with the original input signal. A resonant circuit is required if the output is to be a single-frequency signal.

Feedback from the output to the input is responsible for generating a negative resistance between terminals, and if a resonant circuit can be formed by the parasitics or other elements, the amplifier becomes an oscillator. A circuit that often becomes an unintentional high-frequency oscillator is the emitter follower of Fig. 6.24, specifically when the base is driven from a capacitor with sufficient lead inductance. When analyzing this with a linear CAD tool, a negative resistance is evidenced from base to ground and a negative resistance from emitter to ground.

Following is a derivation of the negative resistance responsible for the oscillation. As shown in Fig. 6.24, the bipolar transistor and the two capacitors will

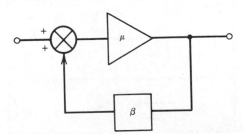

Figure 6.18 Feedback amplifier.

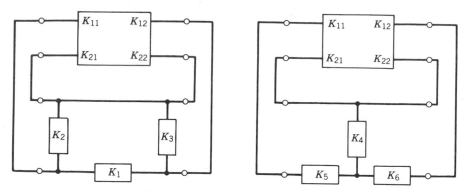

Figure 6.19   Two oscillator circuits.

generate a negative resistance. The negative resistance is responsible for the cancellation of the resonator and load losses, and oscillation can be obtained. To see how a negative resistance is realized, the input impedance of the circuit will be derived.

If h_{oe} is sufficiently small ($h_{oe} \ll 1/R_L$), the equivalent circuit is as shown in Fig. 6.24. The steady-state loop equations are

$$
\begin{aligned}
V_{in} &= I_{in}\,(X_{C1} + X_{C2}) - I_b\,(X_{C1} - \beta X_{C2}) \\
0 &= -I_{in}\,(X_{C1}) + I_b\,(X_{C1} + h_{ie})
\end{aligned}
\tag{6.55}
$$

After I_b is eliminated from these two equations, Z_{in} is obtained as

$$
Z_{in} = \frac{V_{in}}{I_{in}} = \frac{(1 + \beta)X_{C1}\,X_{C2} + h_{ie}\,(X_{C1} + X_{C2})}{X_{C1} + h_{ie}}
\tag{6.56}
$$

If $X_{C1} \ll h_{ie}$, the input impedance is approximately equal to

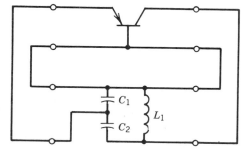

Figure 6.20   Feedback oscillator using capacitive voltage divider; Colpitts oscillator.

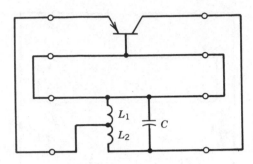

Figure 6.21 Feedback oscillator using inductive voltage divider; Hartley oscillator.

$$Z_{in} \approx \frac{1 + \beta}{h_{ie}} X_{C1} X_{C2} + (X_{C1} + X_{C2})$$

$$\approx \frac{-g_m}{\omega^2 C_1 C_2} + \frac{1}{j\omega[C_1 C_2/(C_1 + C_2)]} \qquad (6.57)$$

That is, the input impedance of the circuit shown in Fig. 6.24 is a negative resistor

$$R = -g_m/\omega^2 C_1 C_2 \qquad (6.58)$$

in series with a capacitor

$$C_{in} = C_1 C_2/(C_1 + C_2) \qquad (6.59)$$

which is the series combination of the two capacitors. With an inductor L (with the series resistance R_s) connected across the input, it is clear that the condition for sustained oscillation is

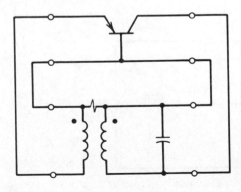

Figure 6.22 Feedback oscillator using mutual coupling; Armstrong oscillator.

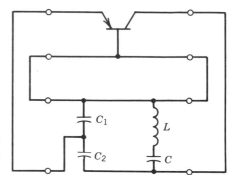

Figure 6.23 Feedback oscillator using series resonant circuit; Clapp–Gouriet oscillator.

$$R_s = g_m / \omega^2 C_1 C_2 \qquad (6.60)$$

and the frequency of oscillation

$$F_0 = \frac{1}{2\pi \sqrt{L \left(C_1 C_2 / (C_1 + C_2) \right)}} \qquad (6.61)$$

This interpretation of the oscillator readily provides several guidelines that can be used in the design. First of all, C_1 should be as large as possible so that

$$X_{C1} \ll h_{ie} \qquad (6.62)$$

and C_2 is to be large, so that

$$X_{C2} \ll 1/h_{oe} \qquad (6.63)$$

When these two capacitors are large, the transistor base-to-emitter and collector-to-emitter capacitances will have a negligible effect on the circuit's performance. However, (6.58) limits the maximum value of the capacitances since

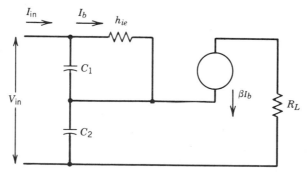

Figure 6.24 Negative input impedance generated by capacitive feedback.

$$R_s \leq g_m/\omega^2 C_1 C_2 \leq G/\omega^2 C_1 C_2 \tag{6.64}$$

where G is the maximum value of g_m. For a given product of C_1 and C_2, the series capacitance is a maximum when $C_1 = C_2 = C_m$. Thus (6.64) can be written

$$1/\omega C_m > \sqrt{R_s/g_m} \tag{6.65}$$

This equation is important because it shows that for oscillations to be maintained, the minimum permissible reactance $1/\omega C_m$ is a function of the resistance of the inductor and the transistor's transconductance g_m.

An oscillator circuit known as the Clapp circuit or Clapp–Gouriet circuit is shown in Fig. 6.25. This oscillator is equivalent to the one just discussed, but it has the practical advantage of being able to provide another degree of design freedom by making C_v much smaller than C_1 and C_2. It is possible to use C_1 and C_2 to satisfy the condition of (6.64) and then adjust C_v for the desired frequency of oscillation ω_0, which is determined from

$$\omega_0 L - \frac{1}{\omega_0 C_v} - \frac{1}{\omega_0 C_1} - \frac{1}{\omega_0 C_2} = 0 \tag{6.66}$$

Consider a Clapp–Gouriet VCO example, as shown in Fig. 6.26. The transistor operated at 1 mA has a $G_{\max}$ of 40 mS and the dynamic g_m is found to be 20 mS. The reduction of $G_{\max}$ is explained in Section 6.2. The inductor, a quarter-wavelength transmission line, has a Q_0 of 200. The oscillator is to be designed for 7.5 GHz.

The feedback is provided by the voltage divider C_1 and C_2, where C_1 is C_{te} and about 2 pF and C_2 is C_{ce} and about 0.5 pF. The transistor is operated at 1 mA and has an F_t of 5 GHz and an ac current gain h_{fe} of 125.

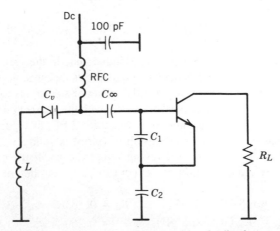

Figure 6.25   VCO using capacitive feedback.

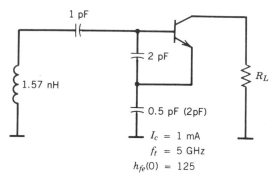

Figure 6.26   Practical valued for VCO at 7.5 GHz.

First we use (6.64) to determine the negative resistance under feedback conditions. From

$$r = -gm/(\omega^2 C_1 C_2) \tag{6.67}$$

we obtain $r = -9\ \Omega$. This means that the loss resistance r' of the combination of L and C_0 must be less then $9\ \Omega$. With an unloaded Q_0 of 200 and a series capacitor of 1 pF as a typical value for the tuning diode, we need an inductance of 1.57 nH. This value can be computed from

$$1/C = 1/C_1 + 1/C_2 + 1/C_v = 1/0.2857\ \text{pF} \tag{6.68}$$

and

$$L = 1/\omega^2 C = 1.57\ \text{nH} \tag{6.69}$$

Finally, we compute $R_s = \omega L_s / Q_0 = 0.26\Omega$ with

$$L_s = (X_1 - X_v)/\omega = 1.12\ \text{nH} \qquad L_v = 0.45\ \text{nH} \tag{6.70}$$

where X_v accounts for 0.45 nH of package inductance in the varactor. The use of C_{ce} as feedback capacitor is somewhat dangerous because its value is small compared to C_{te} or C_1 and may vary greatly. A better solution is to add an external capacitor which also makes it possible to control the temperature response. By adding 1.5 pF to C_{ce} the new value of C_2 becomes 2 pF, or C_1 and C_2 are equal. We have to make sure that the relationship of (6.65) is valid.

Finally, we must see over what tuning range the oscillator can be used. By rearranging (6.65) and calculating the effect of C_v, we can determine

$$
\begin{aligned}
f_{\min} &= 6.5\ \text{GHz}\ (2\ \text{pF}), & R_s &= 0.183\ \Omega < |-3|\ \Omega & (6.71)\\
f_{\max} &= 9.2\ \text{GHz}\ (0.5\ \text{pF}), & R_s &= 0.260\ \Omega < |-1.5|\ \Omega & (6.72)
\end{aligned}
$$

where $L = 0.902$ nH. A practical tuning diode, however, can have a much lower Q then assumed for the inductor. Some diodes exhibit values such as 5, in which case the system has to be reevaluated.

For $F_0 = 7.5$ GHz and the same working condition we calculate

$$R_v = \frac{1/\omega C}{Q} = \frac{21.2}{5} = 4.24 \ \Omega \neq \left| -2.25 \right|$$

The oscillator will not work. There is no change of C_1 and/or C_2 possible for which oscillation can be resumed since

$$1/\omega C < \sqrt{2.25/0.02} = 10.61 \quad \text{or} \quad C > 2.00 \text{ pF}$$

However, by increasing the dc bias of the transistor to about 2 to 3 mA and adjusting for $g_m = 0.042$, we find that

$$R_s = 4.24 < \left| -4.728 \right|$$

For higher frequencies or at about 9 GHz, the oscillator requires more gain and the dc bias point must be moved to 7 mA in order to obtain constant working conditions over temperature and semiconductor tolerances. The power dissipation must not be higher than the manufacturer's specifications.

We have briefly mentioned that the oscillator amplitude stabilizes due to the nonlinear performance of the transistor. There are various mechanisms involved, and depending on the circuit, several of them simultaneously may be responsible for the performance of an oscillator. Under most circumstances, the transistor is operated in an area where the dc bias voltages are substantially larger than the ac voltages. Therefore, the theory describing the transistor performance under these conditions is called "small-signal theory." In a microwave transistor oscillator, however, we are dealing with a feedback circuit that applies positive feedback. The energy that is generated by the initial switch-on of the circuit is fed back to the input of the circuit, amplified, and returned to the input again until oscillation starts. The oscillation would theoretically increase in value unless some limiting or stabilization occurs. In transistor circuits, we have two basic phenomena responsible for limiting the amplitude of oscillation:

1. Limiting because of gain saturation and reduction of open-loop gain.
2. Automatic bias generated by the rectifying mechanism of either the *pn* junction in the bipolar transistor or in the junction field-effect transistor.

A third factor, external AGC, will not be considered here.

The oscillators we discuss here are self-limiting oscillators. The self-limiting process, by generating a dc offset bias that moves the operating point into a region of less gain, is generally noisy. For very low noise oscillators, this operation is not recommended. We deal here only with the so-called negative resistance oscillator, where a negative resistance is generated due to feedback and is used to start

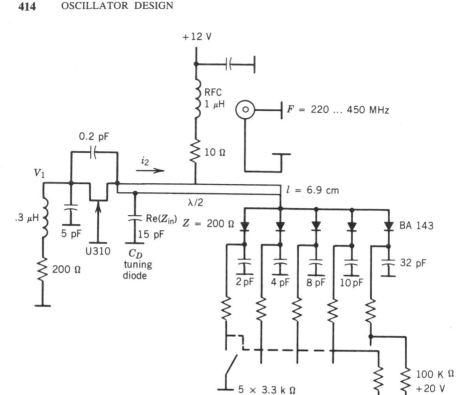

Figure 6.27 Half-wavelength oscillator using transmission line as resonator.

oscillation into the passive load. Here we look at what happens inside the transistor
that is responsible for amplitude stabilization, and therefore, we will be in a position
to make a prediction regarding the available energy and the harmonic content.

Figure 6.27 shows a UHF half-wavelength FET oscillator. A complete analysis
of this oscillator will show that the circuit is amplitude stabilized by the variation
in dc bias from gate to source [6.3]. This circuit uses a U310 JFET with a g_m of
20 ms. The circuit can be tuned 200 to 400 MHz by switching the capacitive
termination at the end of the half-wave resonator. Fine tuning is provided by the
15-pF varactor at the other end of the resonator. The computed result for the oscillator negative impedance is given in Fig. 6.28 for a tuning capacitance of 3 to 44
pF, where an impedance higher than $-50 \, \Omega$ is calculated over 220 to 400 MHz.
This circuit is used in a commercial signal generator, a Rohde & Schwarz model.

6.6 OSCILLATOR Q AND OUTPUT POWER

The oscillator will deliver power at frequencies near f_0 as the load is changed from
the ideal condition of a 50-Ω termination. This effect can easily be measured by
finding the maximum frequency shift for a load mismatch of all phases.

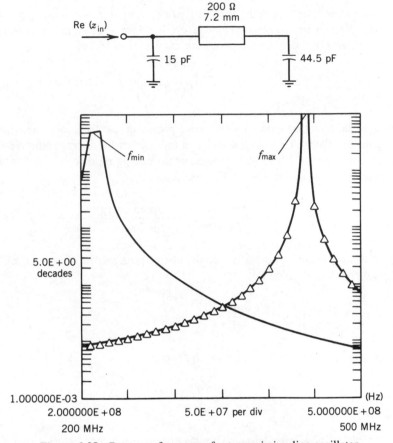

Figure 6.28 Resonant frequency for transmission-line oscillator.

For example, consider a 6-dB attenuator terminated in a sliding short circuit, which can give a complete phase rotation around the Smith chart. Since the load VSWR is 1.67:1, the change in frequency will give the oscillator Q or Q_{ext} by [6.18]

$$Q_{ext} = \frac{f_0}{2\Delta f}\left(S - \frac{1}{S}\right) \qquad (6.73)$$

For a 10-GHz oscillator with a 10-MHz frequency deviation, we have

$$Q_{ext} = \frac{10}{2(0.01)}\left(1.67 - \frac{1}{1.67}\right) = 536$$

The load VSWR must not be too large for this measurement since this could cause the oscillation to stop.

Another method for measuring the oscillator Q is by injection locking the free-running oscillator with a known signal level. If a low-level signal P_i is injected into the oscillator at a distance Δf from the carrier, then [6.19]

$$Q_{\text{ext}} = \frac{2f_0}{\Delta f}\sqrt{P_{\text{in}}/P_{\text{out}}} = \frac{2f_0}{\Delta f}\frac{1}{\sqrt{G_i}} \qquad (6.74)$$

where G_i is the injection gain. Both of these measurement techniques are used to measure the oscillator external Q, which is a measure of the average stored energy in the oscillator circuit. This figure of merit is the energy ratio delivered externally to all dissipative loads. The Q_{ext} of the oscillator is

$$Q_{\text{ext}} = 2\pi\,\frac{\text{time-averaged stored energy}}{\text{energy delivered to the load per cycle}} \qquad (6.75)$$

The load that accepts this power is calculated from the loaded Q of the oscillator at the load port and is given by

$$\frac{1}{Q_L} = \frac{1}{Q_u} + \frac{1}{Q_e} \qquad (6.76)$$

For a parallel resonant load this is given by

$$\frac{\omega L}{R_T} = \frac{\omega L}{R_u} + \frac{\omega L}{R_e} \qquad (6.77)$$

where the useful power is delivered to R_e, the external load. Notice that Q_e is different from Q_{ext}, although these parameters often have the same name. The loaded Q is also given by

$$\frac{1}{Q_L} = \frac{1 + \beta}{Q_u} \qquad (6.78)$$

where

$$\beta = Q_u/Q_e \qquad (6.79)$$

The oscillator output power is difficult to predict, but we can expect it to be less than the saturated power of the same transistor in large-signal amplifier applications. The available power from the transistor must be absorbed in (1) losses in the tuning elements and resonator (Q_u), and (2) power to the load. An estimate for the output power for GaAs MESFETS has been derived by Johnson [6.8].

It is helpful to use the common-source amplifier to compute the oscillator output power. For oscillators, the objective is to maximize $(P_{\text{out}} - P_{\text{in}})$ of the amplifier, which is the useful power to the load. An empirical expression for the common-source amplifier output power is

$$P_{out} = P_{sat}\left(1 - \exp\frac{-GP_{in}}{P_{sat}}\right) \tag{6.80}$$

where P_{sat} is the saturated output power of the amplifier and G is the tuned small-signal common-source transducer gain of the amplifier, which is identical to $|S_{21}|^2$. Since the objective is to maximize $P_{out} - P_{in}$,

$$d(P_{out} - P_{in}) = 0 \tag{6.81}$$

$$\partial P_{out}/\partial P_{in} = 1 \tag{6.82}$$

$$\partial P_{out}/\partial P_{in} = G\exp\left(-GP_{in}/P_{sat}\right) = 1 \tag{6.83}$$

$$\exp\left(GP_{in}/P_{sat}\right) = G \tag{6.84}$$

$$P_{in}/P_{sat} = \ln G/G \tag{6.85}$$

At the maximum value of $P_{out} - P_{in}$, the amplifier output is

$$P_{out} = P_{sat}\left(1 - 1/G\right) \tag{6.86}$$

and the maximum oscillator output power is

$$P_{osc} = P_{out} - P_{in}$$
$$= P_{sat}\left(1 - 1/G - \ln G/G\right) \tag{6.87}$$

Thus the maximum oscillator output power can be predicted from the common-source amplifier saturated output power and the small-signal common source transducer gain G. A plot of (6.87) is given in Fig. 6.29, which shows the importance of high gain for a high oscillator output power. Another gain that is useful for large-signal amplifier or oscillator design is the maximum efficient gain, defined by

$$G_{ME} = \frac{P_{out} - P_{in}}{P_{in}} \tag{6.88}$$

For maximum oscillator power the maximum efficient gain is, from (6.85) and (6.86),

$$G_{MEmax} = \frac{G - 1}{\ln G} \tag{6.89}$$

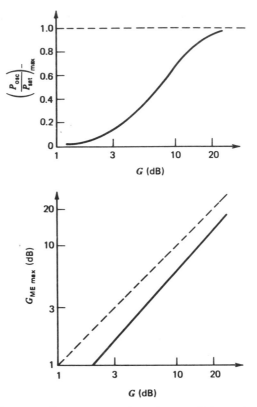

Figure 6.29 Maximum oscillator power and maximum efficient gain versus small-signal transducer power gain.

This gain is also plotted in Fig. 6.29, showing a considerably smaller value of G_{MEmax} compared to G, the small-signal gain. From these results, the oscillator output power will approach P_{sat} at low frequencies where the gain is large. As the gain approaches unity, the oscillator power approaches zero.

6.7 NOISE IN OSCILLATORS

Noise generated by the transistor and passive devices shows up at the output signal of amplifiers. In the case of an oscillator, which is a nonlinear device, these noise voltages and currents are modulating the signal produced by the oscillator. While introducing the concept of noise in oscillators, we first discuss noise measurement techniques and then calculate some of the noise voltages and currents that are modulated onto the carrier. This allows us finally to use Leeson's model [6.4] to obtain an expression of the normalized (in 1-Hz bandwidth) single-sideband noise ($\mathcal{L}$ in dBc/Hz) power. Finally, we analyze the various contribution to the noise and calculate the noise of VCOs.

We start by looking at different noise test techniques [6.20–6.26] that give rise to noise descriptions that are related to one another. The following equations are common definitions of oscillator spectral purity:

$$S_\theta(f_m) = \text{spectral density of phase fluctuation}$$

$$S_\theta(f_m) = \Delta\theta^2_{\text{rms}} \tag{6.90}$$

$$S_{\dot\theta}(f_m) = \text{spectral density of frequency fluctuations}$$

$$S_{\dot\theta}(f_m) = \Delta f^2_{\text{rms}} \tag{6.91}$$

$$\mathcal{L}(f_m) = \frac{\text{ratio of noise power in a 1-Hz bandwidth}}{\text{at } f_m \text{ offset from carrier to carrier signal power}}$$

$$\mathcal{L}(f_m) = \frac{N(1\text{-Hz BW})}{C} \tag{6.92}$$

6.7.1 Using a Spectrum Analyzer

The easiest technique for measuring oscillator noise is to view the oscillator spectrum directly on a spectrum analyzer, giving a display as in Fig. 6.30. This method allows direct measurement of $\mathcal{L}(f_m)$. The oscillator output power is read off the screen in dBm. The noise at a frequency offset f_m away from the carrier may also be read directly. Noise measured in this way will usually require correc-

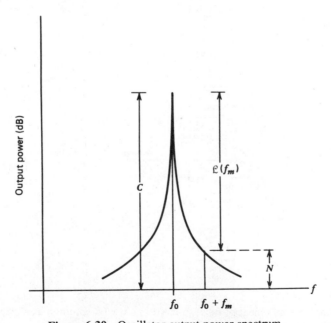

Figure 6.30 Oscillator output power spectrum.

tion factors, since the detector of the analyzer is ordinarily an envelope rather than a true rms detector, the log amplifiers amplify noise peaks less, and the bandpass filters may be Gaussian or trapezoidal in shape, which requires correction to a square bandpass. Additionally, since 1-Hz bandpass filters are uncommon, this results in measurement of the noise in a wider bandwidth which must be corrected to 1 Hz by reducing the noise measured by 10 dB for every decade by which the filter is wider than 1 Hz.

After applying these corrections, $\mathcal{L}(f_m)$ is equal to

$$\mathcal{L}(f_m) = \frac{\text{noise power with corrections at } f_m}{\text{carrier power}} = \frac{N(1\text{-Hz BW})}{C} \quad (6.93)$$

Certain precautions must be taken when measuring $\mathcal{L}(f_m)$ in this fashion. The technique is most useful when it can be determined that the noise of the oscillator being measured is worse than that of the local oscillator of the spectrum analyzer. The reason for this is apparent from Fig. 6.31a, which shows a spectrum analyzer from the front end. The noiseless local oscillator translates the oscillator under test to an IF frequency where the amplitude and noise can be analyzed with narrow fixed filters. The spectrum analyzer cannot distinguish between noise from its own

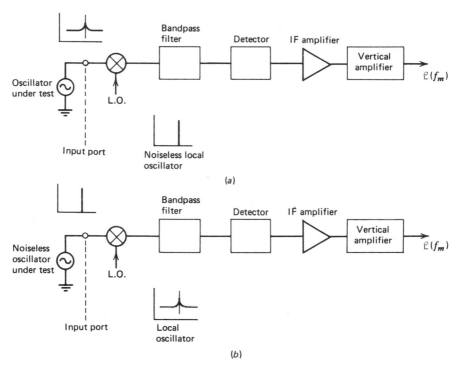

Figure 6.31 Measurement of noise-to-carrier ratio with spectrum analyzer: (a) noiseless local oscillator; (b) noiseless oscillator under test.

local oscillator and that from an oscillator under test (which may be better) as in Fig. 6.31*b*. This situation frequently occurs at microwave frequencies where multiplied, low-noise oscillators often outperform the commonly used YIG-tuned oscillator in spectrum analyzers.

An important point that should be made is that the bulk of oscillator noise particularly close to the carrier is phase or FM noise. Oscillator limiting mechanisms, whether self-limiting or automatic gain control type, tend to eliminate AM noise. Under these conditions $\mathcal{L}(f_m)$ can be related to phase modulation in the following way. A table of Bessel functions will reveal that if a carrier is phase modulated (for a small modulation index $\Delta\theta_{peak} \ll \pi/2$), the ratio of the first-order sideband to the carrier J_0 is

$$\frac{J_1}{J_0} \simeq \frac{1}{2}\,\Delta\theta_{peak} \simeq \frac{1}{2}\,\sqrt{2}\,\theta_{rms} \tag{6.94}$$

Since $\mathcal{L}(f_m)$ is the ratio of noise power (J_1^2) to carrier power (J_0^2),

$$\mathcal{L}(f_m) = \frac{N}{C} = \left(\frac{J_1}{J_0}\right)^2 = \frac{1}{2}\,\theta_{rms}^2 \tag{6.95}$$

This description of $\mathcal{L}(f_m)$ holds only where it can be assumed that the $f_0 \pm \Delta f$ noise sidebands are correlated (i.e., caused by the same modulation source). This is not true in the additive noise region, where the noise at $f_0 \pm \Delta f$ is not correlated (i.e., independent thermal noise generation at $\pm \Delta f$).

6.7.2 Two-Oscillator Method

Frequency Discriminator. A more sensitive technique is to measure $S_\theta(f_m)$. Figure 6.32 shows a common oscillator noise measuring approach that gives $S_\theta(f_m)$.

Describe oscillator 1 as

$$v_1 = V_1 \cos\left[\omega_1 t + \theta_1(t)\right] \tag{6.96}$$

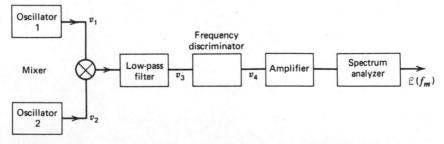

Figure 6.32 Measurement of spectral density of frequency fluctuations with frequency discriminator.

and oscillator 2 as

$$v_2 = V_2 \cos \left[\omega_2 t + \theta_2(t) \right] \tag{6.97}$$

Then mixing these oscillators together gives an IF frequency low enough to apply to the frequency discriminator.

$$v_3 = V_1 V_2 \cos \left[\omega_1 t + d\theta_1(t) \right] \cos \left[\omega_2 t + d\theta_2(t) \right]$$

$$= \frac{V_1 V_2}{2} \cos \left\{ (\omega_1 - \omega_2)t + \left[d\theta_1(t) - d\theta_2(t) \right] \right\} \tag{6.98}$$

The sum frequency term is eliminated by a low-pass filter. The output from the discriminator is

$$v_4 = K\omega_{\text{in}} \tag{6.99}$$

The frequency of the foregoing signal is

$$\omega_4 = (\omega_1 - \omega_2) + \frac{d\theta_1(t) - d\theta_2(t)}{dt} \tag{6.100}$$

where $(\omega_1 - \omega_2)$ is a constant and $[d\theta_1(t) - d\theta_2(t)]/dt$ represents the sum of the frequency fluctuations in ω_1 and ω_2.

The output of the discriminator will be

$$v_4(t) = K \left[(\omega_1 - \omega_2) + \frac{d\theta_1 - d\theta_2}{dt} \right] = K_2 + K\frac{d\theta_1 - d\theta_2}{dt} \tag{6.101}$$

A high-pass filter will remove the constant term, leaving

$$v_4(t) = K\frac{d\theta_1 - d\theta_2}{dt} = K\frac{d\theta_1}{dt} - K\frac{d\theta_2}{dt} = Kd\omega_1 - Kd\omega_2$$

$$= 2\pi K(df_1 - df_2) \tag{6.102}$$

This time function is then applied to a low-frequency spectrum analyzer. In the transform domain or frequency domain

$$v_4(f_m) = 2\pi K \left[dF_1(f_m) - dF_2(f_m) \right] \tag{6.103}$$

Since $dF_1(f_m)$ and $dF_2(f_m)$ are uncorrelated, they combine as follows:

$$\Delta f_{\text{rms}} = dF_1(f_m) + dF_2(f_m) = \sqrt{(dF_1)^2 + (dF_2)^2} \tag{6.104}$$

$$v_4(f_m) = 2\pi K \sqrt{\overline{(dF_1)^2} + \overline{(dF_2)^2}} \tag{6.105}$$

Since spectrum analyzers normally display power rather than voltage, the display represents

$$[v_4(f_m)]^2 = (2\pi K)^2 \left[\overline{(dF_1)^2} + \overline{(dF_2)^2}\right] \tag{6.106}$$

It will be recognized that $\overline{dF_1^2} = S_\theta(f_m)$ for oscillator 1 and $\overline{dF_2^2} = S_{\theta 2}(f_m)$ for oscillator 2.

The spectrum analyzer display is proportional to the sum of the spectral densities $S_\theta(f_m)$ for oscillator 1 and oscillator 2.

$$[v_4(f_m)]^2 = (2\pi K)^2 \left[S_{\theta 1}(f_m) + S_{\theta 2}(f_m)\right] \tag{6.107}$$

$\mathcal{L}(f_m)$ now represents noise sideband power to carrier power caused by phase fluctuations as a function of frequency from the carrier. $S_\theta(f_m)$ represents FM deviation squared as a function of frequency offset from the carrier.

The two parameters may be related as follows:

$$S_\theta(f_m) = \overline{\Delta f(f_m)^2} \tag{6.108}$$

$$df(t) = \frac{1}{2\pi} d\omega(t) = \frac{1}{2\pi} \frac{d\theta(t)}{dt} \tag{6.109}$$

Then in the transform domain

$$df(f_m) = \frac{1}{2\pi}(s) d\theta(f_m) \tag{6.110}$$

$$S_\theta(f_m) = \overline{df(f_m)^2} = \left(\frac{s}{2\pi}\right)^2 \overline{d\theta(f_m)^2} \tag{6.111}$$

$$\overline{d\theta(f_m)^2} = \left(\frac{2\pi}{s}\right)^2 S_\theta(f_m) = \frac{1}{f_m^2} S_\theta(f_m) \tag{6.112}$$

$$\mathcal{L}(f_m) = \frac{1}{2} \overline{\theta(f_m)^2} = \frac{1}{2f_m^2} S_\theta(f_m) \tag{6.113}$$

This technique affords better sensitivity than direct measurement of $\mathcal{L}(f_m)$ at microwave frequencies, since the translation down to low RF frequencies permits the use of spectrum analyzers with lower-noise local oscillators or fast Fourier transform analyzers. In general, the sensitivity of this system is limited by the internal noise of the frequency discriminator.

Double-Balanced Mixer. A more sensitive scheme removes the frequency discriminator as shown in Fig. 6.33. We assume that the oscillators are or can be

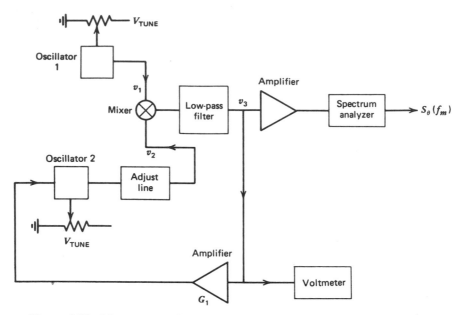

Figure 6.33 Measurement of spectral density of phase fluctuations with a mixer.

adapted so that one can be phase-locked to the other. In Fig. 6.33 the oscillators are set so that they are at approximately the same frequency. Oscillator 1 and oscillator 2 then mix to produce sum and difference frequencies. The sum frequencies are removed by the low-pass filter. The difference frequency error signal is sent back to lock oscillator 2 to oscillator 1. Inside the loop bandwidth, which can be adjusted by varying the gain of G_1, the noise of oscillator 2 tracks that of oscillator 1. Outside the loop bandwidth, the noise of the two oscillators shows no correlation.

The mixer is usually a double-balanced mixer consisting of four diodes. The IF port is dc-coupled to provide the phase-locked dc signal. This phase-locked dc signal is adjusted to be 0 V on the voltmeter, since the sensitivity $dv/d\theta$ is maximum for this condition. This is done by adjusting a line length such that the phases of the two oscillators are 90° apart.

Figure 6.34 shows the typical sensitivity of the mixer. Beyond the loop bandwidth, the output of the mixer may be described as follows:

$$v_1 = V_1 \cos(\omega t + \theta_{n1}) \tag{6.114}$$

$$v_2 = V_2 \cos(\omega t + \theta_{n2} - \pi/2) \tag{6.115}$$

$$v_3 = V_1 V_2 \cos(\omega t + \theta_{n1}) \cos(\omega t + \theta_{n2} - \pi/2)$$

$$= \frac{V_1 V_2}{2} \cos\left(\theta_{n1} - \theta_{n2} + \frac{\pi}{2}\right) \tag{6.116}$$

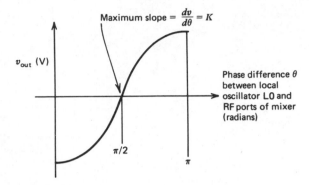

Maximum slope $= \dfrac{dv}{d\theta} = K$

v_{out} (V)

Phase difference θ
between local
oscillator LO and
RF ports of mixer
(radians)

$\pi/2$

π

Figure 6.34 Output voltage of doubly balanced mixer versus phase difference between local oscillator and RF signal ports.

The θ_{n1} and θ_{n2} terms are rms phase noise, which can be combined as

$$\theta_{nT} = \sqrt{\overline{\theta_{n1}^2} + \overline{\theta_{n2}^2}} \tag{6.117}$$

$$v_3 = \frac{V_1 V_2}{2} \cos\left(\theta_{n1} - \theta_{n2} + \frac{\pi}{2}\right) = \frac{V_1 V_2}{2} \cos\left(\theta_{nT} + \frac{\pi}{2}\right)$$

$$= -\frac{V_1 V_2}{2} \sin \theta_{nT} \tag{6.118}$$

For θ_{nT} very small,

$$\sin \theta_{nT} \simeq \theta_{nT} = \sqrt{\overline{\theta_{n1}^2} + \overline{\theta_{n2}^2}} \tag{6.119}$$

Since the spectrum analyzer displays power, it will show the square of the term

$$-\frac{V_1 V_2}{2} \sqrt{\overline{\theta_{n1}^2} + \overline{\theta_{n2}^2}} \quad \text{or} \quad \left(\frac{V_1 V_2}{2}\right)^2 \left(\overline{\theta_{n1}^2} + \overline{\theta_{n2}^2}\right)$$

$$\overline{\theta_{n1}^2} = S_\theta(f_m) \text{ of oscillator 1} \tag{6.120}$$

$$\overline{\theta_{n2}^2} = S_\theta(f_m) \text{ of oscillator 2} \tag{6.121}$$

If the spectral densities have equal power distribution but are not correlated, the mixer output is 3 dB greater than either one alone. This technique yields the sum of the $S_\theta(f_m)$ for oscillator 1 and oscillator 2.

$S_\theta(f_m)$ can now be related to $\mathcal{L}(f_m)$. $S_\theta(f_m)$ is equal to $\mathcal{L}(f_m)$ folded about itself. Therefore, $S_\theta(f_m) = 2\mathcal{L}(f_m)$ if the noise sidebands about f_1 are correlated and $S_\theta(f_m) = \sqrt{2}\mathcal{L}(f_m)$ if they are not correlated.

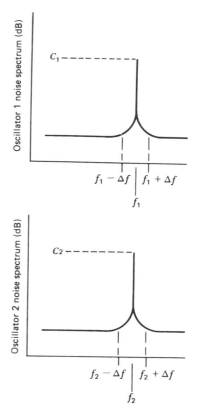

Figure 6.35 Noise spectrum of two oscillators at f_1 and f_2 carrier frequency.

In Fig. 6.35 the noise below $f_0 - \Delta f$ is assumed to be uncorrelated to the noise above $f_0 + \Delta f$ in oscillator 1 and in oscillator 2. Closer than $f_0 \pm \Delta f$ the assumption is that there is correlation of the noise above and below the carrier in both oscillators. Beyond $\pm \Delta f$ we assume that this is the noise floor of the device. Closer than $\pm \Delta f$ we assume that the noise is caused by phase modulation mechanisms in the device or other components that generate related sidebands above and below the carrier. When these two spectrums are mixed together, the following occurs: If $f_1 = f_2$, then (if we ignore the sum frequency components, which are eliminated by the low-pass filter), $f_1 - f_2 = 0$; f_1 then mixes against the noise spectrum of f_2. This causes the noise spectrum of f_2 to fold upon itself. For instance, f_1 mixing against $f_2 \pm \Delta f_x$ will yield two correlated noise components at Δf_x which add in power to cause a 6-dB increase as in Fig. 6.36. However, if f_1 mixes against $f_2 \pm \Delta f_y$, there is only a 3-dB increase, since the noise at $f_2 - \Delta f_y$ is not correlated to that at $f_2 + \Delta f_y$.

The reverse also occurs: f_2 can mix with the noise of f_1 at $f_1 \pm \Delta f_x$ and $f_1 \pm \Delta f_y$ to cause an additional 3-dB increase in noise measured at the mixer's output. This increase occurs because this reverse process generates another spectrum identical

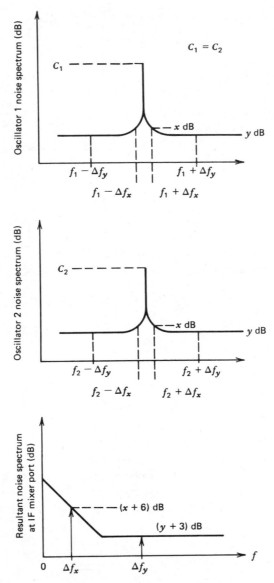

Figure 6.36 Resultant noise spectrum due to foldover of spectrum about the carrier for one oscillator.

in amplitude to that in Fig. 6.36; however, the noise of the two oscillators is not correlated except within the phase-locked-loop bandwidth. The mixer takes these two uncorrelated spectrums and adds them at its output, causing an additional 3-dB increase in noise, as shown in Fig. 6.37. $\mathcal{L}(f_m)$ can be obtained from this spectrum by subtracting 9 dB from the part where the upper and lower noise

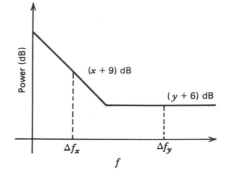

Figure 6.37 Power spectrum of mixer IF port as displayed on a spectrum analyzer due to the combined effects of foldover and addition of 3 dB for noise spectrum of two uncorrelated oscillators.

sidebands are correlated and by subtracting 6 dB from the area where no correlation exists.

It is possible to go back to Fig. 6.36 before the addition of 3 dB (due to two uncorrelated oscillators) to see how $S_\theta(f_m) = \Delta\theta^2_{rms}$ is related to $\mathcal{L}(f_m)$. Since

$$\mathcal{L}(f) = \frac{1}{2}\Delta\theta^2_{rms} \qquad (6.122)$$

and

$$S_\theta(f_m) = \Delta\theta^2_{rms} \qquad (6.123)$$

we see that the folded-over spectrum of a single oscillator at Δf_x or where the upper and lower sidebands of f_1 are correlated is equal to $S_\theta(f_m)$.

The noise spectrum of an amplifier would appear as in Fig. 6.38. For a moment, it is of interest to discuss the $1/f$ noise spectrum near dc. Noise in amplifiers is often modeled as in Fig. 6.39, which was also discussed in Chapter 2 (see Section

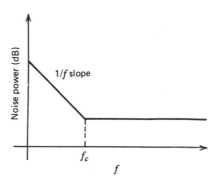

Figure 6.38 Noise power versus frequency of a transistor amplifier.

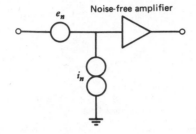

Noise-free amplifier

Figure 6.39 Equivalent noise sources at the input of an amplifier.

2.3). In bipolar amplifiers, e_n is related to the thermal noise of the base spreading resistance:

$$e_n = \sqrt{4kTr_bB} \qquad (6.124)$$

This noise source has a relatively flat frequency response. The i_n noise source is associated with the shot noise in the base current:

$$i_n = \sqrt{2qI_bB} \qquad (6.125)$$

This i_n noise generator has associated with it a $1/f$ noise mechanism.

In FET devices, the situation is reversed. The e_n noise generator has a $1/f$ noise component where i_n shows none. It is interesting to note that in general, the $1/f$ noise corner of bipolar silicon devices is lower than that of silicon JFETs. Silicon JFETs are less noisy than silicon MOSFETS. GaAs MESFETs usually have the highest $1/f$ corner frequencies, which can extend to several hundred megahertz. Carefully selected bipolar devices can have $1/f$ noise corners below 100 Hz.

There are various instruments that can measure e_n and i_n directly with no carrier signal present. These measurement methods would provide a noise plot as in Fig. 6.38. However, if a carrier signal is applied to the amplifier, the noise plot would be modified as in Fig. 6.40. The low-frequency noise sources can effect the phase shift through the amplifier, causing the $1/f$ phase noise spectrum about the carrier.

6.7.3 Leeson's Oscillator Model

Since an oscillator can be viewed as an amplifier with feedback [6.4] it is helpful to examine the phase noise added to an amplifier that has a noise figure F. With F defined by [see (2.1)]

$$F = \frac{(S/N)_{\text{in}}}{(S/N)_{\text{out}}} = \frac{N_{\text{out}}}{N_{\text{in}}G} = \frac{N_{\text{out}}}{GkTB} \qquad (6.126)$$

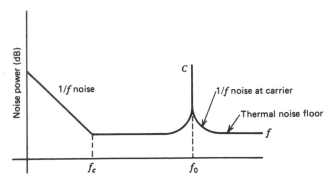

Figure 6.40   Noise power versus frequency of a transistor amplifier with an input signal applied.

$$N_{\text{out}} = FGkTB \qquad (6.127)$$

$$N_{\text{in}} = kTB \qquad (6.128)$$

where N_{in} is the total input noise power to a noise-free amplifier. The input phase noise in a 1-Hz bandwidth at any frequency $f_0 + f_m$ from the carrier produces a phase deviation given by (Fig. 6.41)

$$\Delta\theta_{\text{peak}} = \frac{V_{\text{nRMS1}}}{V_{\text{avsRMS}}} = \sqrt{\frac{FkT}{P_{\text{avs}}}} \qquad (6.129)$$

$$\Delta\theta_{1\text{RMS}} = \frac{1}{\sqrt{2}}\sqrt{\frac{FkT}{P_{\text{avs}}}} \qquad (6.130)$$

Since a correlated random phase relation exists at $f_0 - f_m$, the total phase deviation becomes

$$\Delta\theta_{\text{RMS total}} = \sqrt{FkT/P_{\text{avs}}} \qquad (6.131)$$

The spectral density of phase noise becomes

$$S_\theta(f_m) = \Delta\theta_{\text{RMS}}^2 = FkTB/P_{\text{avs}} \qquad (6.132)$$

where $B = 1$ for a 1-Hz bandwidth. Using

$$kTB = -174 \text{ dBm/Hz} \qquad (B = 1) \qquad (6.133)$$

allows a calculation of the spectral density of phase noise that is far removed from the carrier (i.e., at large values of f_m). This noise is the theoretical noise floor of

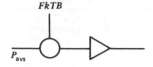

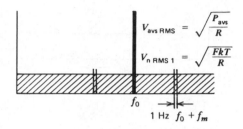

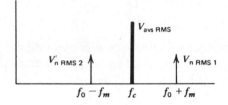

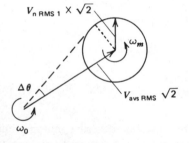

Figure 6.41 Phase noise added to carrier.

the amplifier. For example, an amplifier with $+10$ dBm power at the input and a noise figure of 6 dB gives

$$S_\theta(f_m > f_c) = -174 \text{ dBm} + 6 \text{ dB} - 10 \text{ dBm} = -178 \text{ dB}$$

For a modulation frequency close to the carrier, $S_\theta(f_m)$ shows a flicker or $1/f$ component which is empirically described by the corner frequency f_c. The phase noise can be modeled by a noise-free amplifier and a phase modulator at the input as shown in Fig. 6.42. The purity of the signal is degraded by the flicker noise at frequencies close to the carrier. The spectral phase noise can be described by

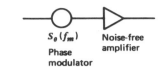

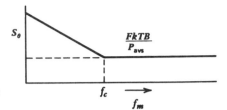

Figure 6.42 Phase noise modeled by a
noise-free amplifier and a phase modulator.

$$S_\theta(f_m) = \frac{FkTB}{P_{\text{avs}}}\left(1 + \frac{f_c}{f_m}\right) \quad (B = 1) \tag{6.134}$$

The oscillator may be modeled as an amplifier with feedback as shown in Fig. 6.43. The phase noise at the input of the amplifier is affected by the bandwidth of the resonator in the oscillator circuit in the following way. The tank circuit or bandpass resonator has a low-pass transfer function

$$L(\omega_m) = \frac{1}{1 + j(2Q_L\omega_m/\omega_0)} \tag{6.135}$$

where

$$\omega_0/2Q_L = B/2 \tag{6.136}$$

is the half-bandwidth of the resonator. These equations describe the amplitude response of the bandpass resonator; the phase noise is transferred unattenuated through the resonator up to the half bandwidth. The closed loop response of the phase feedback loop is given by

$$\Delta\theta_{\text{out}}(f_m) = \left(1 + \frac{\omega_0}{j2Q_L\omega_m}\right)\Delta\theta_{\text{in}}(f_m) \tag{6.137}$$

The power transfer becomes the phase spectral density

$$S_{\theta\,\text{out}}(f_m) = \left[1 + \frac{1}{f_m^2}\left(\frac{f_0}{2Q_L}\right)^2\right]S_{\theta\,\text{in}}(f_m) \tag{6.138}$$

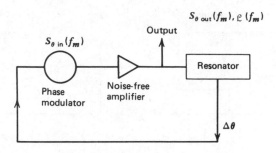

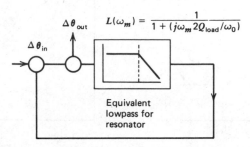

Figure 6.43   Equivalent feedback models of oscillator phase noise.

where $S_{\theta\text{in}}$ was given by (6.134). Finally, $\mathcal{L}(f_m)$ is

$$\mathcal{L}(f_m) = \frac{1}{2}\left[1 + \frac{1}{f_m^2}\left(\frac{f_0}{2Q_L}\right)^2\right]S_{\theta\text{in}}(f_m) \qquad (6.139)$$

This equation describes the phase noise at the output of the amplifier. The phase pertubation $S_{\theta\text{in}}$ at the input of the amplifier is enhanced by the positive phase feedback within the half bandwidth of the resonator, $f_0/2Q_L$.

Depending on the relation between f_c and $f_0/2Q_L$, there are two cases of interest as shown in Fig. 6.44. For the low-Q case, the spectral phase noise is unaffected by the Q of the resonator, but the $\mathcal{L}(f_m)$ spectral density will show a $1/f^3$ and $1/f^2$ dependence close to the carrier. For the high-Q case, a region of $1/f^3$ and $1/f$ should be observed near the carrier. Substituting (6.134) in (6.139) gives an overall noise of

$$\mathcal{L}(f_m) = \frac{1}{2}\left[1 + \frac{1}{f_m^2}\left(\frac{f}{2Q_L}\right)^2\right]\frac{FkT}{P_{\text{avs}}}\left(1 + \frac{f_c}{f_m}\right)$$

$$= \frac{FkTB}{2P_{\text{avs}}}\left[\frac{1}{f_m^3}\frac{f^2 f_c}{4Q_L^2} + \frac{1}{f_m^2}\left(\frac{f}{2Q_L}\right)^2 + \frac{f_c}{f_m} + 1\right] \quad \text{(dBc/Hz)} \quad (6.140)$$

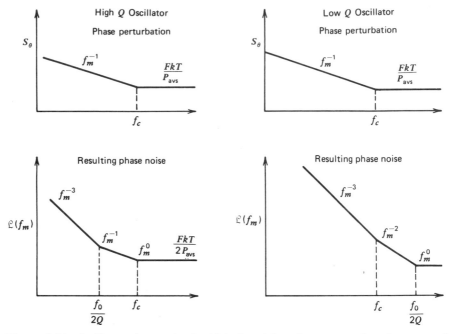

Figure 6.44 Oscillator phase noise for high-Q and low-Q resonator viewed as spectral phase noise and as noise-to-carrier ratio versus frequency from the carrier.

Examining (6.140) gives the four major causes of oscillator noise: the up-converted $1/f$ noise or flicker FM noise, the thermal FM noise, the flicker phase noise, and the thermal noise floor, respectively [6.27, 6.28].

6.7.4 Low-Noise Design

By rearranging the equation and evaluating each term, we obtain for a 1-Hz bandwidth,

$$\mathcal{L}(f_m) = \frac{1}{2}\left[1 + \frac{\omega_o^2}{4\omega_m^2}\left(\frac{P_{in}}{\omega_0 W_e} + \frac{1}{Q_u} + \frac{P_{sig}}{\omega_0 W_e}\right)^2\right]\left(1 + \frac{\omega_c}{\omega_m}\right)\frac{FkT_o}{P_{avs}} \quad (6.141)$$

input power over reactive power

resonator Q

signal power over reactive power

phase perturbation

flicker effect

This equation is extremely significant because it contains most of the causes of phase noise in oscillators. To minimize the phase noise, the following design rules apply:

1. Maximize the unloaded Q.
2. Maximize the reactive energy by means of a high RF voltage across the resonator and obtain a low LC ratio. The limits are set by breakdown voltages of the active devices and the tuning diodes and the forward-bias condition of the tuning diodes.
3. Avoid saturation at all cost, and try to either have limiting or AGC without degradation of Q. Isolate the tuned circuit from the limiter or AGC circuit. Use antiparallel tuning diode connections to avoid forward bias.
4. Choose an active device with the lowest noise figure. Currently, the best VHF bipolar transistor is the Siemens BFT66 and the lowest noise field-effect transistors are U310 and 2N5397 up to 500 MHz. The noise figure of interest is the noise figure obtained at the actual impedance at which the device is operated. For the microwave range up to 10 GHz, low-noise silicon bipolar transistors include the AT 60535 and AT 41435, described in Chapter 3. For even higher frequencies the GaAs MESFETs described in Chapter 3 are also recommended for oscillator applications. Usually, the lowest-noise microwave oscillators use bipolar transistors and a multiplier if needed. As GaAs MESFETs improve with lower $1/f$ noise, this choice between transistors must be reevaluated continuously.
5. Phase perturbation can be minimized by using high-impedance devices such as field-effect transistors, where the signal-to-noise ratio or the signal voltage relative to the equivalent noise voltage can be made very high. This also indicates that in the case of a limiter, the limited voltage should be as high as possible.
6. Choose an active device with low flicker noise. The effect of flicker noise can be reduced by RF feedback. An unbypassed emitter resistor of 10 to 30 Ω in a bipolar circuit can improve the flicker noise by as much as 40 dB [6.21]. The proper bias point of the active device is important, and precautions should be taken to prevent modulation of the input and output dynamic capacitance of the active device, which will cause amplitude-to-phase conversion and therefore introduce noise.
7. The energy should be coupled from the resonator rather than another portion of the active device so that the resonator limits the bandwidth because the resonator is also used as a filter. A dielectric-resonator oscillator using this principle is described later.

The loading effect of a tuning diode is due to losses, and these losses can be described by a resistor parallel to the tuned circuit.

It is possible to define an equivalent noise $R_{a,\mathrm{eq}}$ that, inserted in Nyquist's equation

$$V_n = \sqrt{4kT_oR\Delta f} \tag{6.142}$$

where $kT_o = 4.2 \times 10^{-21}$ J at about 300 K, R is the equivalent noise resistor, and Δf is the bandwidth, determines an open noise voltage across the tuning diode. Practical values of $R_{a,\mathrm{eq}}$ for carefully selected tuning diodes are in the vicinity of 1000 Ω to 50 kΩ. If we now determine the noise voltage $V_n = \sqrt{4 \times 4.2 \times 10^{-21} \times 10,000}$, the resulting voltage value is 1.265×10^{-8} $\frac{V}{\sqrt{Hz}}$.

This noise voltage generated from the tuning diode is now multiplied with the modulation sensitivity, resulting in the rms frequency deviation

$$(\Delta f_{\mathrm{rms}}) = K_0 \times (1.265 \times 10^{-8} \text{ V}) \text{ in 1-Hz bandwidth} \tag{6.143}$$

In order to translate this into the equivalent peak phase deviation,

$$\theta_d = \frac{K_o\sqrt{2}}{f_m}(1.265 \times 10^{-8} \text{ rad}) \text{ in 1-Hz bandwidth}$$

or for a typical modulation sensitivity of 100 kHz/V,

$$\theta_d = \frac{0.00179}{f_m} \text{ rad in 1-Hz bandwidth}$$

For $f_m = 25$ kHz (typical spacing for adjacent channel measurements for FM mobile radios), the $\theta_d = 7.17 \times 10^{-8}$. This can be converted now into the SSB signal-to-noise ratio

$$\mathcal{L}(f_m) = 20 \log_{10} \frac{\theta_d}{2}$$
$$= -149 \text{ dBc/Hz} \tag{6.144}$$

This is the value typically achieved in the Rohde & Schwarz SMDU or with the Hewlett-Packard 8640 signal generator and considered state of the art for a free-running oscillator. It should be noted that both signal generators use a slightly different tuned circuit; the Rohde & Schwarz generator uses a helical resonator, whereas the Hewlett-Packard generator uses an electrically shortened quarter-wavelength cavity. Both generators are mechanically pretuned and the tuning diode with a gain of about 100 kHz/V is used for frequency-modulation purposes or for the AFC input. It is apparent that, because of the nonlinearity of the tuning diode, the gain is different for low dc voltages than for high dc voltages. The impact of this is that the noise varies within the tuning range.

Let us now examine some test results. If we go back to (6.141), Fig. 6.45 shows the noise sideband performance as a function of Q, whereby the top curve with Q_L

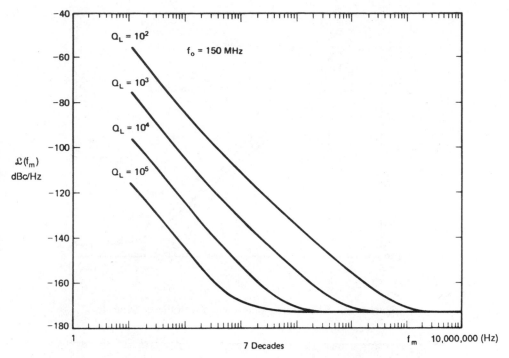

Figure 6.45 Noise sideband of an oscillator at 150 MHz as a function of the loaded Q of the resonator.

= 100 represents a somewhat poor oscillator and the lowest curve with Q_L = 100,000 probably represents a crystal oscillator where the unloaded Q of the crystal was in the vicinity of 3×10^6. Figure 6.46 shows the influence of flicker noise.

Corner frequencies of 1 Hz to 10 kHz have been selected, and it becomes apparent that around 1 kHz the influence is fairly dramatic, whereas the influence at 20 kHz off the carrier is not significant. Finally, Fig. 6.47 shows the influence of the tuning diodes on a high-Q oscillator.

Curve A uses a lightly coupled tuning diode with a K_o of 10 kHz/V; the lower curve is the noise performance without any diode. As a result, the two curves are almost identical, which can be seen from the somewhat smeared form of the graph. Curve B shows the influence of a tuning diode at 100 kHz/V and represents a value of −143 dBc/Hz from −155 dBc/Hz, already some deterioration. Curve C shows the noise if the tuning diode results at a 1-MHz/V modulation sensitivity, and the noise sideband at 25 kHz has now deteriorated to −123 dBc/Hz. These curves speak for themselves.

It is of interest to compare various oscillators. Figure 6.48 shows the performance of a 10-MHz crystal oscillator, a 40-MHz *LC* oscillator, the 8640 cavity tuned oscillator at 500 MHz, the 310- to 640-MHz switched reactance oscillator of the 8662 oscillator, and a 2- to 6-GHz YIG oscillator at 6 GHz.

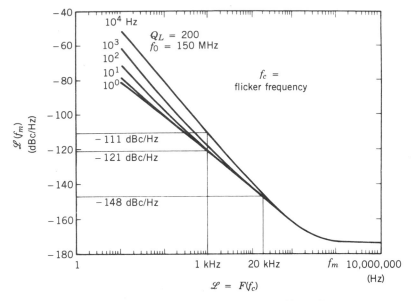

Figure 6.46   Noise sideband as a function of flicker frequency.

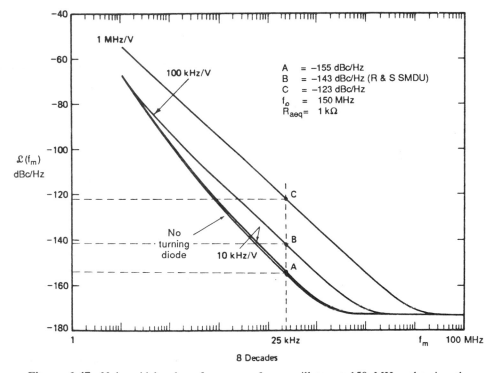

Figure 6.47   Noise sideband performance of an oscillator at 150 MHz, showing the influence of various tuning diodes and increased noise as a function of modulation sensitivity.

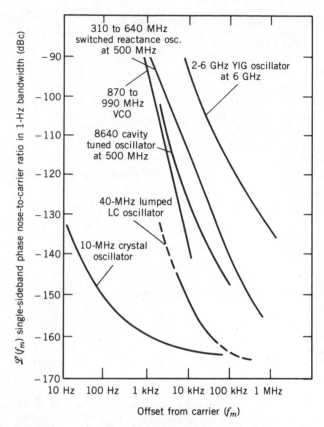

Figure 6.48 Comparison of noise sideband performance of a crystal oscillator, *LC* oscillator, cavity-tuned oscillator, switched reactance oscillator, and YIG oscillator.

Neither linear nor nonlinear CAD programs can handle the SSB phase noise prediction. The PLL Design Kit [6.29] has been specifically written to handle low-noise VCO design.

The following is a short synthesis of a VCO using the PLL design kit. The starting parameters are:

$$\text{Tuning range} \quad 3.8 \text{ to } 4.2 \text{ GHz}$$

$$\text{Tuning diode hyper-abrubt} \quad 0.5 \text{ to } 2.5 \text{ pF}$$

$$\text{Bipolar transistor} \quad f_T = 6 \text{ GHz}, I_c = 15 \text{ mA}, h_{fe} = 150$$

$$K_{\text{VCO}} \quad 40 \text{ MHz/V}$$

Based on these parameters the following data in Table 6.5 were calculated. Table 6.5 shows a calculation of the elements used to cover the specified range. Based on the dc input power we have calculated the output power. The SSB phase noise

TABLE 6.5   Calculation of VCO Tuning Range

F_{min} = 3800 MHz
F_{max} = 4200 MHz
Center range = 4000 MHz
Tuning ratio = 1.105
C_{min} (at V_{max}) of tuning diode = 0.5 pF
C_{max} (at V_{min}) of tuning diode = 2 pF
Bipolar transistor chosen:
   Transistor is operated at I_c = 6 mA, V_c = 8 V
   Cutoff frequency of transistor = 12 GHz at 7 mA
Theoretical output power based on Fourier analysis = 21 mW or 13 dBm
Board stray capacitance = 1 pF
Stripline oscillator used
Transmission line Z = 50 Ω
Half-wave circuit:
   C_{in} = 2 pF, C_{out} (min) = 0.5 pF
   Center conductor = 15.3 mm
   C_{out} (max) = 0.774 pF
Frequency-compensated microstrip calculation:
   Substrate ϵ = 2.3
   Substrate thickness = 1.6 mm
   Width = 4.76 mm
   Mechanical length = 10.1 mm

was calculated first for a MESFET with a flicker frequency of 10 MHz and a noise figure of 10 dB, plotted in Fig. 6.49. The first portion of the phase plot is horizontal and indicates the incidental noise of the oscillator, which is 40 Hz. On a spectrum analyzer this would look similar in linear form. There is no modulation effect of pickup shown. Figure 6.50 shows the phase noise for the same oscillator using a TI bipolar heterojunction transistor described in Fig. 3.9 (f_c = 5 kHz, F = 10 dB).

Using capacitive feedback and loose coupling, Fig. 6.51 shows the oscillator in a grounded base configuration and inductive output coupling. Other configurations can be used as seen in Fig. 6.52. It is a narrowband VCO using the grounded collector circuit introduced earlier in this chapter. Figure 6.53 shows the wideband version operating from 2 GHz to about 3.5 GHz. It has two outputs for different applications, such as one to the digital divider chain and the other to the output amplifier.

A 700- to 900-MHz VCO used in cellular telephones is shown in Fig. 6.54. It combines best phase noise with best microphonic suppression. The oscillator transistor is a junction FET and the band switching and coarse steering are accomplished by diode switching. The design is similar to that of Fig. 6-27.

Two power oscillators are shown in Figs. 6.55 and 6.56. These oscillators are mechanically tuned. A special form of this is the cavity-tuned oscillator shown in Fig. 6.57. The transistor is loosely coupled to the quarter-wave resonator and

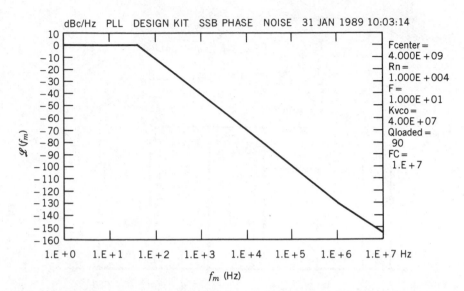

Figure 6.49 Single-sideband phase noise for a 4-GHz oscillator using element values as shown in Table 6.5 using a MESFET.

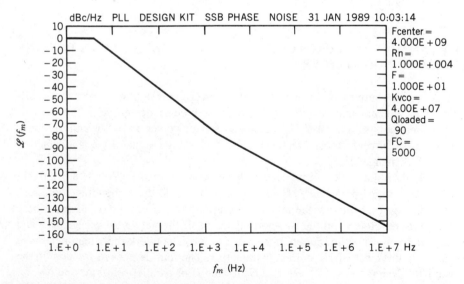

Figure 6.50 Single-sideband phase noise for a 4-GHz oscillator with a bipolar heterojunction transistor (HBT) using Table 6.5.

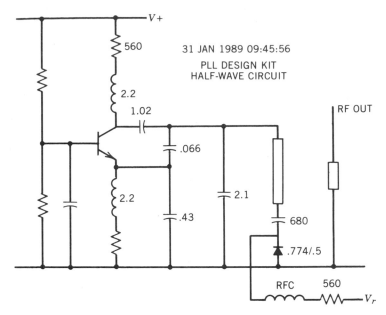

Figure 6.51   Circuit of a 4-GHz oscillator using Table 6.5.

provides a negative impedance to the circuit. This mechanically tuned system has been very popular for extremely low noise oscillators such as the HP8640. Covering the range 2 to 4 GHz, these oscillators can drive a frequency doubler or a frequency divider. Figure 6.58 shows a close-up of the cavity.

6.8 ANALYTIC APPROACH TO OPTIMUM OSCILLATOR DESIGN USING *S* PARAMETERS

Since an oscillator is often operating at the maximum output power, the small-signal parameters may not be accurate for a precise design. For this reason, designers may use a large-signal parameter set for power amplifier and oscillator designs. Usually, the most significant effect under large-signal drive is a reduction in S_{21} and changes in S_{22}.

Several oscillator design procedures have been reported [6.7–6.9, 6.30, 6.31] using large-signal parameters. Using y or z parameters, the embedding conditions given in Fig. 6.59 predict the maximum output power from the oscillator. These solutions are found from the two-port equations of the networks in Fig. 6.19 and the condition of maximum output power. For the Π network the y parameters are used, and the T network requires z parameters. The four degrees of freedom in the oscillator design are B_1, B_2, B_3, and G_n for the Π or parallel solutions and X_1, X_2, X_3, and R_n for the T or series solutions. This gives six oscillator designs for the active two-port without considering the load value or stability factor. We are forcing

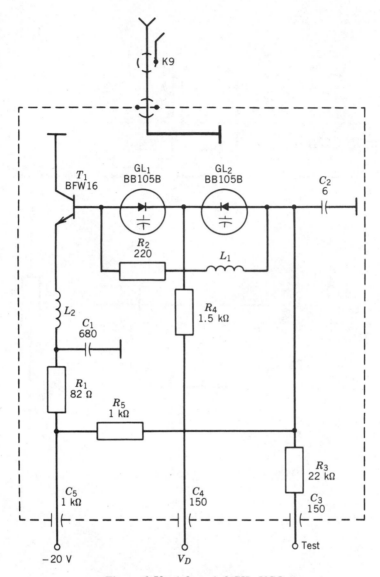

Figure 6.52 1.3- to 1.6-GHz VCO.

the circuit to be resonant and deliver power to a resistor (probably different from 50 Ω). A new software product, SONATA, was recently introduced by Compact Software to perform this oscillator synthesis [6.32].

An example of this calculation for a 500-μm GaAs MESFET at $V_{DS} = 8$ V, $I_{DS} = 50$ mA (DXL 3501A) is given in Fig. 6.60 using lumped elements at 10 GHz. The S parameters for this example are

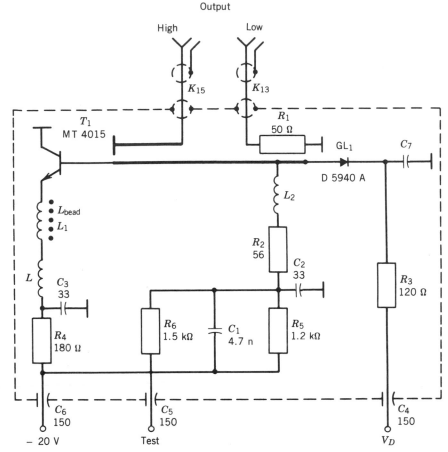

Figure 6.53   2- to 3.46-GHz power VCO.

$$S_{11} = 0.66 \; \underline{/-143^\circ}$$

$$S_{12} = 0.071 \; \underline{/117^\circ}$$

$$S_{21} = 1.26 \; \underline{/46^\circ}$$

$$S_{22} = 0.74 \; \underline{/-59^\circ}$$

The series resonant case with the resistor in the gate gives nearly 50 Ω (case 4), but there is no requirement on the load resistor. Another description of these six oscillators is power out the gate, drain, or source and power out the gate–source, drain–source, or drain–gate.

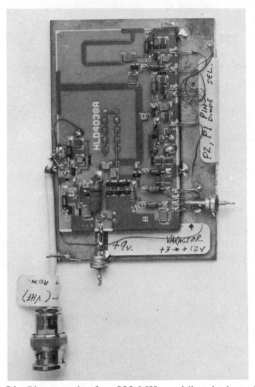

Figure 6.54 Photograph of an 800-MHz mobile telephone VCO.

Another analytic approach due to Gilmore [6.9] uses large-signal S parameters. The upper portion of Fig. 6.61 shows a two-port network described by large-signal S parameters which represent the active element used in the oscillator. The quantity V_1^+ is the power incident on port 1 of the device; V_2^+ is the power incident on port 2.

V_1^+ is fixed at the outset of the oscillator design to be equal to the amplitude used during measurement of S_{11} and S_{21}. It may be set at the point of maximum power added efficiency as described by Pucel [6.33] or Johnson [6.8], or through simulations as was done here. Similarly, V_2^+ is unknown at the outset. However, at optimum power, V_2^+ should be minimized. Hence a small level of V_2^+ should be assumed in specifying $S_{12}(V_2^+)$ and $S_{22}(V_2^+)$ at the outset. This can be checked in the final design.

The complex voltage V_2^- is a free parameter. By defining the gain

$$A = V_2^-/V_1^+ = A_R + jA_I \qquad (6.145)$$

the power delivered to the external network can be maximized as a function of A.

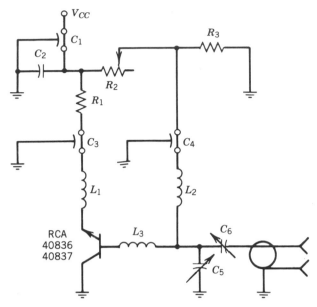

C_1, C_2, C_4 = 470-pF, Feed-through
Allen-Bradley FA5C, or
equivalent
C_2 = 0.2 μF, disk ceramic
C_5, C_6 = 0.35 to 3.5-pF, Johanson
4702 or equivalent

L_1, L_3 = RF choke, 0.5 in. (12.70 mm) length of
No. 32 wire
L_3 = Copper strip:
  0.005 in. (0.127 mm) thick
  0.18 in. (0.457 mm) wide
  0.3 in. (0.76 mm) long
R_1 = 3 to 10 Ω 1/2 W
R_2 = 0 to 500 Ω 2 W
R_3 = 1200 Ω 1/2 W

Figure 6.55 2-GHz power oscillator.

Consider the device described by its large-signal S parameters embedded in an external network as shown in Fig. 6.61. The condition for oscillation is

$$(Z_{in})\text{device}\big|_1 = -(Z_{in})\text{network}\big|_1$$
$$(Z_{in})\text{device}\big|_2 = -(Z_{in})\text{network}\big|_2$$

(6.146)

corresponding to the condition

$$V_{1N}^+ = V_1^- \quad \text{and} \quad V_{1N}^- = V_1^+$$

(6.147a)

$$V_{2N}^+ = V_2 \quad \text{and} \quad V_{2N}^- = V_2^+$$

(6.147b)

where (Z_{in})device $\big|_1$ is the device input impedance at port 1, and (Z_{in})network $\big|_1$ is the external embedding network input impedance at port 1. For example, port 1 might represent the gate of a common source FET, and port 2 the drain.

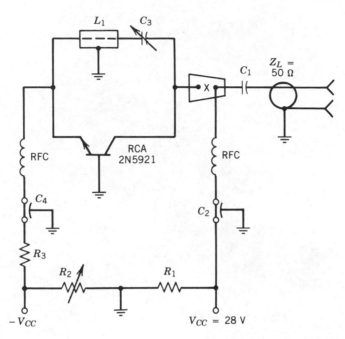

C_1 = 300 pF, disk ceramic
C_3, C_4 = 470 pF, feed-through type,
  Allen-Bradley FA5C, or equivalent
C_3 = 0.3-3.5 pF, Johanson 4702, or equivalent
L_1 = 1.3 in. (33.02 mm) length of 50-ohm coaxial
  line
R_1 = 1200 Ω
R_2 = 0-250 Ω
R_3 = 5 Ω

RFC = 3 turns, No. 29 wire, 0.06 in.
  (1.59 mm) I.D., 0.18 in. (4.77 mm) lng.
X = Tapered microstripline—
  0.1 in. (2.54 mm) wide, input end
  0.24 in. (6.09 mm) wide, output end
  0.475 in. (12.06 mm) long
  0.005 in. (0.13 mm) thick, copper
Dielectric Material = DuPont 5-mil Kapton
  H, or equivalent

Figure 6.56 1.3-GHz 4-W oscillator.

The device is described by its large-signal S parameters, each assumed to be a function of a single variable:

$$V_1^- = S_{11}(V_1^+)V_1^+ + S_{12}(V_2^+)V_2^+ \qquad (6.148a)$$

$$V_2^- = S_{21}(V_1^+)V_1^+ + S_{22}(V_2^+)V_2^+ \qquad (6.148b)$$

The linear embedding network is best described by inverse S parameters. Let S^{-1} represent the inverse S matrix of the embedding network. Then

$$V_{1N}^+ = S_{11}^{-1}V_{1N}^- + S_{12}^{-1}V_{2N}^-$$
$$V_{2N}^+ = S_{21}^{-1}V_{1N}^- + S_{22}^{-1}V_{2N}^- \qquad (6.149)$$

which becomes, using the conditions for oscillation (6.147),

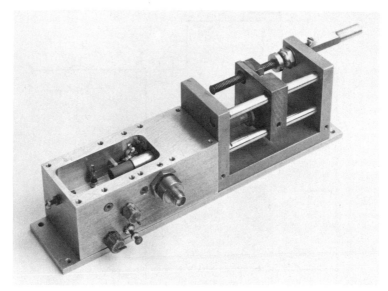

Figure 6.57   4- to 8-GHz cavity-tuned oscillator.

$$V_1^- = S_{11}^{-1}V_1^+ + S_{12}^{-1}V_2^+$$
$$V_2^- = S_{21}^{-1}V_1^+ + S_{22}^{-1}V_2^+$$
$$(6.150)$$

Substituting $V_2^- = AV_1^+$ into (6.148b) gives

$$V_2^+ = \frac{A - S_{21}(V_1^+)}{S_{22}(V_2^+)} V_1^+ \qquad (6.151)$$

Equating (6.148) and (6.150) and using (6.151) gives the design equations

$$S_{22}S_{11}^{-1} + S_{12}^{-1}(A - S_{21}) - S_{11}S_{22} + S_{12}S_{21} - AS_{12} = 0$$
$$S_{22}S_{21}^{-1} + S_{22}^{-1}(A - S_{21}) - AS_{22} = 0$$
$$(6.152)$$

This set of four equations (using real and imaginary parts) describes the condition for oscillation. Provided that A is suitably chosen so that the device generates power, the required network conditions are completely described by (6.152).

If it is desired to extend the reference planes of the device past the terminals at which the device was characterized (e.g., the FET bond wires), the extension is easily incorporated, and can, if desired, be optimized for a given set of load conditions.

Let the length of the line added at the gate be θ_1, and at the drain θ_2. If the characteristic impedances of the lines is Z_0, the incident waves V_1^+ and V_2^+, and hence the large-signal device parameters, normalized to Z_0, are unaffected. The S parameters that should be used in (6.152) for the device are then

Figure 6.58   Close-up of Fig. 6.57.

$$[S] = \begin{bmatrix} S_{11}e^{-j2\theta_1} & S_{12}e^{-j(\theta_1+\theta_2)} \\ S_{21}e^{-j(\theta_1+\theta_2)} & S_{22}e^{-j2\theta_2} \end{bmatrix} \tag{6.153}$$

where θ_1 and θ_2 are the electrical lengths of the lines and can be chosen as variables (if desired).

The S parameters of the linear network are most easily found by first cascading the transmission matrices of the three component elements, normalized to Z_0. In the example considered here, the external network is a Π (shunt)-topology, as shown in Fig. 6.61. The series oscillator, in which the external circuit has a T-topology, can be analyzed in the same way.

For the shunt network,

$$\begin{bmatrix} A & B \\ C & D \end{bmatrix} = \begin{bmatrix} 1 & 0 \\ Y_1 & 1 \end{bmatrix} \begin{bmatrix} 1 & Z_3 \\ 0 & 1 \end{bmatrix} \begin{bmatrix} 1 & 0 \\ Y_2 & 1 \end{bmatrix} = \begin{bmatrix} 1 + Z_3Y_2 & Z_3 \\ Y_1 + Y_2 + Y_1Y_2Z_3 & 1 + Y_1Z_3 \end{bmatrix}$$

$$\tag{6.154}$$

By conventional conversion to the S matrix and simple matrix inversion, the inverse S matrix of the embedding Π network is then

$$S^{-1} = \frac{1}{(Y_1 + Y_2) + Z_3(1 - Y_2)(1 - Y_1) - 2}$$

$$\begin{bmatrix} Z_3(1 + Y_1)(1 - Y_2) - (Y_1 + Y_2) & -2 \\ -2 & Z_3(1 - Y_1)(1 + Y_2) - (Y_1 + Y_2) \end{bmatrix} \tag{6.155}$$

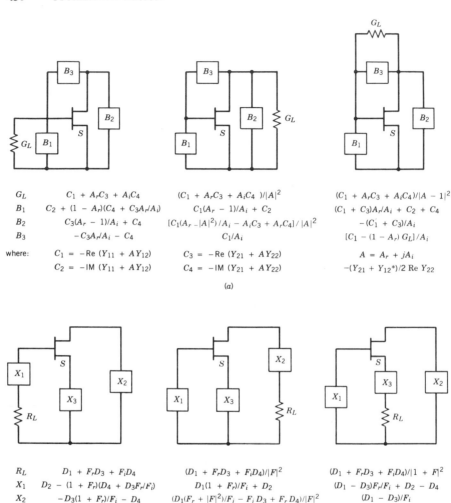

Figure 6.59 Optimum embedding elements for six oscillator structures: (a) three shunt oscillators; (b) three series oscillators. (From Ref. 6.7.)

On substitution of $[S]$ and $[S^{-1}]$ into the four equations (6.152), four of the eight unknowns ($G_1 + jB_1$, $G_2 + jB_2$, $R_3 + jX_3$, θ_1, θ_2) are determined. By arbitrarily determining the other four unknowns by circuit constraints (such as specifying the load impedance, taking the embedding elements reactive, and physical realizability constraints), the circuit can be optimized through choice of A to deliver maximum power into the load. The circuit is found simply through solution of (6.152) using standard nonlinear root finding methods.

Optimization of $A = V_2^-/V_1^+$, keeping V_1^+ constant, requires that V_2^- be varied

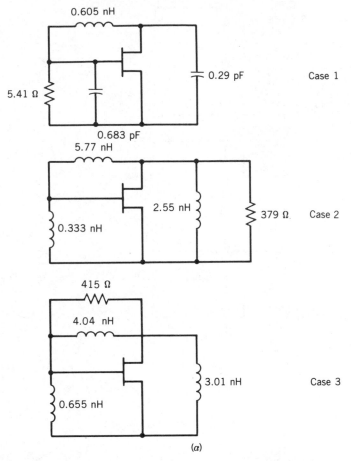

Figure 6.60 Six oscillator structures at 10 GHz for DXL-3501A GaAs MESFET: (a) three shunt oscillators; (b) three series oscillators.

in both magnitude and phase until the power delivered to the external network is a maximum. Referring to Figure 6.61, the power delivered to the load is given by

$$P = (|V_{1N}^+|^2 - |V_{1N}^-|^2) + (|V_{2N}^+|^2 - |V_{2N}^-|^2)$$

$$P = |V_1^-|^2 - |V_1^+|^2 + |V_2^-|^2 - |V_2^+|^2$$

$$= |V_1^+|^2 \left\{ |S_{11}|^2 + 2 \operatorname{Re}\left(\frac{S_{11}^* S_{12}(A - S_{21})}{S_{22}}\right) \right.$$

$$\left. + \frac{|S_{12}|^2 - 1}{|S_{22}|^2}\left(|A - S_{21}|^2\right) + |A|^2 - 1 \right\} \tag{6.156}$$

Using $|x|^2 = x \cdot x^*$, $A = A_R + jA_I$, and (6.151), equation (6.156) becomes

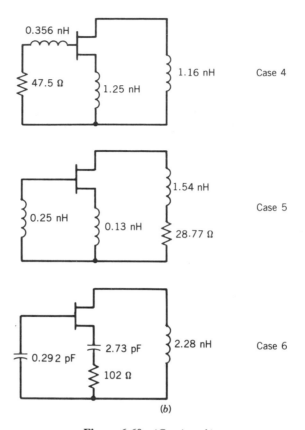

Figure 6.60 (*Continued*)

$$P = \left|V_1^+\right|^2 \left(\begin{aligned} &\left|S_{11}\right|^2 + \frac{S_{11}S_{12}^*}{S_{22}^*}(A_R - jA_I - S_{21}^*) + \frac{S_{12}S_{11}^*}{S_{22}} \\ &\cdot (A_R + jA_I - S_{21}) + A_R^2 + A_I^2 - 1 \\ &+ \frac{\left|S_{12}\right|^2 - 1}{\left|S_{22}\right|^2} \times \left(A_R^2 - A_R S_{21}^* + A_I^2 - jA_I S_{21}^* \right. \\ &- A_R S_{21} + jA_I S_{21} + \left|S_{21}\right|^2 \Biggr) \end{aligned} \right)$$

(6.157)

The power has a local turning point at $\partial P / \partial A_R = 0$ and $\partial P / \partial A_I = 0$, giving

$$\mathrm{Re}\left[\frac{S_{12}^* S_{11}}{S_{22}^*}\right] + A_R\left[1 + \frac{\left|S_{12}\right|^2 - 1}{\left|S_{22}\right|^2}\right] - \left[\frac{\left|S_{12}\right|^2 - 1}{\left|S_{22}\right|^2}\right] \mathrm{Re}\, S_{21} = 0$$

$$\mathrm{Im}\left[\frac{S_{12}^* S_{11}}{S_{22}^*}\right] + A_I\left[1 + \frac{\left|S_{12}\right|^2 - 1}{\left|S_{22}\right|^2}\right] - \left[\frac{\left|S_{12}\right|^2 - 1}{\left|S_{22}\right|^2}\right] \mathrm{Im}\, S_{21} = 0 \quad (6.158)$$

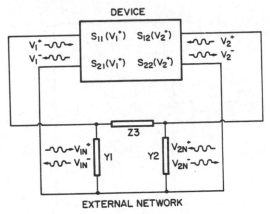

Figure 6.61 Shunt oscillator topology. The device is represented by large-signal S-parameters. The incident and reflected voltage waves are shown. (From Ref. 6.9 © IEEE 1983.)

so that

$$A_{opt} = \frac{1}{|S_{22}|^2 + |S_{12}|^2 - 1} \left(|S_{12}|^2 S_{21} - S_{21} - S_{22}S_{11}S_{12}^* \right) \quad (6.159)$$

Now, P is a maximum if the second derivatives are negative, which is true if

$$|S_{12}|^2 + |S_{22}|^2 < 1 \quad\quad (6.160)$$

This condition will almost always be satisfied for the FET.

It will be noted in (6.159) on substituting for $[S]$ from (6.153) that the magnitude of A_{opt} is independent of the line lengths θ_1 or θ_2. Similarly, the coefficient of $|V_1^+|$ in (6.151) and of $|V_1^+|^2$ in (6.156) is unchanged by the addition of lengths of line on either side of the transistor. Thus, for a given transistor, the optimum gain and power out is set solely by the FET S parameters. As would be expected intuitively, the line lengths serve only as impedance transformers.

For typical GaAs MESFETS, the equation for A_{opt} can be further simplified to

$$A_{opt} \simeq \frac{S_{21}}{1 - |S_{22}|^2} \quad\quad (6.161)$$

when S_{12} is a small number. This result states the optimum voltage gain is the S_{21} or 50-Ω voltage gain divided by the output mismatch factor, $1 - |S_{22}|^2$. For a transistor with S_{22} nearly zero, A_{opt} is simply S_{21}.

A simple computer program was written to optimize, with respect to power, the design of an FET oscillator into a 50-Ω load. For reactive embedding elements

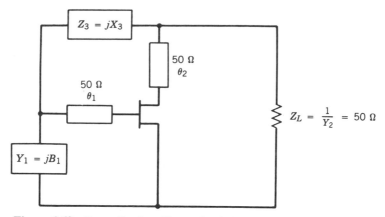

Figure 6.62 Generalized oscillator circuit for the design example given.

$(G_1 = R_3 = 0)$ and $Y_2 = 1$, the unknown quantities are θ_1, θ_2, B_1, and X_3. Such a case is completely constrained and is illustrated in Fig. 6.62.

This design is practical from the viewpoint of a carrier-mounted FET with external feedback, in which the feedback points are movable. Note further that no output load transformer is required, since Y_2 is specified directly to represent 50 Ω. A short computer program was written to calculate the device S parameters for assumed values of θ_1 and θ_2 using (6.153), and the network inverse S matrix for assumed values of B_1 and X_3 using (6.155). These values were then substituted into (6.152) and the equations solved using a standard quasi-Newton routine, as a function of the variables θ_1, θ_2, B_1, and X_3.

From the viewpoint of implementation, it is easiest to treat the feedback element Z_3 as a transmission line of some suitable characteristic impedance. The software was modified so that the unknown parameter X_3 was replaced by l_3, the length of the required feedback transmission line. The network inverse S matrix (6.155) is still easily calculable, although less simply, due to the more complicated form of the transmission matrix for a transmission line rather than a series element.

The design process then proceeds as follows:

1. The large-signal device S parameters are modeled (or measured) at a selection of incident powers, at both ports.
2. A value for V_1^+ is chosen. This fixes $S_{11}(V_1^+)$ and $S_{21}(V_1^+)$. A value of V_2^+ is estimated, to set $S_{12}(V_2^+)$ and $S_{22}(V_2^+)$.
3. A_{opt} is calculated from (6.159) using only the chosen device S parameters; the ratio $|V_2^+|/|V_1^+|$ can then be found from (6.151) and the ratio $Z_0 P / |V_1^+|^2$ from (6.156). All these quantities are independent of the device reference planes (i.e., of the transmission line lengths ultimately chosen). This, then, enables the designer:

a. To check that the output S parameters $S_{12}(V_2^+)$ and $S_{22}(V_2^+)$ that were used correspond to the actual value of V_2^+ calculated.

b. To calculate the oscillator output power at the selected value of V_1^+.

By varying the value of V_1^+ selected in (2), a curve of output power versus incident input power (as done by Johnson [6.8] or Pucel [6.33]) may be generated. Each such output power is the maximum deliverable power to the load at the selected value of V_1^+.

4. The value of V_1^+ that gives the peak output power is used (with corresponding V_2^+) to set the S parameters at the device operating point. The external element values are then found through solution of (6.152).

A large-signal model [6.8] was used to predict the S parameters of the NEC 869177 FET employed in the construction of two oscillators, which operated at 5 GHz. This transistor has a nominal I_{DSS} of 330 mA and a pinch-off voltage around 5 V. The transistor S parameters are given in Table 6.6.

Table 6.7 illustrates these steps in the design of a 5-GHz oscillator. The first column groups blocks of data according to the values of V_1^+ (incident input power) selected. The final line in each block, indicated by a check, is that for which the correct value of V_2^+ (incident power at the output) has been obtained. Thus in the first line of block 1, an incident input power of 17.7 dBm was chosen and an incident output power of 19.9 dBm estimated. This estimate for the power incident on the output port might be based on a transistor with a gain of 6 dB, operating into a net output reflection coefficient of 0.65. The S parameters are thus defined since the incident powers are known, and step 3 of the design process can be performed. From (6.159), A_{opt} is $2.77 + j1.56$; using this in (6.151) gives V_2^+ corresponding to 24.6 dBm. Since our initial estimate of V_2^+ was only 19.9 dBm, the initial guess for incident power at port 2 was modified upward in the second line of block 1, changing $S_{12}(V_2^+)$ and $S_{22}(V_2^+)$, and the cycle repeated; after recalculation, the new incident output power is found to be close to that initially assumed. As shown, the magnitude of A is 2.27 in this case, and the ratio $Z_0P/|V_1^+|^2$, from (6.156), is 4.00, giving a power of 235 mW delivered to the 50-Ω load, for the selected V_1^+ of 17.7 dBm. This is the maximum available power from the device under the chosen terminal conditions (i.e., for the given S parameters used).

Blocks 2, 3 and 4 then vary the terminal conditions by selecting higher values of incident input power. The output power does not continue to increase beyond

TABLE 6.6   S-Parameters for NEC 869177[a]

$S_{11} = 0.73 \underline{/-116°}$
$S_{21} = 1.95 \underline{/85°}$
$S_{12} = 0.048 \underline{/-45°}$
$S_{22} = 0.50 \underline{/-42°}$

[a]At $V_{DS} = 7.5$ V, $V_{GS} = -3.0$ V, $I_{DSS} = 330$ mA, and $V_p = -5$ V.

TABLE 6.7   Design Values for the 5-GHz Oscillator

Block		Input Power Selected $\|V_1^+\|^2$ dBm	Incident Power at Output, $\|V_2^+\|^2$ dBm		A_{OPT}	$\dfrac{Z_0 P_{LOAD}}{\|V_1^+\|^2}$	P_{LOAD} calc. mW
			Estimated	Calculated			
1	x	17.7	19.9	24.6	$2.77 + j1.56$	5.16	—
	✓	17.7	21.2	20.9	$2.07 + j1.37$	4.00	235
2	x	19.3	19.9	26.0	$2.67 + j1.54$	4.86	—
	✓	19.3	22.0	21.1	$1.82 + j1.35$	3.52	300
3	x	20.7	26.4	16.4	$0.678 + j1.45$	2.05	—
	x	20.7	19.9	26.1	$2.30 + j1.36$	3.55	—
	x	20.7	23.4	19.1	$1.27 + j1.24$	2.28	—
	✓	20.7	22.0	21.3	$1.51 + j1.25$	2.55	300
4	x	22.0	23.4	19.3	$0.998 + j1.17$	1.61	—
	✓	22.0	22.0	21.4	$1.25 + j1.15$	1.82	288

bound with V_1^+, but reaches a peak of 300 mW in blocks 2 and 3. This is then the desired operating point. Block 2 was used instead of block 3 because of the higher accuracy in the application of S parameters at the lower input power level, since the FET is not saturated as much. The reason for the peaking of output power can be seen in Fig. 6.63, which shows the measured gain–saturation characteristics of the unmatched transistor chip. The net available power from the device, $P_{out} - P_{in}$, shown by the lower curve, indicates that the available output power will peak for some value of incident power. The shape of this experimentally measured curve correlates well with the column labeled P_{LOAD} in Table 6.7, even though the matching conditions are different in the two cases.

The values of the S parameters at this operating point are used in a root-finding

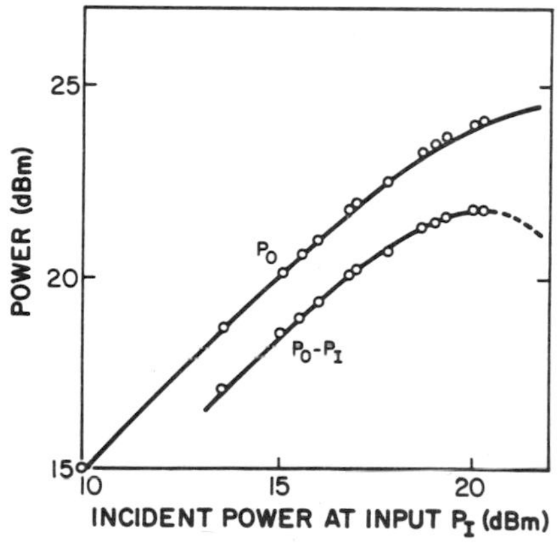

Figure 6.63 Measured gain–saturation characteristics of the FET chip. (From Ref. 6.9. © IEEE 1983.)

routine to solve (6.152) for these elements. The routine used here was a standard IMSL FORTRAN routine, ZXMIN. It was found that convergence to a solution was highly dependent on the initial starting guess, and that multiple solutions are possible. The approximate CPU cost of one design was $0.10.

The design attained was readily realizable and is shown in Fig. 6.64. The driving-point impedance of $-50\ \Omega$ was verified using Super Compact at the load port.

The characteristic impedance of the feedback line was selected through realizability considerations. Doubling Z_0 of this line approximately halved its length, which made it too short to allow connection to the tap points. This line was implemented through a piece of copper ribbon suspended close to the substrate. The gate inductance, which has a very high admittance (and hence can be thought of as the oscillator resonator), was implemented by a short piece of copper ribbon to ground. A photograph of the oscillator is shown in Fig. 6.65. The gate is on the right; the bias leads can be seen coming in at the edges of the picture; the feedback loop, shunt inductance, and chip capacitors are easily discernible.

The oscillator was operated with a gate bias of -3 V and drain voltage of 7.5 V. By slightly changing the length of the feedback loop, the frequency was adjusted to 5350 MHz. Without any tuning of the output, the power into the (designed) 50-Ω load was 23.4 dBm. By tuning around the output connector, the power out increased to 23.9 dBm, compared with the predicted output power of 24.8 dBm. The efficiency obtained was 34.7%, which was the maximum efficiency over all operating points. Maximum power of 25 dBm was obtained by raising the drain voltage to 9.56 V.

Figure 6.66 plots efficiency and power out as a function of drain and gate bias. Oscillations started to build up at a drain voltage of 2.2 V and were observed for gate voltages higher than -4.4 V. It can be seen that even near peak power the efficiency is still very high. Figure 6.67 shows the frequency pushing observed

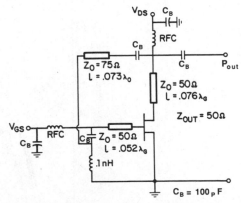

Figure 6.64 Optimized oscillator circuit for the topology chosen. (From Ref. 6.9 © IEEE 1983.)

Figure 6.65 Photograph of 5.350-GHz oscillator. (From Ref. 6.9 © 1983 IEEE.)

due to changes in the bias voltage. Sensitivity to gate voltage is about 150 MHz/V, while frequency is relatively insensitive to drain voltage variations. Although the oscillator was designed using S parameters at a fixed gate bias of -3 V, oscillation was still achieved over a wide range of bias voltages, indicating the usefulness of S parameters (which are relatively bias insensitive) in oscillator design.

Figure 6.66 Oscillator efficiency and output power. (From Ref. 6.9 © 1983 IEEE.)

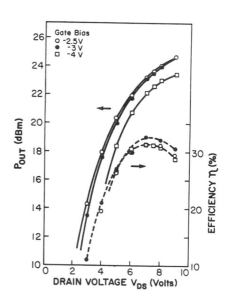

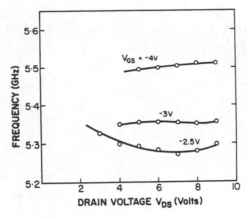

Figure 6.67 Oscillator frequency pushing characteristics. (From Ref. 6.9 © 1983 IEEE.)

Temperature variations in power and frequency at the design point are plotted in Fig. 6.68. When the temperature was raised from $-50°C$ to $+125°C$, the power decreased by 1 dB. The frequency dropped by 74 MHz, giving an average temperature sensitivity of -0.42 MHz/°C.

The frequency noise was also measured in a 300-Hz bandwidth, from 1 to 50 kHz off the carrier. The noise is plotted in decibels below the carrier in Fig. 6.69. The gate bias was held at -2.5 V and the drain voltage at 9 V, corresponding to

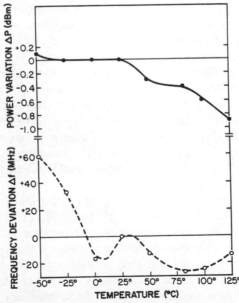

Figure 6.68 Oscillator frequency temperature variation (From Ref. 6.9 © IEEE 1983.)

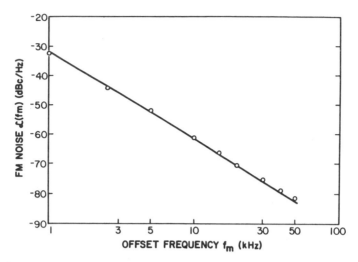

Figure 6.69   Oscillator FM noise. (From Ref. 6.9 © IEEE 1983.)

an output power of 24.9 dBm, close to peak power. The external Q of the oscillator (see Section 6.6) was found to be 29 through load-pull measurements (a sliding load with VSWR of 1.5 was varied over all phases; frequency pulling was 74 MHz). Because the noise measurements are within the resonator bandwidth, the FM noise slope over frequency was 30 dB/decade.

The oscillator was also operated in the self-biased mode, with the gate RFC terminated by a resistor. As the drain voltage is raised, the drain current increases very rapidly at first until oscillation begins and a negative bias develops on the gate. Performance under these conditions was excellent, with peak power of 25 dBm being obtained at a drain voltage of 9.5 V, at an efficiency of 30.5%. A peak efficiency of 35% was achieved at an output power of 24 dBm. Because of the self-limiting oscillation process, the harmonic power was high. However, at peak power the total harmonics were 18.4 dB down, although at lower powers the ratio was much higher. Output power, harmonic power, and efficiency are plotted in Fig. 6.70. The parameter is the external resistor used to develop the self-bias. Figure 6.71 presents the frequency and developed gate bias of the oscillator as a function of externally applied drain voltage. The frequency pushing is reasonably linear (+50 MHz/drain volt). The oscillator could also be used as a frequency modulator, by varying the gate voltage.

Finally, to test the sensitivity of the design to the transistor S parameters, a second NEC-869177 FET was inserted into an identical circuit. The $I-V$ characteristics of this transistor were substantially different from the first (up to 30% higher drain current was observed at identical bias points). By tuning only the output connector on the drain, a slightly smaller output power of 22.7 dBm was obtained at the design bias point; the frequency was 5230 MHz. However, a much lower efficiency of 22% was recorded due to the substantially worse dc character-

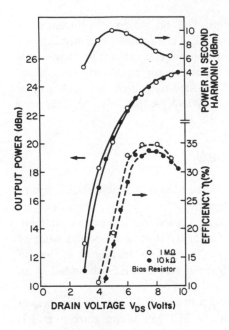

Figure 6.70 Oscillator characteristics in self-bias operation. (From Ref. 6.9 © IEEE 1983.)

istics. In self-bias operation, a peak power of 24.2 dBm was obtained within the safe thermal limits of device operation.

This systematic method for oscillator design using a GaAs MESFET permits an embedding network to be derived analytically that will deliver specified power into a required load at a given frequency. The advantages of this approach are that it is a true two-port design method which requires no creation of an equivalent one-port circuit, the output load is directly specified, and power is automatically

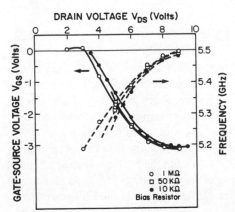

Figure 6.71 Oscillator frequency and gate bias in self-bias operation. (From Ref. 6.9 © IEEE 1983.)

maximized for the device operating conditions. Thus no computer optimization of the circuit is necessary. Furthermore, the device is completely described only by its large-signal S parameters. No assumptions are required about the form of the nonlinearity in order to set the required load impedance for maximum power output as long as the device S parameters have the required functional dependence to be defined by large-signal methods. The program SONATA [6.32] has recently been introduced to generate such oscillators.

6.9 NONLINEAR OSCILLATOR MODEL

A nonlinear large-signal model of the FET can be used with practical circuitry to predict such performance parameters as frequency of oscillation, power output, and efficiency. For this purpose a nonlinear model based on the frequency domain rather than the time domain is preferable. The analysis is based on the work of Tajima [6.10, 6.11]. The validity of this model was established by its success in predicting the large-signal (oscillating) performance of an FET embedded in an actual circuit.

The circuit employed for this study is shown in Fig. 6.72. An FET is located in the middle of a microstrip circuit, wherein the FET gate is connected to an RF choke by means of a wire inductance, while the source is bonded to a long stripline which is terminated by another RF choke. The dc drain bias is fed through a bias tee at the drain port. The gate bias is applied through the gate choke. The source choke is connected to ground for the dc return. RF output power is obtained from the drain port via a matching circuit. The circuits connected to the source and gate terminals function together as a resonator and a feedback circuit, respectively.

This oscillator circuit for purposes of analysis can be divided into two parts at the plane where the drain is connected to the output matching circuit. The admittances at this plane "looking" to the right and left are denoted by Y_{ld} and Y_a. The admittance Y_{ld} represents the load and its matching circuit. The admittance Y_a includes the FET and its source and gate terminations. The oscillator condition is determined by the equation

$$Y_a(A, \omega) + Y_{ld}(\omega) = 0 \qquad (6.162)$$

where A represents the RF voltage amplitude dependence of Y_a. The frequency dependence is represented by ω. The amplitude dependence of Y_a, of course, stems from the nonlinear behavior of the FET. The frequency behavior of the linear admittance Y_{ld} is easy to establish given the circuit configuration. However, the amplitude and frequency dependence of $Y_a(A, \omega)$ is more difficult to determine.

The first step in this procedure is to establish the nonlinear behavior of $Y_a(A, \omega)$ and to obtain the correlation between $Y_a(A, \omega)$ and the FET equivalent circuit. To do this, the FET is represented in terms of its equivalent circuit, as shown in Fig. 6.72b. Under large-signal operation, the element values of the FET equivalent circuit vary with time because at large drive levels they become dependent on

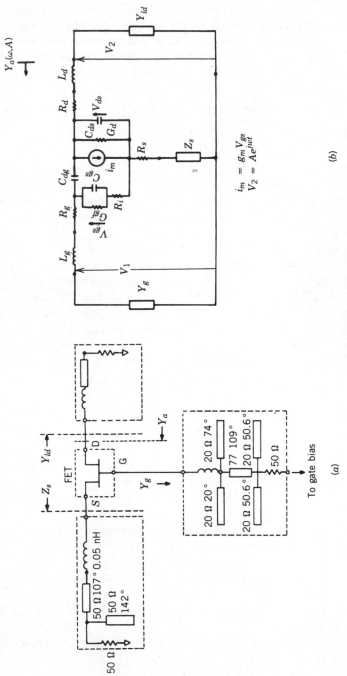

$$i_m = g_m V_{gs}$$
$$V_2 = Ae^{j\omega t}$$

(b)

(a)

Figure 6.72 Schematic of oscillator circuit and equivalent circuit of FET used in oscillator analysis. (From Ref. 6.34.)

terminal voltages. We may consider two of the terminal voltages to be independent and choose the set V_{gs} and V_{ds}, V_{gs} being the voltage across the gate capacitance and V_{ds} that across the drain conductance. If we restrict our interest to the signal frequency and ignore the effects due to higher harmonic components, these voltages can be written as

$$V_{gs} = V_{gso} + v_{gs} \cos(\omega t + \phi)$$
$$V_{ds} = V_{dso} + v_{ds} \cos \omega t \tag{6.163}$$

where V_{gso} and V_{dso} are the dc bias voltages, v_{gs} and v_{ds} the amplitudes of signal frequency components, and ϕ the phase difference between the gate and drain voltages. The equivalent circuit for the signal frequency can now be expressed as a function of the following parameters, which are independent of time: V_{gso}, V_{dso}, v_{gs}, v_{ds}, ω, and ϕ.

To avoid unnecessary complexity of calculations, we limit the nonlinear behavior to five elements, gate forward conductance, G_{gf}, gate capacitance C_{gs}, gate charging resistance R_i, transconductance g_m, and drain conductance G_d. This is justifiable. Here G_{gf} represents the effect of the forward-rectified current across the gate junction under large-signal operation. No voltage dependence was assumed for the parasitic elements, that is, the lead inductances (L_g, L_d, L_s) and contact resistances (R_g, R_d, R_s). Also ignored was the small voltage dependence of the drain channel capacitance C_{ds} and feedback capacitance C_{dg} because of their small values.

6.9.1 Expressions for g_m and G_d

Transconductance g_m and drain conductance G_d are defined as

$$g_m = \left(\frac{i_{ds}}{v_{gs}} \right)_{v_{ds}=0} \qquad G_d = \left(\frac{i_{ds}}{v_{ds}} \right)_{v_{gs}=0} \tag{6.164}$$

where i_{ds} is the RF drain current amplitude. The instantaneous drain current can be written in terms of g_m and G_d as

$$I_{ds}(t) = I_{dso} + g_m v_{gs} \cos(\omega t + \phi) + G_d v_{ds} \cos \omega t \tag{6.165}$$

where I_{dso} is the dc drain current. In this expression linear superposition of the dc and RF currents is assumed.

Now, if we have a function that can simulate the nonlinear dependence of the drain current I_{ds} on V_{gs} and V_{ds}, as

$$I_{ds} = I_{ds}(V_{gs}, V_{ds}) \tag{6.166}$$

then under large-signal conditions, the instantaneous current $I_{ds}(t)$ can be obtained by inserting (6.163) into (6.166). By multiplying sin ωt by (6.165) and integrating over a complete period, g_m is obtained as

$$g_m = -\frac{\omega}{\pi v_{gs} \sin \phi} \int_0^{2\pi/\omega} I_{ds} \sin \omega t \, dt \qquad (6.167)$$

Similarly, G_d is obtained as

$$G_d = \frac{\omega}{\pi v_{ds} \sin \phi} \int_0^{2\pi/\omega} I_{ds} \sin(\omega t + \phi) \, dt \qquad (6.168)$$

Equations (6.167) and (6.168) are now functions of RF amplitudes v_{gs} and v_{ds}, as well as of bias voltages V_{gso} and V_{dso}. We turn now to a more detailed discussion of the nonlinear relation (6.166).

The functional relation $I_{ds}(V_{gs}, V_{ds})$ was established empirically by simulating the dc I–V characteristics by a nonlinear function given by

$$I_{ds}(V_{ds}, V_{gs}) = I_{d1}I_{d2} \qquad (6.169)$$

$$I_{d1} = \frac{1}{k} \left\{ 1 + \frac{V'_{gs}}{V_p} - \frac{1}{m} + \frac{1}{m} \exp\left[-m\left(1 + \frac{V'_{gs}}{V_p}\right)\right]\right\}$$

$$I_{d2} = I_{dsp} \left\{ 1 - \exp\left[\frac{-V_{ds}}{V_{dss}} - a\left(\frac{V_{ds}}{V_{dss}}\right)^2 - b\left(\frac{V_{ds}}{V_{dss}}\right)^3\right]\right\}$$

$$k = 1 - \frac{1}{m}[1 - \exp(-m)]$$

$$V_p = V_{po} + pV_{ds} + V_\phi$$

$$V'_{gs} = V_{gs} - V_\phi$$

where V_{po} (> 0) = pinch-off voltage at $V_{ds} \approx 0$
V_{dss} = drain current saturation voltage
V_ϕ = built-in potential of the Schottky barrier
I_{dsp} = drain current when $V_{gs} = V_\phi$

and a, b, m, and p are fitting factors that can be varied from device to device.

6.9.2 Nonlinear Expressions for C_{gs}, G_{gf}, and R_i

Although the gate junction is also a function of V_{gs} and V_{ds}, we assume here that it can be approximated by a Schottky barrier diode between gate and source, with

V_{gs} as the sole voltage parameter. Gate capacitance C'_{gs} and forward gate current i_{gf} can be found from Schottky barrier theory as

$$C'_{gs} = \frac{C'_{gso}}{\sqrt{1 - V_{gs}/V_\phi}} \quad (-V_p \leq V_{gs}) \tag{6.170}$$

or

$$C'_{gs} = \frac{C'_{gso}}{\sqrt{1 + V_p/V_\phi}} \quad (-V_p \geq V_{gs}) \tag{6.171}$$

$$i_{gf} = i_s \exp(\alpha V_{gs} - 1) \tag{6.172}$$

where C'_{gso} is the zero-bias gate capacitance, i_s the saturation current of the Schottky barrier, and $\alpha = q/nkT$.

When V_{gs} varies according to (6.163), the effective gate capacitance C_{gs} and gate forward conductance G_{gf} for the signal frequency are obtained from (6.170)–(6.172) as

$$C_{gs} = \frac{1}{\pi v_{gs}} \int_0^{2\pi} \left(\int^{V_{gs}} C'_{gs} \, dv \right) \cos \omega t \, d(\omega t) \tag{6.173}$$

$$G_{gf} = 2i_s \exp(\alpha V_{gso}) \frac{I_1(\alpha v_{gs})}{v_{gs}} \tag{6.174}$$

where $I_1(x)$ is the modified Bessel function of the first order.

The gate-charging resistance R_i was assumed to vary in such a way that the charging time constant was invariant, with bias.

$$R_i C_{gs} = \tau_i (\text{constant}) \tag{6.175}$$

Thus all nonlinear element values of the equivalent circuit can be expressed in terms of the terminal RF amplitudes and their relative phase. One may now determine more precisely the admittance $Y_a(A, \omega)$, or $Y_a(v_{gs}, v_{ds}, \omega)$, by an iteration method such as the following.

First, starting values for v_{gs} and the equivalent-circuit parameters are assumed. For the latter, small-signal values based on measured S parameters are suitable. With these parameters specified, the output voltage v_{ds} and its phase can be calculated in a straightforward manner. With the resultant value of v_{ds}, ϕ, and the initially assumed V_{gs}, the "first-cut" evaluation of the equivalent-circuit elements can be made with the help of (6.167), (6.168), and (6.173)–(6.175). The procedure above is then repeated, each time using the most recently evaluated values of v_{ds}, ϕ, and v_{gs}, until convergence is obtained. The process converges when successive iterations reproduce the equivalent circuit parameters to within some specified

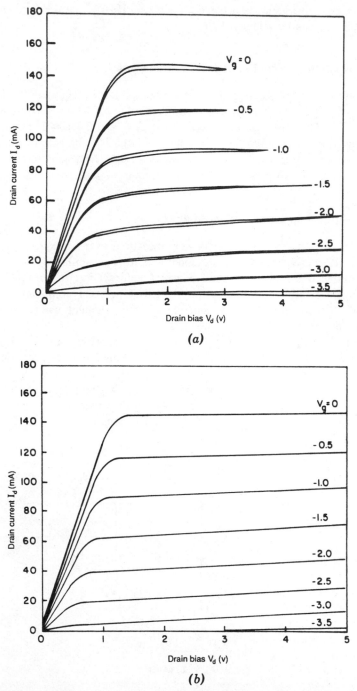

Figure 6.73   (a) Measured and (b) simulated I–V characteristic used in nonlinear oscillator analyzer. (From Ref. 6.34.)

error. Once convergence is achieved, such oscillator properties as power output
and efficiency can be calculated.

6.9.3   Analytic Simulation of I–V Characteristics

The analysis begins by applying the analytic expressions [equation (6.169)] to the
set of measured I–V characteristics shown in Fig. 6.73a for a 1 μm $\times$ 400 μm
FET [6.34]. The fitting parameters $a = -0.2$, $b = 0.6$, $m = 3$, and $p = 0.2$ in
these equations were determined, and the simulated I–V characteristics calculated.
Figure 6.72b is the result of this simulation. Note the excellent agreement (of
course, the hysteresis shown in Fig. 6.73a cannot be represented) [6.34].

6.9.4   Equivalent-Circuit Derivation

Next, the small-signal S parameters were measured over a broad frequency range
(2 to 12 GHz) at the operating bias conditions for the oscillator. These were used
to determine the equivalent-circuit element values (Fig. 6.74). These element
values were determined by using the SUPERCOMPACT computer-aided design
program, which optimizes the equivalent-circuit element values to provide a
"good" fit to the measured S parameters.

Figure 6.75 illustrates the excellent agreement between measured S parameters
and those calculated from the equivalent circuit. This establishes confidence in the
equivalent-circuit element values.

The equivalent-circuit elements are used for two purposes: (1) to determine
what range of circuit terminations are necessary to initiate oscillations (i.e., establish instability); and (2) to establish initial conditions for the nonlinear analysis.

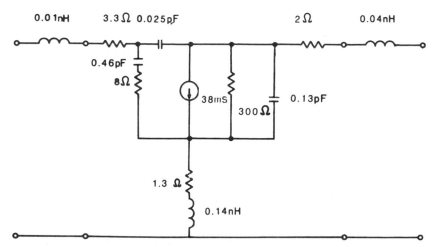

Figure 6.74   Equivalent circuit of FET based on measured S parameters. (From Ref. 6.34.)

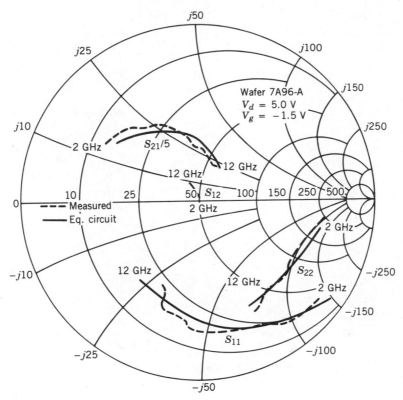

Figure 6.75 Comparison of measured S parameters and calcuated S parameters based on equivalent circuit ($Z = 400$ μm, $L_G = 1$ μm). (From Ref. 6.34.)

6.9.5 Determination of Oscillation Conditions

The oscillation conditions, that is, the load conditions at the drain terminals necessary for oscillations to start, are delineated by the shaded regions in Fig. 6.76. Shown is a plot of $-Y_a/Y_0$, where $Y_0 = 0.02$ S is the characteristic line admittance. This plot was obtained for the conditions where the source and gate terminals were terminated by the oscillator circuit elements established earlier (Fig. 6.72a). The unstable regions where oscillation is possible were determined by the Nyquist criterion. It shows that oscillation is most likely to occur close to 10 GHz, but with greater mismatch, oscillation could also occur at lower frequencies, 6 to 8 GHz. Past measurements with similar FETs have shown a tendency to hop in frequency as the circuit was tuned.

6.9.6 Nonlinear Analysis

Having established the oscillation conditions, we now apply the equivalent-circuit elements and the nonlinear equations from the dc $I-V$ simulation to determine the

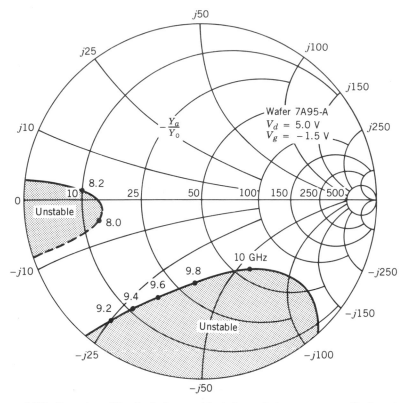

Figure 6.76 Domains of load admittance (shaded areas) that support oscillations in FET circuit. Reference plane for load is that marked Y_a in Fig. 6.72a. (From Ref. 6.34.)

oscillator properties under steady-state oscillation conditions for the permissible range of load-terminating conditions. The result of the nonlinear analysis is shown in Fig. 6.77. Shown are closed constant-output power contours (in dBm) as a function of load (drain) terminating conditions. Also shown are intersecting loci of constant-frequency contours. For example, the 10-GHz contour shows the predicted power output at various terminating admittance levels. The power levels indicated are in the range obtained experimentally, as shown by the measured data.

6.9.7 Conclusion

A large-signal model of the FET has been derived. This model has been applied to an FET embedded in an actual oscillator circuit, and the predicted performance has been shown to be consistent with experimental results [6.34].

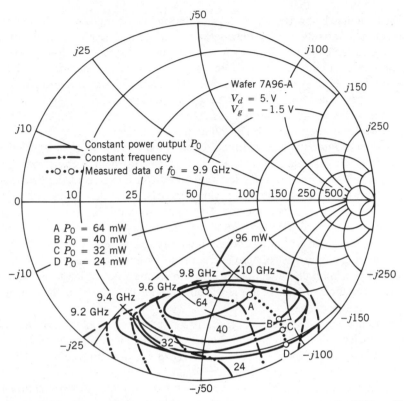

Figure 6.77 Calculated constant frequency and power output contours for FET oscillator circuit. These contours represent load admittance conditions at the drain which are necessary to yield the stated power output at the given frequency. All shown are measured oscillator data. (From Ref. 6.34.)

6.10 OSCILLATOR DESIGN USING NONLINEAR CAD TOOLS

The recent introduction of nonlinear computer tools such as Sonata [6.32] and Microwave Harmonica [6.35] allows a design engineer to produce theoretical oscillator designs to achieve given specifications. Oscillator design is an autonomous problem in which there is no external driving source. This complicates the design process considerably, as the frequency of an oscillator is then an additional degree of freedom in the circuit that must be accounted for.

Microwave Harmonica is a nonlinear CAD package using the harmonic balance technique. Because of its nonlinear optimization capabilities, it is uniquely suited to the task of oscillator design [6.36]. The example that follows, from the work of Rizzoli et al. [6.37], illustrates the design of a 5-GHz dielectric resonator oscillator using Microwave Harmonica.

The harmonic balance method seeks a solution to a steady-state nonlinear design problem by iteratively solving for a set of variables, referred to as state variables. The state variables can typically be chosen as the voltages at the linear–nonlinear interface in a circuit partitioned into linear and nonlinear segments. They are expressed as the phasor components, and their harmonics, of a sinusoidal excitation frequency. The state variables are usually found iteratively by a gradient-based technique which seeks a simultaneous solution for Kirchhoff's equations applied to the linear and nonlinear sides of the network separately.

For the nonautonomous, or driven, circuit, the driving frequency is known a priori and Kirchhoff's equations are a well-determined system of equations in which the phase and amplitude of the excitation appear on one side of the equations as forcing terms. In an autonomous circuit, the only excitation terms appearing in Kirchhoff's equations are dc sources. One stable solution to these equations for a circuit with no applied RF drive will always be the dc (or degenerate) solution, as all phasor terms at an arbitrarily chosen frequency can be set to zero and still satisfy the RF driving conditions (of zero excitation). For an autonomous circuit to have a solution to Kirchhoff's equations at nonzero frequencies, at least one additional degree of freedom is required, as there is one additional unknown in the equations—the oscillation frequency.

For oscillator analysis purposes (i.e., for a fixed circuit topology), the additional free parameter is just the unknown frequency of oscillation. For oscillator design purposes, the frequency is fixed as a design goal, and the required degree of freedom must be represented by a free circuit variable, such as a bias voltage or other tuning element. In this way, a solution can be found to the conditions for oscillation (which are just Kirchhoff's equations), by varying some circuit parameter. This parameter must be adjusted so that the equations can be satisfied at the oscillation frequency, with some set of (solution) state variables, which are determined at the same time.

The harmonic balance problem must then allow for the simultaneous solution of both the state variables and the circuit elements to satisfy Kirchhoff's equations under the chosen conditions (i.e., at the design frequency). In fact, Microwave Harmonica allows the user to set additional circuit parameters to be variables in order to optimize for other circuit responses, such as output power, efficiency, spectral purity, or distortion, while simultaneously satisfying Kirchhoff's equations. Consequently, the methodology of introducing additional degrees of freedom into the harmonic-balance problem allows not only for the solution of autonomous designs, but also the optimization of all types of circuits for desired response.

As an example, Figure 6.78 shows a DRO design using the Plessey GAT6 FET (I_{DSS} = 40 mA) embedded in a simple microstrip circuit. The specified optimization goals were 20 mW output power at a frequency of 5 GHz. In the design, the additional degrees of freedom for the optimization problem and the oscillator synthesis were provided by the resonator diameter and the microstrip lengths, together with the FET bias voltages. Prior to circuit optimization, the program automatically optimizes the DRO dimensions to resonate at the $TE_{01\delta}$ mode at 5 GHz. The preoptimization is needed to ensure that the initial starting point for the

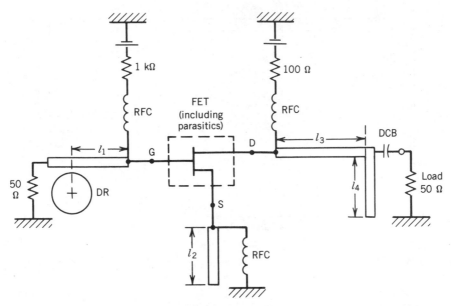

Figure 6.78 DRO design example. (From Ref. 6.37 with permission.)

design equations is within the operating regime. A sequence of optimizations was then performed by increasing the lower bounds on the output power. The maximum power available was found to be 29 mW with an efficiency of 14.2%. The bias point was $V_D = 7.95$ V, $V_G = -0.77$ V, and $I_0 = 26$ mA. Output harmonics were 22.3 dB below the fundamental.

The circuit can also be tuned by moving the metal tuning plate over the resonator. To tune to a frequency of 5.05 GHz, the circuit must be reoptimized with the plate distance s as the only tuning variable. This change to the single circuit parameter is needed to allow there to be a solution to Kirchhoff's equations at the new oscillation frequency. The values of the state variables are, of course, also different at the new point, and the output power is reduced correspondingly.

An oscillator analysis can be performed by repetitively reoptimizing the tuning parameter s at a series of frequencies and constructing a tuning curve of oscillation frequency versus s. For an oscillator of unknown frequency, the actual oscillation frequency can be determined by entering the tuning curve at the known value of s. Note that in the case of an oscillator analysis, only a single tuning variable is adjusted at each frequency point, so that a tuning curve can be constructed to give a single, unique relationship between the design frequency and the actual circuit parameter value in the circuit being analyzed.

Although the harmonic balance approach ensures that a steady-state solution exists, the buildup of oscillations and the stability of the steady-state operating point also need to be addressed. One way to provide a numerical solution to this problem is to use the principles of bifurcation theory [6.38]. Using the frequency

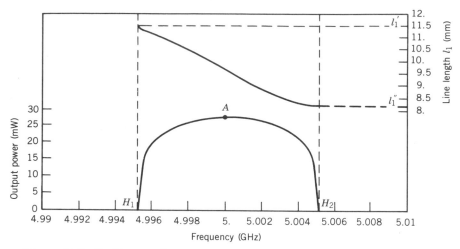

Figure 6.79 Oscillator performance versus l_1. (From Ref. 6.37 with permission.)

of oscillation as a continuation parameter [6.39], and choosing the output power as a parameter to describe the circuit state, a solution path may be built by stepping the frequency through a prescribed range and performing a sequence of circuit optimizations with respect to the state variables and some other free circuit parameter.

Figure 6.79 shows the solution path for the DRO depicted in Fig. 6.78, with the distance l_1 between the DR plane and the FET gate chosen as the free circuit variable. Both the output power and the tuning (continuation) parameter are plotted against frequency. Point A is the nominal operating point obtained by circuit optimization as just described. The critical points H_1 and H_2 are Hopf bifurcations [6.40], and all states belonging to the bifurcated branch H_1AH_2 are stable. On the other hand, a local stability analysis about the dc bias point reveals that each dc state between H_1 and H_2 has two natural frequencies with positive real parts, and is thus unstable. This guarantees oscillation buildup whenever the DRO is biased with the distance between the DR plane and the FET gate set to any values between l_1' and l_1''. The bifurcation diagram is a tuning diagram that provides full information on the DRO tuning range with respect to the circuit variable, and on the power-to-frequency relationship within this range.

Figure 6.80 shows the results of a similar analysis, with the tuning plate position s now being chosen as the circuit variable. In the range of all positive s, there is only one Hopf bifurcation at point H_1, corresponding to $s = s_1$. This bifurcation is supercritical, so that the circuit is dc stable below H_1 and dc unstable above H_1 due to a couple of complex-conjugate natural frequencies with positive real part. Thus, whenever the FET is biased with the tuning plate set to any position above s_1, oscillation buildup will take place. In this particular case, stable oscillation is possible even if the plate is suppressed (s goes to infinity).

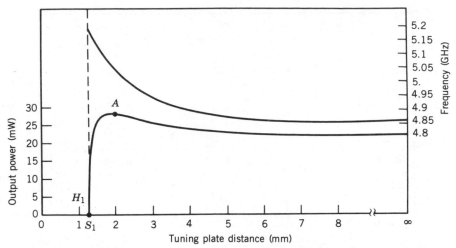

Figure 6.80 Oscillator performance versus tuning plate distance. (From Ref. 6.37 with permission.)

Finally, the DRO may be electronically tuned by inserting a series varactor in the gate feedback circuit as shown in Fig. 6.81. An inductance was also introduced into the gate to resonate the varactor capacitance at the nominal operating frequency of 5 GHz. Figure 6.82 shows the results of the bifurcation analysis for the DRO, with the varactor bias chosen as the free circuit parameter. A comparison of Figs.

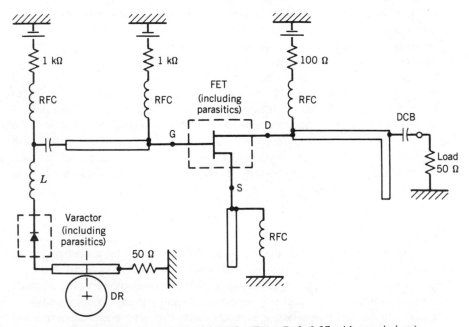

Figure 6.81 Varactor-tuned DRO. (From Ref. 6.37 with permission.)

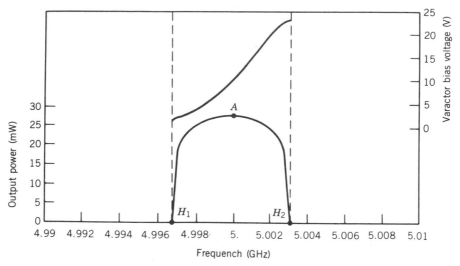

Figure 6.82 Oscillator performance versus varactor tuning voltage. (From Ref. 6.37 with permission.)

6.79, 6.80, and 6.82 clearly displays the superior tuning performance obtained by changing one of the resonator geometrical parameters over that of other types of mechanical or electronic tuning.

In summary, the advent of nonlinear CAD programs now allows the designer to verify many aspects of circuit operation not previously obtainable through linear programs alone. In this way, the levels of output power, harmonic content, dc-to-RF efficiency, device currents, and load pushing and pulling can be examined before the oscillator is constructed.

A second oscillator example described in the remainder of this section will develop the methodology needed from commencement of the nonlinear design to its completion. The $I\text{--}V$ and small-signal S-parameter data are used with a parameter extraction routine to characterize the FET for the nonlinear model used in the harmonic-balance simulator. The overall design philosophy is demonstrated with the design of an oscillator–amplifier subsystem. The steps of design specification, device modeling, circuit analysis, and system optimization are illustrated by this buffered oscillator example.

6.10.1 Parameter Extraction Method

Parameter extraction of an accurate nonlinear model plays an essential part in nonlinear simulation. Conventionally, small-signal parameters are extracted from S parameters measured at a single bias point. Designers relying on this approach are often frustrated by nonunique solutions. Usually, there exists a family of solutions, all of which produce a similar match between model response and measurement. As a consequence, the particular solution obtained depends on the initial guess; using a different starting point will probably result in a different solution. Additional difficulties arise for large-signal nonlinear modeling, since we need to determine parameters that may vary with bias, such as the transconductance. It is obvious that small-signal S-parameter measurements at a single bias point are not adequate for extracting bias-dependent parameters.

A common practice in an attempt at large-signal modeling is to extract the bias-dependent parameters and bias-independent parameters separately. The bias-dependent parameters are extracted by curve fitting, while the bias-independent parameters are extracted from S parameters measured at a single bias point. Such an approach can be an improvement over conventional small-signal modeling since the extraction of the bias-independent parameters reduces the number of unknowns. However, two problems may plague such an approach: the results may not be unique, and bias-dependent parameters extracted from the dc data alone (transconductance, output conductance) may not be valid at the operating microwave frequencies.

A novel approach to nonlinear parameter extraction [6.41, 6.42] simultaneously extracts the dc, bias-dependent, and bias-independent parameters. The motivation of this method is to strengthen the model identifiability and to enforce a unique solution. S-parameter measurements at multiple bias points can be utilized to achieve a robust solution applicable to dc, small-signal, and large-signal modeling.

A software package called RoMPE (Robust Model Parameter Extractor) [6.43] is commercially available to perform simultaneous extraction of dc, bias-dependent, and bias-independent parameters. The program accounts for the dependence, explicit and implicit, of small-signal parameters on the bias. It implements the theoretically established relationship between small-signal parameters and bias-dependent parameters, such as between C_{gs} and C_0 (C_{gs} at zero gate voltage) and V_B (built-in potential voltage). Extraction is performed through one of two available state-of-the-art optimizers: the l_1 or l_2 optimizers.

To illustrate the usefulness of the approach, an example will be presented. The large-signal FET model [6.44] used appears in Fig. 6.83. The bias-independent parameters are identified as

$$\phi_a = [L_g, L_d, L_s, R_g, R_d, R_s, C_{dg}, C_{ds}] \tag{6.176}$$

and the bias-dependent parameters as

$$\phi_b = [C_{gs}, I_{ds}] \tag{6.177}$$

The constraints imposed on the ϕ_b are

$$C_{gs} = \frac{C_0}{\sqrt{1 - V_g/V_b}} \tag{6.178}$$

$$I_{ds} = I_{dss} \left(1 - \frac{V_g}{V_p}\right)^2 \tanh \frac{\alpha_d V_d}{V_g - V_p}$$

$$V_p = V_{p0} + \gamma V_d \tag{6.179}$$

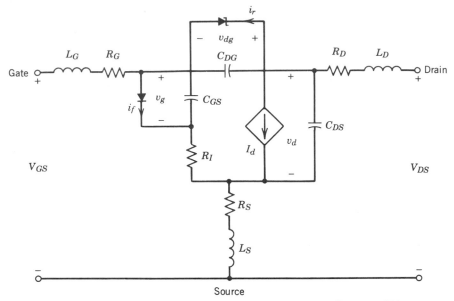

Figure 6.83   Large-signal FET model. (From Ref. 6.41 © IEEE 1988.)

The diode currents are given by

$$i_f = I_s \exp\left(\alpha_s v_g - 1\right) \qquad (6.180)$$

$$i_r = I_{sr} \exp\left[\alpha_{Sr}(V_d - V_g - V_{BR})\right] \qquad (6.181)$$

The model parameters were then optimized such that the S-parameters matched those reported in Ref. 6.41 from 2 to 18 GHz at three bias points. Starting values and model values at solution are listed in Table 6.8. The optimization required 35 iterations of the l_1 optimizer and took 55 CPU minutes on 8-MHz IBM PC–compatible equipment (a total of 51 frequencies, 18 optimizable variables). Figure 6.84 displays the relative S parameters over the entire frequency range for the second bias point ($V_{gs} = 1.74$ V, $V_{ds} = 4$ V). The agreement between measured and modeled responses is superb. The match at the other bias points is slightly degraded.

As a comparison to the conventional approach of parameter extraction, the dc parameters alone were first extracted to give a good match to the measured bias points. Next, the ac parameters (C_0, V_b) for the bias-dependent gate–source capacitance (C_{gs}) were extracted using (6.178). Then the bias-independent ac parameters were extracted at the second bias point and gave an excellent fit to the measured S parameters. However, when the entire model was simulated (using these bias-dependent and bias-independent parameters) the response was not acceptable, as shown in Fig. 6.85. It is also worth mentioning that the parameters extracted in this way are significantly different from those extracted using the simultaneous method.

TABLE 6.8 Parameter Values of the FET Model

Parameter		Bias 1		Bias 2	Bias 3
		Start	Solution	Solution	Solution
R_g	(OH)	1.0	0.0119	0.0119	0.0119
R_d	(OH)	1.0	0.0006	0.0006	0.0006
G_{ds}	(s)	a	0.004647	0.005843	0.006382
R_i	(OH)	5.0	5.855	4.164	3.642
R_s	OH	1.0	0.3514	0.3514	0.3514
L_s	(nH)	0.01	0.0107	0.0107	0.0107
C_{gs}	(pF)	a	0.7568	0.3997	0.3194
C_{dg}	(pF)	0.1	0.04226	0.04226	0.04226
C_{ds}	(pF)	0.3	0.1958	0.1958	0.1958
G	(s)	a	0.05888	0.04467	0.03048
T	(ps)	5.0	3.654	3.654	3.654
L_g	(nH)	0.05	0.1257	0.1257	0.1257
L_d	(nH)	0.05	0.0719	0.0719	0.0719

Coefficient		Start	Solution	Coefficient		Start	Solution
I_s	(nA)	0.5	0.5	α_d		4.0	3.039
α_s	(/v)	20.0	20.0	V_{po}	(V)	−4.0	−4.275
I_{sr}	(NA)	0.5	0.5	γ		−0.2	−0.3912
α_{sr}	(/v)	1.0	1.0	C_0	(pF)	1.0	0.7961
I_{dss}	(A)	0.2	0.191	V_b	(V)	1.0	0.5975

^{a}Values determined by constraints on ϕ_b.

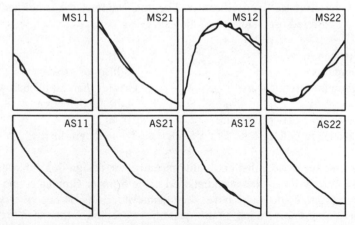

Figure 6.84 S-parameter match using simultaneous ac and dc parameter extraction.

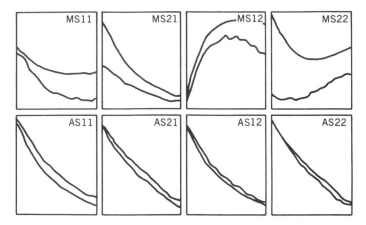

Figure 6.85 *S*-parameter match using conventional parameter extraction. (From Ref. 6.41 © IEEE 1988.)

A test of robustness was performed [6.41] by perturbing the starting point of some variables by 20 to 200%. All the variables converged to virtually the same solution. This demonstrates the uniqueness of the solution when dc and ac parameters are extracted simultaneously.

6.10.2   Example of Nonlinear Design Methodology: 4-GHz Oscillator–Amplifier

A simple example to illustrate the design steps needed to generate a buffered oscillator subsystem is presented. The steps are listed methodically to best demonstrate the overall concept of a nonlinear design rather than the details.

1. Design Specification. A key to determining product quality is to measure its degree of conformance to specification. For this example, the simplified specification for a buffered oscillator is as follows:

Frequency	4 GHz
Output power	20 dBm or greater
Harmonic frequencies	Greater than 20 dB below the fundamental
Mean time between failure	10^5 hr
Frequency pulling (into 2 : 1 VSWR)	To be determined

It is instructive to note that until quite recently, the design of a subsystem such as this could only have been accomplished using a linear simulator, and that the oscillator design could only have been achieved using a negative-resistance approach. Furthermore, none of the quantities specified (other than frequency) could be determined until the design is completed and built. The advent of nonlinear simulators allows verification of these at the design stage. This is crucial for quality considerations, as it allows conformance to specification to be designed in.

2. Design Modeling. A parameter extraction program, such as RoMPE, can be used to fit the coefficients used in the model of Materka and Kacprzak [6.44], as modified in Microwave Harmonica, to the available measurement data. After selection of the FET and measurements of *I–V* data and *S*-parameter data, the parasitic elements and coefficients used in the nonlinear Microwave Harmonica model can be determined. An example of the modeled *I–V* response of the FET is shown in Fig. 6.86.

3. Component Design. The specification calls for the development of both an amplifier (for buffering) and an oscillator. The development of an amplifier is relatively simple. Figure 6.86 shows a circuit topology developed using linear circuit techniques (i.e., optimized for gain and bandwidth). For the Microwave Harmonica analysis, bias and RF sources must be added, and the FET represented by the model parameters found in step 2. When driven by a 4-GHz input level of 15 dBm, the load line in Fig. 6.87 results. The amplifier output power that results is 20 dBm.

The oscillator design is more difficult, as this type of circuit is now autonomous (i.e., has no external RF drive). Although the designer might derive the circuit topology shown in Fig. 6.88 as a first step from linear analysis, there is no guarantee that the circuit will oscillate at precisely 4 GHz. Some additional degree of freedom (such as the 2.2-pF tuning capacitor) must be adjusted to tune the frequency.

The design of oscillators using the harmonic balance method also requires this additional degree of freedom. The frequency, as in the specification, must first be

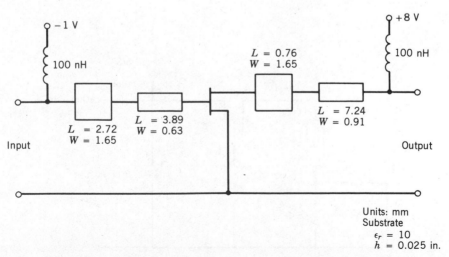

Figure 6.86 Amplifier design at 4 GHz. (From Ref. 6.36 © IEEE 1988.)

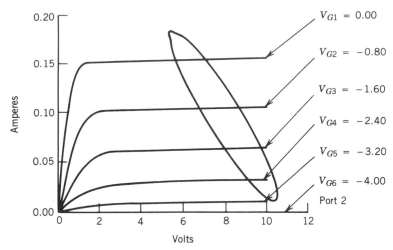

Figure 6.87   Amplifier load line with $P_{in} = 15$ dBm at 4 GHz. (From Ref. 6.36 © IEEE 1988.)

imposed on the problem. At least one corresponding degree of freedom, such as a circuit parameter, must then be adjusted to ensure oscillation at that frequency. In Microwave Harmonica, the degree of freedom is introduced as an optimization variable, with the variable adjusted so that the harmonic-balance equations (i.e., the conditions for oscillation) have a solution at the frequency imposed. Other variables and other optimization criteria, such as maximum output power, maximum spectral purity, or maximum efficiency, may also be imposed.

If a solution exists, oscillations will occur at the imposed frequency with an amplitude determined by the final values of the state variables. The output spectrum

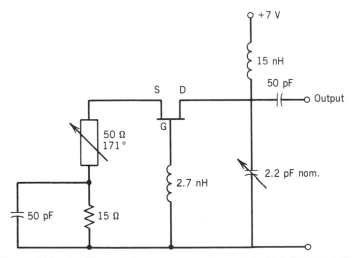

Figure 6.88   Oscillator design at 4 GHz. (From Ref. 6.36 © IEEE 1988.)

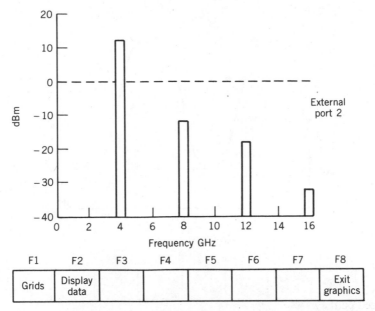

Figure 6.89 Oscillator power output spectrum. (From Ref. 6.36 © IEEE 1988.)

in Fig. 6.89 results, with the corresponding drain current waveform shown in Fig. 6.90 (over two cycles of oscillation). The load line for the oscillation is shown in Fig. 6.91 and now represents a true limit cycle, with the drain current confined between slightly greater than I_{DSS} and pinch-off. The elliptical shape results because the load impedance is now reactive.

4. Subsystem Design. The resulting oscillator–amplifier subsystem is shown in Fig. 6.92. The total system is also autonomous but will no longer oscillate at precisely 4 GHz because the amplifier input loads the oscillator differently from the 50 Ω used in the initial oscillator design. Microwave Harmonica can again be used to set the frequency to the desired specification frequency by adjustment of a single tuning parameter. The tuning capacitor assumes a final value of 3.14 pF after "optimization." Optimization in this case is nothing more than allowing the program's optimizer to adjust the value of the tuning element until the harmonic-balance equations can be solved, at the desired frequency, with some set of (solution) state variables.

The output waveform is shown in Fig. 6.93 and corresponding output spectra in Fig. 6.94. The specified output power and harmonic content can be determined from this figure, and additional optimization criteria added to improve the performance further. The gate current waveform in Fig. 6.95 reveals substantial harmonic content with a dc offset; this would have negative implications for system reliability.

Finally, an oscillator analysis (rather than synthesis) is required in order to determine the effect of load pull. In Microwave Harmonica this can be performed

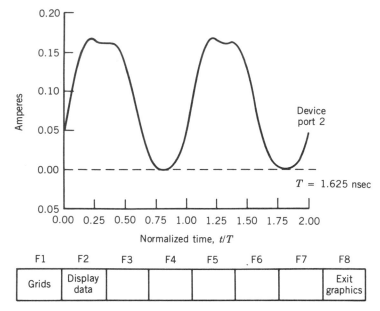

Figure 6.90   Oscillator drain current versus time (two cycles). (From Ref. 6.36 © IEEE 1988.)

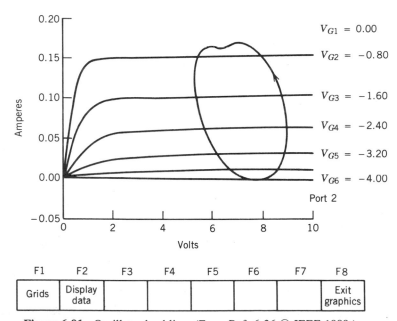

Figure 6.91   Oscillator load line. (From Ref. 6.36 © IEEE 1988.)

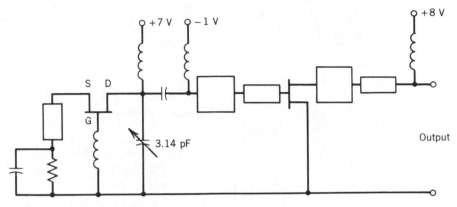

Figure 6.92 Buffered oscillator design. (From Ref. 6.36 © IEEE 1988.)

relatively easily by automatically sweeping the frequency and reoptimizing the tuning element at each step of the sweep. In this way, a tuning curve of frequency versus element value can be constructed, and the frequency determined from the (fixed) value of the tuning element that is used in the actual circuit. By repeating this process for different values of the load, the effect of various loads on the oscillation frequency can be determined.

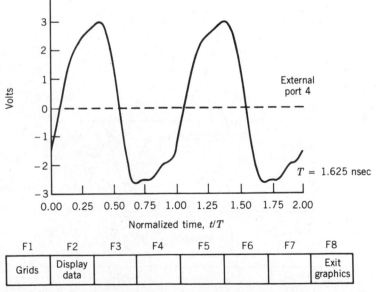

Figure 6.93 Buffered oscillator load voltage versus time (two cycles). (From Ref. 6.36 © IEEE 1988.)

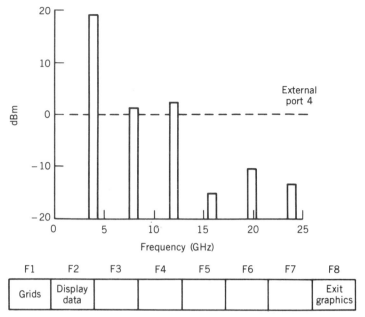

Figure 6.94 Buffered oscillator power output spectrum. (From Ref. 6.36 © IEEE 1988.)

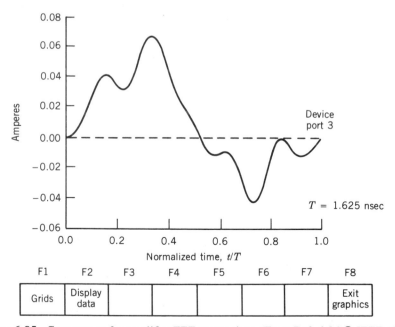

Figure 6.95 Gate current for amplifier FET versus time. (From Ref. 6.36 © IEEE 1988.)

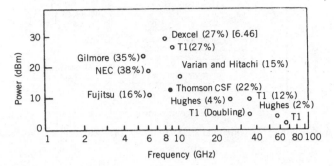

Figure 6.96 Survey of narrowband FET oscillator performance. (From Ref. 6.45.)

6.10.3 Conclusion

To effectively utilize any harmonic-balance program, a time-domain model for the nonlinear device is required. The extraction of parameters to describe devices according to the notation used in available models has always been a factor limiting the usefulness of nonlinear circuit simulators. Programs are now available which extract these parameters from I–V data and small-signal S parameters measured at one or more bias conditions. Simultaneous extraction of the parasitic network and the nonlinear model parameters provides consistency between the data sets and the model. Because the equations used for device modeling are known and fixed, the adjoint technique can be used to determine sensitivities of the model elements for fitting purposes. This results in very fast execution times.

The recent advances in modeling and circuit simulation makes it possible to optimize circuits for given nonlinear responses. Mixers, oscillators, power amplifiers, limiters, and PIN attenuators are among the types of circuits that can be quantitatively designed using the harmonic balance method. The ability to predict compressed output powers, spectral purity, dc-to-RF and RF-to-RF efficiencies, and transducer and conversion gains under varying drive is likely to result in tremendous productivity gains as the design and optimization of nonlinear components are transferred from the computer to the workbench.

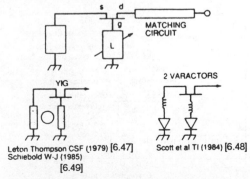

Figure 6.97 Wideband tuning of an FET oscillator. (From Ref. 6.45.)

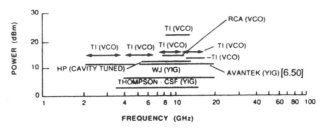

Figure 6.98   Survey of wideband FET oscillator performance. (From Ref. 6.45.)

6.11   1988 MICROWAVE OSCILLATORS

In this chapter we have looked at various aspects of oscillators. It may be useful to give readers some reference points regarding oscillator performance. Figure 6.96 shows a survey of narrowband FET oscillators. Power output at the frequency of operation and dc-to-RF efficiency are listed. These are narrowband VCO, DRO, or cavity-tuned circuits. For example, the 1-W oscillator at 8 GHz uses a 4000-μm-gate-width DXL 4640A-P100 GaAs MESFET on a duroid substrate with microstrip resonators [6.46].

For wideband operation, Fig. 6.97 illustrates some implementations using either varactor diodes or YIG resonators. Figure 6.98 provides an overview of the output power and tuning range for some selected oscillators. The Avantek 2- to 18-GHz YIG oscillator uses the AT22000 silicon bipolar transistor described in Chapter 3 [6.50].

The noise performance is another important parameter, as we have learned, and Fig. 6.99 is a graphic presentation of the $\mathcal{L}$ (f_m) at $f_m = 10$ kHz. There are many ways to couple a resonator to an oscillator for frequency stabilization, as shown in Fig. 6.100. A very low noise DRO at 1.3 GHz using a silicon bipolar transistor gave -142 dBc/Hz at 10 kHz from the carrier [6.59]. This oscillator uses the resonator in the feedback loop and as a filter. This is similar to the crystal oscillator

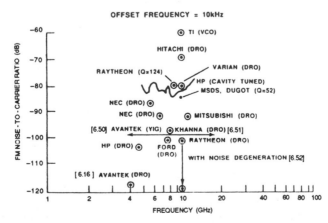

Figure 6.99   Survey of FET oscillator noise performance. (From Ref. 6.45.)

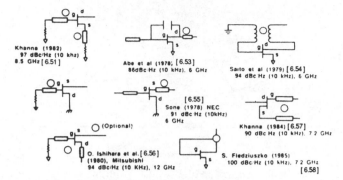

Figure 6.100 Methods of frequency stabilization. (From Ref. 6.45.)

reported by Rohde [6.3, p. 198]. Perhaps the lowest oscillator phase noise of a 4-GHz DRO was −130 dBc/Hz at 10 kHz from the carrier [6.60] using a low-noise GaAs FET amplifier with parallel feedback. This oscillator used a two-stage GaAs amplifier with 20 dB gain and produced 11.5 dBm of output power.

The FM noise of an X-band microstrip GaAs FET oscillator is typically −65 dBc/Hz at 10 kHz offset from the carrier. Use of a dielectric resonator will reduce this noise to −100 dBc/Hz at the same offset frequency. The noise can be degenerated to a lower level by means of a frequency-locked loop, and a novel realization of such a circuit is presented in Fig. 6.101.

The uniqueness of this circuit lies in its utilization of one dielectric resonator in both the basic oscillator and the discriminator within the frequency-locked loop. The resulting noise performance is given in Fig. 6.102, both with and without noise degeneration. This method seems to have achieved the best carrier-to-noise ratio of an FET oscillator at X-band (10 GHz).

Figure 6.103 is a photograph of an ultrafast switching DRO with three resonators [6.61]. This is a single GaAs MESFET switched between three resonator ports, which gives very low post-tuning drift. Also, the oscillator has no spurious outputs at the unwanted oscillator frequencies.

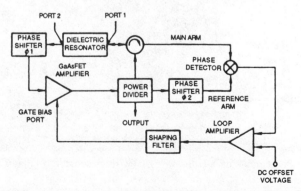

Figure 6.101 Circuit for noise degeneration in FET oscillators. (From Ref. 6.52.)

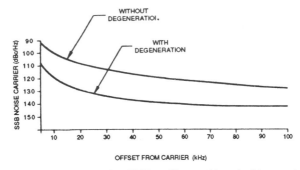

Figure 6.102 Noise performance of an FET oscillator with and without noise degeneration. (From Ref. 6.52.)

Figure 6.103 Photograph of an ultrafast dielectric resonator oscillator. (Courtesy of Avantek.)

Some recent wideband results reported on GaAs MMIC structures indicate a bandwidth of 2.5 to 6 GHz using off-chip varactors [6.62], and 6 to 12 GHz and 10 to 20 GHz low-power oscillators with multiple integrated varactor structures [6.63].

REFERENCES

6.1   G. R. Basawapatna and R. B. Stancliff, "A Unified Approach to the Design of Wide-Band Microwave Solid-State Oscillators," *IEEE Transactions on Microwave Theory and Techniques*, Vol. MTT-27, May 1979, pp. 379–385.

6.2   A. P. S. Khanna and J. Obregon, "Microwave Oscillator Analysis," *IEEE Transactions on Microwave Theory and Techniques*, Vol. MTT-29, June 1981, pp. 606–607.

6.3   Ulrich L. Rohde, *Digital PLL Frequency Synthesizers Theory and Design*, Prentice-Hall, Englewood Cliffs, N.J., 1983. Copies available from Compact Software, Inc., 483 McLean Boulevard, Paterson, N.J. 07504.

6.4   D. B. Leeson, "A Simple Model of Feedback Oscillator Noise Spectrum," *Proceedings of the IEEE*, Vol. 54, February 1966, pp. 329–330.

6.5   K. Kurokawa, "Noise in Synchronized Oscillators," *IEEE Transactions on Microwave Theory and Techniques*, Vol. MTT-16, April 1968, pp. 234–240.

6.6   S. Hamilton, "FM and AM Noise in Microwave Oscillators," *Microwave Journal*, June 1978, pp. 105–109.

6.7   K. L. Kotzebue and W. J. Parrish, "The Use of Large Signal S-Parameters in Microwave Oscillator Design," *Proceedings of the International IEEE Microwave Symposium on Circuits and Systems*, 1975.

6.8   K. M. Johnson, "Large Signal GaAs MESFET Oscillator Design," *IEEE Transactions on Microwave Theory and Techniques*, Vol. MTT-27, March 1979, pp. 217–227.

6.9   R. J. Gilmore and F. J. Rosenbaum, "An Analytic Approach to Optimum Oscillator Design Using S-Parameters," *IEEE Transactions on Microwave Theory and Techniques*, Vol. MTT-31, August 1983, pp. 633–639.

6.10 Y. Tajima and B. Wrona, "GaAs FET Large Signal Model and Design Applications," *Technical Digest of IEDM*, 1980, pp. 122–125.

6.11 Y. Tajima, B. Wrona, and K. Mishima, "GaAs FET Large Signal Model and Its Application to Circuit Designs," *IEEE Transactions on Electron Devices*, Vol. ED-28, February 1981, pp. 171–175.

6.12 I. Bahl and P. Bhartia, *Microwave Solid State Circuit Design*, Wiley, New York, 1988, Chapters 3 and 9.

6.13 A. P. S. Khanna and Y. Garault, "Determination of Loaded, Unloaded, and External Quality Factors of a Dielectric Resonator Coupled to a Microstrip Line," *IEEE Transactions on Microwave Theory and Techniques*, Vol. MTT-31, March 1983, pp. 261–264.

6.14 D. Kajfez and P. Guillon, *Dielectric Resonators*, Artech, Dedham, Mass., 1986.

6.15 N. K. Osbrink, "YIG-Tuned Oscillator Fundamentals," *Microwave Systems News*, Vol. 13, November 1983, pp. 207–225.

6.16 G. D. Vendelin, W. C. Mueller, A. P. S. Khanna, and R. Soohoo, "A 4 GHz DRO," *Microwave Journal*, June 1986, pp. 151–152.

6.17   I. Kipnis, "Silicon Bipolar MMIC for Frequency-Conversion Applications up to 20 GHz," *IEEE MTT-S Digest*, 1987, pp. 855–858.

6.18   J. Obregon and A. P. S. Khanna, "Exact Derivation of the Non-linear Negative Resistance Oscillator Pulling Figure," *IEEE Transactions on Microwave Theory and Techniques*, Vol. MTT-30, July 1982, pp. 1109–1111.

6.19   K. Kurokawa, "Injection Locking of Microwave Solid State Oscillators," *Proceedings of the IEEE*, Vol. 61, 1973, p. 1386.

6.20   J. A. Barnes, A. R. Chie, and L. S. Cutter, "Characterization of Frequency Stability," *IEEE Transactions on Instrumentation and Measurement*, Vol. IM-20, No. 2, May 1971, pp. 105–120.

6.21   D. J. Healey III, "Flicker of Frequency and Phase and White Frequency and Phase Fluctuations in Frequent Sources," *Proceedings of the 26th Annual Symposium on Frequency Control*, Fort Monmouth, N.J., June 1972, pp. 43–49.

6.22   Patrick Lesage and Claude Audoin, "Characterization of Frequency Stability: Uncertainty due to the Finite Number of Measurements," *IEEE Transactions on Instrumentation and Measurement*, Vol. IM-22, No. 2, June 1973, pp. 157–161.

6.23   Jacques Rutman, "Characterization of Frequency Stability: A Transfer Function Approach and Its Application to Measure via Filtering of Phase Noise," *IEEE Transactions on Instrumentation and Measurement*, Vol. IM-22, 1974, pp. 40–48.

6.24   "Understanding and Measuring Phase Noise in the Frequency Domain," *Hewlett-Packard Application Note 207*, October 1976.

6.25   Chuck Reynolds, "Measure Phase Noise," *Electronic Design*, February 15, 1977, pp. 106–108.

6.26   A. L. Lance, W. D. Seal, F. G. Mendozo, and N. W. Hudson, "Automating Phase Noise Measurements in the Frequency Domain," *Proceedings of the 31st Annual Symposium on Frequency Control*, June 1977.

6.27   D. Scherer, "Learn About Low-Noise Design," *Microwaves*, April 1979, pp. 120–122.

6.28   "Phase Noise Characterization of Microwave Oscillators," *Hewlett-Packard Product Note 11729C*.

6.29   PLL Design Kit, Compact Software, Inc., Paterson, N.J.

6.30   M. Maeda, K. Kimura, and H. Kodera, "Design and Performance of X-Band Oscillators with GaAs Schottky-Gate Field-Effect Transistors," *IEEE Transactions on Microwave Theory and Techniques*, Vol. MTT-23, August 1975, pp. 661–667.

6.31   M. Vehovec, L. Houselander, and R. Spence, "On Oscillator Design for Maximum Power," *IEEE Transactions on Circuit Theory*, Vol. CT-15, September 1968, pp. 281–283.

6.32   *Sonata User's Manual*, Compact Software, Paterson, N.J.

6.33   R. A. Pucel, R. Bera, and D. Masse, "Experiments on Integrated Gallium-Arsenide FET Oscillators at X-Band," *Electronics Letters*, Vol. 11, May 15, 1975, pp. 219–220.

6.34   Raytheon Company, Research Division, Waltham, Mass., "FET Noise Studies," *Final Technical Report F49620-79-C-0024*, Air Force Office of Scientific Research, March 1981.

6.35   *Microwave Harmonica Manual*, Compact Software, Inc., Paterson, N.J.

6.36   R. J. Gilmore, J. Gerber, and M. Eron, "Design Methodologies for Nonlinear Circuit Simulation and Optimization," *32nd ARFTG Symposium Digest*, Phoenix, Ariz., December 1988.

6.37 V. Rizzoli, A. Neri, and A. Costanzo, "Analysis and Optimization of DROs Using a General Purpose CAD Program," *Alta Frequenza*, Vol. 57, September 1988, pp. 389–398.

6.38 A. Neri and V. Rizzoli, "Global Stability Analysis of Microwave Circuits by a Frequency-Domain Approach," *IEEE MTT-S International Microwave Symposium Digest*, June 1987, pp. 689–692.

6.39 H. Wacker, *Continuation Methods*, Academic Press, New York, 1978.

6.40 G. Ioos and D. D. Joseph, *Elementary Stability and Bifurcation Theory*, Springer-Verlag, New York, 1980.

6.41 J. W. Bandler, S. H. Chen, S. Ye, and Q. J. Zhang, "Robust Model Parameter Extraction Using Large Scale Optimization Concepts," *IEEE MTT-S International Microwave Symposium Digest*, 1988, pp. 319–322.

6.42 J. W. Bandler, S. H. Chen, S. Diajavad, and K. Madsen, "Efficient Optimization with Integral Gradient Approximations," *IEEE Transactions on Microwave Theory and Techniques*, Vol. MTT-36, No. 2, February 1988, pp. 444–455.

6.43 RoMPE, Compact Software, Inc., Paterson, N.J.

6.44 A. Materka and T. Kacprzak, "Computer Calculation of Large-Signal GaAs FET Amplifier Characteristics," *IEEE Transactions on Microwave Theory and Techniques*, Vol. MTT-33, February 1985, pp. 129–135.

6.45 R. A. Pucel, "The GaAs FET Oscillator, Its Signal and Noise Performance," *40th Annual Frequency Control Symposium*, 1986, pp. 385–391.

6.46 R. M. Rector and G. D. Vendelin, "A 1.0 Watt GaAs MESFET Oscillator at X-Band," *IEEE MTT-S International Microwave Symposium Digest*, June 1978, pp. 145–146.

6.47 Y. Leton, S. Barvet, and J. Obregon, "Multioctave FET Oscillators Double Tuned by a Single YIG," *ISSCC Digest of Technical Papers*, 1979.

6.48 B. N. Scott, M. Wurtele, and B. B. Cregger, "A Family of Four Monolithic VCO MICs Covering 2 to 18 GHz," *IEEE Microwave and Millimeter-Wave Monolithic Circuits Symposium Digest*, May 1984, pp. 58–61.

6.49 C. F. Schiebold, "An Approach to Realizing Multi-octave Performance in GaAs-FET YIG-Tuned Oscillators," *IEEE MTT Digest*, 1985, pp. 261–263.

6.50 C. C. Leung, C. P. Snapp, and V. Grande, "A 0.5-μm Silicon Bipolar Transistor for Low Phase Noise Oscillator Applications up to 20 GHz," *IEEE MTT-S International Microwave Symposium Digest*, 1985, pp. 383–386.

6.51 A. P. S. Khanna, J. Obregon, and Y. Garault, "Efficient Low Noise Three Port X-Band FET Oscillator Using Two Dielectric Resonators," *IEEE MTT-S International Microwave Symposium Digest*, 1982, pp. 277–279.

6.52 Z. Galani, M. J. Bianchini, R. C. Waterman, R. DiBiase, R. W. Laton, and J. B. Cole, "Analysis and Design of a Single-Resonator GaAs FET Oscillator with Noise Degeneration," *IEEE Transactions on Microwave Theory and Techniques*, Vol. MTT-32, December 1984, pp. 1556–1564.

6.53 H. Abe, Y. Takayama, A. Higshisaka, and H. Takamizawa, "A Highly-Stabilized Low-Noise GaAs FET Integrated Oscillator with a Dielectric Resonator in the C-Band," *IEEE Transactions on Microwave Theory and Techniques*, Vol. MTT-26, March 1978, pp. 156–162.

6.54 T. Saito, Y. Arai, H. Komizo, Y. Itoh, and T. Nishikawa, "A 6 GHz Highly Stabilized GaAs FET Oscillator Using a Dielectric Resonator," *IEEE MTT Digest*, 1979, pp. 197–199.

6.55 J. Sone and Y. Takayama, "A 7 GHz Common-Drain GaAs FET Oscillator Stabilized with a Dielectric Resonator," *NEC R&D*, No. 49, April 1978, pp. 1–8.

6.56 O. Ishihara, T. Mori, H. Sawano, and M. Nakatani, "A Highly Stabilized GaAs FET Oscillator Using a Dielectric Resonator Feedback Circuit in 9-14 GHz," *IEEE Transactions on Microwave Theory and Techniques*, Vol. MTT-28, August 1980, pp. 817–824.

6.57 A. P. S. Khanna, "Parallel Feedback FETDRO Using 3-Port *S*-Parameters," *IEEE MTT-S International Microwave Symposium Digest*, 1984, pp. 181–183.

6.58 S. J. Fiedziuszko, "Miniature FET Oscillator Stabilized by a Dual Mode Dielectric Resonator," *IEEE MTT Digest*, 1985, pp. 264–265.

6.59 E. C. Niehenke and P. A. Green, "A Low-Noise L-Band Dielectric Resonator Stabilized Microstrip Oscillator," *IEEE MTT-S International Microwave Symposium Digest*, 1987, pp. 193–196.

6.60 G. Lan, D. Kalokitis, E. Mykietyn, E. Hoffman, and F. Sechi, "Highly Stabilized, Ultra-Low Noise FET Oscillator with Dielectric Resonator," *IEEE MTT-S International Microwave Symposium Digest*, 1986, pp. 83–86.

6.61 A. P. S. Khanna and R. Sohoo, "Single DRO Rapidly Switches Three Frequencies," *Microwaves & RF*, July 1987, pp. 142–145.

6.62 J. E. Andrews, R. J. Holden, K. W. Lee, and A. F. Podell, "2.5–6.0 GHz Broadband GaAs MMIC VCO," *IEEE MTT Digest*, 1988, pp. 491–494.

6.63 E. Reese and J. M. Beall, "Optimized X and Ku Band GaAs MMIC Varactor Tuned FET Oscillators," *IEEE MTT Digest*, 1988, pp. 487–490.

6.64 S. R. Lesage, M. Madihian, N. Nayama, and K. Honjo, "15.6 GHz HBT Microstrip Oscillator," *Electronics Letters*, Vol. 24, February 1988, pp. 230–232.

BIBLIOGRAPHY

Adler, R., "A Study of Locking Phenomena in Oscillators," *Proceedings of the IEEE*, Vol. 61, No. 10, October 1973, pp. 1380–1385.

Alley, G. D., and H. C. Wang, "An Ultra-Low Noise Microwave Synthesizer," *IEEE Transactions on Microwave Theory and Techniques*, Vol. MTT-27, December 1979, pp. 969–974.

Bierman, H., "DRO's Meet EW and Countermeasure System Needs," *Microwave Journal*, Vol. 30, October 1987, pp. 44–66.

Debney, B. T., and J. S. Joshi, "A Theory of Noise in GaAs FET Microwave Oscillators and Its Experimental Verification," *IEEE Transactions on Electron Devices*, Vol. ED-30, July 1983, pp. 769–776.

Everard, J. K. A., King's College London, "Minimum Sideband Noise in Oscillators," *IEEE 40th Annual Frequency Control Symposium*, 1986.

Fiedziuszko, S. J., "Microwave Dielectric Resonators," *Microwave Journal*, Vol. 29, September 1986, pp. 189–200.

Finlay, H. J., J. S. Joshi, and S. C. Cripps, "An X Band F.E.T. Oscillator with Low F.M. Noise," *Electronics Letters*, Vol. 14, March 16, 1978, pp. 198–199.

Graffeuil, J., K. Tantrarongroj, and J. F. Sautereau, "Low Frequency Noise Physical Analysis for the Improvement of the Spectral Purity of GaAs FETs Oscillators," *Solid-State Electronics*, Vol. 25, 1982, pp. 367–374.

Khanna, A. P. S., "Fast Settling Low Noise Ku Band Fundamental Bipolar VCO," *IEEE MTT-S International Microwave Symposium Digest*, 1987, pp. 579–581.

Khanna, A. P. S., "Q Measurement of Microstrip-Coupled Dielectric Resonators," *Microwaves & RF*, January 1984, pp. 81–84.

Khanna, A. P. S., "Review of Dielectric Resonator Oscillator Technology," *Proceedings of the Frequency Control Symposium*, Philadelphia, May 1987, pp. 478–486.

Khanna, A. P. S., and J. Obregon, "Direct Measurement of the Non-linear MIC Oscillator Characteristics Using Injection Locking Polar Diagram," *IEEE MTT-S International Microwave Symposium Digest*, 1983, pp. 501–503.

Khanna, A. P. S., and J. Obregon, "Network Analyzer Doubles as Oscillator Diagnostician," *Microwaves & RF*, Vol. 23, July 1984, pp. 106–112.

Kurokawa, K., *An Introduction to the Theory of Microwave Circuits*, Academic Press, New York, 1969, Chapter 9.

Kurokawa, K., "Microwave Solid State Oscillator Circuits," in *Microwave Devices*, M. J. Howes and D. V. Morgan, Eds., Wiley, New York, 1976.

Kurokawa, K., "Some Basic Characteristics of Broadband Negative Resistance Oscillator Circuits," *Bell System Technical Journal*, Vol. 48, July–August 1969, pp. 1937–1955.

Niehenke, E. C., and R. D. Hess, "A Microstrip Low-Noise X-Band Voltage Controlled Oscillator," *IEEE Transactions on Microwave Theory and Techniques*, Vol. MTT-27, December 1979, pp. 1075–1079.

Ollivier, P., "Microwave YIG-Tuned Oscillator," *IEEE Journal of Solid-State Circuits*, Vol. SC-7, February 1972, pp. 50–60.

Papp, J. C., and Y. Y. Koyano, "An 8-18 GHz YIG-Tuned FET Oscillator," *IEEE Transactions on Microwave Theory and Techniques*, Vol. MTT-28, July 1980, pp. 762–766.

Penfield, P., "Circuit Theory of Periodically Driven Nonlinear Systems," *Proceedings of the IEEE*, Vol. 54, February 1966, pp. 266–280.

Podcameni, A., and L. A. Bermudez, "Stabilised Oscillator with Input Dielectric Resonator: Large Signal Design," *Electronics Letters*, Vol. 17, January 1981, pp. 44–45.

Pucel, R. A., and J. Curtis, "Near-Carrier Noise in FET Oscillators," *IEEE MTT-S International Microwave Symposium Digest*, 1983, pp. 282–284.

Rauscher, C., "Large-Signal Technique for Designing Single-Frequency and Voltage-Controlled GaAs FET Oscillators," *IEEE Trans. Microwave Theory and Techniques*, Vol. MTT-29, April 1981, pp. 293–304.

Riddle, A. N., and R. J. Trew, "A New Method of Reducing Phase Noise in GaAs FET Oscillators," *IEEE MTT-S International Microwave Symposium Digest*, 1984, pp. 274–276.

Rohde, U. L., "Crystal Oscillator Provides Low Noise," *Electronic Design*, October 11, 1975.

Rohde, U. L., "Effects of Noise in Receiving Systems," *Ham Radio*, Vol. 11, 1977, p. 34.

Rohde, U. L., "Evaluating Noise Sideband Performance in Oscillators," *Ham Radio*, Vol. 10, 1978, p. 51.

Rohde, U. L., "Mathematical Analysis and Design of an Ultra Stable Low Noise 100 MHz Crystal Oscillator with Differential Limiter and Its Possibilities in Frequency Standards," Presented at the 32nd Annual Frequency Symposium, Fort Monmouth, N.J., June 1978, p. 409.

Trew, R. J., "Design Theory for Broadband YIG-Tuned FET Oscillators," *IEEE Transactions on Microwave Theory and Techniques*, Vol. MTT-27, January 1979, pp. 8–14.

Tserng, H. Q., and H. M. Macksey, "Wide-Band Varactor-Tuned GaAs MESFET Oscillators at X- and Ku-Bands," *IEEE MTT-S International Microwave Symposium Digest*, 1977, pp. 267–269.

Wagner, W., "Oscillator Design by Device Line Measurement," *Microwave Journal*, February 1979, pp. 43–48.

Winch, R. G., and J. L. Matson, "Ku-Band MIC Bipolar VCO," *Electronics Letters*, Vol. 17, April 1981, pp. 296–298.

Zensius, D., J. Hauptman, and N. Osbrink, "GaAs FET YIG Oscillator Tunes from 26 to 40 GHz," *Microwaves & RF*, Vol. 22, October 1983, pp. 129–139.

PROBLEMS

6.1. The S parameters are given below for a HXTR-5001 silicon bipolar transistor, including parasitic bonding inductances. ($L_B = 0.3$ nH, $L_E = 0.1$ nH.) Design an oscillator at 5 GHz which delivers power to a 50-Ω load. Design the dc bias circuit. Give the complete RF and dc schematic. Assume that a 28-V dc supply is available.

S parameters at $f = 5.0$ GHz ($V_{CE} = 18$ V, $I_C = 30$ mA):

Oscillator	S_{11}	S_{21}	S_{12}	S_{22}	k
Common emitter	$0.54\ \underline{/167°}$	$1.45\ \underline{/52°}$	$0.120\ \underline{/44°}$	$0.49\ \underline{/42°}$	1.41
Common base	$0.83\ \underline{/141°}$	$1.58\ \underline{/-87°}$	$0.162\ \underline{/115°}$	$1.10\ \underline{/-53°}$	-0.45

6.2. The S parameters of the NE567 bipolar transistor are given for the common-emitter and common-base oscillators ($V_{CE} = 10$ V, $I_{CE} = 40$ mA), $f = 8$ GHz.

(a) Design a common-emitter oscillator with a power-out collector.

(b) Design a common-base oscillator with a power-out collector.

$$\text{Common Emitter}$$
$$S = \begin{bmatrix} 0.85\ \underline{/117°} & 0.142\ \underline{/88°} \\ 0.87\ \underline{/24°} & 0.68\ \underline{/-59°} \end{bmatrix}$$
$$k = 0.33$$
$$G_{ms} = 7.9 \text{ dB}$$

$$\text{Common Base}$$
$$S = \begin{bmatrix} 1.32\ \underline{/88°} & 0.595\ \underline{/99°} \\ 1.47\ \underline{/172°} & 1.03\ \underline{/-96°} \end{bmatrix}$$
$$k = 0.24$$
$$G_{ms} = 3.9 \text{ dB}$$

Use lumped elements for these designs.

6.3. Design a common-base bipolar oscillator at 4 GHz using lumped elements.

$$S = \begin{bmatrix} 0.707 \ \underline{/-30°} & 0.35 \ \underline{/45°} \\ 1.414 \ \underline{/45°} & 0.50 \ \underline{/-60°} \end{bmatrix}$$

$$V_{CE} = 10 \text{ V}$$
$$I_C = 10 \text{ mA}$$
$$(h_{FE})_{min} = 30$$

If the supply voltage is 15 V, give dc bias circuit. Draw a complete RF and dc schematic.

6.4. The S parameters of a common-base bipolar transistor are given and the input resonator has been selected. Design the load circuit for an oscillator ($f = 10$ GHZ) using distributed elements.

$$S = \begin{bmatrix} \sqrt{2} \ \underline{/90°} & 0.707 \ \underline{/90°} \\ \sqrt{2} \ \underline{/180°} & 1 \ \underline{/-90°} \end{bmatrix} \qquad \Gamma_G = 1 \ \underline{/-90°}$$

6.5. Design an oscillator using a common-collector NE219 bipolar transistor at $f = 4$ GHz, $V_{CE} = 8$ V, $I_{CE} = 40$ mA.

$$S = \begin{bmatrix} 0.91 \ \underline{/-135°} & 0.67 \ \underline{/-30°} \\ 1.41 \ \underline{/-90°} & 0.60 \ \underline{/90°} \end{bmatrix} \qquad k = 0.389$$

Give the RF design and schematic using lumped elements (ignore dc design).

6.6. A common-collector NEC 645 bipolar transistor (NPN) was selected for an oscillator design at $f = 10$ GHz, $V_{CE} = 8$ V, $I_{CE} = 20$ mA. The power supply is $+15$ V.

$$S_{11} = 0.56 \ \underline{/92°} \qquad S_{12} = 0.89 \ \underline{/-119°} \qquad k = 0.49$$

$$S_{21} = 0.92 \ \underline{/177°} \qquad S_{22} = 0.47 \ \underline{/-56°}$$

Draw the complete RF and dc schematic using 50-Ω microstrip-line elements for the RF design.

6.7. Given S parameters of a common-source FET at $f = 6$ GHz,

$$S = \begin{bmatrix} 0.95 \ \underline{/-45^\circ} & 0.25 \ \underline{/45^\circ} \\ 1.414 \ \underline{/45^\circ} & 0.50 \ \underline{/-45^\circ} \end{bmatrix}$$

$V_{DD} = 12 \ V$

$V_{DS} = 6 \ V$

$V_{GS} = -1 \ V$

$I_{DS} = 10 \ mA$

(a) Calculate k.

(b) Give the RF design of an oscillator with a 50-Ω load; use lumped elements.

(c) Give the dc design.

(d) Prepare a complete schematic.

6.8. Given the S parameters of a common-source or common-gate GaAs FET at $f = 12$ GHz (DXL-2502A chip).

FET	S_{11}	S_{21}	S_{12}	S_{22}	k
Common source	$0.52 \ \underline{/-139^\circ}$	$1.47 \ \underline{/60^\circ}$	$0.039 \ \underline{/140^\circ}$	$0.75 \ \underline{/-40^\circ}$	2.44
Common gate	$0.32 \ \underline{/155^\circ}$	$1.38 \ \underline{/-75^\circ}$	$0.226 \ \underline{/50^\circ}$	$1.19 \ \underline{/-34^\circ}$	-0.12

Design an oscillator using distributed 50-Ω microstrip-line elements which delivers power to a 50-Ω load. Assuming a +15-V power supply, design the dc bias network for the operating point

$$I_{DS} = 0.30 \ A$$

$$V_{DS} = 6.0 \ V$$

$$V_{GS} = -1.0 \ V$$

Finally, draw the entire oscillator schematic (RF and dc).

6.9. Given the following S parameters of a common-source GaAs MESFET at 2 GHz:

$$S_{11} = 0.93 \ \underline{/-43^\circ} \qquad k = 0.53$$

$$S_{21} = 2.72 \ \underline{/146^\circ} \qquad V_{DS} = 5.0 \ V$$

$$S_{12} = 0.022 \ \underline{/69^\circ} \qquad I_{DS} = 40 \ mA$$

$$S_{22} = 0.77 \ \underline{/-9^\circ} \qquad V_{GS} = -2 \ V$$

$$V_{DD} = +12 \ V$$

Data on stability circles at 2 GHz:

	Γ_G Plane			Γ_L Plane	
Center	Radius	Stable Region	Center	Radius	Stable Region
$1.10\underline{/50°}$	0.17	Outside	$1.60\underline{/38°}$	0.83	Outside

(a) Design an oscillator that delivers power from the drain–source port to a 50-Ω load.

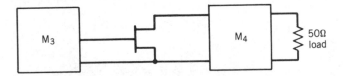

(b) Design the dc bias circuit.

(c) Draw the complete RF and dc circuit schematic.

6.10. Use a silicon varactor diode modeled by

$$C = C_{\min} \left(\frac{V_B + \phi}{V_R + \phi}\right)^{\gamma}$$

where

$$\phi = 0.7 \text{ V (for Si)} \qquad V_B = 20 \text{ V}$$

$$\gamma = 0.71 \qquad\qquad V_R = \text{reverse voltage applied}$$

$$C_{\min} = 0.35 \text{ pF}$$

The capacitance versus voltage is shown on the following plot, which shows a capacitance range of better than $10:1$ over 0.7 to 20.7 V. Design a 2- to 6 GHz with this diode and the silicon bipolar transistor shown below. Suggested topology: a common base with varactor at input port.

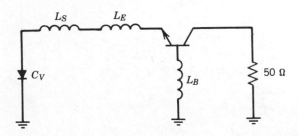

Problem 6.10

Freq. (GHz)	S_{11} Mag.	S_{11} Angle	S_{21} Mag.	S_{21} Angle	S_{12} Mag.	S_{12} Angle	S_{22} Mag.	S_{22} Angle
0.1	0.56	−60	39.07	152	0.009	69	0.87	−18
0.5	0.54	−145	15.00	104	0.023	56	0.49	−28
1.0	0.54	−170	8.03	90	0.033	65	0.42	−23
1.5	0.55	179	5.30	82	0.045	72	0.41	−22
2.0	0.56	170	4.04	76	0.058	75	0.41	−23
2.5	0.56	165	3.24	72	0.070	78	0.40	−23
3.0	0.58	159	2.75	65	0.083	79	0.40	−25
3.5	0.59	154	2.37	62	0.096	82	0.41	−26
4.0	0.60	149	2.06	57	0.108	83	0.42	−28
4.5	0.61	145	1.87	53	0.124	84	0.42	−33
5.0	0.62	142	1.67	49	0.136	83	0.43	−36
5.5	0.64	137	1.54	44	0.150	85	0.42	−40
6.0	0.65	134	1.40	41	0.165	84	0.44	−45

S parameters, CE AT-41400, silicon bipolar transistor chip ($V_{CE} = 8$ V, I_C = 25 mA, $L_B \approx 0.5$ nH, $L_E \approx 0.2$ nH):

6.11. Design a wideband buffered VCO using the AT-10600 FET at 6 to 12 GHz. The perfect varactor would be

$$C = C_{min} \left(\frac{V_B + \phi}{V_R + \phi} \right)^{\gamma}$$

where

$$\gamma = 0.71 \qquad V_B = 30 \text{ V}$$

$$C_{min} = 0.10 \text{ pF} \qquad \phi = 1.0 \text{ (for GaAs)}$$

Suggested topology: a common-drain oscillator and common-source amplifier. Use 3 V, 10 mA for oscillator; 5 V, 30 mA for amplifier. Give the tuning voltage versus frequency curve.

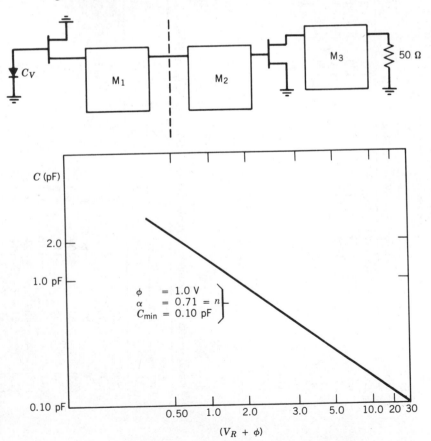

Typical scattering parameters, common source ($V_{DS} = 3$ V, $I_{DS} = 10$ mA):

Freq. (GHz)	S_{11}		S_{21}			S_{12}			S_{22}	
	Mag.	Angle	dB	Mag.	Angle	dB	Mag.	Angle	Mag.	Angle
5.0	.90	−47	6.9	2.22	131	−21.5	.084	69	.61	−13
6.0	.86	−59	7.1	2.27	120	−19.9	.101	64	.57	−17
7.0	.80	−72	7.2	2.29	110	−18.6	.117	59	.52	−23
8.0	.74	−83	7.3	2.32	100	−17.7	.130	54	.48	−28
9.0	.68	−97	7.3	2.33	89	−16.9	.143	48	.43	−36
10.0	.61	−114	7.3	2.31	77	−16.1	.156	42	.38	−46
11.0	.55	−133	7.1	2.27	66	−15.6	.166	35	.33	−56
12.0	.52	−153	6.8	2.18	54	−15.2	.174	28	.28	−66
13.0	.49	−174	6.4	2.10	44	−14.9	.180	21	.24	−78

Typical scattering parameters, common source ($V_{DS} = 5$ V, $I_{DS} = 30$ mA):

Freq. (GHz)	S_{11}		S_{21}			S_{12}			S_{22}	
	Mag.	Angle	dB	Mag.	Angle	dB	Mag.	Angle	Mag.	Angle
5.0	.81	−59	9.7	3.05	124	−25.8	.051	76	.65	−11
6.0	.74	−72	9.5	2.98	112	−24.7	.058	74	.62	−14
7.0	.68	−86	9.1	2.86	101	−23.4	.068	72	.60	−18
8.0	.63	−100	8.9	2.80	91	−22.5	.075	70	.57	−22
9.0	.56	−113	8.8	2.76	82	−21.7	.082	69	.55	−26
10.0	.51	−130	8.4	2.62	72	−20.6	.096	68	.53	−30
11.0	.46	−143	8.0	2.52	62	−20.0	.100	66	.51	−40
12.0	.42	−156	7.8	2.45	54	−19.6	.105	64	.47	−49
13.0	.41	−168	7.5	2.37	46	−18.9	.114	60	.46	−54

6.12. Find Γ_L and S_{22}' for the 15.6-GHz oscillator shown below. Also find Γ_G and S_{11}' for the oscillator, assume a common-base configuration.

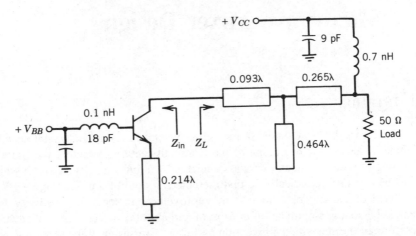

7 Microwave Mixer Design

7.0 INTRODUCTION

For many years the key element in receiving systems has been the crystal detector or diode mixer. At the beginning of the twentieth century, detectors were crude, consisting of a semiconductor crystal contacted by a fine wire ("whisker") which had to be adjusted periodically so that the detector would keep functioning. With the advent of the triode, a significant improvement in receiver sensitivity was obtained by adding amplification in front of and after the detector. A real advance in performance came with the invention by Edwin Armstrong of the super regenerative receiver. Armstrong was also the first to use a vacuum tube as a frequency converter (mixer) to shift the frequency of an incoming signal to an intermediate frequency (IF), where it could be amplified with good selectivity and low noise, and later detected. The superheterodyne receiver, which is the major advance in receiver architecture to date, is still employed in virtually every receiving system.

The development of microwave mixers was fostered during World War II with the development of radar. At the beginning of the war, single-diode mixers exhibited poor noise-figure performance, but by the 1950s, noise figures of 8 dB could be obtained. Today, single-diode mixers exhibit this type of performance at frequencies in excess of 100 GHz. The burden of establishing receiver sensitivity is still largely dependent on the mixer throughout the microwave frequency range. Recently, low-noise GaAs FET amplifiers are being used below 50 GHz to improve the system noise figure, but above this frequency, the diode is the only device that can be used for frequency-conversion applications.

The mixer, which can consist of any device capable of exhibiting nonlinear performance, is essentially a multiplier. That is, if at least two signals are present, their product will be produced at the output of the mixer. This concept is illustrated in Fig. 7.1. The RF signal applied has a carrier frequency of ω_s with modulation $M(t)$, and the local oscillator signal (LO or pump) applied has a pure sinusoidal frequency of ω_p. From basic trigonometry we know that the product of two sinusoids produces a sum and difference frequency. Either of these frequencies can be selected with the IF filter. Unfortunately, no physical nonlinear device is a perfect multiplier. Thus they contribute noise and produce a vast number of spurious frequency components.

For example, the voltage–current relationship for a diode can be described as an infinite power series,

$$I = a_0 + a_1 V + a_2 V^2 + a_3 V^3 + \cdots \tag{7.1}$$

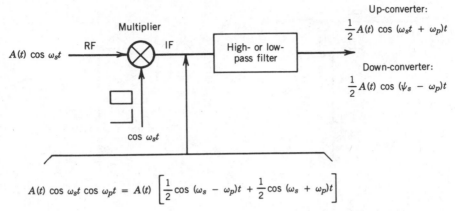

$$A(t) \cos \omega_s t \cos \omega_p t = A(t) \left[\frac{1}{2} \cos (\omega_s - \omega_p)t + \frac{1}{2} \cos (\omega_s + \omega_p)t \right]$$

Figure 7.1 Ideal multiplier mixer model showing both up- and down-converter performance.

where V is the sum of both input signals and I is the total signal current. If the RF signal is substantially smaller than the LO signal and modulation is ignored, the frequency components of the current I are

$$\omega_d = n\omega_p \pm \omega_s \tag{7.2}$$

As mentioned above, the desired component is usually the difference frequency ($|\omega_p + \omega_s|$ or $|f_p - f_s|$), but sometimes the sum frequency ($f_s + f_p$) is desired when building an up-converter, or a product related to a harmonic of the LO, as in a subharmonically pumped mixer, can be selected.

A mixer can also be analyzed as a switch that is commutated at a frequency equal to the pump frequency ω_p. This is a good first-order approximation of the mixing process for a diode since it is driven from the low-resistance state (forward bias) to the high-resistance state (reverse bias) by a high-level LO signal. The simplified diode model is shown in Fig. 7.2. With this switching action in mind, a single-ended mixer can be represented by the circuit in Fig. 7.3a. In this example the RF signal appearing at the IF load is interrupted by the switching action of the diode, which is caused by the pump. From the modulation theorem it can be shown

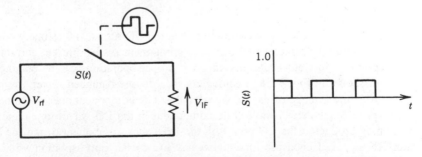

Figure 7.2 Single-ended mixer employing diode switching model.

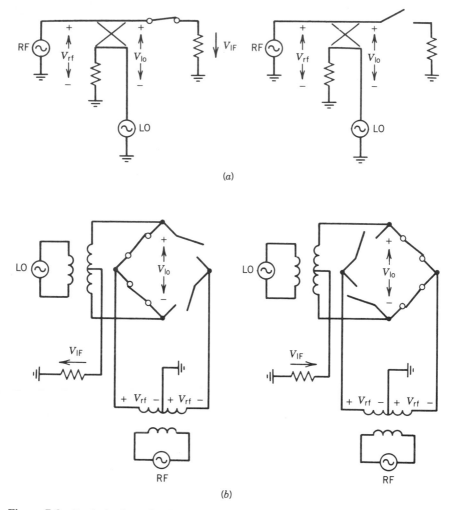

Figure 7.3 Typical mixer circuits employing diode switching model depicting IF voltage (or current) as a function of LO polarity; (a) single-ended mixer; (b) double-balanced mixer.

that the sum and difference frequencies appear at the IF port as well as many other products. It should be remembered that a dc component is also present and must not be suppressed in a physical diode mixer if proper operation is to be obtained. The circuit shown in Fig. 7.3b is equivalent to a double-balanced mixer. In this instance, the time average of the RF signal and LO dc component does not appear at the IF port. Since there is no LO dc component in the LO waveform, there is no switching product at the LO port with the frequency component of the fundamental RF signal. Hence the mixer also has LO to RF port isolation without requiring filters as in the single-ended case.

The concept of the switching mixer model can also be applied to field-effect transistors used as voltage-controlled resistors. In this mode, the drain-to-source resistance can be changed from a few ohms to many thousands of ohms simply by changing the gate-to-source potential. At frequencies below 1 GHz, virtually no LO power is required to switch the FET, and since no dc drain bias is required, the resulting FET mixer is passive. However, as the operating frequency is raised above 1 GHz, passive FET mixers require LO drive powers comparable to diode or active FET designs.

Regardless of the nonlinear or switching element employed, mixers can be divided into several classes: (1) single ended, (2) single balanced, or (3) double balanced. Depending on the application and fabrication constraints, one topology can exhibit advantages over the other types. The simplest topology, which is shown in Fig. 7.4a, consists of a single diode and filter network. Although there is no isolation inherent in the structure (balance), if the RF, LO, and IF frequencies are sufficiently separated, the filter (or diplexer) networks can provide the necessary isolation. In addition to simplicity, single-diode mixers have several advantages over other configurations. Typically, the best conversion loss is possible with a single device, especially at frequencies were balun or transformer construction is difficult or impractical. These excellent conversion-loss characteristics have been demonstrated numerous times in the literature, especially at millimeter wave frequencies or with image-enhanced waveguide microwave designs. The LO requirements are also minimal since only a single diode is employed and dc biasing can easily be accomplished to reduce drive requirements further. The disadvantages of the topology are: (1) sensitivity to terminations; (2) no spurious response suppression; (3) minimal tolerance to large signals; and (4) narrow bandwidth due to spacing between the RF filter and mixer diode.

The next topology commonly used is the single balanced structure shown in Fig. 7.4b. These structures tend to exhibit slightly higher conversion loss than that of a single-ended design, but since the RF signal is divided between two diodes, the signal power-handling ability is better. However, more diodes require more LO power. Since the structure is balanced, some isolation between ports is obtained and there is some spurious suppression for RF or LO products, depending on which is balanced.

The double-balanced structure is the topology most commonly employed between 2 and 18 GHz. It exhibits the best large signal-handling capability, port-to-port isolation, and spurious rejection. Alas, double-balanced mixers usually exhibit the poorest conversion loss characteristics and require the most LO drive. However, in strong signal environments such as the EW arena, spurious rejection and large-signal performance usually outweigh the 1 dB or so loss in sensitivity. Some high-level mixer designs can employ multiple-diode rings with several diodes per leg in order to achieve the ultimate in large-signal performance. Such designs can easily require hundreds of milliwatts of pump power. A general performance comparison for various mixer topologies is shown in Table 7.1. It should be noted that these performance traits are quite general and are highly dependent on balun design, diode quality, and operating frequency.

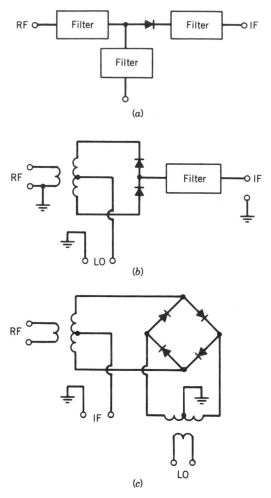

Figure 7.4 Common mixer topologies; (*a*) single-ended; (*b*) single-balanced; (*c*) double-balanced.

7.1 DIODE MIXER THEORY

The simple metal-semiconductor junction, first investigated by Braun in 1874, exhibits a nonlinear impedance as a function of voltage, making it an ideal candidate for mixer applications. Although other semiconductor junctions, such as the *p-n* junction, also exhibit nonlinear behavior, the metal-semiconductor diode (Schottky barrier) is primarily a majority carrier device, making it essentially free of minority carrier effects such as reverse recovery-time problems and high charge-storage capacitance. Because of the inherently low junction capacitance and high switching speed, Schottky diodes [7.1], typically the point contact type, operate well into the millimeter wave frequency range with cutoff frequencies exceeding

TABLE 7.1 Mixer Topology Performance Considerations

	Single-ended	Single Balanced	Double Balanced
Conversion gain	High	Moderate	Low
Spurious performance	None	Moderate	High
Dynamic range	Low	Moderate	High
Isolation	None	Moderate	High
Pump power	Low	Moderate	High
Complexity	Low	Moderate	High
Bandwidth	Narrow	Wide	Wide

2000 GHz. Most practical diodes employ either Si (silicon) or GaAs (gallium arsenide) as the semiconductor material, with the most common metals being Cu (copper), Pt (platinum), Ag (silver), Al (aluminum), Ti (titanium), and Au (gold). Since n-type GaAs exhibits a mobility many times greater than that of p-type material, n-type structures using Pt, Au, or Ti are most prevalent.

There have been many models explaining the operation of metal-semiconductor junctions, but the original model described by W. Schottky in 1942 [7.2] has endured and amply describes diode operation. As mentioned above, diode operation is based on majority carrier injection into the metal (anode) from the semiconductor (cathode), made possible because of free electrons present in the doped semiconductor. Diode operation is easily illustrated by first considering the metal and semiconductor properties separately and then combining the two.

Figure 7.5 depicts various energy levels for both the isolated metal and n-type semiconductor at equilibrium. From the electron gas theory for metals, we know that the average energy to remove an electron from the Fermi level and place it at rest in free space is $e\psi_m$, where ψ_m is the thermionic or vacuum work function.

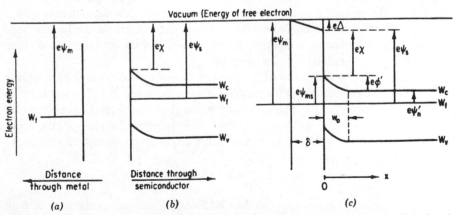

Figure 7.5 Energy levels for a metal and a semiconductor; (*a*) isolated metal; (*b*) isolated semiconductor; (*c*) metal and semiconductor in contact.

Values for ψ_m are on the order of several volts and are unique for each metal. However, the work function can vary depending on the metal surface conditions. Similarly, a work function ψ_s exists for the semiconductor, which not only is a function of surface conditions, but is influenced by the position of the Fermi level ψ_{fs}, which is dependent on doping level. The quantities ψ_v and ψ_c denote the positions of the valance and conduction band energy levels. Depending on the distribution and type of surface states of the semiconductor, a net positive or negative charge may exist at the surface, thus distorting the conduction and valance band energy levels as shown in Fig. 7.5b. The energy required to remove an electron from the conduction band to free space is $e\chi$, where χ is the electron affinity, which is a constant for each material and does not vary with doping level.

As the metal and semiconductor, which are joined by an external conductor, are brought together, a shift in energy levels must occur because equilibrium must be maintained for the combined system. Thus, from thermodynamics, the Fermi levels for both the metal and semiconductor must then coincide. Hence a potential difference Δ between the metal and semiconductor, due to the difference between their work functions, will result. This is known as the barrier potential or contact potential and varies depending on the materials used. As δ becomes sufficiently small (at or near contact), charge will be transferred between the two materials. In this case, electrons will be transferred from the semiconductor to the metal because of its lower Fermi level, leaving behind a positively charged depletion region and creating a negatively charged metal surface; hence a junction capacitance is created via the contact potential.

However, it should be remembered that the barrier height, which is defined as

$$\psi_{ms} = \psi_m - X' \tag{7.3}$$

where $X' = X - \Delta$, is not predicted by the original Schottky theory, since surface states, which imply that W is nonzero, were not considered. With this assumption the barrier height would be simply the difference between ψ_m and X. This is the case when the surface state density is very low, but for materials such as GaAs, ψ_{ms} is almost metal independent but has a strong dependence on the surface state density. It should also be noted that the model above does not take into account image forces on the electron at the junction.

Several other characteristics of the junction can also be determined by solving Poisson's equation throughout the junction and depletion regions. The solution can begin by assuming that the charge density in the depletion region can be approximated by $\rho = eN_d$ for $x = 0$ to $x = w_0$, and zero elsewhere, where N_d is the donor density. The electric field, which is a simple triangle function, can be written as

$$E(x) = -(eN_d/\epsilon_s)(w_0 - x) \tag{7.4}$$

The voltage across the junction can then be found by integrating the electric field and applying the boundary conditions at $x = 0$. At that point, we know that the voltage must then be equal to $-\phi_{bi}$. Hence the voltage can be written as

$$V(x) = (eN_d/\epsilon_s)[w_0x - (0.5)x^2] - \psi_{ms} \tag{7.5}$$

It can then be shown that the quantity ϕ_{bi}, which is the degree of band bending in the semiconductor (built-in potential), can be related to the depletion width as

$$\phi_{bi} = eN_dw_0^2/2\epsilon_s \tag{7.6}$$

When a voltage V is applied to the diode (Fig. 7.6), equation (7.6) is still valid provided that ϕ_{bi} is replaced by $\phi_{bi} - V$, and w_0 is replaced by w. Hence the depletion width is a function of applied voltage. Diode conduction properties can also be illustrated by examining Fig. 7.6.

During forward-bias conditions, the barrier height is lowered by an amount equal to the applied voltage. Thus it becomes easier for electrons to travel from the n-type semiconductor to the metal. However, the potential barrier for electrons traveling in the reverse direction is unaffected. The greater the applied voltage, the easier it becomes for forward charge flow, and the thinner the depletion region becomes. When reverse bias is applied, the potential barrier for forward-traveling electrons becomes large; hence there is a small probability that an electron with sufficient thermal energy will cross the junction.

The capacitance of the diode is also a function of applied voltage and can be found by first determining the total junction charge, which is $eN_d \times A$ (area) $\times w$ (depletion width). Using the relation

$$w = [(2(\phi_{bi} - V)\epsilon_s)/(eN_d)]^{1/2} \tag{7.7}$$
$$\text{junction charge} = [2(\phi_{bi} - V)eN_d\epsilon_s]^{1/2} \tag{7.8}$$

Then by taking the partial derivative of the charge ψ_1 with respect to the applied voltage, the capacitance becomes

$$C(V) = A[(2\epsilon_sN_d)/(2(\phi_{bi} - V))]^{1/2} \tag{7.9}$$

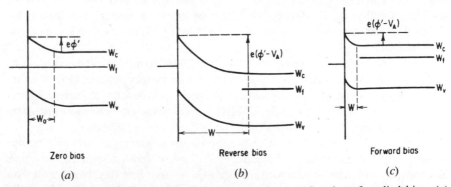

Zero bias

(a)

Reverse bias

(b)

Forward bias

(c)

Figure 7.6 Energy-level diagram of Schottky barrier as a function of applied bias: (a) zero bias; (b) reverse bias; (c) forward bias.

which can be put in the form

$$C(V) = C_{j0}/(1 - V/\phi_{bi})^{1/2} \qquad (7.10)$$

where C_{j0} is the capacitance at zero bias.

The current–voltage relationship for the metal-semiconductor diode can be derived from a variety of theories which lead to the same basic exponential relationship. The model formulated by Bethe [7.3], called the diode or thermionic emission model, assumes that the junction depletion region is small and electrons do not suffer collisions when traversing the junction. Hence the charge carriers are affected primarily by the barrier potential and traverse the junction only if the carriers possess sufficient thermal energy (velocity) to overcome the barrier height. It is also assumed that at zero bias, equal numbers of electrons cross the junction in both directions, yielding a net current of zero.

The other common approach to determining the diode's I–V characteristics is based on the diffusion theory proposed by Schottky. In this approach, the depletion region length is assumed to be large and the charge carriers (electrons) suffer numerous collisions. Passage across the barrier is determined partly by diffusion and carrier concentrations are assumed to be independent of current.

As mentioned above, both approaches yield essentially the same I–V relationship, which has the dominant characteristic

$$I(V) = I_0[\exp(eV/kT) - 1] \qquad (7.11)$$

where k is Boltzmann's constant ($1.37 \times 10^{-23} \, J/K$), T is the absolute temperature, and V is the applied voltage. However, to account for the nonideal behavior of real diodes, the ideal diode equation of (7.11) can be modified by adding the factor n to the relationship as follows:

$$I(V) = I_0[\exp(eV/nkT) - 1] \qquad (7.12)$$

where n is a number close to unity, usually between 1.05 and 1.4. The factor n, sometimes called the diode ideality factor or slope parameter, can usually be selected so that the I–V relationship obtained from (7.12) matches measured diode performance.

There are a variety of reasons why physical diodes do not follow the ideal diode equation, such as imperfections in fabrication and factors not included in either simple model. Some of these factors are (1) series resistance, (2) surface imperfections, (3) image forces, (4) edge effects, and (5) tunneling. The most important of these with respect to mixer performance is diode series resistance.

Unfortunately, the fabrication requirements for producing a Schottky barrier diode somewhat contradict the requirements for low series resistance, which is essential for optimum mixer performance. Generally, Schottky barriers require lightly doped semiconductors, but fabricating diode contacts with low series resistance requires highly doped material. Therefore, diodes are usually fabricated by growing a lightly doped epitaxial layer on top of a highly doped substrate, to achieve the best junction versus ohmic contact performance. Series resistance is minimized, but there is still a contribution from the lightly doped epitaxial layer, which must be made thick enough to contain the depletion region during reverse-bias conditions. These conflicting requirements make it difficult to fabricate high-Q diodes using an ion implantation process, which is becoming very popular in the manufacture of high-volume MMICs.

RF skin resistance is the other main contributor to the total series resistance of the diode. This component of resistance cannot be measured at dc as is most manufacturers' data, but must be measured at RF frequencies. The skin effects of connecting beams or wires further complicate measurements and estimates of the true value of R_s. The DeLoach [7.4, 7.5] method, described in Chapter 5, can be used to determine diode Q, but cannot separate the values of C_j and R_s. However, by measuring or estimating C_j at low frequencies, a good approximation of R_s can be obtained.

Neither theory presented above includes the effects of image forces on the electrons in the depletion region. This force arises from the electron's negative charge, which is positively imaged in the metal, thus attracting charge carriers to the metal side of the junction, effectively lowering the barrier height. This phenomenon produces a voltage-dependent deviation from the ideal diode current characteristic.

Quantum mechanical tunneling can also cause charge carriers to traverse the junction by tunneling through the barrier. Tunneling is more prevalent at low temperatures, where thermionic emission has been reduced, and when doping levels are high, can sometimes degrade the noise performance of some mixers. Up to this point, imperfections in diode performance were attributed to material and fabrication quantities, with little regard to operating temperature. However, at commercial or military temperature ranges, diode characteristics change enough to affect mixer performance. As the temperature is lowered, the diode becomes more sensitive to applied voltage, with the diode knee increasing in voltage. If the LO power is marginal and no bias control is employed, large changes in conversion loss can occur. If LO power is varied as a function of temperature, mixer perfor-mance can be held constant, but in broadband designs, which are typically unbiased with wide variations in LO power as a function of frequency, the controlling of LO power as a function of temperature is completely impractical. Stable perfor-mance can be accomplished by overdriving the mixer diodes. This technique can usually limit mixer conversion loss variations to within 0.5 dB at any given frequency.

The junction capacitance of a diode remains essentially constant as a function of temperature but can increase slightly because of shifts in knee voltage. This may cause a slight degradation in conversion-loss performance of a mixer at the low end of its temperature range. However, operation of a properly designed mixer at very low temperatures can result in extremely good noise-figure performance. Mixers for very low noise radio-astronomy applications are commonly operated at cryogenic temperatures as low as 4.8 K.

Now that the junction current and capacitance characteristic are known, a large-signal model of the pumped diode can be formulated. As we have learned, when a diode is pumped with a LO signal, an infinite number of products, including a dc component, are generated. If a second signal is added, the simple set of LO and harmonic frequencies becomes much larger, since products of the signal and pump, as well as products that include harmonics of both frequencies, are generated. If we now assume that the signal amplitude is substantially smaller than the pump (LO), which is commonly the case, a small-signal mixing spectrum (Fig. 7.7) is generated [7.6]. It should be remembered that all mixing product frequencies exist in the diode and it does not matter which of the two signals is the larger. Short-circuiting the diode at a particular mixing product frequency eliminates the product's voltage component across the diode terminals, but a current component can still exist. In a similar manner, open-circuiting a voltage component of a mixing product across the diode terminals eliminates the current component, but the product's voltage is still present. However, it will be seen later that a performance difference will arise for a particular circuit application whether a mixing product is open circuited or short circuited.

We can now investigate the effects of pumping on junction current and capacitance and combine these effects to form both a large- and a small-signal diode model (Fig. 7.8). From equation (7.12) for diode current as a function of voltage, the I–V relationship for the ideal diode component of the large-signal model is obtained, since that expression is valid for any junction voltage. Similarly, the expression for junction capacitance (7.9) was obtained from the charge in the depletion region by the relation

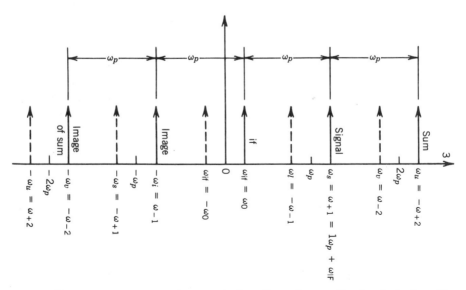

Figure 7.7 Modulation spectra for a pumped nonlinear element allowing for both positive and negative frequencies.

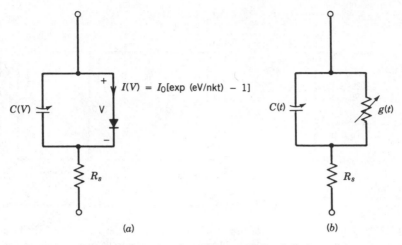

Figure 7.8 (a) Large-signal and (b) small-signal diode models.

$$C(V) = dQ_d/dV \qquad (7.13)$$

The current in the capacitor is defined as

$$I_c(t) = \frac{dQ_d}{dt} = \frac{dQ_d}{dV}\bigg|_{V=V_j(t)} \frac{dV_j(t)}{dt} \qquad (7.14)$$

$$= C(V_j(t))[dV_j(t)/dt] \qquad (7.15)$$

where $V_j(t)$ is the large-signal junction voltage.

It is this quantity, $V_j(t)$, that must be determined prior to the solution of the small-signal problem. The solution for $V_j(t)$ must include the effects of the diode embedding network at both the pump frequency ω_p and its harmonics, as well as at dc. The analysis assumes that the RF signal is negligible and circuit performance is determined solely by the LO or pump signal. A circuit representation of the LO analysis is shown in Fig. 7.9 [7.7]. The circuit performance of the diode model and embedding network shown in the figure can also be described in terms of the Fourier coefficients [7.8–7.11] of the diode's junction voltage and circuit current as

$$V_j(t) = \sum_{k=-\infty}^{\infty} V_k e^{jk\omega_p t} \qquad (7.16)$$

where $V_k = V^*_{-k}$, and

$$I_e(t) = I_c(t) + I_d(t) = \sum_{k=-\infty}^{\infty} I_{ek} e^{jk\omega_p t} \qquad (7.17)$$

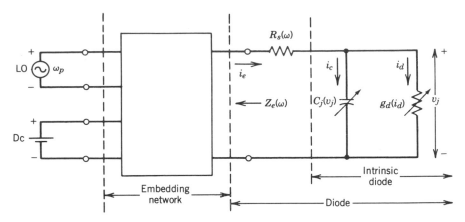

Figure 7.9 Equivalent circuit for mixer LO analysis with the large-signal diode model characterized in the time domain and with the series resistance R_s and embedding network Z_e represented in the frequency domain.

where $I_{ek} = I^*_{e-k}$. The circuit solution must also satisfy the boundary conditions. The first condition is best applied in the time domain and is imposed by the diode currents I_d and I_c. The second condition, which is imposed by the embedding network, can be described as

$$V_k = -I_{ek}[Z_e(k\omega_p) + R_s(k\omega_p)] \tag{7.18}$$

where $k = \pm 1, \pm 2, \ldots$. Equation (7.18) can then be written as

$$V_{\pm 1} = V_p - I_{e\pm 1}(Z_{10}) \tag{7.19}$$
$$Z_{10} = Z_e(\pm\omega_p) + R_s(\pm\omega_p)$$

at the pump frequency (ω_p), where V_p is the LO voltage, and as

$$V_0 = V_{dc} - I_{dc}[Z_e(0) + R_s(0)] \tag{7.20}$$

where V_{dc} is the dc bias voltage. The boundary conditions for the circuit relationships above are best applied in the frequency domain, where the embedding network, which is usually composed of linear lumped and distributed elements, is easily described by analytical functions not requiring differentiation or integration. However, to solve the harmonic balance diode/network problem efficiently, the number of LO harmonics must be truncated. Typically, a value of $n = 5$ provides a good compromise between execution speed and numerical efficiency.

Once the LO voltage waveform has been determined, a small-signal conductance and capacitance can be defined. The diode incremental conductance is obtained from the expression for current (7.12). Thus

$$g(t) = \frac{dI_d}{dV_j} = \frac{I_0 e V_j(t)}{nKT} \exp\left[\frac{e V_j(t)}{nKT}\right] \simeq \frac{e V_j(t)}{nKT} I_d(t) \qquad (7.21)$$

Hence the diode conductance presented to a small-signal (RF) for any instant of time during the LO cycle can be found. Similarly, for a small-signal analysis, the junction capacitance can be treated as a linear time-variant capacitance. With this assumption,

$$C(t) = C(V_j(t)) \qquad (7.22)$$

and

$$i_c(t) = \frac{d[C(t)v(t)]}{dt}$$
$$= C(t)\frac{dv(t)}{dt} + v(t)\frac{dC(t)}{dt} \qquad (7.23)$$

where $v(t)$ is the small-signal junction voltage.

The mixing process in a diode mixer is due to the periodic modulation of junction conductance and reactance by the pump signal. Although the variation in diode conductance can change by several orders of magnitude during the LO cycle, the 3 or 4:1 variation in capacitive reactance that occurs can still influence mixer performance and must be included in the small-signal analysis.

As in the large-signal model, the small-signal admittance components can be expressed by their Fourier coefficients. Thus

$$g_d(t) = \sum_{k=-\infty}^{\infty} G_k e^{jk\omega_p t} \qquad (7.24)$$

$$c_j(t) = \sum_{k=-\infty}^{\infty} C_k e^{jk\omega_p t} \qquad (7.25)$$

and

$$i_d(t) = \sum_{k=-\infty}^{\infty} I_k e^{jk\omega_p t} \qquad (7.26)$$

where $G_k = G^*_{-k}$, $C_k = C^*_{-k}$, and $I_k = I^*_{-k}$. These quantities, in conjunction with the embedding impedance $Z_e(\omega)$, can be used to determine the small-signal mixer characteristics and used to develop a conversion matrix for the diode.

The components of the conversion matrix, which relates the various small-signal components of voltage to current at each sideband frequency, can be constructed from the foregoing nonlinearities. For the intrinsic diode (Fig. 7.10), the admittance matrix Y relates the current and voltage at port m, corresponding to a sideband frequency of $(\omega_0 + m\omega_p)$, to the current and voltage at port n, which corresponds to a sideband frequency of $(\omega_0 + n\omega_p)$. Although these ports are not physical, the

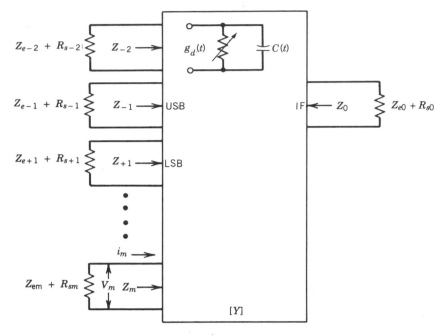

Figure 7.10 Multiport model of pumped intrinsic diode.

model can be treated as a multiport circuit since each port is at a different frequency. The ports, instead of being different sets of terminals at the same frequency, are the same set of terminals at different sideband frequencies.

The square conversion matrix $[Y]$, which is of the form

$$
\begin{bmatrix}
 & \vdots & \vdots & \vdots & \\
\cdots & Y_{11} & Y_{10} & Y_{1-1} & \cdots \\
\cdots & Y_{01} & Y_{00} & Y_{0-1} & \cdots \\
\cdots & Y_{-11} & Y_{-10} & Y_{-1-1} & \cdots \\
 & \vdots & \vdots & \vdots &
\end{bmatrix}
$$

is the admittance matrix of the intrinsic diode; hence the current–voltage relationship for the circuit must obey Ohm's law. Thus

$$[i] - [Y][v] \tag{7.27}$$

where $[i] = [\ldots, i_1, i_0, i_{-1}, \ldots]^T$ and $[v] = [\ldots, v_1, v_0, v_{-1}, \ldots]^T$ are the small-signal voltage and current components at each port. For the intrinsic diode, it can be shown that the elements of the $[Y]$ matrix are given by

$$Y_{mn} = G_{m-n} + j(\omega_0 + m\omega_p)C_{m-n} \tag{7.28}$$

where G_{m-n} and C_{m-n} are the Fourier coefficients of the diode's small-signal conductance and capacitance defined in (7.24) and (7.25). The series resistance and the effects of the embedding network, as shown in Fig. 7.11, can now easily be added. This can be done by forming a new conversion matrix, $[Y']$, which is the admittance of the total mixer. This augmented network (Fig. 7.12) for the pumped mixer allows one external source, v_1, which is the RF input. However, the augmented matrix includes all external terminating impedances Z_{em}. The new matrix, which is of the form

$$
\begin{bmatrix}
& \vdots & \vdots & \vdots & \\
\cdots & Y'_{11} & Y'_{10} & Y'_{1-1} & \cdots \\
\cdots & Y'_{01} & Y'_{00} & Y'_{0-1} & \cdots \\
\cdots & Y'_{-11} & Y'_{-10} & Y'_{-1-1} & \cdots \\
& \vdots & \vdots & \vdots &
\end{bmatrix}
$$

and can be defined as

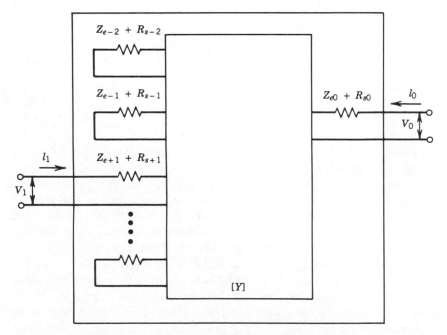

Figure 7.11 Augmented Y' matrix for complete mixer.

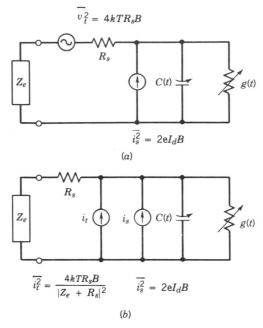

$$\overline{v_t^2} = 4kTR_sB$$

$$\overline{i_s^2} = 2eI_dB$$

(a)

$$\overline{i_t^2} = \frac{4kTR_sB}{|Z_e + R_s|^2} \qquad \overline{i_s^2} = 2eI_dB$$

(b)

Figure 7.12 Mixer diode noise model: (*a*) voltage representation for thermal noise source in R_s; (*b*) noise source converted to a current source via Thévenin's theorem.

$$[Y'] = [Y] + \text{diag}\left[\frac{1}{Z_{em} + R_{sm}}\right] \qquad (7.29)$$

assuming that

$$[i'] = [Y'][v] \qquad (7.30)$$

At this point the entire mixer circuit can be evaluated since the embedding impedances are determined from the mixer topology; but only two ports are of interest, the signal and the IF output. All other ports are terminated and contained within the new matrix $[Y']$. The addition to the intrinsic diode admittance matrix $[Y]$ of the diagonal matrix above is the first step in converting the $(2n + 1)$-port network into a two-port. To complete the conversion, the voltages at the unwanted ports are set to zero, effectively placing the embedding impedances Z_{em} in series with their respective ports. However, the IF load impedance Z_{e0} is not defined or absorbed, but in the physical mixer can be adjusted for optimum performance by matching. Hence a conjugately matched load can be assumed.

The conversion matrix can also be expressed in terms of impedance by inverting $[Y']$; thus

$$[Z'] = [Y']^{-1} \qquad (7.31)$$

With this in mind, we can now define the ratio of available power transferred from the RF source Z_{e1} to the IF load Z_{01} (conversion loss). The conversion loss property of a real mixer can be considered to be composed of three main components; (1) the conversion loss of the intrinsic diode without R_s, (2) the loss contribution at the IF by R_s, and (3) the loss associated with R_s at the signal frequency. The embedding network and matching network are assumed to be lossless. It should be noted how important the diode parameter R_s is to mixer performance since it contributes loss twice, once at the RF and again at the IF frequency. Thus it is also important in determining noise performance.

For the case when the RF signal is defined as ω_1 and the IF signal is defined as ω_0, the input and output impedances, as well as the conversion loss characteristics, can be calculated from two-port network theory. The conversion loss contribution due to R_s at the IF and RF frequencies can be shown to be

$$K_0 = \frac{\mathrm{Re}[Z_{e0} + R_{s0}]}{\mathrm{Re}[Z_{e0}]} \tag{7.32}$$

and

$$K_1 = \frac{\mathrm{Re}[Z_{e1} + R_{s1}]}{\mathrm{Re}[Z_{e1}]} \tag{7.33}$$

while the conversion-loss component from the intrinsic diode is

$$L' = \frac{1}{4|Z_{01}'|^2} \frac{|Z_{e0} + R_{s0}|^2}{\mathrm{Re}[Z_{e0} + R_{s0}]} \frac{|Z_{e1} + R_{s1}|^2}{\mathrm{Re}[Z_{e1} + R_{s1}]} \tag{7.34}$$

From (7.32)–(7.34), the total conversion loss becomes

$$L = \frac{1}{4|Z_{01}'|^2} \frac{|Z_{e0} + R_{s0}|^2}{\mathrm{Re}[Z_{e0}]} \frac{|Z_{e1} + R_{s1}|^2}{\mathrm{Re}[Z_{e1}]} \tag{7.35}$$

We may now generalize the expression above to obtain the conversion loss from any sideband j to any other sideband i. Thus (7.35) becomes

$$L_{ij} = \frac{1}{4|Z_{ij}'|^2} \frac{|Z_{ei} + R_{si}|^2}{\mathrm{Re}[Z_{ei}]} \frac{|Z_{ej} + R_{sj}|^2}{\mathrm{Re}[Z_{ej}]} \tag{7.35a}$$

The noise properties of microwave mixers, starting from the earliest Schottky diodes and later to vacuum tubes, were well verified and understood for many years. But until the 1970s, an accurate noise model did not exist [7.12]. An accurate mixer noise model must include not only the effecs of shot noise, thermal noise, and phonon scattering, but must also encompass circuit interactions, correlation properties from other mixing products, and nonlinear junction capacitance effects. The theory that will be presented was developed by Held and Kerr [7.9, 7.10] and agrees well with experimental results.

As mentioned above, the main noise sources in Schottky mixers are (1) the thermal noise contribution from the diode series resistance; (2) shot noise generated by the random flow of charge carriers across the barrier, which is analogous to shot noise in vacuum tubes; and (3) noise due to phonon scattering and in the case of GaAs, intervalley scattering. The first two noise sources are almost always the dominant contributors, with scattering noise becoming more pronounced at very high frequencies such as for radio astronomy applications at several hundred gigahertz.

The thermal noise component of the mixer diode model, which is present when any power-dissipating element is involved at a temperature above absolute zero, is determined by the total diode series resistance. The noise, which is attributed to the random motion of charge carriers (in this case electrons) can be assumed to be frequency independent and linearly related to temperature. Hence the noise power from a resistor can be modeled as a noiseless resistor of value R with a noise voltage source in series with it which has a magnitude of

$$\overline{v_t^2} = 4kTBR \tag{7.36}$$

The noise power generated in bandwidth B is then

$$P_n = kTB \tag{7.37}$$

where k is Boltzmann's constant and T is temperature in kelvin. Shot noise, which is also significant, occurs because the current through the Schottky barrier is not constant. This is true even for the dc-biased case since the charge carriers are discrete points of charge (electrons) and are not continuous. This concept can easily be understood if the current is assumed to be a series of random impulses of charge traversing the junction. That is, at any instant of time, the current is made up of discrete charge carriers that randomly transit the junction, giving rise to different values of current at any instant of time. However, the average number of discrete pulses of charge per unit time, provided that the time unit is substantially longer than the junction transit time, is proportional to a constant dc current. The fluctuations in the diode current that result from this random impulse process cause a noise current component with a mean-squared magnitude proportional to dc junction current. Thus the noise current in a forward-biased diode can be shown to be

$$\overline{i_s^2} = 2eI_dB \tag{7.38}$$

where e is the electron charge, B the bandwidth, and I_d the junction current.

A diode noise model can now be formulated by adding these Gaussian noise sources to the diode model shown in Fig. 7.9. The new diode equivalent circuit is shown in Fig. 7.12a. To be more compatible with the mixer circuit shown in Fig. 7.10, the thermal noise voltage source can be converted to a current source by using Thévenin's theorem. The current source model (Fig. 7.12b) is more practical for noise analysis.

The thermal and shot noise sources, although both Gaussian in nature, act quite differently when LO power is applied to the diode. The thermal noise components from R_s when down-converted from the desired and undesired products to the IF are uncorrelated since R_s is assumed to be time invariant. With just a dc current component in the diode, the noises centered at mixing frequencies of $\omega_0 + n\omega_p$ are independent, and therefore when mixed down to the IF frequency of ω_0 will be uncorrelated (Fig. 7.13a). However, when an LO voltage is applied to the diode, the shot-noise components at each mixing product frequency are translated to the IF band (Fig. 7.13b). In addition, the noise components from any mixing frequency include up- and down-converted noise components from other mixing frequencies since they are all related to the same LO and its harmonics. The corre-

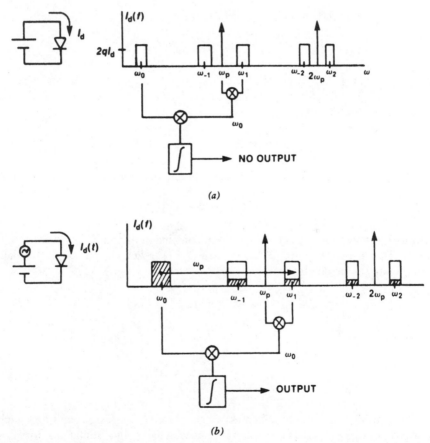

Figure 7.13 Noise correlation illustration: (a) IF output noise for dc-biased diode; (b) down-converted correlated noise for LO driven diode.

lation occurs because the converted noise modulation is due to the same random process.

By using an analysis method similar to the previous small-signal conversion analysis, correlation matrices for thermal and shot noise can be found. For thermal noise, which is uncorrelated, the matrix is simply a diagonal. Thus

$$
C_{tmn} = \begin{bmatrix} \dfrac{4kTR_sB}{(Z_{em} + R_s)^2} & m = n \\[2ex] \dfrac{4kTR_sB}{(Z_0)^2} & m = 0 \\[2ex] 0 & \text{elsewhere} \end{bmatrix} \tag{7.39}
$$

Similarly, the correlation matrix for shot noise is

$$
C_{smn} = 2eI_{m-n}B \tag{7.40}
$$

where I_{m-n} is the Fourier coefficient of the $(m - n)$th term for the series representing the LO diode current. It can then be shown that the noise power dissipated in the output termination Z_{em} is

$$
P_m = |V_m|^2 \, \text{Re} \, [Z_{em}] / |Z_{em} + R_s|^2 \tag{7.41}
$$

where the voltage at ω_m is the sum of the shot and thermal noise and is defined as

$$
\langle V_m^2 \rangle = Z_m(C_s + C_t)Z_m^*t \tag{7.42}
$$

It should be noted that Z_m is the row of the conversion matrix Z_e corresponding to ω_m. If the conversion matrix Z' defined in (7.31) is known, the single-sideband noise figure can be defined as

$$
F_{ssb} = 1 + T_m/T_0 \tag{7.43}
$$

where $T_0 = 290$ K and

$$
T_m = \frac{\langle V_0^2 \rangle}{4kB} \frac{|Z_{e1} + R_{s1}|^2}{|Z_{01}'|^2 \, \text{Re} \, [Z_{e1}]} \tag{7.44}
$$

The analysis techniques outlined above should give the design engineer insight into the conversion-loss and noise-figure mechanisms and limitations for most practical mixers. However, design engineers cannot be expected to develop computer algorithms to evaluate conversion and correlation matrices in order to conduct their daily design activities. Therefore, several design methods will be presented in the following sections which can be executed with commercially available software. Designs conducted in this manner can give excellent results, especially when diodes rather than transistors are used as the nonlinear element. This is in part due to the fact that diodes are very forgiving; that is, they are very tolerant of matching errors, LO drive variations, and temperature excursions. Also, most mixer requirements do not demand state-of-the-art performance in regard to noise-figure and conversion-loss characteristics.

7.2 SINGLE-DIODE MIXERS

The single-diode mixer, although fondly remembered for its use as an AM "crystal" radio or radar detector during World War II, has become less popular due to demanding broadband and high dynamic range requirements encountered at frequencies below 20 GHz. However, there are still many applications at milli-meter wave frequencies, as well as consumer applications in the microwave portion of the spectrum, which are adequately served by single-ended designs. The design techniques presented can also be applied to single-ended or balanced mixers using planar or waveguide approaches.

In this chapter we focus on a more "hands-on" design approach than previously presented, with emphasis on microstrip applications. Matching considerations, circuit approximations, and design philosophy for all types of mixers are developed. Commercially available design tools will also be used to determine nonlinear and linear circuit and diode characteristics.

The design of single-diode mixers can be approached in the same manner as multiport network design. The multiport network contains all mixing-product frequencies regardless of whether they are ported to external terminations or terminated internally. With simple mixers, the network's main function is frequency-component separation; impedance-matching requirements are secondary. Hence, in the simplest approach, the network must be capable of selecting the LO, RF, and IF frequencies (Fig. 7.14).

However, before a network can be designed, the impedance presented to the network by the diode at various frequencies must be determined. Unfortunately, the diode is a nonlinear device; hence, determining its characteristics is more involved than determining an unknown impedance with a network analyzer. Since the diode impedance is time varying, it is not readily apparent that a stationary impedance can be found. Stationary impedance values for the RF, LO, and IF frequencies can be measured or determined if sufficient care in analysis or evaluation is taken.

The first impedance to measure or determine is the diode's LO impedance. As we have learned, this impedance is a function of LO drive power, dc bias, frequency, and physical diode characteristics; thus impedance measurements must be conducted at the correct bias and drive power, which correspond to the expected mixer environment. Typically, S_{11} (large signal) is measured as a function of frequency for several values of LO power. This information allows the design

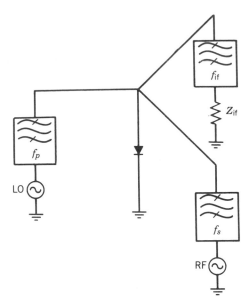

Figure 7.14   Three main frequency components and filtering requirements for single-diode mixer.

engineer to optimize the network characteristics to the expected LO power variation so that the resulting mixer will exhibit the best performance traits, such as conversion loss, noise figure, VSWR, and so on. The large-signal LO impedance can also be simulated numerically with a harmonic balance or time-domain simulator. Figure 7.15 illustrates typical LO impedance characteristics for a beam-lead diode as a function of frequency and pump power. It should be remembered that when measuring the diode's LO impedance, a dc return path must be present even if fixed dc bias is not employed (Fig. 7.16).

Measuring or determining the RF and IF impedances is somewhat more difficult in that the LO drive must be present. When computer simulation is employed, the problem is numerically difficult but conceptually easy for the user. By adding a second signal to the analysis, and evaluating the voltages and currents in the circuit, the conjugate matched impedance values for the embedding network can be determined. A computer program such as microwave SPICE is ideal for such analyses. Measuring such parameters is a bit more tricky.

An approximate value for RF impedance can be obtained with the measurement setup shown in Fig. 7.17. With this arrangement, LO power is injected by means of a directional coupler. The RF impedance is measured by using a second source and network analyzer. Sometimes there are problems with this approach, in that the LO signal interferes with the network analyzer measurements. Usually, this is when the LO and RF signals are very close in frequency. As mentioned above, measurements of this type are only approximate, because the diode is not presented with the terminating impedances that it will see in the final mixer circuit at all

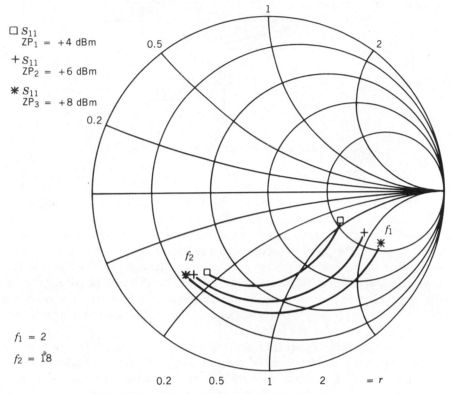

$\square\ S_{11}$
  $ZP_1\ =\ +4$ dBm

$+\ S_{11}$
  $ZP_2\ =\ +6$ dBm

$\ast\ S_{11}$
  $ZP_3\ =\ +8$ dBm

$f_1\ =\ 2$

$f_2\ =\ 18$

Figure 7.15   LO impedance as a function of pump power.

product frequencies. The IF impedance can be measured in a similar manner, or if the IF frequency is sufficiently far removed from the RF or LO frequencies, a slide screw tuner can be placed at the diode's IF output port and adjusted for optimum conversion loss. The tuner and load can then be measured. The tuner/ load combination will be the conjugate of the diode's IF impedance. This measurement configuration is shown in Fig. 7.18. Although the measured impedance values

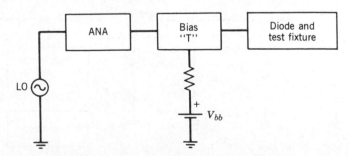

Figure 7.16   Measurement setup for determining diode LO impedance.

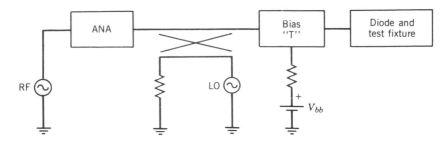

Figure 7.17   Diode RF impedance measurement setup.

determined by the foregoing method are only approximate, they are typically sufficiently accurate for most design problems.

With the diode impedance data obtained, we can now continue with the network design problem. The problem proceeds as a multiport network design where key impedance terminations and sources are external to the network. Each port of the network is at a different frequency, not necessarily a different physical location (Fig. 7.19). Hence as one of the source frequencies is changed, a new set of network conditions result. When broadband performance is desired, this analysis method can become a bit cumbersome.

When both the large-signal (LO) and small-signal (RF and IF) impedances are known, the design problem can be reduced to the solution of linear two-port networks. That is, a two-port network can be constructed for each frequency component, such as the LO-to-diode matching network, image termination-to-diode matching network, and so on. This concept is illustrated in Fig. 7.20. The dashed lines between each network in the figure denote the fact that each network is not independent, but composed of at least some of the identical elements. Thus during computer optimization of the network performance, a change, for example, in the value of a capacitor that is common to all signal paths must be made identical from network to network. This is easily accomplished by employing the variable element block when using either Super-Compact or Touchstone.

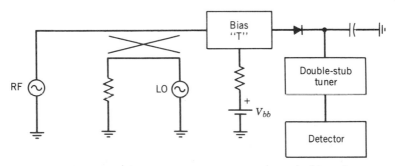

Figure 7.18   IF impedance measurement setup using a double-stub tuner.

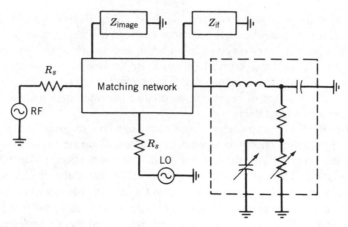

Figure 7.19 Matching circuit for single-diode mixer with external ports at separate frequencies.

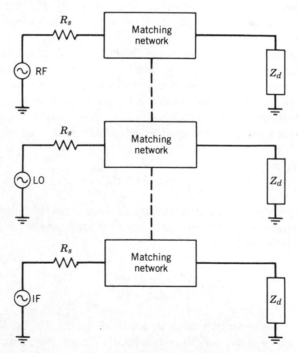

Figure 7.20 Multiport matching network for single-diode mixer reduced to three coupled two-port networks. A linear simulator can yield a good design approximation if the small- and large-signal impedances are substituted for Z_d in the appropriate analyses.

If a nonlinear simulator is available, such as Libra Microwave SPICE, and a diode model is formulated using the methods outlined in Chapter 5 for GaAs FETs, the complete problem can be solved using a single network. Provided that a sufficient number of LO and signal harmonics or sufficiently small time step is employed, excellent simulation accuracy can be obtained. The nonlinear simulator has the added advantage of being able to predict intermodulation and conversion gain compression characteristics.

Although the design approach outlined above can be straightforward and is well defined, the problems encountered when selecting a diode are sometimes ignored. Selecting the proper mixer diode for a particular application is dependent on a variety of factors, such as conversion loss, cost, intermodulation characteristics, LO drive power, frequency of operation, and the desired transmission-line media.

The actual mechanical characteristic of the diode is probably the first parameter to be considered. This parameter is not only influenced by the transmission-line media involved, but is influenced by performance, cost, and the final application environment. For example, better mixer performance will almost always be obtained when unpackaged devices are employed rather than chip or beam-lead diodes, due to package parasitics and losses. However, packaged diodes are much easier to handle, and product assembly can be accomplished without expensive soldering, welding, or bonding equipment. A good illustration of the use of packaged diodes for low-cost products are the vast variety of commercially available broadband mixers constructed with ribbon-lead diodes on soft substrate material.

As the frequency of operation becomes higher and soft substrate materials are replaced by fused silica or alumina, performance requirements usually dictate the use of either beam-lead or chip devices. However, there are vast performance differences between diodes, which are due to semiconductor type, barrier potential, and junction area. Typically, a figure of merit is defined for a diode, analogous to the parameter f_t for an active device, called the cutoff frequency f_c, which is defined as

$$f_c = 1/2\pi R_s C_{j0} \tag{7.45}$$

The cutoff frequency is an important parameter since it can be shown that a degradation in conversion-loss performance δ_1 from the ideal case at the RF frequency [7.13] can be defined as

$$\delta_1 = 1 + \frac{R_s}{r_{dr}} + \frac{r_{dr}}{R_s} \left(f/f_c\right)^2 \tag{7.46}$$

where r_{dr} is the nonlinear impedance of the diode junction at the RF frequency. At the IF frequency, a finite R_s also degrades conversion loss; thus a degradation factor δ_2 can also be defined as

$$\delta_2 = 2R_s/r_{di} + 1 \tag{7.47}$$

where r_{di} in this case is the nonlinear impedance of the diode junction at the IF frequency. Equation (7.46) can also be written in the form

$$\delta_1 = 1 + R_s/r_{dr} + R_s r_{dr}(\omega C_{j0})^2 \qquad (7.48)$$

Hence it becomes easy to see that as the R_s-C_{j0} product becomes large, the conversion loss degrades rapidly. Equation (7.48) also illustrates that an optimum diode size exists for a given frequency. This can be seen if one examines the last term in (7.48). At low frequencies, C_{j0} cannot be made arbitrarily large in an attempt to minimize R_s. Also, as the operating frequency is increased, the $(\omega)^2$ term begins to dominate. Unfortunately, the relationship between R_s and C_{j0} for any diode geometry is inversely related; that is, as the junction area is reduced so that C_{j0} can be minimized, R_s increases. Thus a minimum in conversion loss occurs for a given frequency as the diode junction area is varied. As the operating frequency is raised, the minimum occurs at smaller and smaller junction areas; the absolute value of conversion loss deteriorates with frequency. In Fig. 7.21a the parasitic conversion-loss contribution due to the diode parasitics for diodes fabricated using a high-quality silicon process is shown. A similar plot of conversion-loss degradation for ion-implanted GaAs diodes fabricated using a conventional FET process is illustrated in Fig. 7.21b. For both cases it was assumed that the $R_s C_{j0}$ product remains constant as the diode size is scaled. This problem is always a challenge to the diode designer, who invariably is asked to fabricate a diode with minimum C_{j0} and minimum R_s. It should be noted that when selecting a diode for a particular frequency range, the largest C_{j0} should be chosen. This will minimize performance sensitivity due to diode variations and will usually give the best overall results.

The junction area and the semiconductor material also influence the nonlinear diode impedance at various product frequencies. The diode size can be chosen for ease in matching, rather than for conversion-loss performance as is sometimes done in waveguide designs or broadband applications. Generally, GaAs diodes exhibit higher impedances than comparably sized silicon devices [7.14] and typically have lower values of R_s, but are usually more costly. Two diode models, for Si and GaAs devices driven into LO saturation, are shown in Fig. 7.22. As can be seen the junction resistance is quite different even though the parasitic element values are similar.

The final major consideration in diode selection is intermodulation performance. This key characteristic is directly related to LO power and barrier potential. Most commercial diode manufacturers offer silicon mixer diodes with various barrier potentials. Low-barrier Si devices require only about 0.3 V for a 1-mA diode current, while high-barrier diodes may require as much as 0.7 V for the same current. If the LO impedances are similar, about 6 dB more LO power is required with high-barrier devices. As the input RF signal to the diode approaches the pump power level, the converted product levels begin to saturate; that is, the converted signal level is no longer related to the input RF signal power. The 1-dB compression point occurs when the input signal power is approximately 6 dB less

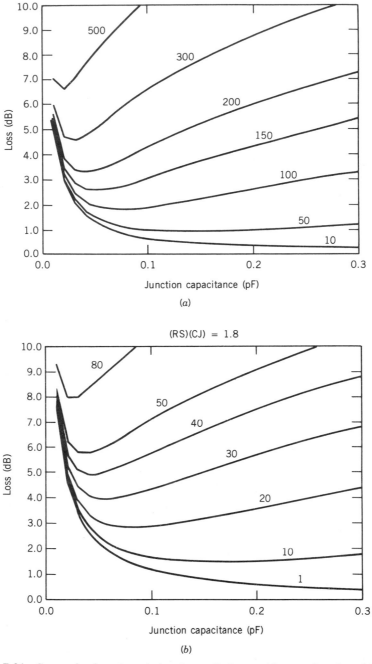

Figure 7.21 Conversion loss degradation due to diode parasitics as a function of junction capacitance and frequency (GHz): (*a*) high-quality silicon diode; (*b*) GaAs ion-implanted planar diode.

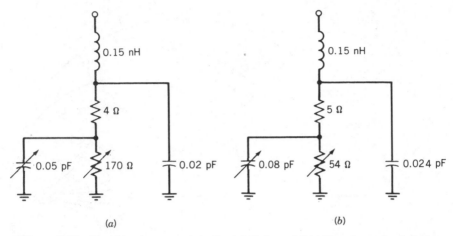

Figure 7.22 Parasitic element values for (*a*) GaAs and (*b*) silicon beam-lead diodes.

than the LO power level. Increasing the LO or pump power increases the compression point until the diode is fully saturated. This occurs when the LO waveform drives the diode well into the conduction region, which for a high-barrier Si diode is about 0.7 to 0.8 V of peak forward voltage. For GaAs diodes the saturation level may even be slightly greater. The third-order intercept point is also related to the barrier height and LO power in the same manner. A common "rule of thumb" is that the third-order intercept point (IP_3) is 10 dB above the 1-dB compression point; however, the designer should not be surprised if the IP_3 level of a particular mixer is on the order of the LO power. Product terminations, especially the image termination, greatly influence the distortion characteristics of any mixer.

To illustrate the foregoing design concepts and circuit interactions, the design of a single-diode microstrip mixer for the frequency range (RF) 9.0 to 10.2 with an IF frequency of 500 MHz will now be described. If we select alumina as the substrate material and desire a simple matching structure capable of reasonable bandwidth performance, either a beam-lead or chip diode should be employed. In addition, for X-band operation, the diode cutoff frequency should be in excess of 100 GHz, and R_s should be less than 10 Ω to assure good mixer performance. C_{j0} must also be less than 0.3 pF so that optimum performance can be obtained. LO pump power is minimized by selecting a low-barrier Si diode.

The design begins by determining the LO, RF, and IF impedances at the desired LO power level for the diode. This can be accomplished by using numerical techniques or laboratory measurements. Figure 7.23 illustrates the LO, RF, and IF impedances of a typical silicon beam-lead diode driven at the onset of LO saturation. The element values for a simple diode model are shown in Fig. 7.24.

The unmatched return-loss impedance characteristics for LO and RF signals in the frequency range 9.0 to 10.7 GHz are shown in Fig. 7.25. Because of the semiconductor material (Si) and diode size, the impedances are relatively close to

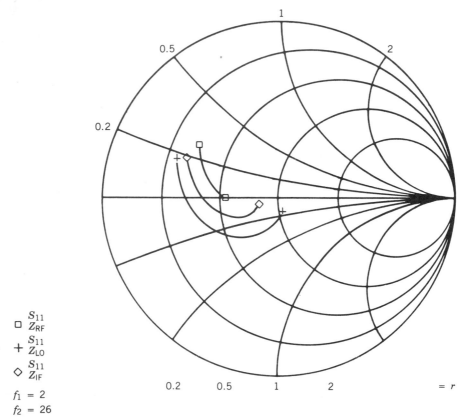

$\square$ $\dfrac{S_{11}}{Z_{RF}}$

$+$ $\dfrac{S_{11}}{Z_{LO}}$

$\Diamond$ $\dfrac{S_{11}}{Z_{IF}}$

$f_1 = 2$
$f_2 = 26$

Figure 7.23 LO, RF, and IF impedance of typical silicon beam-lead diode.

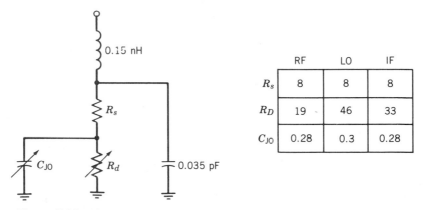

	RF	LO	IF
R_s	8	8	8
R_D	19	46	33
C_{JO}	0.28	0.3	0.28

Figure 7.24 Element values for silicon beam-lead diode used in mixer example.

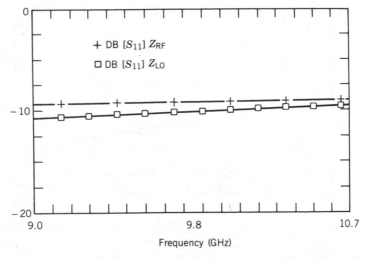

Figure 7.25 Return loss of LO and RF impedance of unmatched diode.

$50\times$. However, some improvement in conversion-loss performance can be obtained by matching. A simple matching network topology that separates the IF signal from the LO and RF signals and provides a dc and IF return path is shown in Fig. 7.26. The new RF and LO return loss performances for the total structure are shown in Fig. 7.27. It was assumed the the LO and RF signal separation would be accomplished with the aid of a directional coupler. It should also be noted that although the RF return loss is greater than 20 dB, the LO return loss was slightly degraded from the unmatched case. This type of VSWR trade-off in favor of the RF port is common. IF port matching at 500 MHz can be accomplished with lumped-element techniques.

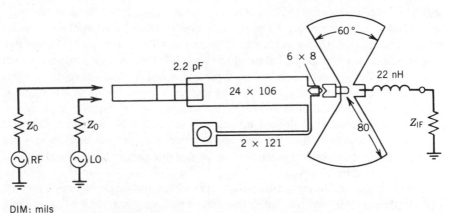

DIM: mils
SUB: $H = 15$, $\epsilon_r = 9.9$ $T = 0.1$

Figure 7.26 Single-diode mixer microstrip circuit layout with beam-lead silicon diode.

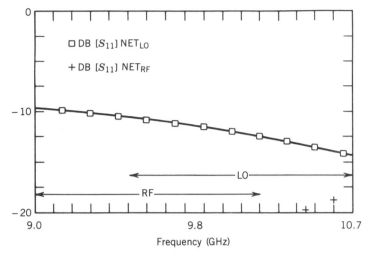

Figure 7.27 LO and RF return loss performance of single diode and matching network.

Although the example above is somewhat simple, it does illustrate the design methodology, device selection, and circuit trade-offs. The single-diode mixer design problem also forms the basis for many other topologies which are illustrated in the following sections.

7.3 SINGLE-BALANCED MIXERS

The simplicity and performance of the single-diode mixer presented in Section 7.2 make it very attractive in terms of cost, producibility, and conversion loss, but it exhibits some serious drawbacks for broadband high-dynamic-range applications. Probably its greatest disadvantage is the difficulty in injecting LO energy while still being able to separate the LO, RF, and IF signals in the embedding network. Without the aid of some form of balun or hybrid structure, this problem becomes even more difficult as the mixer operating bandwidth is increased. Aside from circuit design considerations, balanced mixers offer some unique advantages over single-ended designs such as LO noise suppression and rejection of some spurious products. The dynamic range can also be greater because the input RF signal is divided between several diodes, but this advantage is at the expense of increased pump power. However, balanced mixers tend to exhibit higher conversion loss and are more complex. Both the increase in complexity and conversion loss can be attributed to the hybrid or balun, and to the fact that perfect balance and lossless operation cannot be achieved.

There are essentially only two design approaches for single-balanced mixers; one employs a 180° hybrid, while the other employs some form of quadrature structure. The numerous variations found in the industry are related to the trans-mission-line media employed and the ingenuity involved in the design of the hybrid structure. The most common designs for the microwave frequency range employ either a branch-line, Lange, or "rat-race" hybrid structure. At frequencies below about 5 GHz, broadband transformers are very common, while at frequencies above 40 GHz, waveguide structures become prevalent, although GaAs monolithics are beginning to encroach in the 44-GHz, 60-GHz, and 94-GHz frequency ranges.

Before we can analyze or design a single-balanced mixer, we must understand the operation and realization of the 180° or 90° hybrid. Ideal representations illustrating the performance characteristics of both hybrid types are shown in Fig. 7.28. The ideal performance of the quadrature hybrid illustrated in Fig. 7.28a can be described with the aid of an S-parameter matrix. Assuming that the ports are defined as in the figure, the matrix can be defined as

$$S_{90} = \frac{1}{(2)^{1/2}} \begin{bmatrix} 0 & -j & 0 & 1 \\ -j & 0 & 1 & 0 \\ 0 & 1 & 0 & -j \\ 1 & 0 & -j & 0 \end{bmatrix} \qquad (7.49)$$

Similarly, the performance of an ideal 180° hybrid (or balun) can be described as

$$S_{180} = \frac{1}{(2)^{1/2}} \begin{bmatrix} 0 & 0 & 1 & 1 \\ 0 & 0 & 1 & -1 \\ 1 & 1 & 0 & 0 \\ 1 & -1 & 0 & 0 \end{bmatrix} \qquad (7.50)$$

As can be seen in the relationships above, the zero-valued diagonal terms indicate that the hybrids are perfectly matched, and the nondiagonal zero-valued terms indicate infinite reverse isolation.

A popular form of the quadrature hybrid described above is the Lange coupler, illustrated in Fig. 7.29. The coupler is used in microstrip applications because of its ability, due to the interdigitated structure, to obtain very tight coupling values (approximately -6 to -1.25 dB). Because of this ability, couplers are commonly designed overcoupled (about -2.5 dB), thus achieving very broadband performance for a single-section structure. Typical broadband amplitude and phase performance for a four-strip coupler fabricated on an alumina substrate is shown in Fig. 7.30. As can be seen, the phase performance is very close to 90° throughout a considerable frequency range. The amplitude balance, although not perfect, is also very respectable. Couplers with similar performance can also be realized in three-layer stripline by using parallel-plate coupling. Quadrature coupler performance can also be obtained in stripline or microstrip over a limited bandwidth with a branch-line coupler (Fig. 7.31). The amplitude and phase performances for a typical coupler are shown in Figure 7.32. Although the single-section branch-line coupler exhibits narrowband performance, multisection couplers can be constructed for 40% bandwidth applications.

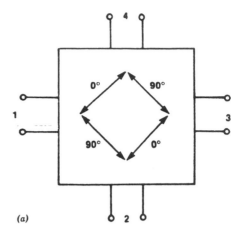

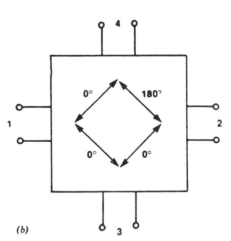

Figure 7.28 Ideal hybrid performance: (a) quadrature coupler (90°); (b) Balun (180°).

Realizing a 180° hybrid at microwave frequencies is somewhat more difficult. However, the ring or "rat-race" hybrid, illustrated in Fig. 7.33, is easy to construct and offers good performance for narrowband applications. The coupler ring is realized with a 70.7-Ω transmission line which is 1.5 wavelengths in circumference. The four 50-Ω ports are connected to the ring in such a manner that two of them are separated by 0.75 wavelength, with the remaining two being separated by 0.25 wavelength.

The operation of the 180° hybrid is simple and is illustrated in Fig. 7.34. If a signal is injected into port 1, the voltage appearing at port 2 is zero, since the path lengths differ by 0.5 wavelength; thus port 2 can be treated as a virtual ground. Hence the transmission-line portions of the ring between ports 3 and 2, and ports 4 and 2, act as short-circuited stubs connected across the loads presented at ports

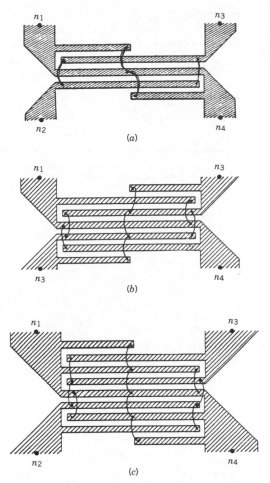

Figure 7.29 Microstrip Lange coupler configurations: (*a*) four-strip; (*b*) six-strip; (*c*) eight-strip.

3 and 4. For center-frequency operation, these stubs appear as open circuits. Similarly, the transmission-line lengths between ports 1 and 3, and ports 1 and 4, transform the 50-Ω load impedances (Z_0) at ports 3 and 4 to 100 Ω ($2Z_0$). When combined at port 1, these transformed impedances produce the 50-Ω input impedance seen at port 1. A similar analysis can be applied at each port, thus showing that the hybrid exhibits a matched impedance of 50 Ω or Z_0 at all nodes. It should be noted that when port 1 is driven, the outputs at ports 3 and 4 are equal and in phase, while ideally there is no signal at port 2. However, when port 2 is driven, the output signals appearing at ports 3 and 4 are equal but exactly out of phase. Also, no signal appears at port 1. Hence ports 1 and 2 are isolated. Unfortunately, the hybrid performance is narrowband, although modifications to the ring by replacing 0.5 wavelengths of transmission line between ports 2 and 4 by a constant

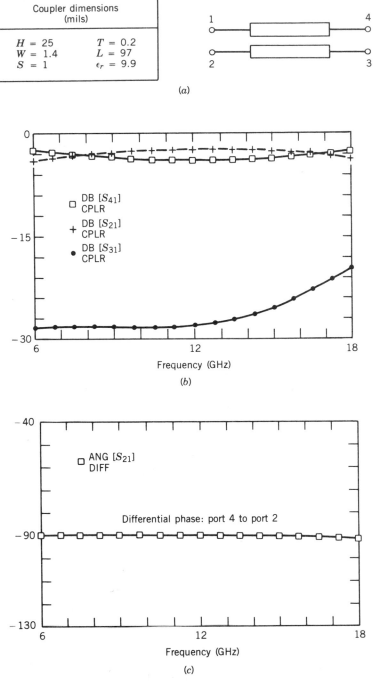

Figure 7.30 Microstrip Lange coupler fabricated on an alumina substrate: (*a*) coupler strip dimensions with nodel connections; (*b*) amplitude response; (*c*) phase performance.

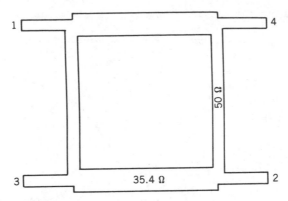

Figure 7.31 Branchline hybrid for 50-Ω system.

180° phase "flip" (Fig. 7.35) does improve the bandwidth considerably, but realizing the necessary transmission line, since it must be very tightly coupled, is difficult. Parallel suspended coupled strips or a Lange coupler can be used for this application, with good results. The performances of both hybrids are shown in Figs. 7.36 and 7.37.

Another form of a four-port 180° hybrid is the trifilar wound center-tapped transformer [7.15]. These types of transformers, which are typically wound on ferrite cores (beads or toroids), can be designed to exhibit extremely broadband performance. Although the transformer performance is influenced by a variety of factors, such as wire size and winding length, core material, and size and aspect ratio, transformers that can operate from 10 mHz to 4 GHz can be constructed. Typical construction of a trifilar wound transformer with a 2:1 turns ratio is illustrated in Fig. 7.38.

The performance of a transformer hybrid is similar to the "rat race" described previously if it is connected as shown in Fig. 7.39, with the exception that all four ports do not have the same driving-point impedance. This condition will always exist regardless of the turns ratio of the transformer because the secondary winding will always be center tapped. The operation of the transformer is easily explained with the aid of Fig. 7.40.

If a voltage source of 2 V with impedance Z_0 is connected to the transformer's primary winding (port 4), the current I will induce a current of $I/2$ in the secondary with a voltage of V across each winding. When ports 1 and 2 are terminated with an impedance of $2Z_0$, voltage V and $-V$ appear across the outputs. By summing the voltages around the loop, it is found that no voltage appears at port 3; thus port 3 is a virtual ground.

A similar analysis can be used to describe the hybrid's operation when port 3 is driven. In this case, because of symmetry in the transformer, the current is divided equally between each secondary winding. However, because the currents in the windings are in opposite directions, no current is induced in the primary winding. Instead, equal output voltages of magnitude V appear across the loads at

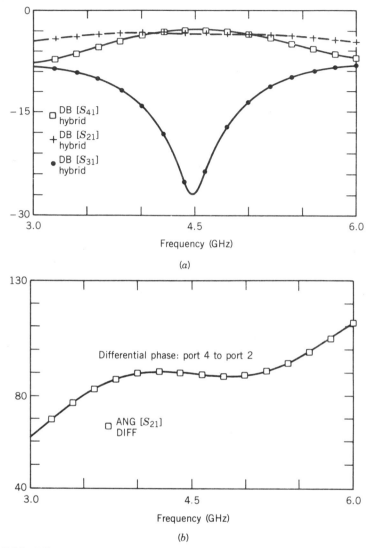

Figure 7.32 Microstrip branchline coupler performance: (*a*) amplitude response; (*b*) differential phase performance.

ports 1 and 2. It should be noted that with this excitation, the output voltages are in phase and the input power is divided equally.

When either port 1 or 2 is excited, operation of the hybrid is even more interesting. For example, if port 2 is driven, the current in the secondary winding nearest to the excited node must equal I; thus the voltage induced in that winding must equal V. Since the transformer is ideal, the current in the primary winding must also equal I; hence a voltage V will be present at port 4. Since there is a

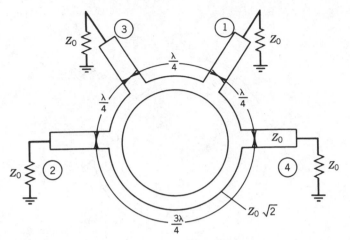

Figure 7.33 Ring or ''ratrace'' hybrid circuit topology for microstrip or stripline applications.

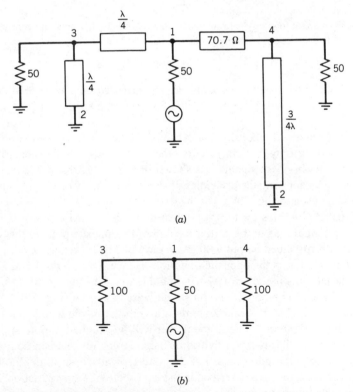

Figure 7.34 Equivalent circuit of ring hybrid with port 1 excited and with ports 3 and 4 as outputs; (*a*) transmission-line model with port 2 as a virtual ground; (*b*) equivalent circuit at center frequency.

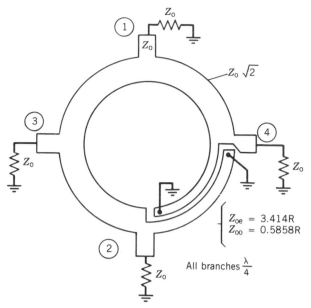

Figure 7.35   Circuit topology for broadband ring hybrid employing an inversely connected transmission line in order to achieve an additional 180° phase "flip."

voltage V across the primary winding, there must also be a voltage V developed in the remaining winding. From Kirchhoff's laws, the voltage at port 1 must be zero.

A fourth excitation mode, shown later to be the key to the operation of balanced mixers employing transformers, occurs when both ports 1 and 2 are driven with equal-amplitude in-phase signals. As shown in Fig. 7.41, this mode of operation can be solved by applying the principles of superposition. By employing an analysis similar to that used in Fig. 7.40c for the case when port 2 was excited, the current and voltage relationships for the case when port 1 is excited can also be obtained (Fig. 7.40d). Again, as in the earlier example, the opposite port, in this case port 2, exhibits a zero-valued output voltage. A current I is, however, developed in the primary winding but in the opposite direction. Hence, when ports 1 and 2 are both excited, the currents in the primary winding due to each source cancel, and all the power is summed to port 3 with no power delivered to port 4 (Fig. 7.41). It should also be noted that if ports 1 and 2 are driven with out-of-phase signals, no power will be delivered to port 3 and all the power will be summed to port 4.

There are also other types of hybrid or balun structures that can be used in the design of single-balanced mixers, but they are typically used in double-balanced designs; therefore, they are described in the next section. Many variations of transformer hybrids involving multiple cores and coaxial windings also exist, but they are typically employed at frequencies below several hundred megahertz and when very low loss or high power-handling capability is desired.

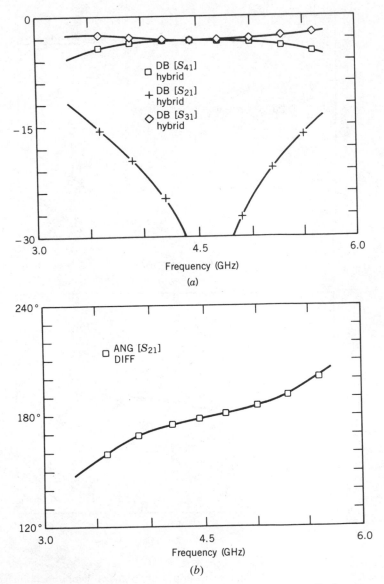

Figure 7.36 Microstrip ring (ratrace) hybrid performance: (*a*) amplitude response; (*b*) differential phase performance.

᠈ Now that the operation of hybrids is understood, the design philosophy for single-balanced mixers can be described. Regardless of the frequency range involved and the method of construction, single-balanced mixers can be classified as either of two types: (1) designs utilizing 90° hybrids, and (2) designs based on 180° hybrids. Which approach is taken depends strongly on the system application, since spurious performance and isolation differ considerably between the two

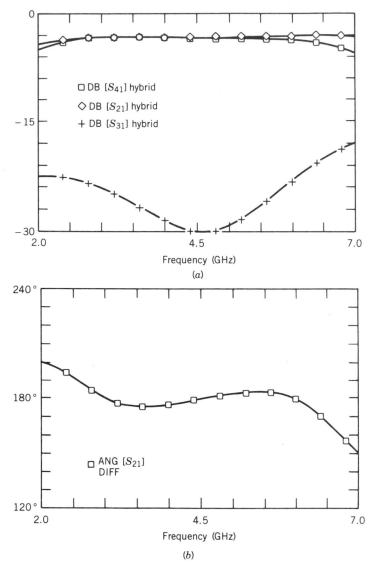

Figure 7.37 Enhanced bandwidth microstrip ring (ratrace) hybrid performance: (*a*) amplitude response; (*b*) differential phase performance.

mixer types. Single-balanced mixers have the added advantage of being able to provide some LO noise rejection in the receiving chain. AM noise rejection would be complete if every component in the mixer were perfect, such as diode matching and hybrid balance and isolation; however, rejection values of 20 to 30 dB can be obtained with practical designs [7.16].

To gain some insight into the performance variations of single-balanced mixers, we will analyze several configurations. The mixer topology illustrated in Fig. 7.42*a*

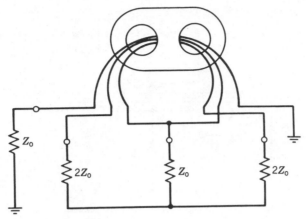

Figure 7.38 Transformer hybrid constructed on a binocular ferrite core with trifilar windings.

employs a 180° hybrid and two series-connected diodes. If the RF signal is applied to the sum port (port 1 of the ring hybrid shown in Fig. 7.33), the signal voltages at the diode ports will be in phase. Similarly, if the LO signal is applied to the delta port (port 2 of the ring hybrid), the pump voltage at the diode ports will be out of phase. In addition, with a perfect hybrid, no LO signal will be present at the RF port (port 1), regardless of whether the diodes are matched to the ring, as long as the diodes are identical. Hence when matched diodes are employed, the L-to-R isolation of the mixer is essentially that of the hybrid. Thus this configuration usually offers superior L-to-R isolation performance to mixers employing 90° hybrids. However, the VSWR at the LO and RF ports of the ring hybrid will be that of the diode impedance mismatch in the system, which is unlike the case of quadrature coupler designs in that some improvement in port VSWR occurs beyond that obtained with the diode matching network.

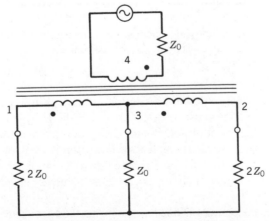

Figure 7.39 Wire diagram of 180° transformer hybrid.

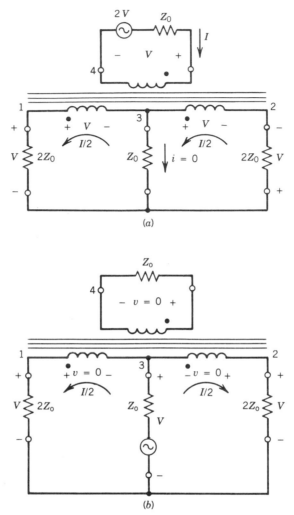

Figure 7.40 Voltage and current conditions in a transformer hybrid for various port excitations: (*a*) voltage source at port 4; (*b*) voltage source at port 3; (*c*) voltage source at port 2; (*d*) voltage source at port 1.

Because of the phase relationships involved between LO and RF signals, the correct diode orientation must be selected so that the IF signal does not cancel. Since one of the diodes is reverse connected from the other, and the LO signals are 180° out of phase, the diode conductance waveforms are in phase. The resultant IF waveforms from each diode are in phase due to the in-phase application of RF voltage. Thus the IF signal may be extracted at the node between the two diodes (Fig. 7.42*b*).

The situation concerning LO noise is somewhat different. The AM noise voltage present on the pump waveform enters the mixer at the delta port and appears at

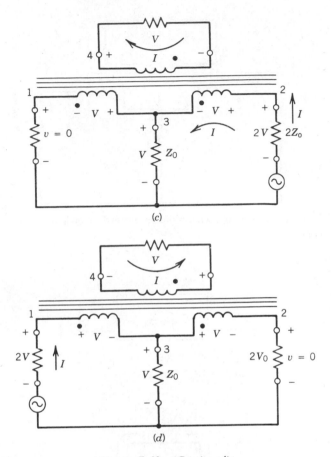

Figure 7.40 (*Continued*)

each diode out of phase. Since the conductance waveforms are in phase, the converted noise-generated products cancel at the mixer's IF port (Fig. 7.42c). In practice, complete cancellation does not occur because of imperfect diode matching and hybrid fabrication.

Determining the spurious performance is a bit more involved, but a qualitative analysis can be used to determine relative spurious characteristics for various topologies. The analysis begins by expressing the diode current, in one of the mixer arms, as an infinite series of the form

$$I_1 = aV_1 + b(V_1)^2 + c(V_1)^3 + d(V_2)^4 + \cdots \qquad (7.51)$$

where V_1 is the total diode voltage ($V_{rf} + V_{lo}$). Reversing a diode is equivalent to reversing the applied voltage, which in turn changes the sign of the odd terms in the current expression. Thus the current for the other diode can be expressed as

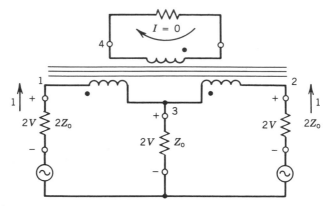

Figure 7.41 Voltage and current conditions in a transformer hybrid when ports 1 and 2 are excited in phase.

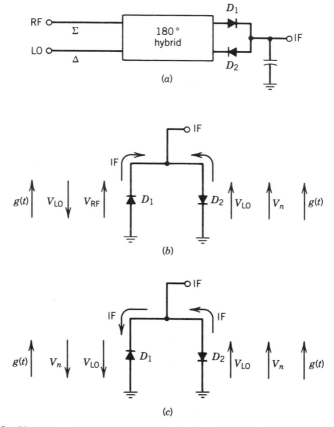

Figure 7.42 Phase relationships between LO, IF, RF, and AM noise voltages in single-balanced mixer: (a) mixer topology; (b) IF current summation; (c) AM noise cancelation.

$$I_2 = -aV_2 + b(V_2)^2 - c(V_2)^3 + d(V_2)^4 - \cdots \qquad (7.52)$$

where V_2 is again the total diode voltage. Remembering that the LO signal is applied to the delta port, the voltages V_1 and V_2 can be expressed as

$$V_1 = V_{lo} \cos \omega_p t + V_{rf} \cos \omega_s t \qquad (7.53)$$

$$V_2 = V_{lo} \cos \omega_p t + V_{rf} \cos \omega_s t \qquad (7.54)$$

Substituting the expressions for diode voltage (7.53) and (7.54) into the expressions for diode current (7.51) and (7.52), and remembering that the current at the IF node is equal to

$$I_{if} = I_1 - I_2 \qquad (7.55)$$

because one of the diodes is reversed, a qualitative spurious performance can be obtained. It can be shown that spurious responses arising from products of $mf_{rf} + nf_{lo}$ where m and n are even [(2,2), (4,4), etc.] are eliminated, and the (2,1) but not the (1,2), where $m = \pm 2$ and $n = \pm 1$, is eliminated. If the roles of the sum and delta ports are reversed (i.e., the LO voltage at the diode ports are in phase and the RF voltage is out of phase, the conversion-loss performance, even-order spurious response, and isolation characteristics will not change. However, the (2,1) spurious product will be suppressed but not the (1,2). Hence the system performance must determine which port is used for LO injection. To determine exact spurious levels, a nonlinear mixer analysis must be performed.

The analysis of a single-balanced mixer realized with a quadrature coupler (Fig. 7.43) may be analyzed in a similar manner. As before, the hybrid is used to inject the LO and RF signals into each diode, which can be treated, as in the preceding example, as separate mixers. However, in this case, the RF and LO signals at the diode ports differ by 90°. This phase relationship is best illustrated with the aid of the phasor diagram of Fig. 7.44. As can be seen in the diagram, at one arm of the hybrid the LO voltage leads the RF voltage, while at the other diode port, the RF voltage leads. But if one of the diodes is reversed, the phase difference between the diode conductance waveform and the applied RF voltage, at any instant of time, is the same for both diodes. Since this phase difference is the same for each diode, the IF currents at the node between both diodes will be summed in phase; thus the IF signal can be extracted at that point in the circuit.

Aside from the differences in isolation and VSWR mentioned previously, the

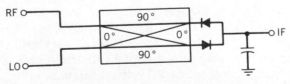

Figure 7.43 Single-balanced mixer employing 90° hybrid.

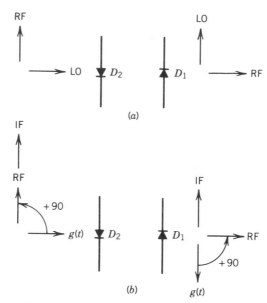

Figure 7.44 Phasor diagram illustrating LO, RF, and IF voltages: (*a*) RF and LO phase relationships at each diode; (*b*) phase relationships between RF, IF, and conductance waveform.

conversion-loss characteristics with this approach are identical to those of a mixer realized with a ring hybrid. The even-order spurious performance is also equivalent, but there is no suppression of either the (1,2) or (2,1) spurious products.

7.4 DOUBLE-BALANCED MIXERS

The most commonly used mixer in the microwave portion of the spectrum is the double-balanced mixer. It usually consists of four diodes and two baluns or hybrids, although a double-ring or double-star design requires eight diodes and three hybrids. The double-balanced mixer has better isolation and spurious performance than the single-balanced designs described previously, but usually requires greater amounts of LO drive power, are more difficult to assemble, and exhibit higher conversion loss. However, they are usually the mixer of choice because of their spurious performance and isolation characteristics.

A typical single-ring mixer with transformer hybrids is shown in Fig. 7.45. With this configuration the LO voltage is applied across the ring at terminals L and L', and the RF voltage is applied across terminals R and R'. As can be seen, if the diodes are identical (matched), nodes R and R' are virtual grounds; thus no LO voltage appears across the secondary of the RF transformer. Similarly, nodes L and L' are virtual grounds for the RF voltage; hence, as before, no RF voltage appears across the secondary of the LO balun. Because of the excellent diode

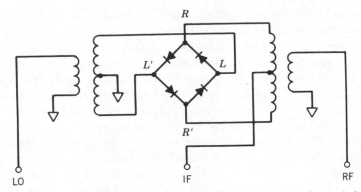

Figure 7.45 Circuit diagram of double-balanced mixer with transformer hybrids.

matching that can be obtained with diode rings fabricated on a single chip, the L-to-R isolation of microwave mixers can be quite good, typically 30 to 40 dB.

We can now analyze the performance of the mixer by applying the techniques described in Section 7.3 for a single-balanced mixers. With the LO and RF excitations shown in Fig. 7.46a, the IF currents for each diode can be qualitatively determined by analyzing the phase relationships between the conductance waveform for each diode and the applied RF voltage.

To illustrate this concept, we will now analyze the performance of each diode. When the LO and RF voltages are applied as shown, the RF voltage and LO conductance waveform for diode D1 are in phase; thus we can define an IF current

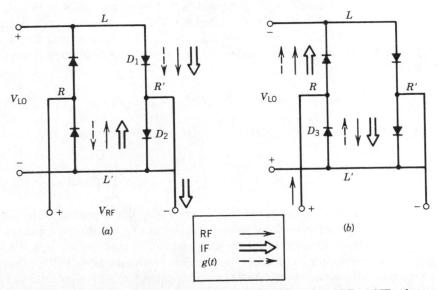

Figure 7.46 Phase relationships between LO, conductance waveform, RF, and IF voltages: (a) diodes D1 and D2 conducting; (b) diodes D3 and D4 conducting with LO signal reversal.

in the direction of the diode symbol. Using the same convention, the RF signal and conductance waveform for diode D2 are out of phase; hence the IF current is reversed from that of the first diode. Since diodes D3 and D4 conduct when the LO voltage polarity is reversed (Fig. 7.46b), their conductance cycles are reversed from that of diodes D1 and D2. Hence the IF currents in diodes D3 and D4 are reversed from the RF signal. With this LO voltage sense, the IF current can be summed at nodes R and R'. By reversing the LO voltage and applying the same analysis convention, the IF current relationships for the diode ring are obtained. This time the IF current can be summed at nodes L and L', or the negative-going current can be summed at nodes R and R'. If the transformer hybrid performs as described in Section 7.3, the IF signal will combine at the center tap of the secondary winding of each balun. The IF signal can be extracted at either tap as long as the opposite tap is used to complete the IF circuit. No IF signal will be present at either LO or RF port because no current is induced in the primary windings since the IF currents enter the arms of the balun in phase (even mode).

The analysis above is also valid if the balun or hybrids used in the mixer design have no center tap, which is usually the case at microwave frequencies above a few gigahertz; but the IF currents must be externally summed and prevented from being terminated by the RF and LO source impedances. This is commonly done in four-diode broadband double-balanced mixers used in the upper portion of the microwave spectrum.

The mixer circuit above can be analyzed further so that some insight into the spurious performance and embedding impedances for the diodes can be obtained. This analysis is easily accomplished because of the inherent isolation in the circuit, which allows the RF and LO analyses to be conducted separately.

If a transformer hybrid is designed with a 2:1 turns ratio (Fig. 7.38), as is commonly done, each half of the secondary winding can be modeled as a voltage source V_{lo} with a source impedance of $2Z_{lo}$ (usually 100 Ω) connected across an antiparallel diode pair as shown in Fig. 7.47a. Thus the LO current in diode D1 can be expressed as

$$I_1 = a[V_{lo}] + b[V_{lo}]^2 + c[V_{lo}]^3 + d[V_{lo}]^4 + \cdots \qquad (7.56)$$

and the LO current in diode D2, since it is reversed from diode D1, can be expressed as

$$I_2 = a[-V_{lo}] + b[-V_{lo}]^2 + c[-V_{lo}]^3 + d[-V_{lo}]^4 + \cdots \qquad (7.57)$$

If we define the total LO current as $I_1 - I_2$, only odd-order terms will be present. However, all harmonics will be present in each diode, but the even harmonic currents will be equal and opposite for each antiparallel diode pair, giving rise to zero-valued embedding impedances for even LO harmonics as well as for the dc component. Of course, the odd harmonic embedding impedances are still determined by the balun. As in the past example of the single-balanced mixer, double-balanced mixers reject spurious responses produced by $mf_{rf} + nf_{lo}$ frequency

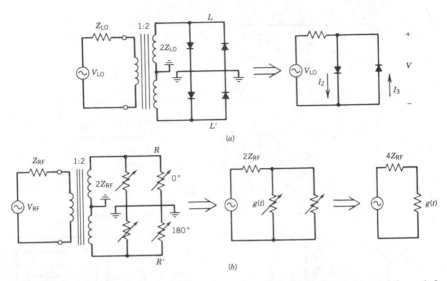

Figure 7.47 LO and RF equivalent circuits for single-ring mixer: (*a*) model used for spurious analysis; (*b*) RF embedding impedance circuit model.

components when m or n are even, and provide no suppression of odd-ordered products. By employing a similar analysis, the embedding impedances at the RF frequency (Fig. 7.47*b*) for each diode can be shown to be $4Z_{rf}$ (typically, 200 Ω). At the IF frequency, all four diodes are connected in parallel; hence the embedding impedance for each diode is $4Z_{if}$ (typically, 200 Ω).

Another version of the double-balanced diode ring mixer, which is essentially its dual, is the star mixer. A transformer hybrid realization of the star mixer is shown in Fig. 7.48. When transformer hybrids are employed in the mixer design, no advantages over conventional approaches are obtained, although some clever mixer circuits have been built with coaxial and microstrip baluns. These microstrip or coaxial mixers are sometimes easier to assemble than conventional realizations.

An extension of the double-balanced ring mixer is the double-ring or double-star double-balanced mixer, referred to occasionally as the double-double-balanced mixer. As can be seen in Fig. 7.49, multiple secondary windings are required on the LO and RF baluns, and an additional IF balun is also needed. Because of the four extra diodes, twice as much pump power is required, but since the RF signal is divided eight ways instead of four, the LO-to-RF signal power ratio at each diode is greater, thus extending the mixer's large-signal saturation level. The conversion loss characteristics for a double-ring mixer are approximately equal to that of single-ring designs; hence a true improvement in receiver dynamic range can be obtained, at the expense of LO power, by using this approach. In addition, the IF balun, although adding to the circuit complexity, does allow for completely independent IF extraction regardless of its frequency range provided that the balun still functions. This is another subtle advantage of double-ring and double-star

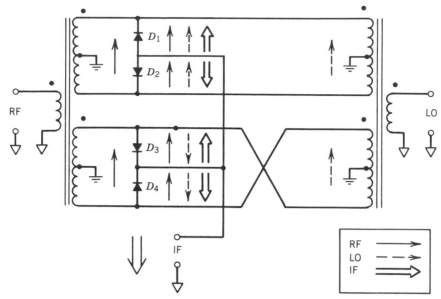

Figure 7.48 Circuit diagram of star mixer with multiple secondary transformer hybrids.

mixers. Occasionally, double-ring mixers are constructed with diode rings that employ multiple series diodes in each leg in order to increase the mixers compression point. Unfortunately, the extra diodes in the ring(s) do degrade the conversion loss, but at frequencies below 30 mHz, where receiver sensitivity is limited by atmospheric noise, and where signal levels at the antenna terminals can reach several hundred millivolts, the extra dynamic range is usually welcomed.

Up until now, we have described the operation of double-balanced mixers that employ transformer hybrids. As the frequency of mixer operation is extended beyond several gigahertz, transformer hybrids can no longer be fabricated. At these frequencies, transmission-line baluns begin to dominate and by time the operating frequency reaches 4 or 5 GHz, a few coupled inductors, fabricated on GaAs substrates, are the only structures resembling transformers that are found.

The simplest of baluns can be realized by employing a length of balanced transmission line 90° in length, connected so that one conductor at the unbalanced end is grounded, with the balanced load (or source) connected across both conductors at the other end (Fig. 7.50a). The structure is indeed a balun, since the current in one terminal at the balanced end is equal and opposite to the current in the other terminal. At the center frequency of the structure, the balanced end is also isolated from ground. Baluns of this type can easily be realized, for example, with insulated twisted wire, such as enameled magnet wire, parallel-plate transmission line, or edge-coupled parallel coplanar strips. Another form of quarter-wavelength balun can be constructed with coaxial cable as shown in Fig. 7.50b. In this case an additional conducting jacket is added around the outside of the cable to form a second transmission line of a noncritical impedance value. If the outer shield is

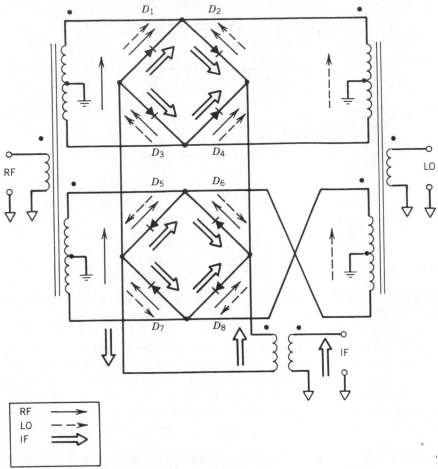

Figure 7.49 Low-frequency representation of double-double-balanced mixer.

connected to the original coaxial cable's jacket 0.25 wavelength away from the balanced end, a short-circuited transmission line is formed. The resultant quarter-wavelength shorted line effectively chokes any current that may tend to exist on the outside of the original transmission line's shield, thus preventing unbalancing and forcing the currents at the balanced end to be equal and opposite in the center conductor and shield. Although these structures are simple, they exhibit limited bandwidth and are of little use in broadband mixer design.

A clever coaxial balun [7.17], which can be designed to exhibit many decades of bandwidth, is the unbalanced-to-balanced coaxial transition (Fig. 7.51a). This structure was first realized by gradually removing the shield along the length of a coaxial cable until only two point connections (terminals) for the center conductor and outer jacket exist at the balanced end. This structure exhibits a very soft low-frequency cutoff point which is reached when the length is approximately 180° at

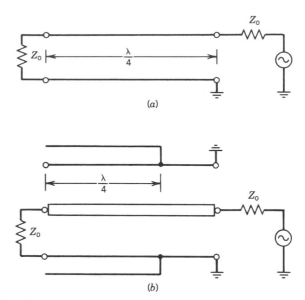

Figure 7.50 Simple balun structures: (*a*) balanced-line quarter-wavelength-long balun; (*b*) coaxial balun.

the lowest frequency of operation. However, good performance can be obtained at frequencies several octaves lower. A very popular version of this type of balun is the microstrip-to-parallel plate line transition [7.18, 7.19] (Fig. 7.51*b*) commonly used in many mixers and antenna feed structures. This structure, which can be built with various impedance contours and ground plane tapering schemes, also exhibits high-pass performance with a very soft low-frequency cutoff characteristic.

Transmission-line structures which are naturally balanced, such as slotline and finline, can also be used as balanced feed in mixer design. However, all of the structures above, and the more complex transmission-line structures to follow, exhibit one major drawback compared to a transformer hybrid: There is no true RF center tap. As will be seen, this deficiency in transmission-line structures, extensively complicates the design of microwave balanced mixers.

The lack of a balun center tap does indeed complicate the extraction of IF energy from the structure, but if the IF frequency is low, diplexing can be employed to ease performance degradation. This concept is illustrated in the following design example of a double-balanced 2- to 12-GHz mixer. It will be assumed that because of the soft-substrate transmission-line media and frequency range, a packaged diode ring with known impedances will be used. For Si diodes in this frequency range, the typical LO impedance range (magnitude) is on the order of 75 Ω, while the RF impedance is approximately 50 Ω. With these values in mind, microstrip-to-parallel plate transmission-line baluns similar to the design illustrated in Fig. 7.51*b* can be fabricated on soft-substrate material. For baluns of this type that require impedance transforming, a Dolph–Tchebycheff contour of impedance values along the balun's

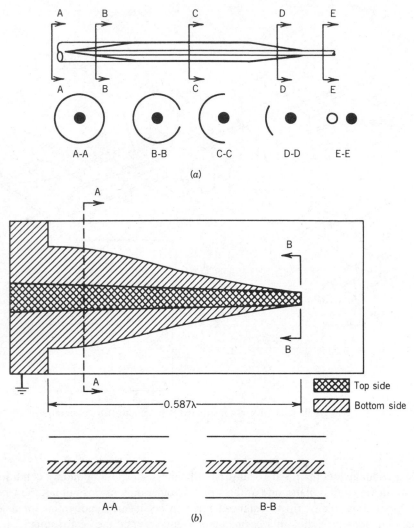

Figure 7.51 Broadband high-pass balun structures: (*a*) coax to balanced line; (*b*) microstrip to parallel-plate line.

length yields excellent broadband performance. These types of tapers can be approximated by dividing the balun into at least 20 segments, with each segment being equally spaced along its length. The cross-section dimensions, which can be analyzed as asymmetrical broadside suspended coupled strips, are adjusted so that the desired impedance contour is achieved. It should be noted that the dimensions at each end of the balun are fixed; that is, the balanced end must terminate in a parallel-plate transmission line, and the unbalanced end must terminate as microstrip. However, the ground-plane width at the microstrip end must only be wide enough to simulate microstrip performance, which can usually be accomplished

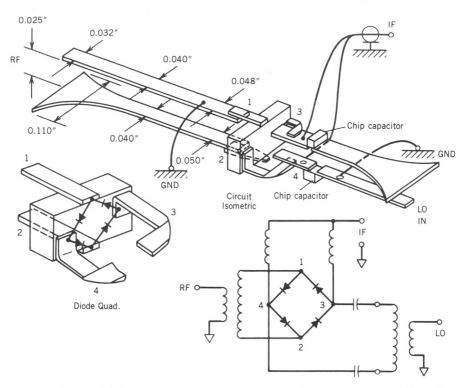

Figure 7.52 Double-balanced mixer constructed with microstrip to parallel-plate line baluns.

with a width greater than 5 to 10 substrate thicknesses. These boundary conditions constrain the ground-plane side of the structure taper dimensions but not the taper. The taper from microstrip to balanced line can be linear, exponential, or determined empirically, although a cosine taper seems to yield the best results.

The completed mixer assembly with package diode ring and its equivalent circuit are shown in Fig. 7.52. As can be seen, both the RF and LO baluns terminate at the diode ring and provide the proper excitation. But since there is no center tap, the IF must be summed from the top and bottom of either balun. This summing is accomplished with bond wires that have high reactances at microwave frequencies but negligible inductances at the IF. Blocking capacitors form the second element in a high-pass filter, preventing the IF energy to be dissipated externally. An IF return path must also be provided at the terminals of the opposite balun. The top conductor side of the balun is grounded with a bond wire, providing a low-impedance path for the IF return and a sufficiently large impedance in shunt with the RF path. The ground-plane side of the balun provides a sufficiently low impedance for the IF return from the bottom side of the diode ring. The balun inductance and blocking capacitor also form a series resonant circuit shunting the IF output; therefore, this resonant frequency must be kept out of the IF passband.

The upper-frequency limit of mixers fabricated using tapered baluns and low parasitic diode packages, along with a lot of care during assembly, can be extended to 20 GHz. Improved "high-end" performance can be obtained by using beam-lead diodes. Although this design technique is very simple, there is little flexibility in obtaining an optimum port VSWR since the baluns are designed to match the magnitude of the diode impedance. The IF frequency response of using this approach is also limited, due to the lack of a balun center tap, to a frequency range below the RF and IF ports.

A slightly more complex balun structure can be formed by combining two quarter-wavelength-long coupled line pairs as illustrated in Fig. 7.53. Without any compensation, this structure can exhibit greater than an octave of bandwidth and can be realized in a variety of media, such as coaxial cable, microstrip, and parallel-plate transmission line. The structures typically demonstrate excellent phase balance between ports, due to the nature of the coupling mechanism and good amplitude balance. The input VSWR on the unbalanced port is the limiting performance factor.

However, when this structure is realized with coaxial cable, currents existing on the outside of the jacket unbalance the system and are unpredictable. An outer conductor can be added, as was done on the simple coaxial balun illustrated in Fig. 7.50b, to eliminate unwanted currents and radiation and to restore balance. The resultant structure, developed by Marchand [7.20] in 1944 and illustrated in Fig. 7.54a, with its circuit model shown in Fig. 7.54b, is sometimes referred to as the compensated Marchand balun. The structure can be realized with as many as four different impedance transmission lines with different lengths. Hence a considerable amount of flexibility in matching is possible since the balun is a

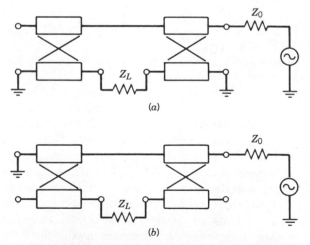

Figure 7.53 Planar-coupled line balun structures.

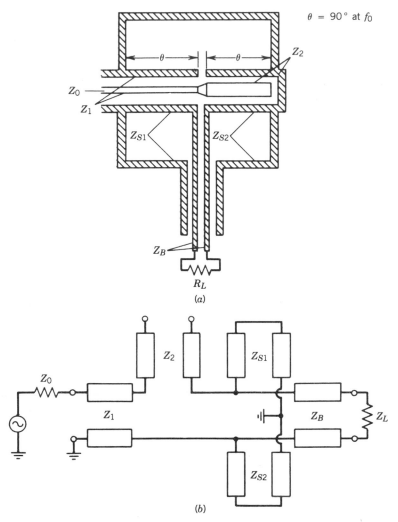

Figure 7.54 Marchand compensated balun: (*a*) coaxial cross section; (*b*) equivalent transmission-line model.

multielement bandpass network. Usually, Z_1 and Z_2 are designed to be of equal value, and Z_{s1} and Z_{s2}, which are effectively in series and then shunted across the balanced load, are made as large as possible. Transmission line Z_b has a characteristic impedance value equal to that of the balanced termination, although it can be used as a matching section. If proper filter synthesis methods are employed in the design of the compensated balun, excellent multioctave performance can be obtained. Coaxial structures are, however, cumbersome to integrate into mixer topologies, but some designs employ simpler versions of the foregoing structure that employ the mixer housing to form the outer balun shield.

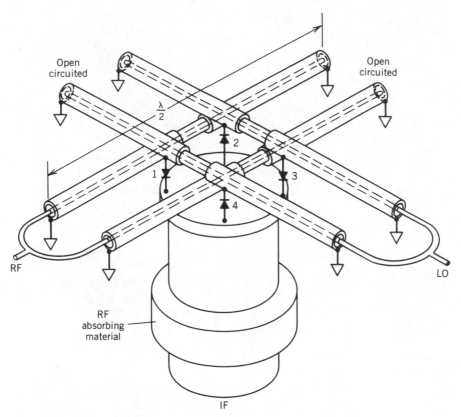

Figure 7.55   Octave bandwidth double-balanced mixer using coaxial compensated balun.

An excellent example of this technique, which can be used in the design of octave or multioctave bandwidth mixers, will now be illustrated. The mixers consist of dual RF and LO baluns fabricated with semirigid coaxial cable and four diodes connected in a star, rather than a ring, formation. The length of each half of the dual balun is 180° at the band center frequency, with the outer jacket of the cable cut in the center to form the balun output (Fig. 7.55). The ends of the balun are connected to ground, while the remainder of the structure is suspended. The IF energy is extracted at the center of the diode star, with the IF return path being formed by the balun outer conductor. It should be noted that the upper-frequency limit of the mixer occurs when the total length of the balun becomes 360°. The IF frequency upper limit occurs at the RF band center frequency, that is, when the IF return path is 90° in length.

Star structures of this type solve the IF overlapping problem encountered with the previous design approach and exhibit excellent octave bandwidth performance. However, extra care must be employed in selecting the diodes and balun imped-ances since there is even less flexibility in obtaining an optimum port VSWR.

A suspended substrate version of the compensated balun is easily realized with

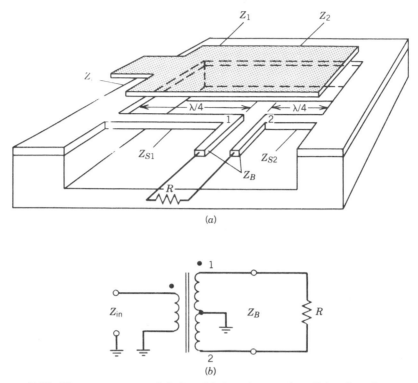

Figure 7.56 Planar compensated balun fabricated on a low-dielectric substrate: (*a*) metallization patern and package floor; (*b*) low-frequency equivalent-circuit model.

a parallel-plate transmission line [7.21]. As in the coaxial case, four distinct distributed elements exist and can be used to the designer's advantage. A typical balun, which is fabricated on a soft-substrate low dielectric medium (not shown) is illustrated in Fig. 7.56*a* and its equivalent circuit is shown in Fig. 7.56*b*. It is evident from the illustration that the balun performance is influenced by the packaging since the impedance of several elements, especially Z_{s1} and Z_{s2}, is dependent on the spacing between the substrate and the ground planes. Although, a center tap is shown in the equivalent circuit, it is a virtual center tap and is completely valid only at dc, as is the case in all distributed baluns. When the IF frequency is low, Z_{s1} and Z_{s2} form an adequate IF return path for most applications.

A dual version of the balun above can easily be constructed by adding a second conducting strip parallel to Z_{s1} and Z_{s2} (Fig. 7.57). Dual baluns [7.22] of this type have many applications, such as providing the power dividing function on a common LO port for a phase-tracking mixer pair, or providing the equivalent of two secondary windings which are required when realizing double-ring or double-star mixers. Little to no degradation in performance occurs by adding a second strip, and when the dual balun is used to feed two mixers, a small amount of isolation is obtained. There is also a subtle advantage with planar baluns of this

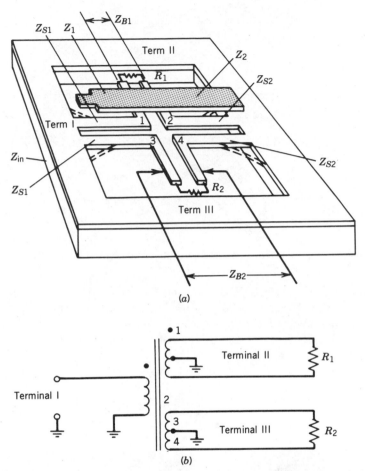

Figure 7.57 Dual planar compensated balun: (*a*) metallization pattern; (*b*) low-frequency equivalent-circuit model.

type in that the balanced output is in the plane of the substrate rather than normal to the substrate as is the case with the microstrip-to-parallel plate balun. A space quadrature relationship of this type can be useful in eliminating crossover connections which are common in multidiode designs.

When hard substrates such as quartz and alumina are used in mixer construction, the added mechanical advantage relative to soft substrates allows the designer to eliminate package parasitics by employing beam-lead or chip diodes. The reduction in parasitics, in conjunction with the elimination of excess diode lead lengths and crossover connections, dramatically extend the upper-frequency limit of the typical mixer structure. For example, the interconnections between diodes and baluns can be virtually eliminated by selecting an LO and RF balun structure with orthogonal orientations of the balanced signal, because the diodes can be mounted directly to the conductor surfaces. A typical diode and balun interconnect arrange-

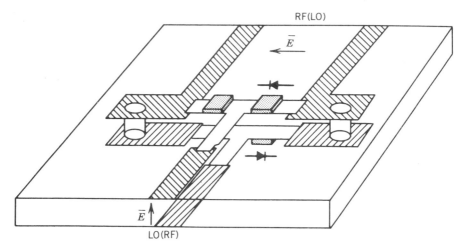

Figure 7.58 Interconnect configuration for planar orthogonal baluns and diode ring.

ment is shown in Fig. 7.58. With this arrangement a diode pair is bonded to the top surface of the substrate, while a second pair is bonded to the bottom surface. Plated-through holes are used to connect the two pairs into a conventional ring. As can be seen, there is less than 0.3 mm of excess length in the structure.

This design technique has been applied to the realization of a 20- to 40-GHz double-balanced mixer [7.23,7.24]. The mixer, which is shown in Fig. 7.59,

Figure 7.59 Double-balance mixer for 20- to 40-GHz operation fabricated on a 0.25-mm-thick quartz substrate (courtesy of Texas Instruments).

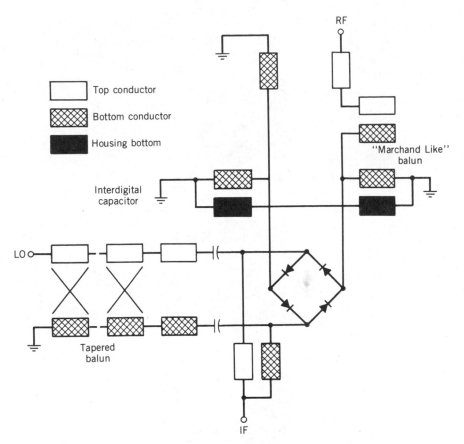

Figure 7.60 Circuit model for 20- to 40-GHz mixer.

consists of a microstrip-to-parallel plate line balun (Fig. 7.51*b*) and a compensated planar balun similar to the design illustrated in Fig. 7.56, except that both baluns are fabricated on a quartz substrate. Interdigitated capacitors were also used in the IF diplexer circuit to eliminate loss and parasitics. The circuit model for the mixer is shown in Fig. 7.60 and is similar to the models described previously. The conversion loss and isolation performance are shown in Fig. 7.61. As can be seen, the performance is comparable to that of the best low-frequency designs.

Planar-compensated baluns can also be used in the design of star mixers, provided that dual secondary windings or their equivalent are available from the balun structure. Dual secondary performance can be obtained, as described above, by adding a second coupled conductor set to the balun feed (Fig. 7.57). If two such structures are orthogonally connected within the plane of the substrate, as depicted in Fig. 7.62, and the feed structures are bridged, the resulting four-node junction will have the proper voltage excitation and phasing for a star mixer. A star mixer employing this technique with glass-packaged diodes is shown in Fig. 7.63. The IF response of mixers fabricated in this manner is quite broadband,

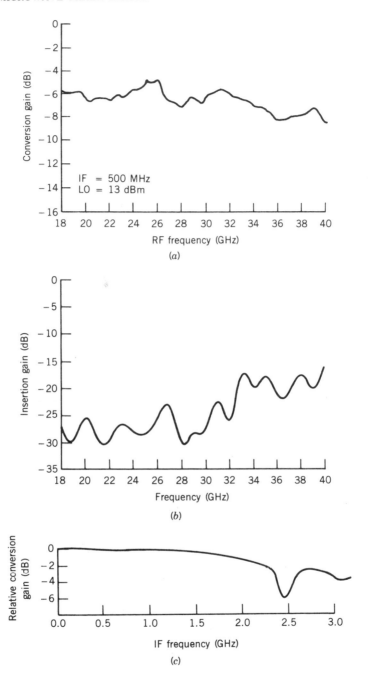

Figure 7.61   20- to 40-GHz mixer performance: (*a*) conversion loss; (*b*) isolation; (*c*) IF response.

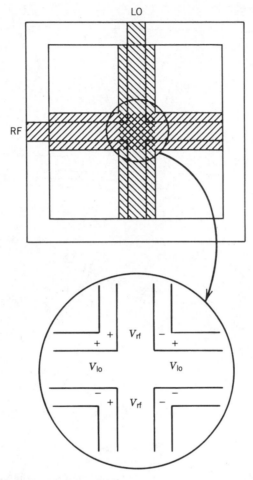

Figure 7.62 Dual balun circuit arrangement for star mixer.

exhibiting a gentle slope as a function of frequency until cutoff is reached (f_0 of balun).

By properly selecting the balun dimensions, an unbalanced source impedance of 50 Ω can be conveniently transformed anywhere within a 50- to 150-Ω impedance range. The circuit bandwidth is almost always greater than an octave and is strongly influenced by the impedance of the open-circuit transmission line in the feed structure (Z_2).

Typically, the lower this impedance can be made, the greater the bandwidth; hence the selection of a thin dielectric will tend to prevent the line widths from becoming unreasonably large when striving for the lowest possible impedance. Typical VSWR (unbalanced port) performance and a transmission-line model for the dual balun is shown in Fig. 7.64.

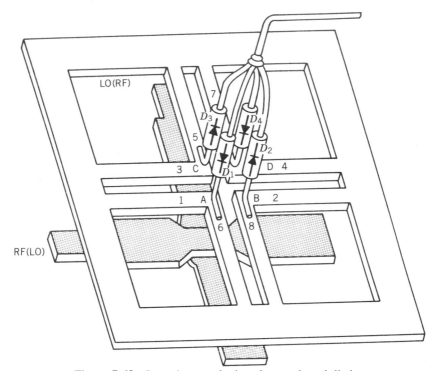

Figure 7.63 Star mixer employing glass packaged diodes.

Double-double-balanced or dual-ring mixers also require balun structures that simulate multiple secondary windings. Microstrip-to-parallel plate transmission-line baluns, similar to the configuration illustrated in Fig. 7.51*b*, if connected in parallel at the unbalanced end, produce the equivalent of the dual planar-compensated balun, with the exception of the balanced output voltage orientation. As in the single-balun case, the balance output of this type of balun is normal to the plane of the substrate. A typical circuit configuration for the balun type above is illustrated in Fig. 7.65. A phase reversal can also be placed in one arm of the dual balun (Fig. 7.65*b*) so that the arm-to-arm voltage orientation is odd. This arrangement can be advantageous, as will be shown later for some mixer topologies.

The mixers described previously, although capable of considerable bandwidth performance, in general exhibit two main drawbacks: (1) IF bandwidth and extraction difficulties, and (2) dynamic range. As we remember from the mixer descriptions above, multiple-ring structures, due to the fact that the IF can be independently extracted at the junction of the two rings, can have overlapping RF, IF, and LO frequency ranges. The extra diode ring, which implies that the mixer will require 3 dB more pump power than will a conventional single-ring design, will also exhibit 3 dB greater compression and distortion characteristics. The major drawback to multiple-ring designs is their added complexity.

When one examines the low-frequency model of a multiple-ring mixer such as

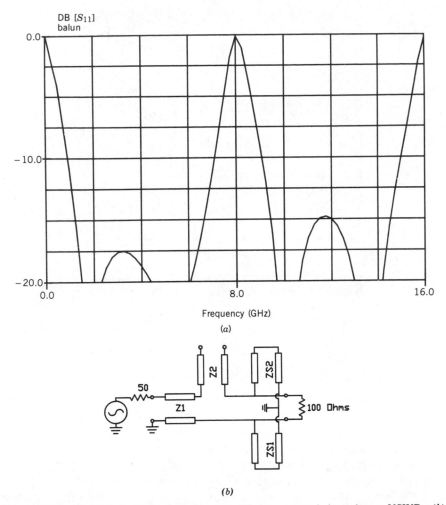

Figure 7.64 Typical dual planar compensated balun: (*a*) unbalanced port VSWR; (*b*) transmission-line model; (*c*) element values.

the one depicted in Fig. 7.49, the actual microwave construction and assembly problems are somewhat elusive. However, when Fig. 7.66 is examined, which illustrates a typical broadband double-ring design employing a variation of the microstrip-to-parallel plate line balun, the complexity becomes apparent. It should be noted that the substrate is not a single piece of material, but two separate circuits assembled orthogonally within the housing. This orientation of the substrates is not a problem when the mixer is assembled in a housing by itself, but a miserable integration results when trying to combine the mixer with other planar components.

The problem is eased somewhat by employing the approach depicted in Fig. 7.67. This structure also uses a modified version of the microstrip-to-parallel plate

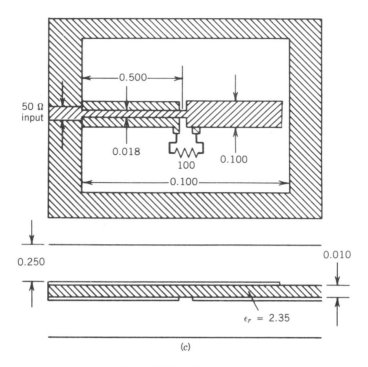

(c)

Figure 7.64 (*Continued*)

line balun described previously, with the addition of a phase reversal at the feed of one of the dual baluns. This reversal allows the substrate to be continuous, although it still needs to be suspended to preserve performance.

Both of the double-ring designs above exhibit conversion-loss characteristics similar to those of single-ring designs (6 to 10 dB) when more pump power is available, with the added advantage of overlapping frequency ports and higher dynamic range. These designs also exhibit the same RF and LO bandwidth capabilities as those of single-ring designs, with several decades of performance not uncommon. However, as one would expect, the IF bandwidth is limited by the IF balun performance. The ferrite baluns shown can be made to operate up to several gigahertz, provided that the total wire length is less than a half-wavelength. If greater bandwidth is desired, a microstrip-to-parallel line balun can be used for microwave IF operation, since its low-frequency cutoff must still be preserved, but unfortunately, the substrate orientation will again, in all likelihood, need to be orthogonal to the main substrate.

The potential for improved dynamic range is the other main advantage of double-ring structures. However, dynamic range is also a strong function of the diode characteristics as well as circuit topology. The high-level performance of any mixer topology can be improved by increasing the diode junction potential. There are a variety of high-barrier diodes available commercially, which have junction potentials greater than 0.6 V. Since most balanced mixer designs employ low-barrier

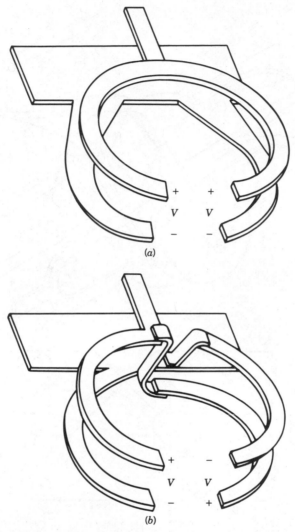

Figure 7.65 Dual microstrip to parallel plate line baluns: (*a*) even excitation; (*b*) odd excitation.

diodes with junction potentials on the order of 0.3 V, an improvement in compression characteristics of 6 dB is feasible, provided that an additional 6 dB of pump power is available.

For example, a typical double-ring design employing conventional high-barrier diodes would require approximately 20 dBm of LO drive; hence the maximum input signal level would be approximately +15 dBm before entering the region of heavy saturation. As the pump power is increased, the compression level of the mixer would also increase until full LO saturation of the diode(s) is reached, which is usually on the order of 23 to 27 dBm. A further increase in LO power would

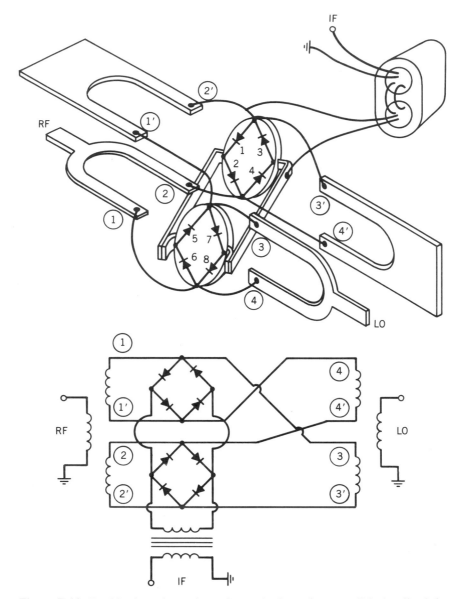

Figure 7.66 Double ring mixer using orthogonal microstrip to parallel-plate line baluns.

result in performance degradation, due partially to the effective increase in diode R_s.

Multiple diode combinations, such as series-connected junctions, can be used to further raise the compression characteristics of a mixer because the effective junction potential of the pair is increased. Various combinations of high- and low-barrier diodes can be used to optimize the mixer's performance for a particular set

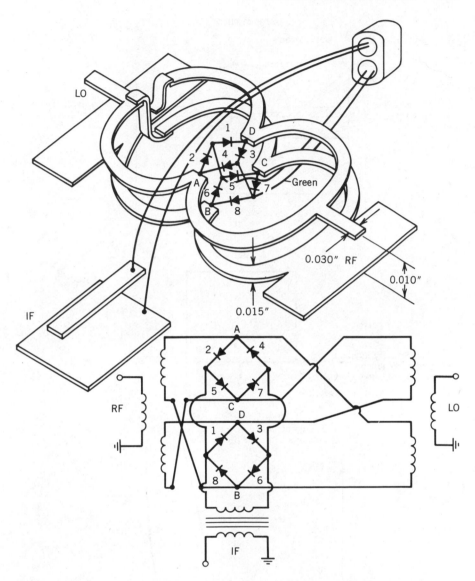

Figure 7.67 Planar double-ring mixer with microstrip to parallel-plate line baluns.

of signal and low-power levels. As in the case of single high-barrier devices, multiple diodes require as much or more LO drive, but the compression point is also raised.

An interesting high-level diode available from MACOM [7.25] is uniquely fabricated with three junctions. The diode consists of three junctions, two *p–n* junctions and a conventional Schottky barrier. A diode cross section depicting its construction and a simplified circuit model are shown in Fig. 7.68*a* and *b*. The

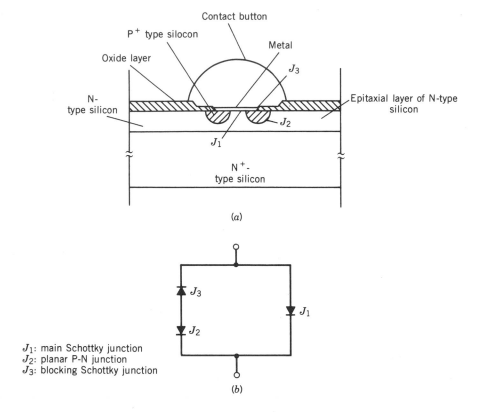

(a)

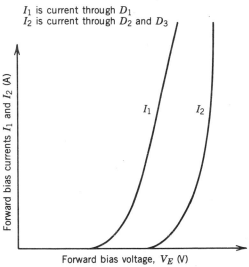

J_1: main Schottky junction
J_2: planar P-N junction
J_3: blocking Schottky junction

(b)

I_1 is current through D_1
I_2 is current through D_2 and D_3

Forward bias currents I_1 and I_2 (A)

I_1

I_2

Forward bias voltage, V_E (V)

Figure 7.68 Triple junction diode: (a) diode cross section; (b) simplified circuit model; (c) relative *I–V* characteristics of junctions J1 and J2/J3 (courtesy of M/A-COM[7-25]).

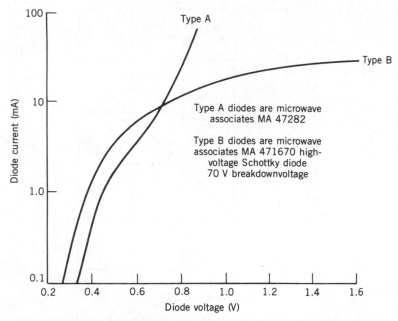

Figure 7.69 Typical diode I–V characteristics: type A, conventional high-barrier diode: type B, triple junction diode (courtesy of M/A-COM[7.25]).

I–V characteristics for junction J2 and the series combination J2 and J3 are compared in Fig. 7.68c. The effective turn-on potential for the J2/J3 combination is considerably higher than for the main junction J1; hence at low bias voltages it is essentially off. A comparison between the I–V characteristics of a typical Schottky high-barrier diode (type b), and the I–V characteristics of the composite triple-junction high-barrier diode (type a) is shown in Fig. 7.69. It is evident from the data that under large-signal conditions, the conventional high-barrier diode (type B) begins to saturate severely at a forward voltage of about 0.5 V, while the triple-junction device is still unsaturated well above a forward voltage of 0.8 V. Hence as LO power is increased, the effective R_s of the type B diode rapidly increases, causing a severe degradation in performance.

A star mixer of the type illustrated in Fig. 7.63, employing othogonally connected planar-compensated baluns, was fabricated and evaluated as a high-level up-converter by Hallford [7.26]. To compare their performance, conventional high-barrier and the triple-junction diodes were evaluated in the same mixer structure. Although the mixer was capable of octave-band performance, it was used as a high-level up-converter in the frequency range 70 MHz to 1500 to 2000 MHz. With conventional high-barrier diodes, a single-sideband output power of 15 dBm was obtained with 28 dBm of LO power and an IF signal level of 22.5 dBm. When triple-junction diodes were used (type A), the performance of the mixer improved dramatically. The compression characteristics of the up-converter are shown in Fig. 7.70 for a LO power of 25 dBm. By increasing the LO power

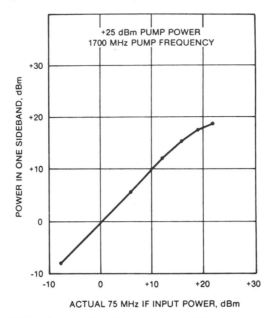

Figure 7.70 Compression characteristics of high-level up-converter.

to 27 dBm, the performance was further enhanced. The two-tone distortion characteristics at this LO drive level are shown in Fig. 7.71.

Although the 35-dBm third-order intercept point is impressive, a careful examination of the measured results reveal a more interesting phenomenon. Note that with an IF signal input level of 22.5 dBm and a LO power level of 27 dBm, a single-sideband output power of 21 dBm was measured. These levels correspond to a double-sideband conversion gain of 1.5 dB in a passive mixer. Although this result is unusual, it should be remembered that the the total power output (DSB) of 250 mW is still substantially less than the total input power (LO + signal) of 680 mW.

It is apparent that some process other than resistive mixing is occurring to account for the conversion performance. Considering that the diode is quite large ($C_{j0} = 1.5$ pF and $R_s = 1.4$) and pumped quite heavily, perhaps some parametric conversion mechanism can account for the added performance. High-level mixers in general exhibit better spurious performance than that of conventional small-signal designs. Unfortunately, calculating or determining the absolute spurious performance is difficult. In fact, the only sure method is to measure the mixer in question embedded in the actual system. Approximate tables and formulas are available throughout the literature, but they assume that the mixer baluns are indeed baluns at harmonics of the LO and RF signals and that broadband resistive terminations are present on all ports. When broadband mixers are used in the lower portion of their operating range, the conditions above may be approximated. But at frequencies above 18 GHz, diode balance, balun performance, and terminations

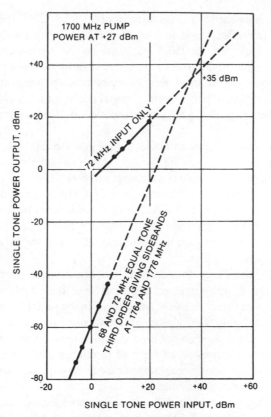

Figure 7.71 Two-tone distortion performance of high-level up-converter.

are surely in question, even when a low-order spurious response such as the (2, 2) is involved.

All the broadband mixer designs that employ suspended substrate techniques exhibit strong interactions with their packages. It is for this reason that it is very common to find the housings of commercially available mixers loaded with lossy ferrite materials. The loading materials not only dampen resonances, but also de-Q the circuit transmission lines. By lowering the circuit Q, the mixer will operate closer to the transmission zeros of the circuit, thus increasing the bandwidth but degrading the overall conversion-loss characteristics. However, if suspended broadband structures are constructed without loading materials, sharp resonances will usually occur somewhere within the mixer passband.

7.5 FET MIXER THEORY

Interest in JFET mixers has been very strong due to their excellent conversion gain and intermodulation characteristics. Numerous commercial products employ JFET mixers, but as the frequency of operation approaches 1 GHz, they begin to disappear. At these frequencies and above, the MESFET can easily accomplish the conversion functions that the JFET performs at low frequencies. However, the performance of active FET mixers reported to date by numerous authors has been somewhat disappointing. In short, they have not lived up to expectations, especially concerning noise-figure performance, conversion gain, and circuit-to-circuit repeatability. Hence, in the hybrid circuit world, most mixer applications are left to the diode.

Recently, growing interest in GaAs monolithic circuits is again beginning to heighten interest in active MESFET mixers. This is indeed fortunate, since properly designed FET mixers offer distinct advantages over their passive counterparts. This is especially true in the case of the dual-gate FET mixer; since the additional port allows for some inherent LO-to-RF isolation, it can at times replace single balanced passive approaches. The possibility of conversion gain rather than loss is also an advantage, since the added gain may eliminate the need for excess amplification, thus reducing system complexity.

Unfortunately, there are some drawbacks when designing active mixers. With diode mixers, the design engineer can make excellent first-order performance approximations with linear analysis; also, there is the practical reality that a diode always mixes reasonably well almost independent of the circuit. In active mixer design, these two conditions do not hold. Simulating performance, especially with a dual-gate device, requires some form of nonlinear analysis tool if any circuit information other than small-signal impedance is desired. An analysis of the noise performance is even more difficult.

In the following section(s), a theory that describes the operation of single- and dual-gate mixers is developed. Emphasis is on the practical implementation of FET mixers for microwave frequency ranges, but techniques for VHF and UHF applications are also described. Special circuit design approaches involving active and passive techniques are stressed, particularly for monolithic circuit applications. Analysis and simulation constraints using commercially available nonlinear solvers are also discussed.

The nonlinear FET modeling described in Chapter 5, combined with the mixer analysis developed for diodes in Section 7.1, can be extended to describe the operation of FET mixers. As we have learned, the dominant nonlinearity of the FET is its transconductance, which is typically (especially with JFETs) a square-law function. Hence it makes a very efficient multiplier with products of reasonably low spuriousness.

The small-signal circuit [7] shown in Fig. 7.72 denotes the principal elements of the FET that must be considered in the model. The parasitic resistances R_g, R_d, and R_s are small compared to R_{ds} and can be considered constant, but they are important in determining the noise performance of the circuit. The mixing products produced by parametric pumping of the capacitances C_{gs}, C_{dg}, and C_{ds} are typically small and add only second-order effects to the total circuit performance. Time-averaged values of these capacitances can be used in circuit simulation with good results.

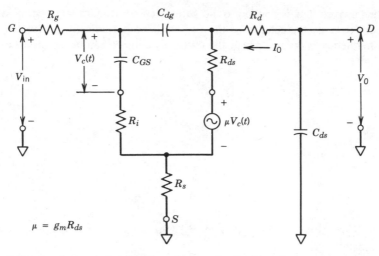

Figure 7.72 Small-signal GaAs FET equivalent circuit with voltage source representation.

The leaves the FET transconductance g_m, which exhibits an extremely strong nonlinear dependence as a function of gate bias. A typical transconductance versus gate bias for a 150-μm low-noise FET is shown in Fig. 7.73. It is evident from the illustration that the greatest change in transconductance occurs near pinch-off, with the most linear change with respect to gate voltage occurring in the center of the bias range. As the FET is biased toward I_{dss}, the transconductance function again becomes nonlinear. It is in these most nonlinear regions that the FET is most efficient as a mixer.

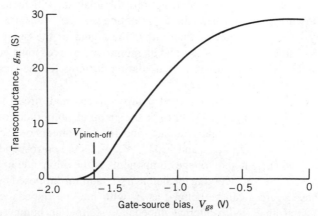

Figure 7.73 Transconductance, as a function of gate bias, for a typical 150-μm gate-width FET.

When a fixed bias is applied to the gate of an FET and a large pump signal is applied, the transconductance of the FET will be periodically modulated. If the pump frequency is defined as ω_p, the transconductance can be expressed as

$$g_m(t) = \sum_{k=-\infty}^{\infty} g_k e^{j\omega_p t} \tag{7.58}$$

where

$$g_k = \frac{1}{2\pi} \int_0^{2\pi} g(t) e^{-j\omega_p t} d\omega_p t \tag{7.59}$$

If we use the time-average value of R_{ds}, the amplification factor $\mu(t)$ can be approximately written as

$$\mu(t) = R_p(g_m(t)) \tag{7.60}$$

where R_p is the time-average value of R_{ds} at the pump frequency.

If we now introduce a second signal, V_c, such that it is substantially smaller than the pump, across the gate-to-source capacitance C_{gs}, the nonlinear action of the transconductance will cause mixing action within the FET. The conventional current source in the FET can be replaced by a voltage source $V_c(t)$ which has mixing product frequencies $|n\omega_p \pm \omega_1|$, where n can be any positive or negative integer. Since V_c is small only, mixing products that are a function of the ω_1, and not its harmonics, need to be considered.

Any practical analysis must include mixing products at both the gate and drain terminal, and at a minimum, allow frequency components in the signal, image, LO, and IF to exist. For ease of conception, the analysis will focus on the FET used as a down-converter where the IF is defined as $|\omega_p - \omega_1|$ and is usually substantially lower in frequency than either the signal or the pump. The image frequency is defined as $|2\omega_p - \omega_1|$ and as mentioned above, must be included in the analysis. The voltage sources and circulating currents for the single-gate FET mixer are shown in Fig. 7.74.

The network above can be analyzed using conventional circuit theory. If we define I_p, I_1, I_2, I_3, and V_p, V_1, V_2, V_3 to be the complex amplitudes of the currents and voltages for the pump, signal, image, and IF components on the gate side of the FET, and define I_p, I_4, I_5, I_6 and V_p, V_4, V_5, V_6 as the corresponding complex amplitudes of the current and voltage components at the drain side of the FET, the problem can be solved by applying the boundary conditions imposed by the external sources. These sources can be defined as E_1 with internal impedance Z_1, for the signal (ω_1), and an E_p with internal impedance Z_p for the pump (ω_p), with all other mixing products, appearing on both sides of the FET, being terminated by

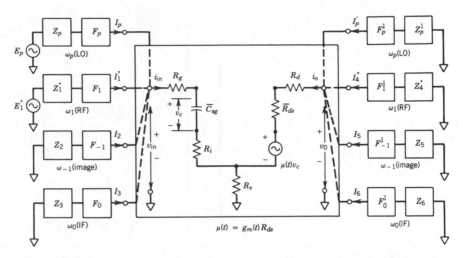

Figure 7.74 Circuit diagram of single-gate FET mixer showing signal, image, and IF circuits (© IEEE 1976).

complex impedances. Thus the voltage and currents at the gate–source and drain–source terminals can be related as

$$V_k = E_k - I_k Z_k \qquad (7.61)$$

where $k = 1, 2, \ldots, 6$ and where $E_k = 0$ for k not equal to unity. For convenience, a lower sideband relationship between the signal and pump was chosen; therefore, I_1 must be replaced by I_1^*. Similarly, I_4 must also be replaced by I_4^*.

Although the filters shown in the illustration above are assumed to be ideal, that is, they only losslessly pass their respective frequency and reject all others, in practice this assumption can be approximated only when the IF frequency is sufficiently spectrally separated form the RF and LO frequencies. In general, the mixer is most efficient when all mixing products except the IF (ω_p) are short-circuited at the drain terminal. It is also desirable to short-circuit the IF at the gate in order to reduce mixer noise. Separating the pump and signal components at the gate terminal is also difficult, and is usually accomplished with the aid of a directional coupler rather than by employing filters.

An equivalent small-signal analysis of the circuit in Fig. 7.74 can be conducted since time-averaged values of the nonlinear elements will be employed. Hence, by applying loop analysis and neglecting any harmonics of the pump frequency, the circuit performance can be represented as

$$[E] = [V] = +[Z_t][I]$$
$$= [Z_m][I] + [Z_t][I] \qquad (7.62)$$

where

$$[E] = \begin{bmatrix} E_1^* \\ 0 \\ 0 \\ 0 \\ 0 \\ 0 \end{bmatrix} \qquad (7.63)$$

$$(V) = \begin{bmatrix} V_1^* \\ V_2 \\ V_3 \\ V_4^* \\ V_5 \\ V_6 \end{bmatrix} \qquad (7.64)$$

$$[I] = \begin{bmatrix} I_1^* \\ I_2 \\ I_3 \\ I_4^* \\ I_5 \\ I_6 \end{bmatrix} \qquad (7.65)$$

The quantities $[Z_m]$ and $[Z_t]$ are the multiport equivalent matrix representation of the mixer proper and the terminating network, which are given by

$$[Z_m] = \begin{bmatrix} Z_{11}^* & 0 & 0 & Z_{14}^* & 0 & 0 \\ 0 & Z_{22} & 0 & 0 & Z_{25} & 0 \\ 0 & 0 & Z_{33} & 0 & 0 & Z_{35} \\ Z_{41}^* & 0 & Z_{43} & Z_{44}^* & 0 & 0 \\ 0 & Z_{52} & Z_{53} & 0 & Z_{55} & 0 \\ Z_{61}^* & Z_{62} & Z_{63} & 0 & 0 & Z_{66} \end{bmatrix} \qquad (7.66)$$

and

$$[Z_t] = \begin{bmatrix} Z_1^* & 0 & 0 & 0 & 0 & 0 \\ 0 & 0 & 0 & 0 & 0 & 0 \\ 0 & 0 & Z_3 & 0 & 0 & 0 \\ 0 & 0 & 0 & Z_4^* & 0 & 0 \\ 0 & 0 & 0 & 0 & Z_5 & 0 \\ 0 & 0 & 0 & 0 & 0 & Z_6 \end{bmatrix} \qquad (7.67)$$

The elements of the matrix $[Z_m]$ are given by

$$Z_{11} = R_g + R_i + R_s + 1/j\omega_1 C \tag{7.68}$$

$$Z_{22} = R_g + R_i + R_s + 1/j\omega_{-1} C \tag{7.69}$$

$$Z_{33} = R_g + R_i + R_s + 1/j\omega_0 C \tag{7.70}$$

$$Z_{44} = R_d + R_{ds} + R_s \tag{7.71}$$

$$Z_{55} = R_d + R_{ds} + R_s \tag{7.72}$$

$$Z_{66} = R_d + R_{ds} + R_s \tag{7.73}$$

$$Z_{14} = Z_{25} = Z_{36} = R_s \tag{7.74}$$

$$Z_{41}^* = \text{conj}(R_s + - g_p R_{ds}/j\omega_1 C) \tag{7.75}$$

$$Z_{61} = R_s + -g_1 R_{ds}/j\omega_1 C \tag{7.76}$$

$$Z_{52} = R_s + -g_p R_{ds}/j\omega_2 C \tag{7.77}$$

$$Z_{62} = -g_1 R_{ds}/j\omega_2 C \tag{7.78}$$

$$Z_{63} = R_s + -g_p R_{ds}/j\omega_3 C \tag{7.79}$$

$$Z_{43} = Z_{53} = -g_1 R_{ds}/j\omega_3 C \tag{7.80}$$

In the equation set above the quantity C represents the time-average value of C_{gs} at the frequencies of ω_k, where k can be either -1, 0, or 1. Similarly, the quantity R_{ds} is the time-average value of the drain-to-source resistance at its respective frequency. The value of R_{ds}, from frequency to frequency, can be quite different, especially if the IF frequency is very low.

The available conversion gain G_{av} [7.27, 7.28], which is of course the quantity of most interest, can be expressed as the ratio of power delivered to the load at port 6 to the power available from the source. Thus

$$G_{av} = \frac{|I_6|^2 \, \text{Re} \, [Z_6]}{|E_1|^2 / \text{Re} \, [Z_1]} \tag{7.81}$$

$$= 4R_g R_L \left| \frac{I_6}{E_1} \right|^2 \tag{7.82}$$

where Z_1 is defined as $R_g + jX_g$, and Z_6 is defined as $R_L + jX_L$. The quantity I_6/E_1 can be obtained by solving (7.62) [7.27]. If the IF frequency is substantially lower than the RF or LO frequencies, the conversion gain expression can be simplified. Thus

$$G_{av} = \frac{2g_1 R_{ds}}{\omega_1 C} \frac{R_g}{(R_g + R_{in}) + (X_g - 1/\omega_1 C)^2} \frac{R_L}{(R_{ds} + R_L)^2 + X_L^2} \tag{7.83}$$

where R_{in} is $R_g + R_i + R_s$. When the source and load impedances are conjugately matched, the conversion gain is a maximum and can be defined as

$$G_c = (g_1)^2 R_{ds}/4(\omega_1)^2 C^2 R_{\text{in}} \tag{7.84}$$

It is interesting to note that the expression for conversion gain is similar to the expression for amplifier gain. The ratio of these two gains, for the same signal frequency, is expressed as

$$G_r = \left(\frac{g_1}{g_m}\right)^2 \left(\frac{C_{ds}}{C}\right)^2 \frac{\tilde{R}_{ds}}{R_{ds}} \tag{7.85}$$

When the IF frequency is low compared to the signal frequency, the gain ratio in (7.85) can be greater than unity even if $g_1 < g_m$. That is because the ratios C_{ds}/C and $\tilde{R}_{ds}/R_{ds}$ are greater than unity when the FET is biased near pinch-off for maximum conversion gain.

   The variation on mixer performance as a function of LO drive is also of prime importance. If we assume that the total gate-to-source voltage is the sum of the LO voltage and DC bias, such that

$$V_{gs} = V_{gdc} + V_p \cos \omega_p t \tag{7.86}$$

then the conversion transconductance, as a function of dc gate bias and peak LO voltage, may be calculated. A plot conversion transconductance versus gate bias for a typical 150-μm FET is shown in Fig. 7.75. The values for g_1 were obtained from a SPICE analysis, but they could have also been obtained by Fourier analysis of $g_m(t)$. As the gate bias was varied, the amplitude of the LO signal V_p was

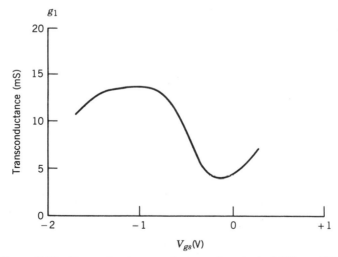

Figure 7.75 Conversion transconductance for a typical 150-μm FET.

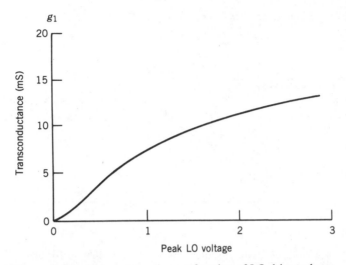

Figure 7.76 Conversion gain as a function of LO drive voltage.

adjusted so that the peak gate voltage was at the onset of forward gate current. Hence, as the gate bias became more negative, the magnitude of V_p must be increased to the point of gate conduction. As can be seen in the illustration, a maximum in conversion gain occurs near pinch-off and a second local maxima occurs near forward conduction. It should be noted that the shape and magnitude of the curve are very dependent on the g_m characteristic of the FET. The dependence of LO drive voltage on conversion gain can be found in a similar manner and is illustrated in Fig. 7.76. The shape of the curve is very similar to that obtained with conventional diode mixers except that the conversion gain saturation is less abrupt.

The dependence of gate capacitance, as a function of gate-to-source voltage, is much less severe. This dependence can be linearly approximated as illustrated in Fig. 7.77. It is evident from the capacitance function above that the values for C_{gs} (static value), C (time-average value at ω_1), and C_p (time-average value at ω_p) can be substantially different.

Unfortunately, the best mixer conversion loss is obtained for large values of LO voltage. With this in mind, and knowing the FET parasitic element values, which can be obtained from small-signal S-parameter measurements, an estimate of the required LO can be made. From conventional circuit theory, the power dissipated in the gate circuit is

$$P_{\text{LO}} = \tfrac{1}{2}(\omega_p \tilde{C}_p V_p)^2 R_{\text{in}} \tag{7.87}$$

where R_{in} is the input resistance, as defined earlier. When the total gate periphery of the FET is small, the optimum LO power will be modest (3 to 6 dBm), but as the FET becomes larger, the term C_p in (7.87) begins to dominate. It is not

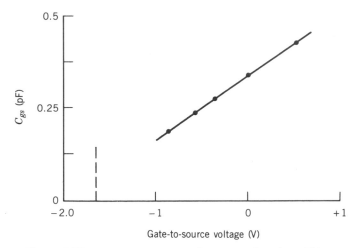

Figure 7.77 Gate-to-source capacitance as a function of bias.

uncommon for a large FET to require 20 dBm of LO drive power. The amount of LO power required for maximum conversion loss at a particular FET size can also be reduced by selecting or designing the FET for the lowest possible value of pinch-off voltage. Since, for a given gate periphery, the values of R_{in} and C_p change little compared to the change in pinch-off voltage obtained when changing the FET doping profile, a dramatic improvement in LO efficiency is obtained when using low-noise FETs with pinch-off voltages in the 1.5-V range as compared to power FETs that exhibit pinch-off voltages in the range 4 to 5 V.

Experimental results for both single- and multiple-FET mixers verify the predictions above for optimum conversion loss, bias, and LO requirements. For example, a single-gate X-band mixer [7.27] using a 500×2.5 μm FET with a pinch-off of 3.3 V was constructed on a 0.5-mm-thick alumina substrate and evaluated. The circuit model for the mixer is shown in Fig. 7.78. The conversion gain performance as a function of LO drive power at a signal frequency of 7.8 GHz and an IF frequency of 30 mHz for the mixer is shown in Fig. 7.79. The calculated performance shown is from the theory above, and the measured data were obtained with the FET biased near pinch-off. It is interesting to note that the conversion gain obtained of 6.4 dB is greater than the 4.7 dB of gain that can be obtained when the same device is used as an amplifier. As can be seen, the calculated results agree quite well with measured performance.

Computer simulation of any proposed mixer circuit can be conducted as long as the FET characteristics are completely known. The computed and measured conversion-loss performance results versus LO power for a similar X-band mixer is shown in Fig. 7.80, and the mixer circuit topology is shown in Fig. 7.81. As can be seen in the illustration above, the computed nonlinear circuit simulation performed using Microwave Harmonica also agrees quite well with the measured results. Computer simulators of this type enable the design engineer, as in this case, to optimize the circuit parameters for best performance. In addition, nonlinear

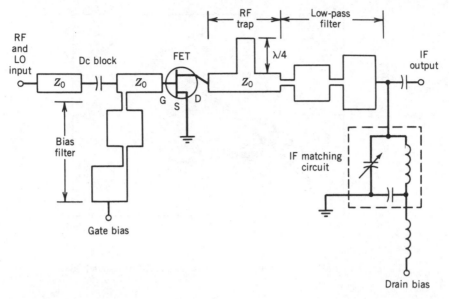

Figure 7.78 X-band single-gate FET mixer reported by Pucel (© IEEE 1976).

simulators allow the designer to examine circuit voltage waveforms and spectral performance. A sample voltage waveform and the spectral response of the circuit above is shown in Figs. 7.82 and 7.83.

The noise performance of a GaAs FET mixer is much less well understood than its companion amplifier. However, the noise-figure performance is related to the intrinsic noise sources of the FET (R_g, R_s, and R_d) as well as to shot noise and noise due to traps in the semiconductor material. The $1/f$ noise spectra of the GaAs FET mixer can extend to several hundred megahertz, but usually, with a well-

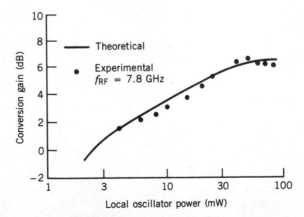

Figure 7.79 Conversion gain of X-band mixer as a function of LO driver power (© IEEE 1976).

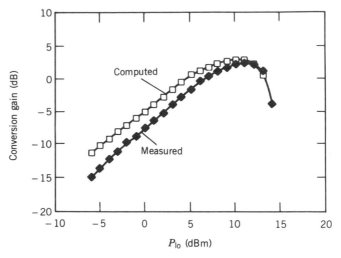

Figure 7.80 Measure versus computed performance of X-band mixer.

designed FET, extends to less than 50 mHz. A variety of single-gate mixers, with operating frequencies extending through X-band, have been reported to have exhibited noise figures less than 5 dB and an associated gain of several decibels. This performance, although not easy to realize, is quite comparable to conventional diode mixers, except that a small amount of gain, not loss, is obtained.

However, a good first-order approximation to the noise-figure performance of a single-gate FET mixer can be determined by applying the accepted noise parameters of MESFET [7.29]. The noise equivalent circuit of an FET is shown in Fig. 7.84, where the noise sources e_g, e_s, and e_d, due to the parasitic resistances R_g, R_s, and R_d, and the noise sources i_d and I_g, due to channel current and induced gate current, are included. The mean-square value of the current sources and their correlation coefficient are given by

$$\overline{I_d^2} = 4kTBg_m P \tag{7.88}$$

$$\overline{I_g^2} = 4kTB \frac{(\omega C_{gs})^2}{g_m} R \tag{7.89}$$

$$C_r = \frac{I_g^* I_d}{(I_g^2 I_d^2)^{1/2}} \tag{7.90}$$

where k = Boltzmann's constant
T = temperature
B = bandwidth
R, P = dimensionless noise parameters of the FET

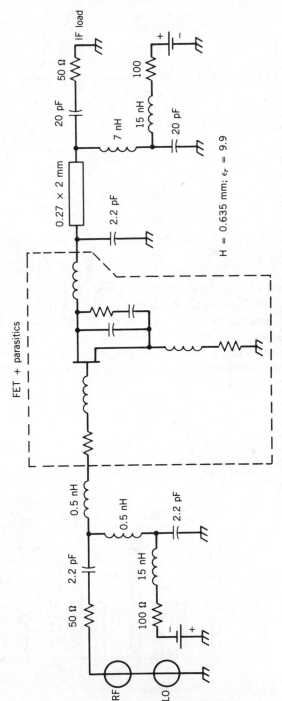

Figure 7.81 Single-gate FET mixer used in harmonic balance analysis.

591

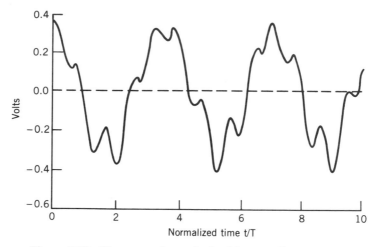

Figure 7.82   IF output voltage obtained from nonlinear analysis.

If we replace the FET model in the mixer model shown in Fig. 7.72 with the noise model of the FET, the mixer noise model shown in Fig. 7.85 results. The noise sources, i_{gn1} and i_{gn0}, in the mixer's input gate circuit are functions of the induced gate noise source at the RF and IF frequencies, while i_{dn1} and i_{dn0} are noise sources in the mixers output circuit that are functions of the drain noise current i_d. The noise voltage sources can also be defined from the general relationship

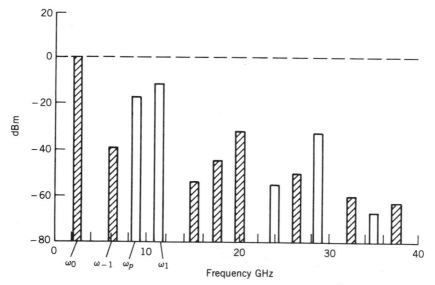

Figure 7.83   Computed mixer spectral performance.

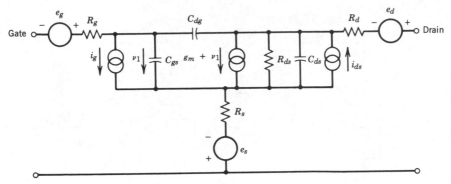

Figure 7.84 Noise equivalent circuit of an FET.

$$\overline{e^2} = 4kTB(\text{Re }[Z])\tag{7.91}$$

The noise figure of the mixer can then be determined by summing all the noise currents within the mixer circuit arriving at the IF output port. This is accomplished with the aid of a formula developed by Friis in 1944 [7.30]. Thus the noise figure can be expressed as

$$F = \frac{\overline{\left| I_{no1} + I_{no0} + I_{no3} + I_{dn0} + (4kTB/R_d)^{1/2} \right|^2}}{\left| I_{no}^2 \right|}\tag{7.92}$$

where I_{no1}, I_{no0}, and I_{no3} are the noise components at the IF output circuit due to the RF and IF noise sources at the FET input and the RF noise source at the FET output. The sources I_{dn0} and $(4kTB/R_d)^{1/2}$ are noise current components generated within the IF drain circuit, and the source I_{no} is the noise at the IF port due to the matching network at the FET gate.

If we now let Y_1, Y_0, and Y_3 be the transfer admittances and G_1, G_0, and G_3 the current gains from the RF and IF input and the RF output ports to the IF output port, respectively, the mixer noise figure can be expressed as

$$\begin{aligned}
F = 1 &+ \frac{R_g + R_s}{R_1} + \frac{|Y_0|^2}{|Y_1|^2}\frac{R_g + R_s + \text{Re }[Z_0]}{R_1}\\[2mm]
&+ \frac{|Y_3|^2}{|Y_1|^2}\frac{R_d + \text{Re }[Z_3]}{R_1} + \frac{1}{|Y_1|^2 R_1 R_d}\\[2mm]
&+ \frac{|G_1|^2}{|Y_1|^2}\frac{\overline{|i_{ng1}^*|^2}}{4kTBR_1} + \frac{|G_3|^2}{|Y_1|^2}\frac{\overline{|i_{nd1}^*|^2}}{4kTBR_1}
\end{aligned}\tag{7.93}$$

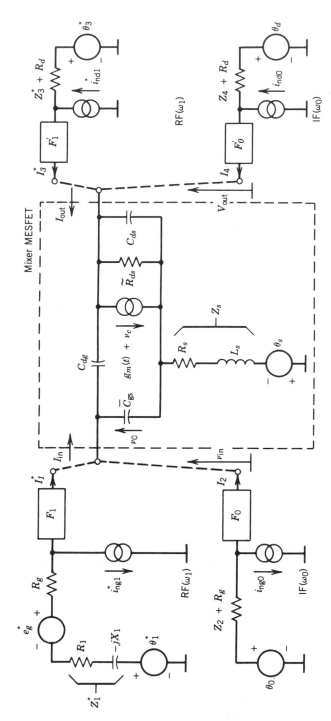

Figure 7.85 Equivalent circuit of single-gate FET mixer used for noise analysis (© IEEE 1983).

$$+ \frac{|G_0|^2}{|Y_1|^2} \frac{\overline{|i_{ng0}|^2}}{4kTBR_1} + \frac{\overline{|i_{nd0}|^2}}{|Y_1|^2\ 4kTBR_1}$$

$$+ \frac{2\{\overline{|i_{ng1}^*|^2}\ \overline{|i_{nd1}^*|^2}\}^{1/2}}{|Y_1|^2\ 4kTBR_1}\ \mathrm{Re}\ \{G_1^*G_3\tilde{C}_{r1}\}$$

$$+ \frac{2\{\overline{|i_{ng0}^*|^2}\ \overline{|i_{nd0}^*|^2}\}^{1/2}}{|Y_1|^2\ 4kTBR_1}\ \mathrm{Re}\ \{G_0^*\tilde{C}_{r0}\}$$

where $\tilde{C}_{r1}$ and $\tilde{C}_{r0}$ are the time-average values of the correlation coefficients between i_{ng1}^* and i_{nd1}^*, and i_{ng0} and i_{nd0}, respectively. The conversion loss is calculated as described previously.

To verify the noise theory, three mixers were constructed with various combinations of RF and IF terminations. Mixer 1 was constructed with an IF short circuit on the gate side of the FET and an RF short at the drain. Mixer 2 was constructed with an IF open circuit at the gate and an RF short at the drain, while mixer 3 used open-circuit terminations at both the gate and drain for the IF and RF, respectively. A comparison between the measured and predicted conversion loss and noise figure performance for the mixers are shown in Figs. 7.86 to 7.88. The

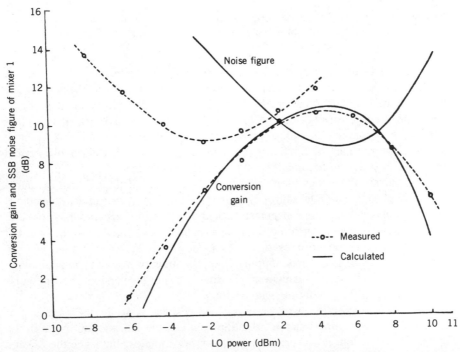

Figure 7.86 Measured and calculated noise figure and conversion gain of mixer 1, in which the IF input and RF output loads are short circuits (© IEEE 1983).

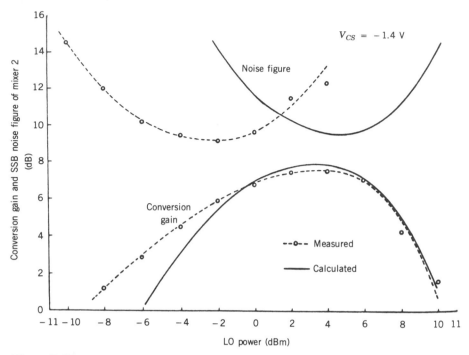

Figure 7.87 Measured and calculated noise figure and conversion gain of mixer 2. The IF input load = high impedance, RF output load = short circuit (© IEEE 1983).

measured and predicted conversion gain performance agree quite well, but the noise-figure performance as a function of LO is in error. However, the noise-figure values themselves are quite accurate and do predict the minimum noise figure obtainable. The error may be due to errors in the values of P, R, and C_r. It is interesting to note that the noise figure was essentially unaffected by the reactive terminations and may be limited by the current gains acting on the noise current components at the signal, image, pump, and IF frequencies.

The analysis above was formulated for gate injection of the LO, but the pump may be applied to either the source or drain terminals. These two injection methods may be preferable when no convenient method of RF and LO separation is possible, such as in applications where no directional coupler is available or when the LO and RF signals are too close together in frequency to be separated by filtering. The performance obtainable when pumping either the drain or source is comparable to gate pumping. However, there are slight differences in conversion gain and noise figure when pumping either the gate or drain.

Since the LO injection, with drain pumped designs, is at the drain terminal of the FET, there is some inherent isolation between the RF and LO ports in the mixer. IF extraction, however, is somewhat more complex, but since the LO and IF are usually separated significantly in frequency, the diplexing task should be easier than diplexing the RF and LO signals as in the gate pumped case.

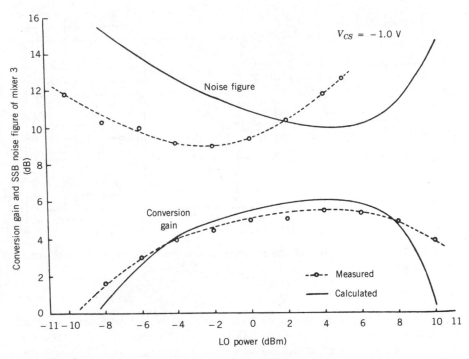

Figure 7.88 Measured and calculated noise figure and conversion gain of mixer 3. The mixer's IF input and RF output loads are high impedances (© IEEE 1983).

Analyzing the drain-pumped configuration is almost identical to the gate-pumped configuration and can be formulated using the circuit model illustrated in Fig. 7.89. The circuit is analyzed as before by writing the loop equations for each mixer product current and treating each frequency component as a separate port in the network. Thus neglecting harmonics of the pump as before, the circuit performance can be described as

$$[E] = [V] = +[Z_t][I]$$

$$= [Z_m][I] + [Z_t][I] \tag{7.94}$$

where the quantities $[Z_m]$ and $[Z_t]$ are the multiport equivalent matrix representation of the mixer proper and the terminating network. With the source and load impedances conjugately matched, the conversion gain can be expressed as

$$G_c = \frac{|g_1 R_{ds1}|^2}{4(\omega_1)^2 C^2 (R_{in})(R_s + R_d + R_{dsp})} \tag{7.95}$$

where R_{ds1} and R_{dsp} are the time-average values of R_{ds} at the signal and pump frequencies. A more detailed discussion of the analysis above was presented by Begemann and Jacob [7.31].

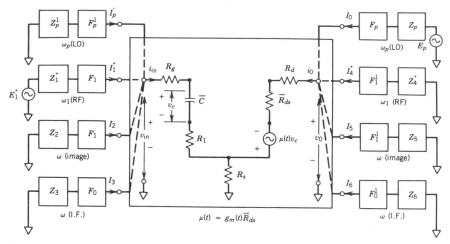

Figure 7.89 Circuit diagram of drain pumped single-gate FET mixer showing signal, image, and IF circuits.

Using the expression above for conversion gain, the performance of an NEC24483 with a signal frequency 4.1 GHz and an IF frequency of 100 MHz was calculated. Figure 7.90 illustrates the conversion performance as a function of gate bias and drain-to-source voltage. The value of the LO amplitude V_p was equal to V_{ds}.

An interesting comparison between the performance of drain- and gate-pumped mixers was shown by Bura and Dikshit [7.32]. Both mixers employed an NEC24483 FET and were pumped at 6 GHz. The RF input frequency range was 5.9 to 6.4 GHz, and the IF band was 3.7 to 4.2 GHz. The conversion gain for both mixers is shown in Fig. 7.91a and the noise-figure performance is shown in Fig. 7.91b. As can be seen, the drain-pumped mixer requires substantially more LO power for optimum conversion loss, but is only slightly lower gain at equivalent pump levels. However, the noise-figure performance for the drain-pumped design is substantially lower and is not too far from the noise figure of a comparable amplifier designed for the same frequency range using the same device type.

Single-ended mixers can also be designed using dual-gate FETs. Dual-gate devices offer several advantages over conventional devices, usch as ease of LO injection, improved isolation, and added gain. They are, however, considerably less stable; hence added care must be used when designing non-self-oscillating mixers.

The operation of a dual-gate FET can easily be understood if thc FET is considered as a cascode-connected FET pair, as was described in Chapter 5. Using this concept for the FET, the drain characteristics for the pair can be approximated by combining the characteristics of each intrinsic FET. This concept is illustrated in Fig. 7.92a. A slightly more convenient representation of the drain characteristics

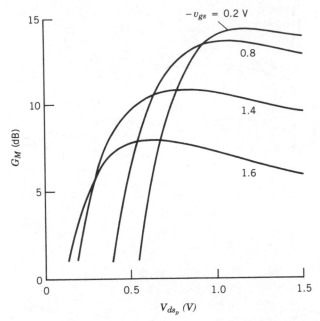

Figure 7.90 Maximum available conversion gain of drain pumped mixer as a function of V_{os} and V_{ds} (© IEEE 1979).

is shown in Fig. 7.92*b*. With this representation the operating point for FET 1 can be found as a function of gate 1 and 2 bias as well as its drain-to-source voltage.

The operating point path can vary significantly depending on how the FET is biased. Typically, gate 1 is used for signal injection with gate 2 biased (V_{gs2} < 0) for FET operation in the low-noise mode (shaded area in Fig. 7.92*b*). Gate 2 is also used for LO injection. Applying the LO at gate 2 is in effect drain pumping the first FET; hence FET 1 is the primary mixing element, while FET 2 acts as a common-gate amplifier. The operation is reversed if a sufficiently high bias voltage (V_{gs2} > 2) is applied to gate 2. With these bias conditions, FET 1 acts as an RF preamplifier, while FET 2 becomes the primary mixing element.

Slightly more insight into the operation of a dual-gate FET mixer can be had by examining the drain current dependence due to gate 1 and 2 bias voltage (Fig. 7.93) and the conversion gain characteristics as a function of LO drive voltage as reported by Tsironis [7.33,7.34] (Fig. 7.94). The shape and characteristics of the curves in the illustration are almost identical to the data obtained from a typical MOSFET (Fig. 7.95). As can be seen, a large percentage change in drain current, as a function of gate bias, occurs in both of the operating regions described above; hence maximum conversion gain should also correspond to the same operating points. Also, as in the case of the single-gate mixer, the conversion gain increases as the LO drive voltage is increased, until the point of LO saturation.

Dual-gate FET mixers can be analyzed as described above or designed with the

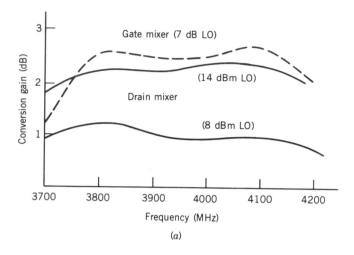

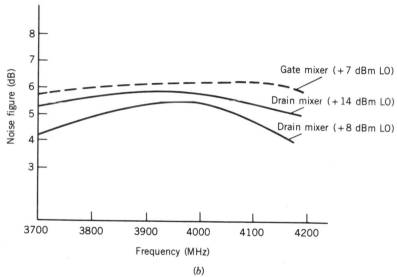

Figure 7.91 Comparison between drain and gate pumped mixers: (*a*) conversion gain performance as a function of LO power; (*b*) single-sideband noise figure performance as a function of LO power (© IEEE 1976).

aid of a nonlinear simulator such as Microwave Harmonica or Microwave SPICE. Conversion gain performance obtainable with dual-gate mixers is comparable to that obtained with conventional devices, with the exception of slightly degraded noise figure and possibly more gain. There is also some suspicion that the intermodulation performance for dual-gate MESFET devices may be better than single-gate FETs; but this idea has not yet been proven.

The distributed amplifier concepts developed in Chapter 5, combined with FET mixer theory, can be applied to the syntheses of distributed mixer structures

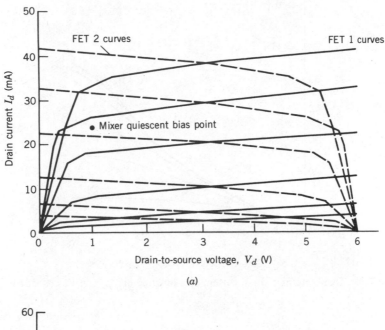

(a)

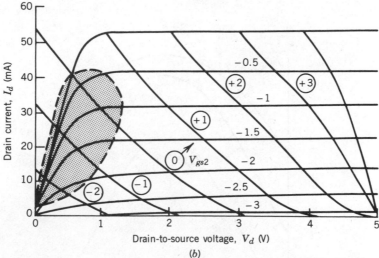

(b)

Figure 7.92 Dual-gate FET I–V characteristics: (a) intrinsic FET I–V characteristics connected "back to back" to simulate a dual-gate FET; (b) dual-gate FET I–V characteristics as a function of V_{gs1} and V_{gs2}.

[7.35, 7.36]. The distributed mixer employs the input capacitance of the FET gates and high-impedance-series transmission lines to realize a lumped-element transmission-line section of impedance Z_0. Several FETs can be cascaded in this manner to form very broadband structures for both the LO and RF mixer ports (Fig. 7.96). Thus low VSWR on both ports and good LO-to-RF isolation can be achieved.

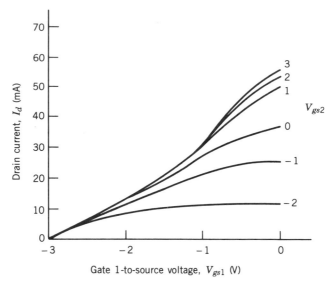

Figure 7.93   Dual-gate FET drain current as a function of V_{gs1} and V_{gs2} (© IEEE 1984).

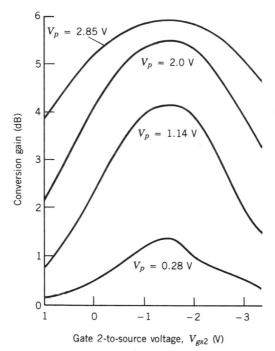

Figure 7.94   Conversion gain characteristics of dual-gate FET as a function of LO peak voltage (© IEEE 1983).

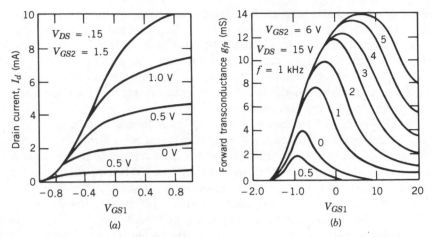

Figure 7.95 Drain current and conversion gain performance of a typical MOSFET: (*a*) drain current as a function of V_{gs1} and V_{gs2}; (*b*) transconductance versus V_{gs1} and V_{gs2}.

These transmission lines must have equal phase shifts as a function of frequency between the FET stages when the mixer is operating as a down-converter with a low IF frequency. Using equal phase shifts between LO and RF signals at each FET yields a constant phase offset at the IF frequency, which allows the IF power to be summed in phase by connecting the drain nodes of the dual-gate FETs together. Higher IF frequencies, or use of this mixer as an up-converter, would require that the drains of the FETs also be connected with a traveling-wave structure.

A variety of design considerations affect the bandwidth and conversion gain of the mixer, such as the number of FETs, gate periphery, gate resistance, LO power

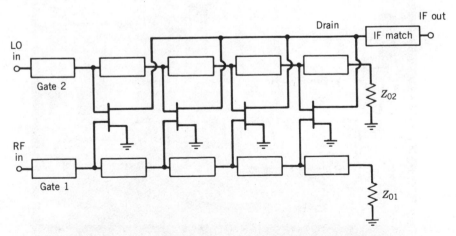

Figure 7.96 Schematic diagram of distributed dual-gate FET mixer.

requirements, and gate length. These design trades can be determined by applying standard distributed amplifier theory and the FET mixer theory discussed above.

The reported mixer circuit was designed using both linear and nonlinear analysis programs. Linear analysis can be used to determine element values for port matching, but conversion analysis can only be performed with a nonlinear solver. Nonlinear analysis allows the design engineer to optimize FET characteristics as well as drive levels and bias conditions.

The monolithic mixer shown in Fig. 7.97 was fabricated on a 0.1-mm-thick semi-insulating GaAs substrate, with the active layers being formed by ion implantation. Since the output impedance at the common drain node of the mixer is on the order of 400 Ω, a source-follower single-gate FET IF amplifier was used for impedance matching. A comparison between the measured and modeled (Microwave SPICE) conversion-loss (gain) performance for the mixer is shown in Fig. 7.98. The conversion gain versus LO power and the conversion gain versus dc bias performances, measured and modeled, are shown in Figs. 7.99 and 7.100. As can be seen, the distributed mixer exhibits similar LO compression characteristics as single FET designs. Although multiple FETs are employed, the mixer is still single-ended and offers no spurious product rejection or AM noise suppression; however, it does offer excellent broadband performance, small size, low port VSWR, and compression characteristics similar to those of multidiode mixers.

7.6 BALANCED FET MIXERS

In section 7.6 we developed the mixer theory for single-ended FET mixers. Combining FETs to form balanced mixer structures and applying the single-ended

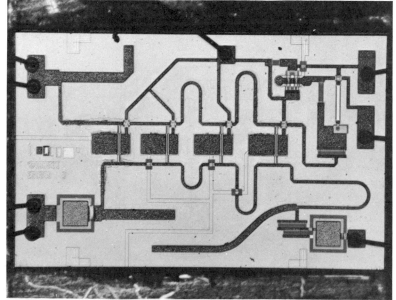

Figure 7.97 Monolithic distributed mixer chip photograph (courtesy of Texas Instruments).

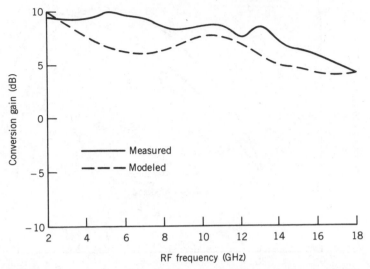

Figure 7.98 Measured versus modeled conversion loss characteristics of distributed mixer.

FET mixer analysis is very straightforward and is easily accomplished by incorporating baluns or hybrids as in the realization of diode mixers. There are, however, some differences and limitations, especially when GaAs monolithic realizations are desired.

In the low-frequency realm, where transformer hybrids are possible, the differences are minor, the main problem being dc bias decoupling. A single-balanced VMOSFET mixer, which illustrates the added decoupling and bias circuitry and

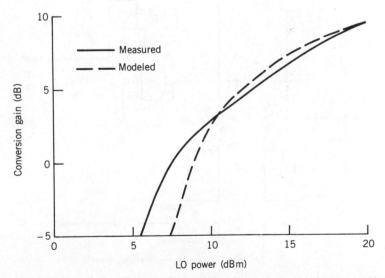

Figure 7.99 Measured and modeled performance of LO saturation characteristics of distributed mixer.

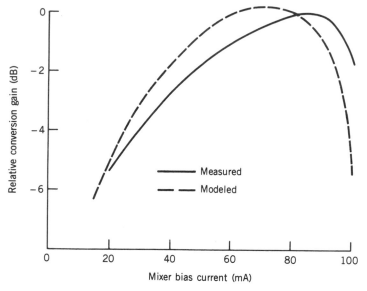

Figure 7.100   Measured versus modeled performance of mixer bias sensitivity.

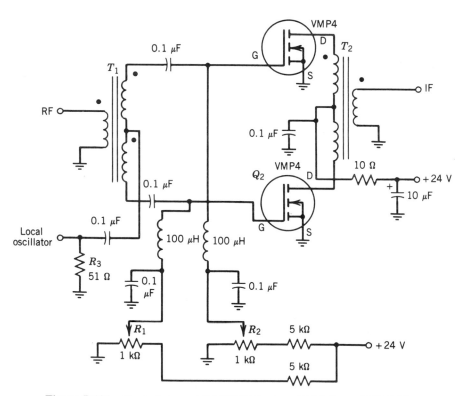

Figure 7.101   Single-balanced VMOSFET mixer (© McGraw-Hill 1988).

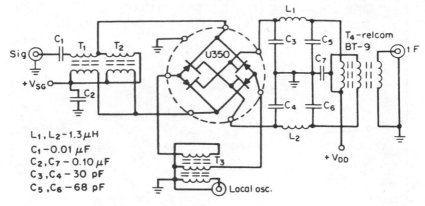

Figure 7.102 Double-balanced JFET mixer employing broadband transformers.

$L_1, L_2 - 1.3 \mu H$
$C_1 - 0.01 \mu F$
$C_2, C_7 - 0.10 \mu F$
$C_3, C_4 - 30$ pF
$C_5, C_6 - 68$ pF

can be designed for HF through UHF operation, is shown in Fig. 7.101. The input signal is applied differentially to the FET gates via the RF balun, while the LO is injected in each gate in phase. Thus the IF signal developed at each FET drain are out of phase and must be summed with an additional balun or hybrid. The extra hybrid is required since we cannot ''flip'' the FET as we can a diode in a conventional signal-balanced mixer. Since the LO is injected in phase, it cancels in the IF balun, but the RF signal, which enters with the proper phase relationship, is summed at the IF port. Hence the mixer exhibits LO-to-RF isolation and LO-to-IF isolation but no RF-to-IF isolation. Typically, a low-pass filter structure is added at the drain to suppress the RF signal. If dual-gate FETs were used in place of the

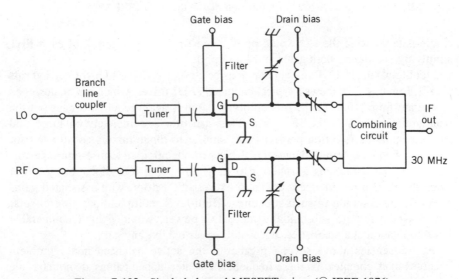

Figure 7.103 Single-balanced MESFET mixer (© IEEE 1976).

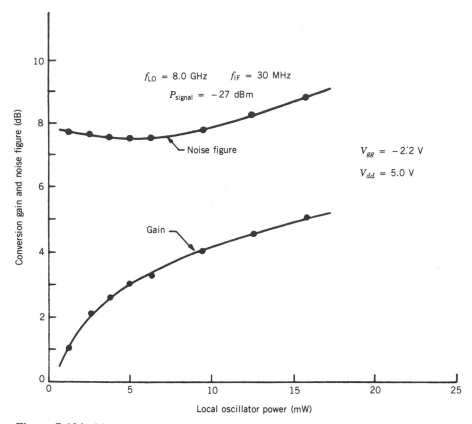

Figure 7.104 Measured conversion gain and double-channel noise figure of a balanced GaAs MESFET mixer at X-band as a function of LO drive (© IEEE 1976).

single-gate devices, the LO would be injected (in phase) at gate 2 of each FET simply by paralleling both gates at the LO port.

Double-balanced FET mixers can also be designed using transformer hybrids [7.37]. Figure 7.102 shows a typical balanced JFET mixer, which can be designed to operate from HF to UHF. An additional balun is again required because of the phase relationships of the IF signal. This structure is completely balanced and exhibits spurious rejection performance, similar to diode mixers constructed for the same frequency range. However, the intermodulation and noise-figure performance of such structures is superior to those of simple four-diode designs. For example, third-order intercept points in excess of 33 dBm, with associated gains of 6 dB, are common in such structures. High-level multiple-diode ring mixers, which would require substantially more LO power, would exhibit comparable intermodulation characteristics, but would never exhibit any gain.

At frequencies above several gigahertz, the active balanced mixer problem becomes more complex. At these frequencies, center-tapped baluns are not possible and it is difficult, because of thermal and other considerations, to integrate

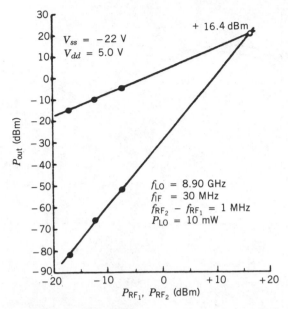

Figure 7.105 Third-order two-tone modulation curves obtained with balanced GaAs MESFET mixer at X-band (© IEEE 1976).

suspended structures with GaAs FETs. Single-balanced FET mixers can, however, be constructed with conventional hybrids and phasing networks.

An X-band single-balanced mixer employing single-gate FETs was reported by Pucel [7.27]. The mixer consisted of two single-gate FETs combined with a quadrature hybrid at the LO and RF ports, and a phasing network to properly sum the 30-mHz IF signal. A representative circuit is shown in Fig. 7.103. The mixer conversion and noise-figure performance, as a function of LO power, are shown in Fig. 7.104. The intermodulation characteristics of the mixer are shown in Fig.

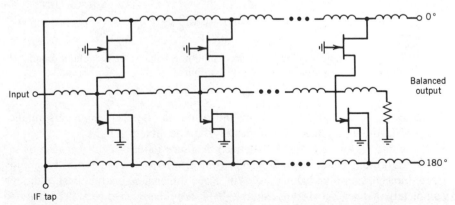

Figure 7.106 Lumped-element equivalent circuit of center-tapped balun.

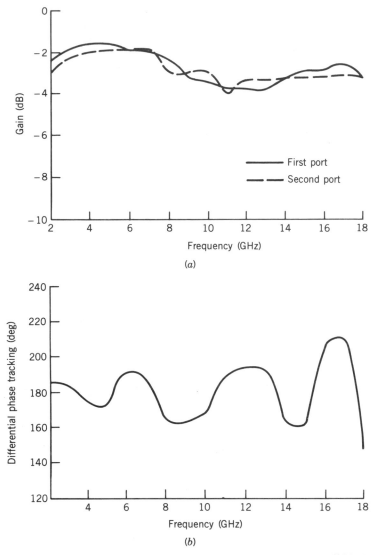

Figure 7.107 (*a*) Frequency response and amplitude balance of monolithic balun; (*b*) differential phase performance of monolithic distributed splitting balun.

7.105. As can be seen, the noise figure and third-order intercept point performance compare favorably to those of well-designed diode mixers.

Designing GaAs FET double-balanced becomes more difficult because of the balun realization. Diode double-balanced mixers, as we recall, typically employ large transmission-line baluns used in three-dimensional structures, although completely planner 2-to 18-GHz double-balanced mixers have been demonstrated [7.38]. Conventional mixer designs such as these are not feasible for monolithic

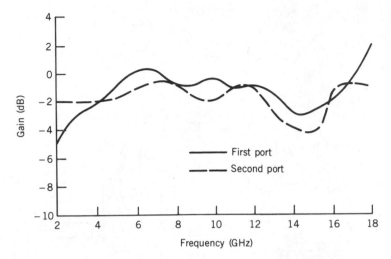

Figure 7.108 Center-tapped balun performance (distributed splitting balun with IF tap).

implementation since their passive elements require excessive GaAs slice area. Hence a completely new design concept must be used to develop double-balanced monolithic GaAs FET mixers.

In the monolithic realm, the balun problem is further constrained by chip area and backside processing requirements. If conventional passive mixer baluns were used, they would be approximately 2 cm in length, which is an order of magnitude too large for a monolithic circuit realization. Thus active baluns or lumped-element transformers are the only feasible options.

A new balun topology that can readily be implemented using monolithic technology has been devised which eliminates the problems described above and provides a virtual center tap. Since the balun uses common-gate and common-source circuit techniques, an ideal 180° phase shift occurs for the signals present between the upper and lower halves of the circuit (Fig. 7.106). Typical broadband amplitude performance for a balun designed with no center tap and resistive terminations at the reverse end of the drain transmission line is shown in Fig. 7.107. As can be seen, the balun exhibits excellent balance through the design band of 2 to 18 GHz. The performance of a center-tapped balun designed for the same frequency range is shown in Fig. 7.108.

Hence if two such baluns are used in conjunction with a diode or FET ring to form a double-balanced mixer, the IF signal appearing at the ring terminals propagates (in phase) down both arms of the balun and can be summed at a common node, thus forming a virtual center tap. This center tap can be used for IF extraction or grounded to complete the IF return path. Since active baluns are not reciprocal, a combining or dividing structure will be needed on the RF port depending on whether the mixer is used as a up- or down-converter (Fig. 7.109).

The frequency limitations of the RF and LO ports are determined by the distributed amplifier-like sections, which can be designed to operate over extremely large

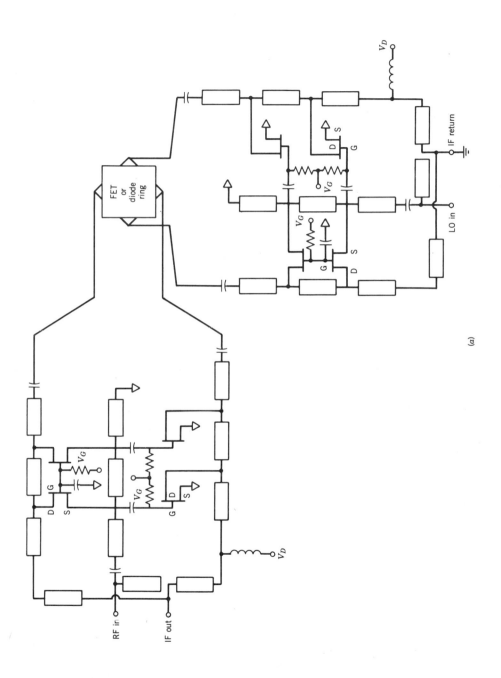

(a)

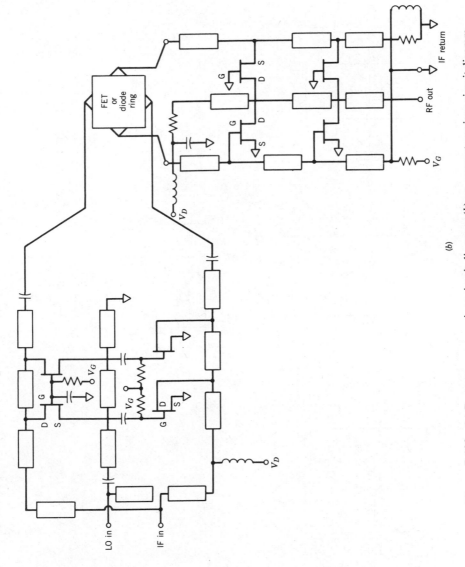

Figure 7.109 (*a*) Down-converter mixer circuit diagram; (*b*) up-converter mixer circuit diagram.

(*b*)

613

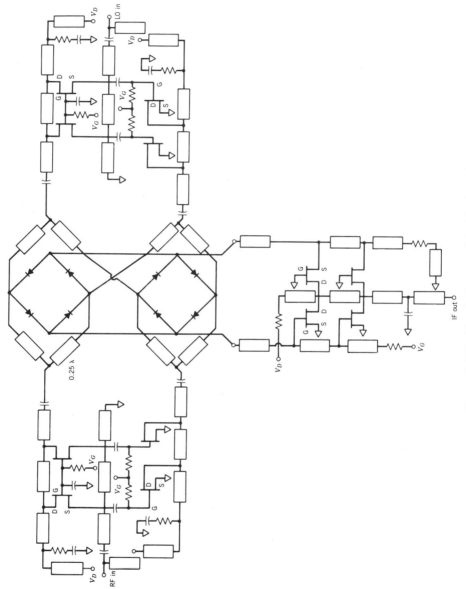

Figure 7.110 Circuit diagram of double-ring mixer.

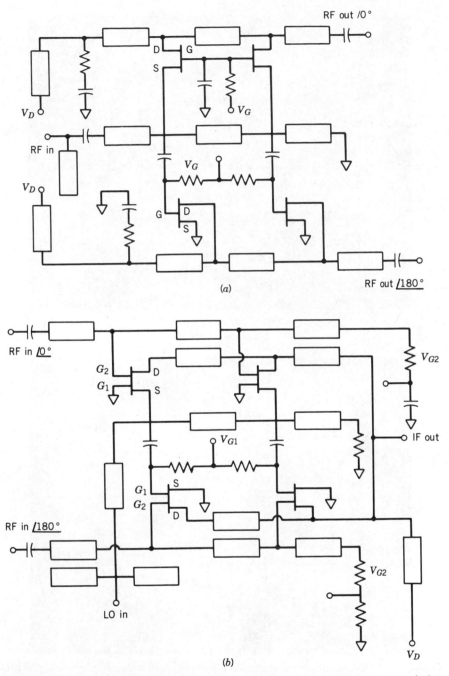

Figure 7.111 (*a*) Transmission-line model of distributed active balun; (*b*) transmission-line model of monolithic double-balanced mixer.

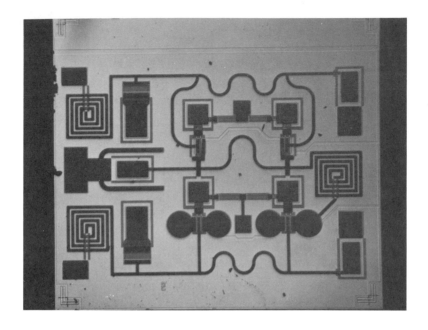

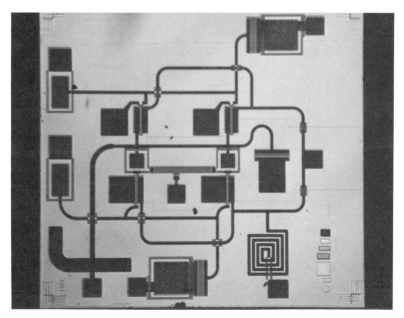

Figure 7.112   Monolithic double-balanced mixer and active balun ICs (courtesy of Texas Instruments).

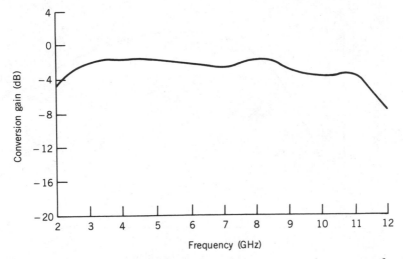

Figure 7.113 Monolithic double-balanced mixer conversion loss performance as a function of frequency.

bandwidths. The IF frequency response can also be designed to exhibit broadband performance. This mixer concept can also be extended to include double-ring mixer topologies (Fig. 7.110). Double-ring approaches have the added advantage of allowing the IF frequency response to overlap the RF and LO frequency bands, thus making IF extraction even easier.

A slightly different topology [7.39] that can readily be implemented using monolithic technology employs active baluns in conjunction with a unique distrib-

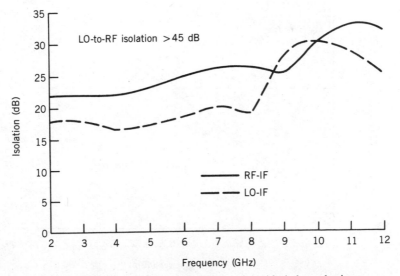

Figure 7.114 Isolation performance of double-balanced mixer.

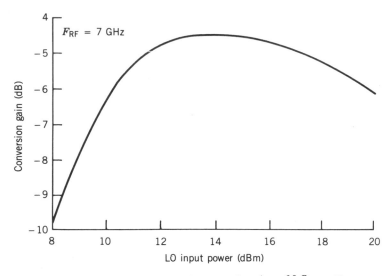

Figure 7.115 Conversion loss as a function of LO power.

uted dual-gate FET mixer structure. The proposed circuit employs a single balun, which can be of either the active or passive lumped-element type (transformer, differential line, etc.) and distributed dual-gate FET mixer sections. Transmission-line models for the balun and mixer are shown in Fig. 7.111. The number of distributed sections employed is somewhat arbitrary and depends on the bandwidth, conversion gain, and impedance-matching requirements. In the design above only

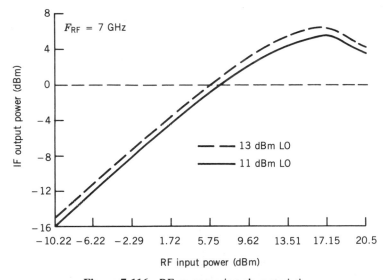

Figure 7.116 RF compression characteristics.

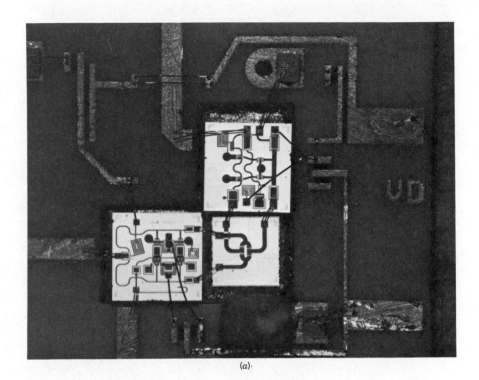

(a).

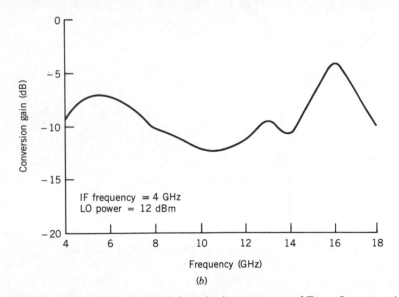

(b)

Figure 7.117 (a) Monolithic double-balanced mixer (courtesy of Texas Instruments); (b) conversion loss performance of down-converter mixer; (c) isolation performance; (d) compression characteristics.

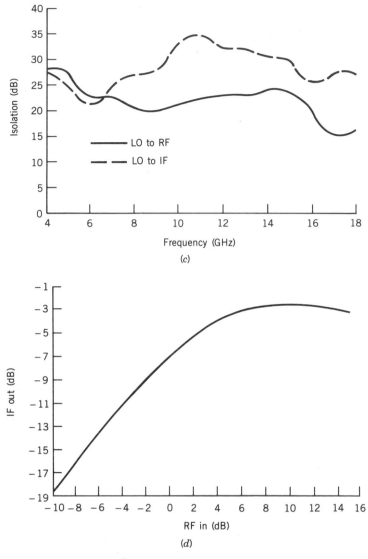

Figure 7.117   (*Continued*)

two sections were required to achieve adequate distributed performance; however, greater conversion gain could probably have been obtained if more sections were employed at the expense of chip complexity.

As can be seen in the circuit diagram, one mixer section employs a common-source topology while the other uses common-gate techniques. Thus an ideal 180° phase shift occurs for the signals present between the upper and lower halves of the circuit, hence eliminating the requirement for a second balun. Both the LO and

RF voltages, which are present at the FET drains of each mixer section, are also out of phase by 180°, while the IF voltages are in phase. By summing the output of both mixer sections, an independent IF port is obtained and the RF and LO signals are canceled. Thus the mixer structure is completely double balanced.

The frequency limitations of the mixing portion of the circuit is determined by the distributed amplifier-like sections, which can be designed to operate over extremely large bandwidths. With the addition of an IF amplifier (active matching) or bandpass matching network, the IF frequency response could be further broadened. Because of the large number of active nonlinear elements, the dynamic range can be made as good as or better than the best conventional diode mixers. The RF balun can also be designed with sufficient gain, thus reducing or eliminating the need for an RF preamplifier in the final receiver.

The distributed monolithic balun and mixer, shown in Fig. 7.112, was designed using 0.5×150 μm dual-gate FETs fabricated on a 0.15-mm-thick GaAs substrate. The FETs were modeled as cascode-connected single-gate FETs with the linear model elements determined from S-parameter measurements. The nonlinear drain current and transconductance characteristics were obtained from $I-V$ curve data obtained at 1 MHz (the technique is described in Chapter 5) and are shown in Fig. 7.92a. The active balun used with the mixer also employed both distributed common-source and common-gate amplifier sections, in order to obtain a broadband differential phase output with good amplitude balance.

The mixer/balun combination was evaluated as a conventional double-balanced mixer with the LO drive applied to the gate 1 circuit. The RF signal was applied to the active balun, which in turn drives the gate 2 circuit. The dc bias on both gates was adjusted for optimum conversion loss; however, since the mixer performance was very insensitive to bias, the second gate voltage was set to zero while

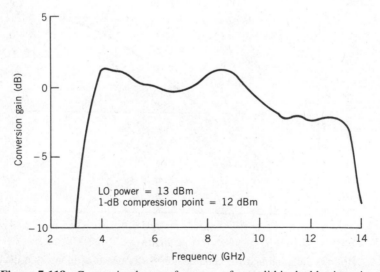

Figure 7.118 Conversion loss performance of monolithic double-ring mixer.

the first gate was biased for a drain current of $I_{dss}/2$. Using the bias conditions above, the conversion loss characteristic shown in Fig. 7.113 was obtained. The RF-to-IF and LO-to-IF isolations, which demonstrate the excellent balance obtained in the design, are shown in Fig. 7.114. Conversion-loss performance as a function of LO power and the RF compression characteristics are shown in Figs. 7.115 and 7.116. As can be seen, the mixer's performance is comparable to hybrid diode designs.

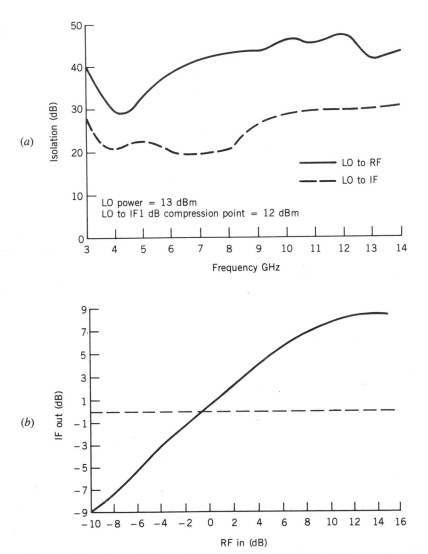

Figure 7.119 (*a*) Isolation and (*b*) compression performance of monolithic double-ring mixer.

This type of structure, with its unique balanced characteristics, can be used as a broadband up-converter as well as a conventional mixer in a variety of receiver applications. In addition, since the mixer is completely balanced, the IF frequency response can overlap the LO or RF responses, which usually can only be accomplished with a double-double-balanced structure.

As mentioned previously, distributed active baluns can be used in conjunction with diode rings to form active/passive double-balanced mixers. Although this approach is not, strictly speaking, a balanced FET mixer, it is an interesting approach and will be presented briefly.

As noted, two distributed baluns can be used in conjunction with a diode ring to form a double-balanced mixer, and a double-double-balanced design can be realized by adding a third balun. By using the above-mentioned technology, both single- and double-ring designs of the types depicted in Fig. 7.109a and 7.110 were fabricated. The single-ring design, with its associated conversion-loss performance at an IF frequency of 4 GHz, is shown in Fig. 7.117. The conversion performance as a function of frequency, which was measured at an IF frequency of 500 mHz, for the double-double-balanced design is shown in Fig. 7.118. Although the mixers employ diodes as the nonlinear element, the conversion loss (gain) of the double-ring design is somewhat greater than that of a conventional structure because of the gain associated with the baluns. The isolation characteristics, which are comparable to those of hybrid designs, are shown in Fig. 7.119.

By using this dual-mode characteristic of distributed broadband baluns in diode mixer topologies, a very compact monolithic circuit which is very process tolerant

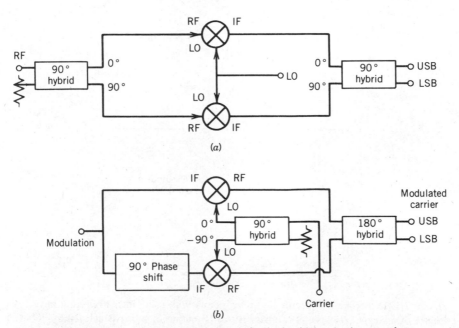

Figure 7.120 (a) Image-rejection and (b) single-sideband mixer topology.

can be designed to operate over a frequency range of several octaves with a performance comparable to that of conventional passive diode mixers.

7.7 SPECIAL MIXER CIRCUITS

There are a variety of interesting mixer topologies in widespread use that perform vital system functions which cannot be simply classified as balanced mixers. Probably the most popular configuration is the image rejection or single-sideband mixer. However, a variety of subharmonically pumped and self-oscillating mixers are in limited use [7.40].

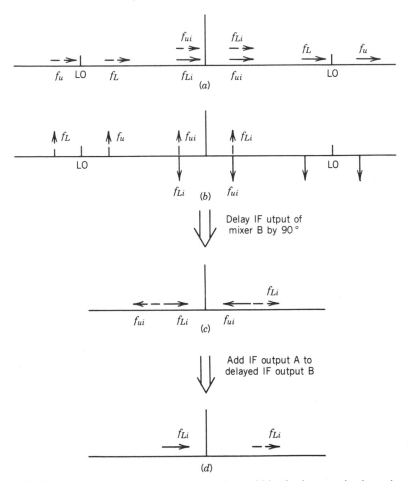

Figure 7.121 Frequency spectra at various points within the image-rejection mixer: (*a*) spectra of mixer A; (*b*) spectra of mixer B; (*c*) IF output of mixer B delayed by 90°; (*d*) composite spectra at LSB IF output.

In many systems it is very advantageous to eliminate or substantially reduce the additional noise power due to the mixer's image response, which is converted to the IF passband. When the IF frequency is low or the operating bandwidth sufficiently large, image filtering cannot be implemented effectively; hence the mixer must be designed with inherent suppression of the unwanted mixing product. This suppression can be obtained by the use of a different type of balanced structure, not unlike a conventional balanced mixer. Unfortunately, no performance enhancement is obtained other than image suppression, as in the case of image-enhanced mixers, since the image energy is not recycled but rather, directed to an unwanted circuit port and dissipated. This is usually not a problem, since many systems may even sacrifice conversion-loss or mixer noise-figure performance to obtain an image-fee response characteristic, since the system noise figure can be restored with additional RF amplification. Also, the image-suppressed response will enable the system to exhibit up to 3 dB of sensitivity improvement beyond what could have been obtained with a conventional broadband mixer regardless of the amount of additional RF amplification.

The classic image-rejection mixer (Fig. 7.120*a*) or single-sideband modulator (Fig. 7.120*b*) topology consists of two mixers with at least one signal applied in quadrature (depending on the application) while the other signal is either combined at the IF output or applied to the IF input in quadrature. Usually, double-balanced

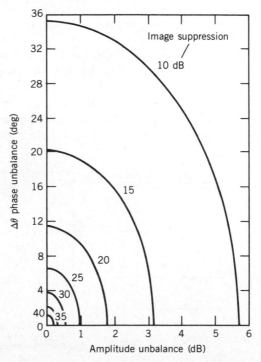

Figure 7.122 Image rejection as a function of circuit amplitude and phase errors.

Figure 7.123 Typical image-rejection mixer consisting of a dual double-balanced mixer and multisection coupler (courtesy of Texas Instruments).

mixers are employed in the circuit to suppress the carrier, but single-balanced designs are sometimes used.

The operation of an image-rejection mixer is easy to understand, but at first glance it may resemble a conventional hybrid combined (balanced) amplifier. The differences lie in the way the upper sideband (USB) and lower sideband (LSB) components for both positive and negative frequency are processed by the circuit. For example, when the positive LSB is down-converted, it becomes a negative frequency IF signal. Similarly, when the negative USB is down-converted, it

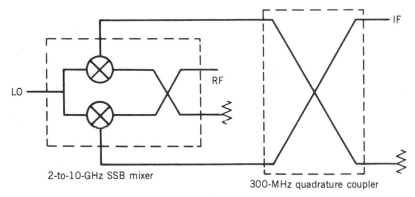

Figure 7.124 Image-rejection mixer circuit configuration.

becomes positive. Hence the positive and negative components are processed differently by the RF and IF couplers. Figure 7.121 illustrates this concept.

With the phase relationships depicted in Fig. 7.120, the IF components due to the positive and negative RF upper and lower sidebands are all down-converted with zero phase shift by mixer A. But if the positive RF signals (USB and LSB) are delayed by 90°, which is equivalent to advancing the negative RF components by 90°, the IF spectra depicted in Fig. 7.121b result. Note that the positive and negative frequency IF components from mixer B due to the upper and lower RF sidebands are 180° out of phase. If we further delay the IF output of mixer B by 90° (IF hybrid), the spectra shown in Fig. 7.121c result. Thus when the IF outputs of mixers A and B are summed in the IF hybrid, the USB components of the IF signal will cancel, leaving only the positive and negative IF LSB components. If the IF outputs of mixers A and B are subtracted, only the USB will be present.

When all the components in the circuit are perfectly matched, image cancellation is complete. Unfortunately, perfection is difficult to obtain, although sometimes demanded; hence the rejection of the image will be finite. This rejection [7.41] which is a measure of the circuit performance, can be expressed as a function

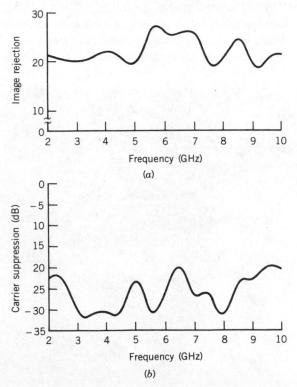

Figure 7.125 Performance of 2- to 10-GHz image-rejection mixer: (a) image rejection; (b) carrier suppression performance.

of the total circuit amplitude and phase imbalance. We then can define the image-rejection as

$$R_i = \left[\frac{1 - 2(A)^{1/2} \cos \Delta\theta + A}{1 + 2(A)^{1/2} \cos \Delta\theta + A} \right] \qquad (7.96)$$

where $\Delta\theta$ is the phase imbalance and A is the amplitude imbalance. Typically, well-designed broadband mixers can achieve image rejections of approximately 20 dB, and narrowband designs can sometimes achieve as much as 30 dB. A suppression of 20 dB corresponds to an amplitude imbalance of 1 dB and phase errors of less than 10°. A convenient graph for determining image rejection based on circuit amplitude and phase errors is shown in Fig. 7.122.

A typical broadband image-rejection mixer (Fig. 7.123) which was fabricated on a 0.5-mm-thick quartz substrate was configured as shown in Fig. 7.124. The image rejection achieved for the 2- to 10-GHz frequency range is shown in Fig. 7.125a and the carrier suppression performance is shown in Fig. 7.125b. The RF quadrature coupler, which is visible in the photograph, was realized with a velocity-compensated multisection coupler [7.42] employing an eight-strip Lange center section. The IF hybrid was external to the mixer.

A similar dual double-balanced mixer and multisection coupler (Fig. 7.126), configured as shown in Fig. 7.127, was evaluated as a broadband SSB modulator. The sideband suppression and carrier suppression for the modulator are shown in

Figure 7.126 Single-sideband 8- to 18-GHz modulator fabricated on a quartz substrate (courtesy of Texas Instruments).

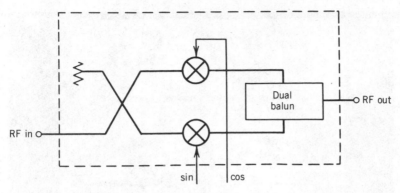

Figure 7.127 Single-sideband modulator circuit configuration.

Fig. 7.128. The mixers and coupler were fabricated on a quartz substrate and the modulation signal was supplied with the proper sin/cos relationship.

Active devices can also be used as nonlinear elements in SSB modulators and image-rejection mixers. A convenient monolithic GaAs FET double-balanced mixer structure, reported by Thompson and Pavio [7.43], is shown in Fig. 7.129. The circuit (Fig. 7.130) consists of two differential pairs with their associated current sources and an external quadrature coupler. The modulation input signal was also supplied from an external source with the proper sin/cos phase relationship.

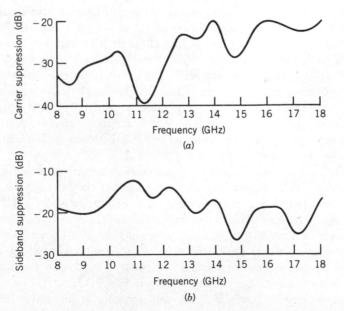

Figure 7.128 Single-sideband modulator performance: (a) carrier suppression; (b) sideband suppression.

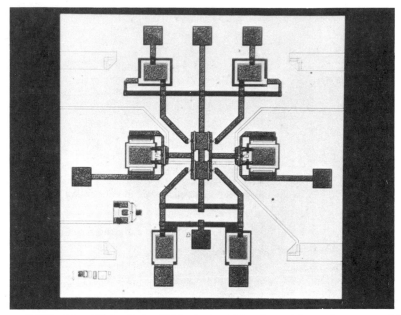

Figure 7.129 Monolithic GaAs FET single-sideband modulator (courtesy of Texas Instruments).

When the RF (carrier) signal is applied to one gate of a differential pair, currents equal in magnitude and opposite in phase are produced in each FET. Since the drains of each FET are connected together, the carrier is canceled. However, since the modulation is applied to each FET differentially, the phase difference between the carrier and modulation in each FET is the same; thus the IF currents developed are summed at the drain terminals of each FET pair. The carrier and modulation are then applied to each pair in quadrature so that only one sideband of the modulator waveform appears at the circuit output. Sideband selection can be accomplished by reversing the phase relationship between the modulation signals ($\sin/\cos$ to $\sin/-\cos$).

A typical output waveform, which was calculated using Microwave SPICE, is shown in Fig. 7.131. It is interesting to note that the number of cycles plotted for the input waveform is one more than the number of cycles of the output waveform for the same time span. Hence the analysis clearly demonstrates lower sideband performance. Although the analysis was conducted at a carrier frequency of 7 GHz, the output voltage waveform of the circuit was verified using a traveling-wave oscilloscope at a carrier frequency of 1 GHz. The measured versus computed spectral performance of the circuit is shown in Fig. 7.132.

Another important structure is the subharmonically pumped mixer. Subharmonic pumping is usually employed when fundamental LO injection is not feasible due to oscillator frequency limitation, noise performance, or economics. These

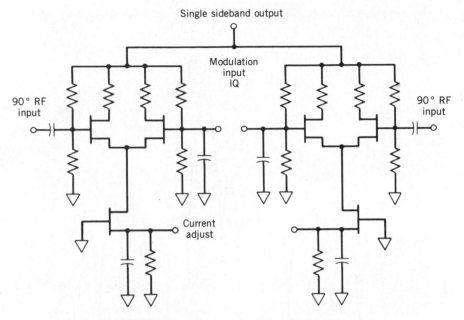

Figure 7.130 FET modulator circuit configuration.

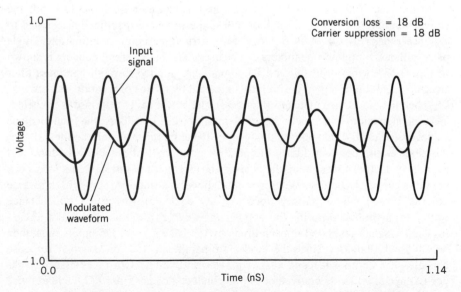

Figure 7.131 Input and output voltage waveforms for SSB modulator.

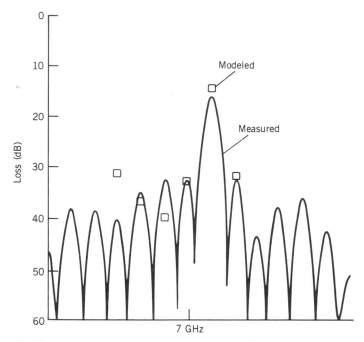

Figure 7.132   Measured versus computed spectral performance of SSB modulator.

techniques are common in the millimeter frequency range, but are sometimes employed in low-cost commercial products.

Any mixer structure can be operated at a subharmonic of the LO, but the conversion efficiency is usually poor unless the mixer is specifically designed to enhance a particular $1 \times m$ product. Most structures employ antiparallel diode pairs, although single-diode mixers can be used. The basic circuit concept is shown in Fig. 7.133, where the LO and RF signals are applied through bandpass filter structures and the IF signal is similarly extracted by means of a bandpass network.

When the diodes are matched (identical), the 1×1 (1,1) response is canceled. This phenomenon can easily be explained if one considers the conductance waveform of each diode and its composite. If the circuit had only a single diode, the diode would conduct during the positive-going half of the LO waveform and be nonconductive throughout the rest of the cycle; thus one conductance peak would occur per LO cycle. When a second diode is introduced, but with a reverse polarity, it too conducts only once per LO cycle. However, its conductance waveform is produced during the opposite half-cycle from that of the first diode.

The IF signal generated in each diode by the LO (f_p) and RF inputs must then be 180° out of phase. Since the diodes are paralleled, the fundamental response cancels. The composite conductance waveform exhibits two conductance peaks per LO cycle, which is equivalent to one conductance peak per LO cycle at twice

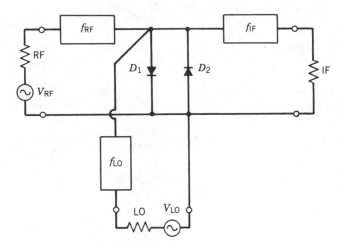

Figure 7.133   Simple subharmonically pumped mixer circuit configuration.

the LO frequency ($2F_p$). This is essentially the same 1×2 $(1,2)$ mixing response that would be obtained with a single diode, except for differences due to the diode's junction capacitance waveform. Relatively efficient mixer performance can also be obtained using the fourth harmonic of the LO, which yields conversion losses in the range 15 to 20 dB.

The previous discussion presented just a few of the special or unique mixer structures that are currently in use. As GaAs FET and bipolar technology mature, many new structures and some old friends from the linear IC world will start to find their way into the microwave region. However, it will still be difficult to surpass the performance of a well-designed diode and diode mixer.

REFERENCES

7.1   W. Schottky. "Halbleitertheorie der Sperrschicht," *Naturwissenschaften*, Vol. 26, 1938, p. 843.

7.2   W. Schottky, *Z. Physik*, Vol. 188, 1942, pp. 539.

7.3   H. A. Bethe, "Theory of the Boundary Layer of Crystal Rectifiers," *MIT Rad. Lab. Report 43–12*, 1942.

7.4   B. C. DeLoach, "A New Technique to Characterize Diodes and an 800 GHz Cutoff Frequency Varactor at Zero Volts Bias," *IEEE Transactions on Microwave Theory and Techniques*, Vol. MTT-12, 1964, p. 15.

7.5   N. Houlding, "Measurement of Varactor Quality," *Microwave Journal*, Vol. 3, 1960, p. 40.

7.6   A. A. M. Saleh, *Theory of Resistive Mixer*, MIT Press, Cambridge, Mass., 1971.

7.7   A. R. Kerr, "A Technique for Determining the Local Oscillator Waveforms in a Microwave Mixer," *IEEE Transactions on Microwave Theory and Techniques*, October 1975.

7.8 S. Egami, "Nonlinear, Linear Analysis and Computer-Aided Design of Resistive Mixers," *IEEE Transactions on Microwave Theory and Techniques*, March 1973.

7.9 D. N. Held and A. R. Kerr, "Conversion Loss and Noise of Microwave and Milli-meter-Wave Mixers: Part 1—Theory," *IEEE Transactions on Microwave Theory and Techniques*, February 1978.

7.10 D. N. Held and A. R. Kerr, "Conversion Loss and Noise of Microwave and Milli-meter-Wave Mixers: Part 2—Experiment," *IEEE Transactions on Microwave Theory and Techniques*, February 1978.

7.11 S. Egami, "Nonlinear, Linear Analysis and Computer-Aided Design of Resistive Mixers," *IEEE Transactions on Microwave Theory and Techniques*, Vol. MTT-22, 1974, p. 270.

7.12 A. R. Kerr, "Shot-Noise in Resistive-Diode Mixers and the Attenuator Noise Model," *IEEE Transactions on Microwave Theory and Techniques*, February 1979.

7.13 W. M. Kelly and G. T. Wrixon, "Conversion Losses in Schottky-Barrier Diode Mixers in the Submillimeter Region," *IEEE Transactions on the Microwave Theory and Techniques*, Vol. MTT-27, 1979, p. 665.

7.14 F. Bernues, H. J. Kuno, and P. A. Crandell, "GaAs or Si: What Makes a Better Diode?" *Microwaves*, March 1981.

7.15 C. L. Ruthroff, "Some Broadband Transformers," *Proceedings of the IRE*, Vol. 47, 1959, p. 1337.

7.16 C. W. Gerst, "New Mixer Designs Boost D/F Performance," *Microwaves*, October 1973.

7.17 J. W. Duncan, "100:1 Bandwidth Balun Transformer," *Proceedings of the IRE*, February 1960, pp. 156–164.

7.18 R. W. Kopfenstein, "A Transmission Line Taper of Improved Design," *Proceedings of the IRE*, January 1956, pp. 31–35.

7.19 M. A. Smith, K. J. Anderson, and A. M. Pavio, "Decade-Band Mixer Covers 3.5 to 35 Ghz," *Microwave Journal*, February 1986, pp. 163–171.

7.20 N. Marchand, "Transmission Line Conversion Transformers," *Electronics*, Vol. 17, No. 12, 1944, p. 142.

7.21 B. R. Hallford, "A Designer's Guide to Planar Mixer Baluns," *Microwaves*, Vol. 18, No. 12, 1979, p. 52.

7.22 B. R. Hallford, "Simple Balun Coupled Mixers," *IEEE MTT-S International Microwave Symposium Digest*, 1981, pp. 304–306.

7.23 A. M. Pavio and M. A. Smith, "An 18–40 GHz Double Balanced Microstrip Mixer," *IEEE International Microwave Symposium Digest*, June 1985, pp. 379–381.

7.24 M. A. Smith, A. M. Pavio, and B. Kim, "A Ka-Band Dual Channel Tracking Receiver Converter," *IEEE International Microwave Symposium Digest*, June 1986, pp. 643–644.

7.25 B. Siegal, "The Triple Barrier Schottky Diode," *Microwave Associates Application Note 2*.

7.26 B. R. Hallford, "Schottky Diodes Upconvert with a Mere 0.25 dB Loss," *Microwaves*, March 1980, pp. 63–70.

7.27 R. A. Pucel, D. Masse, and R. Bera, "Performance of GaAs MESFET Mixers at X-Band," *IEEE Transactions on Microwave Theory and Techniques*, June 1976.

7.28 G. Begemann and A. Hecht, "The Conversion Gain and Stability of MESFET Gate Mixers," *Conference Proceedings of the 9th European Microwave Conference*, 1979.

7.29 G. K. Tie and C. S. Aitchison, "Noise Figure and Associated Conversion Gain of a Microwave MESFET Gate Mixer," *Conference Proceedings of the 13th European Microwave Conference*, 1983.

7.30 H. T. Friis, "Noise Figure of Radio Receivers," *Proceedings of the IRE*, July 1944, pp. 419–422.

7.31 G. Begemann and A. Jacob, "Conversion Gain of MESFET Drain Mixers," *Electronics Letters*, August 30, 1979.

7.32 P. Bura and R. Dikshit, "FET Mixers for Communication Satellite Transponders," *IEEE MTT-S International Microwave Symposium Digest*, 1976.

7.33 C. Tsironis, R. Meierer, and R. Stahlmann, "Dual-Gate MESFET Mixers," *IEEE Transactions on Microwave Theory and Techniques*, Vol. MTT-32, 1984, p. 248.

7.34 C. Tsironis and R. Meierer, "Microwave Wide-Band Model of Dual-Gate MESFETs," *IEEE Transactions on Microwave Theory and Techniques*, Vol. MTT-30, 1982, p. 243.

7.35 T. S. Howard and M. Pavio, "A Distributed Monolithic Dual-Gate FET Mixer," *IEEE Microwave and Millimeter-Wave Monolithic Circuits Symposium Digest*, June 1987, pp. 27–30.

7.36 T. S. Howard and A. M. Pavio, "A Distributed 1–12 GHz Dual-Gate FET Mixer," *IEEE International Microwave Symposium Digest*, June 1986, pp. 329–332.

7.37 A. D. Young, *Designing with Field-Effect Transistors*, McGraw-Hill, New York, 1981.

7.38 R. B. Culbertson and A. M. Pavio, "An Analytic Design Approach for 2–18 GHz Planar Mixer Circuits," *IEEE MTT-S International Microwave Symposium Digest*, 1982, pp. 425–427.

7.39 A. M. Pavio, R. H. Halladay, S. D. Bingham, and C. A. Sapashe, "Monolithic Double-Balanced Diode and GaAs FET Mixers," *IEEE Transactions on Microwave Theory and Techniques,* Vol. MTT-36, December 1988.

7.40 C. Tsironis, E. Stahlmann, and F. Ponse, "A Self-Oscillating Dual Gate MESFET X-Band Mixer with 12 dB Conversion Gain," *Conference Proceedings of the 9th European Microwave Conference*, 1979.

7.41 D. Norgaard, "The Phase-Shift Method of Single Sideband Signal Generation," *Proceedings of the IRE*, December 1956, pp. 1718–1735.

7.42 A. M. Pavio, D. L. Allen, S. D. Thompson, and S. G. Goldman, "A Solid-State 2–10 GHz 1 Watt TWT Replacement Amplifier," *IEEE MTT-S International Microwave Symposium Digest*, 1982, pp. 221–223.

7.43 S. D. Thompson and A. M. Pavio, "A Monolithic Double Balanced Single Sideband Modulator," *IEEE International Microwave Symposium Digest*, June 1987, pp. 898–902.

BIBLIOGRAPHY

Barber, M. R., "Noise Figure and Conversion Loss of the Schottky Barrier Mixer Diode," *IEEE Transactions on Microwave Theory and Techniques*, November 1967.

Carlson, E., M. V. Schneider, and T. F. McMaster, "Subharmonically Pumped Milli-meter-Wave Mixers," *IEEE Transactions on Microwave Theory and Techniques*, vol. MTT-26, p. 706, 1978.

Clifton, B. J., "Schottky Diode Receivers for Operation in the 100–1000 GHz Region," *Radio and Electronic Engineer*, vol. 49, 1979, p. 333.

Collin, R., *Foundations for Microwave Engineering*, McGraw-Hill, New York, 1966.

Cripps, S. C., O. Nielsen, D. Parker, and J. A. Turner, "An Experimental Evaluation of X-Band GaAs FET Mixers Using Single and Dual-Gate Devices," *IEEE MTT-S International Microwave Symposium Digest*, June 1977.

Ernst, R. L., P. Torrione, W. Y. Pan, and M. M. Morris, "Designing Microwave Mixers for Increased Dynamic Range," *IEEE Transactions on Electromagnetic Compatibility*, November 1969.

Evans, Arthur D., *Designing with Field-Effect Transistors*, McGraw-Hill, New York, 1981.

Gardiner, J. G., "An Intermodulation Phenomenon in the Ring Modulator," *The Radio and Electronic Engineer*, April 1970.

Hallford, B. R., "Single-Sideband Mixers for Communications Systems," *IEEE MTT-S International Microwave Symposium Digest*, 1982.

Hicks, R. G., and P. J. Khan, "Numerical Analysis of Subharmonic Mixers Using Accurate and Approximate Models," *IEEE Transactions on Microwave Theory and Techniques*, December 1982.

Hines, M. E., "Image Conversion Effects in Diode Mixers," *IEEE MTT-S International Microwave Symposium Digest*, June 1977, pp. 487–490.

Johnson, K. M., "X-Band Integrated Circuit Mixer with Reactively Terminated Image, *IEEE Transactions on Microwave Theory and Techniques*, July, 1968.

Kerr, A. R., R. J. Mattauch, and J. A. Grange, "A New Mixer Design for 140-220 GHz," *IEEE Transactions Microwave Theory and Techniques* MTT-25, 1977, p. 399.

Maas, Stephen A., *Microwave Mixers*, Artech House, Norwood, MA, 1986.

Maiuzzo, M. A., and S. H. Cameron, "Response Coefficients of a Double-Balanced Diode Mixer," *IEEE Transactions on Electromagnetic Compatibility*, November 1979.

Mania, L., and G. B. Stracca, "Effects of the Diode Junction Capacitance on the Conversion Loss of Microwave Mixers," *IEEE Transactions on Communications*, September 1974.

McColl, M., "Conversion Loss Limitations on Schottky-Barrier Mixers," IEEE Transactions on Microwave Theory and Techniques, January 1977.

O'Neill, H. J., "Image-Frequency Effects in a Microwave Crystal Mixer," Proceedings of IRE, vol. 112, November 1963, pp. 2019–2024.

Orloff, L. M., "Intermodulation Analysis of Crystal Mixer," *Proceedings of the IEEE*, February 1964.

Pavio, Anthony M., "Nonlinear and Special Purpose Circuit Design," *IEEE Galium Arsenide Integrated Circuits Symposium Lecture Series*, October 1987.

Pucel, R., H. A. Haus, and H. Statz, "Signal and Noise Properties of GaAs Microwave Field Effect Transistors," in L. Martin, (ed.), *Advances in Electronics and Electron Physics*, vol. 38, Academic Press, New York, 1975, p. 195.

Rohde, Ulrich L., and T.T.N. Bucher, *Communications Receivers*, McGraw-Hill, New York, 1987.

Scott, J. R., and R. A. Minasian, "A Simplified Microwave Model of the GaAs Dual-Gate MESFET," *IEEE Transactions on Microwave Theory Techniques* MTT-32, 1984, p. 243.

Siegal, Bernard S., "Schottky Diodes—Where We Stand Today," *Microwaves*, April 1971.

Smith, M. A., T. S. Howard, K. J. Anderson, and A. M. Pavio, "RF Nonlinear Device Characterization Yields Improved Modeling Accuracy," *IEEE International Microwave Symposium Digest*, June 1986, pp. 381–384.

Sze, S. M., *Physics of Semiconductor Devices*, Second ed., Wiley, New York, 1981.

Van Der Ziel, A., and R. L. Waters, "Noise in Mixer Tubes," *Proceedings of IRE*, vol. 46, 1958, p. 1426.

Van Der Ziel, A., *Noise: Sources, Characterization, and Measurement*. Prentice-Hall, Englewood Cliffs, NJ, 1970.

Watson, H. A., *Microwave Semiconductor Devices and Their Circuit Applications*, McGraw-Hill, New York, 1969.

Wrixon, G. T., "Select the Best Diode for Millimeter Mixers," *Microwaves*, vol. 15, no. 9, 1976, p. 56.

8 Microwave Computer-Aided Workstations for MMIC Requirements

8.0 INTRODUCTION

Recent changes in technology have led to designs at operating frequencies in the millimeter wave range: above 26 GHz almost to 100 GHz. These designs have been made possible by GaAs foundry services, and there are at least 11 foundries in the United States that provide such services. Many of the microwave/millimeter wave monolithic integrated circuit (MMIC) requirements come from particular high-volume/low-cost designs for military applications. This fact is highlighted by the recent broad agency announcement (BAA) Phase 0. This interest in affordable MMIC activities, together with a reduction in cost of CAD workstations, has provided tools for a MMIC computer-aided design/manufacturing/test workstation. The latest computer workstations using 32-bit microprocessors have increased the computational power to almost true-mainframe performance at PC workstation cost. This has been made possible by 32-bit processors such as the 68020 by Motorola and the 80386 by Intel. In addition, a certain number of technological capabilities are required. Table 8.1 lists required synthesis/analysis capabilities to accommodate the need for MMIC requirements and to provide an integrated microwave CAE/CAM/CAT workstation.

In this chapter we show the Compact Software approach to an integrated CAE/CAM/CAT workstation for general microwave design use, including specific examples and in response to the MMIC goals. The ability to do tolerance analysis or yield optimization is probably most significant in light of the fact that the purpose of the exercise is to be able to provide reliable and cost-effective production. This has not been possible before.

8.0.1 Integrated Microwave Workstation Approach

Figure 8.1 shows a 1989 CAD software workstation approach. This follows the flow of a practical CAD approach for microwave processes or MMIC design, as shown in Fig. 8.2. Starting with system definition and specification, the design is broken up into a variety of subsystems and then further, into circuit blocks. Inside the circuit blocks, active and passive devices must be separated.

A first approach should be to simulate the active devices to get direction for the

TABLE 8.1

1. Linear and nonlinear synthesis programs
   A. Matching networks for single-frequency and wide-frequency bands (e.g., a $4:1$ for complex loads and termination).
   B. Narrowband/wideband lumped and distributed filter synthesis.
   C. Oscillator synthesis from small- or large-signal S parameters. Parallel–series design, determination of all components, determination of efficiency, output power, phase noise, and other relevant data.
   D. Open and closed loop, PLL design, phase noise determination, nonlinear switching, frequency lock phase lock.
   E. Systems analysis and optimization for noise figure IMD performance.

2. Linear and nonlinear analysis program
   A. Analysis of lumped and distributed elements.
   B. Optimization of lumped elements against measured S parameters.
   C. Analysis of linear bipolar and field-effect transistors with provision for temperature and bias (new concept for noise tuning).
   D. Optimization of bipolar and FET models (and HEMT and dual-gate FETs) against measured S parameters.
   E. Analyze and optimize performance of circuits with arbitrary combinations of active and passive devices. These calculations must be available from either electrical or physical parameters.
   F. All proximity effects caused by electrical or magnetic fields must be predictable. Special effects such as multiple coupled lines, cover effects, and multilayer dielectric substances must be included. Arbitrary layouts should be handled using spectral domain techniques. To reduce computational speeds, look-up-table generators should be considered.
   G. A nonlinear analysis of high-frequency circuits can be within limitations using SPICE-like programs. A better way is to use a modified harmonic balanced method that splits the FET and bipolar model into a linear and nonlinear time domain/frequency domain analysis. This technique allows interfacing with a linear program such as SUPER-COMPACT, has the advantage of providing an optimizer, and will distribute elements that are lacking in SPICE programs. In addition, the slow execution speed of SPICE is overcome in this approach.
   H. For accurate device modeling and determination of equivalent circuits special test equipment and modeling software are required that cover both dc and low-frequency analysis. Simulation and verification are part of the CAT portion.
   I. To provide reliable and low-cost designs, yield optimization and sensitivity analysis must be provided in addition to the familiar performance optimization. Traditional software approaches looked only at performance optimization.

type of semiconductor that should be used. This can be approached by using either a library of discrete transistors or GaAs foundry information. Device selection is based on the requirements with respect to low noise, high gain, and/or high output power. The active devices will then be incorporated in a circuit. If synthesis programs are available, a circuit synthesis should be performed, which would be the most cost-effective way. In many cases, unfortunately, designers try to optimize

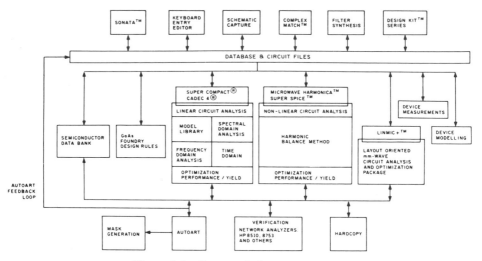

Figure 8.1 Compact Software workstation.

existing circuits for new semiconductors. Most designs do not operate between pure 50-Ω terminations therefore, an overall simulation of the system is essential.

At the systems level, we make a decision whether to accept or reject. If the circuit does not have to be redone, a more detailed analysis, including linear and nonlinear simulation and a first try at layout, including proximity effects, is carried out. If this is acceptable, we look at tests and tolerances and move to the final layout. There we have to check design rules, perform another simulation of the subsystem, and finally, begin fabrication. This is followed by an RF wafer test and total circuit analysis. Unless modifications are required, fabrication can be completed.

Looking again at Fig. 8.1, the first row of programs are essentially synthesis programs. Here we enter electrical specifications for the circuits and obtain circuit element values. The Sonata program is a nonlinear synthesis program for oscillators. The schematic capture program helps to translate a design into a circuit file. Figure 8.3 is a screen picture taken from an Apollo computer showing the schematic capture. The schematic capture program has a library of approximately 145 elements and tracks the program as it is developed. It also contains the synthesis programs shown. These programs write to the data base and output of circuit files, which will then be analyzed. The main block of simulators are for linear circuits and nonlinear circuits. The analysis tools for linear circuits, such as Super-Compact, are based on frequency-domain analysis and provide either performance or yield optimization. Next we introduce to you the concept of MMIC design and foundry consideration.

8.0.2 Nonlinear Tools

The first approach to nonlinear CAD goes back to the development of SPICE (Simulation Program with Integrated Circuit Emphasis) at the University of

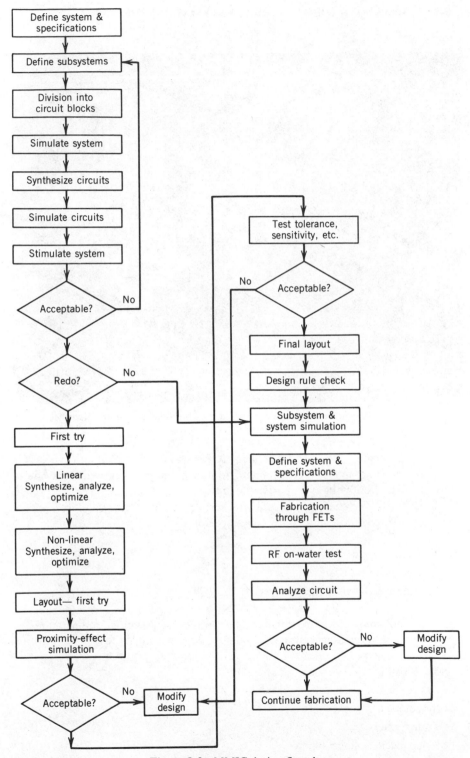

Figure 8.2 MMIC design flowchart.

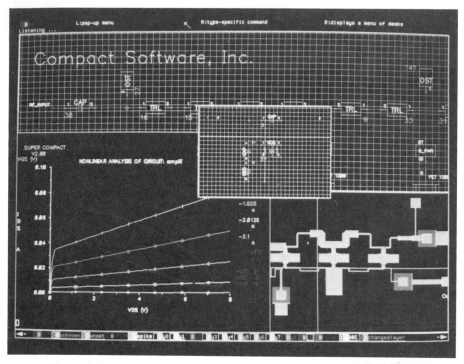

Figure 8.3 Screen dump of Compact Software schematic capture interface.

California–Berkeley. Most commercial SPICE programs are based on version 2G5 or 2G6, which are in the public domain. Although many readers will be aware of this approach, following is a short summary of its capabilities.

Circuit behavior can be simulated with respect to time, frequency, and voltage variation. Three different analyses can be performed:

1. Nonlinear dc
2. Nonlinear transient
3. Linear ac small-signal analysis

Any of these analyses can be conducted at various user-specified temperatures. The following element types are found:

Resistors
Capacitors
Inductors
Transformers
Independent voltage sources

Transmission lines

Diodes

Bipolar junction transistors (BJTs)

Junction field-effect transistors

MOSFETS

Compared to a linear program such as Super Compact, there are some restrictions.

1. There is no optimizer.
2. As the program is a time-domain-only simulator, it is much slower then the frequency-domain approach, and the execution speed depends on the value of the components.
3. Most needed passive microwave elements are missing (see Appendix D).

Table 8.2 shows a typical input form for a SPICE program. Note the data required for describing the bipolar transistor and the FET. To overcome these disadvantages, the harmonic balance method was created. An introduction to this that combines the best of the time- and frequency-domain techniques is given later.

The rest of the chapter is broken down into a series of sections that introduce new CAD-related techniques:

8.1   Introduction to foundry services and CAD

8.2   Introduction to field optimization

8.3   Introduction to the harmonic balance method

8.4   Load-pull technique using programmable microwave tuners

8.5   Introduction to MIMIC considering layout effects

8.6   Introduction to layout-related problems

8.7   Practical design examples and examples for CAD software

8.1   GALLIUM ARSENIDE MONOLITHIC MICROWAVE IC FOUNDRIES: THE ROLE OF CAD

A GaAs monolithic microwave IC (MMIC) foundry produces circuits to customer designs. MMICs are microwave circuits, such as switches, amplifiers, and receivers, which are integrated within one die. This requires the technology to supply low-noise and power FETs, resistors, capacitors, inductors, transmission lines, diodes and so on. Not only must these components be available to the designer, but they must be highly reproducible and have accurate dc and microwave models. The FET technology, in particular, is critical to the high-frequency performance that can be achieved. Today, foundries typically produce FETs that operate to well over 20 GHz and feature minimum sizes of 0.5 μm, requiring the use of deep ultraviolet or electron-beam lithography.

TABLE 8.2

```
*  <41435_S.CIR>
*    AT41435 MODEL
*
.options acct node abstol-10n nopage
.width out-80
.temp 27
.ac  lin 6  .5ghz 3ghz
*
* packaged chip used in circuit + sources
*
vin 52 0      sin (0) 100mv 1ghz      ac 1
r_rs          52 51                   50
c_cs          51 50                   1000pf
1_li          50 53                   1e4nh
vbb           53 0                    0.814
xr            50 54       at          1000pf
c_cout        54 55                   50
r_rl          55 0                    1e4nh
1_lo          54 56                   8
vcc           56 0
   probe
*
*    model for packaged chip — ux (35)
*
.subckt  at       40 49
l_linp    40 41                    0.05nh
t_tip     41 0 42   0               Z0-66      f-1ghz n1-
l_lbp     42 43                     0.3nh      0.178
c_cbep    42 44                     0.03pf
c_cbcp    42 47                     0.04pf
x_qip     43 47 45 q1
t_t3p     44 0   46 0               Z0-25
c_cecp    44 47                     0.03pf     f-1ghz nl-
1_lep     45 44                     0.2nh      2.175
l_lgp     46 0                      0.02nh
t_t2p     47 0   48 0               Z0-65
l_lop     48 49                     0.05nh
.ends                                          f-1ghz nl-
                                               0.023
*
*    equivalent circuit for die
*
.subckt ql        10 14 17
r_riq1     10 11                    r1      1.21
r_r2q2     11 12                    r2      3.12
r_r3q1     12 13                    r3      2.68
d_d1q1     10 15                    dmod    782
```

TABLE 8.2 (*Continued*)

d_d2q2	11 15		dmod	629
d_d3q1	12 15		dmod	366
r_rcq1	14 15		rq	10
r_req1	16 17		rq	0.24
c_ceq1	15 17			0.03pf
c_cbq1	10 15			0.03pf
q_1	13 15	16	qmod	420

```
.ends
*
* include all resistor models and active SPICE
parameter files
*
.inc res.mod
.inc m414.mod
*
.end
```

To design MMICs successfully, a number of important items must be supplied by the foundry:

1. Well-controlled (fixed) IC process(es)
2. Extensive MMIC-component characterizations
3. A design manual
4. A standard components library with data and layout tape
5. Wafer qualification procedures
6. Recommended CAD software and hardware configurations
7. Support engineering during the customer design cycle

For a foundry to be successful, its MMIC process must be well controlled, reproducible, documented, and of high quality. To do this the foundry must have characterized the dc and RF performance of a large number of standard components and their variations with voltage, temperature, and so on. Also, statistical data are needed in the form of means and standard deviations for the component parameters. These data are important, as MMIC designers need to be able to predict circuit performance spreads since unlike conventional hybrid circuits, MMICs are not easy to adjust in performance after they have been processed.

Figures 8.4 and 8.5 illustrate the standard layouts of a 0.5-μm FET and square spiral inductor, respectively. Figure 8.6 is a scanning electron micrograph (SEM) of the inductor. By using data from a design manual and the standard components or derivatives of them, circuits can be designed to meet particular specifications. During this process the foundry usually provides support engineering, since technical questions that may be unique to particular requirements and not covered by the design manual need to be answered. Support engineering continues until

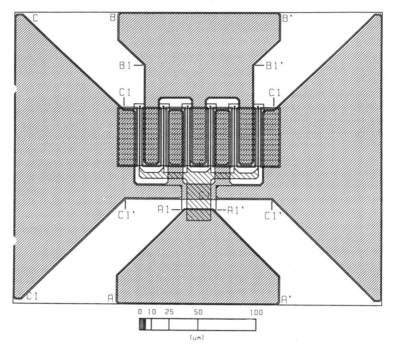

Figure 8.4 Standard 0.5-μm gate length FET layout.

circuit layouts have been completed. Prior to design release, the foundry will check the layout for layer design rule errors and may even check the microwave CAD data files. After design release the masks are procured, the MMIC fabrication completed, and a number of qualification tests made on the wafer lot.

The role of CAD in the design cycle is vitally important. Without modern analysis and optimization tools, MMIC design would be very nearly impossible and, at best, very time consuming! As an example, consider the circuit in Fig. 8.7. This is a microphotograph of a 1 × 1.5 mm 2- to 8-GHz amplifier having 15 dB of gain and 100 mW of output power. It is a simple circuit but typical of today's MMIC foundry circuit complexity. The circuit contains 2 FETs, 11 capacitors, 6 resistors, 5 spiral inductors, 16 transmission lines, and 15 bond pads. The analysis data file for this IC, however, contains 125 data lines, with a typical analysis time on a standard IBM AT PC computer with math coprocessor of 4 minutes. The analysis file is complex because it includes a number of models for monolithic components such as thin-film resistors, inductors, and so on. Analysis and optimization programs are only just starting to put MMIC component models into their routines.

A foundry can use a mixture of their own models and some standard ones. Sometimes these standard models are not sufficient because they do not relate directly to the particular foundry process. One way around this is for the foundry to work with a software developer to produce models and data-bank information that will allow analysis to be more efficient and accurate.

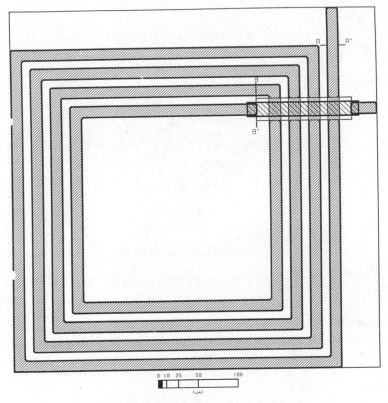

Figure 8.5 Standard four-turn spiral inductor layout.

Circuit analysis and optimization such as that described above can be achieved more efficiently by "pulling" models from a data bank, saving the foundry user the time needed to generate them initially, and the computer the time taken to analyze them. To avoid limitations in this approach during optimization, closed-form equations are needed to allow automatic calculation of full component parameters.

Software such as Linmic+ is available to model more accurately such components as inductors and transformers, and any coupling effects between them can be accounted for. This allows the packing density of MMICs to be increased, which directly affects die costs. This would also allow the generation of the closed-form equations mentioned above without forcing the foundry to go through the time-consuming, tedious, and expensive tasks of component fabrication, test, and modeling.

Of increasing importance to users of foundry services is the need to be able to investigate circuit response as a function of temperature and FET or diode bias currents and voltages. Traditionally, this has been achievable on a computer only by using SPICE, where the accuracy of the simulation is limited by the active component models. However, by introducing bias- and temperature-dependent S

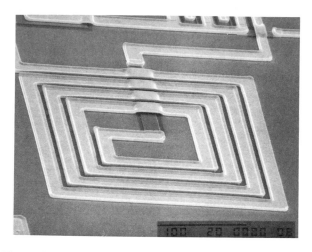

Figure 8.6   Scanning electron micrograph of a four-turn inductor.

parameters into a modern linear analysis program, the designer can quantify these important effects. To be usable, however, this needs to be done in a complete way. It is necessary, for example, to have temperature coefficients for all components. Many of these effects are technology dependent and therefore need to form part of the foundry data bank.

Many foundry users need to develop nonlinear circuits, such as oscillators, multipliers, and mixers for receivers and power amplifiers for transmitters. Traditionally, the characterization of large-signal models for FETs, diodes, and so on, has been very time consuming and, until recently, only approximate. Equally, the

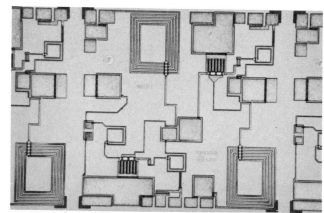

Figure 8.7   Microphotograph of a 2- to 8-GHz MMIC amplifier.

CAD base to the general microwave community for nonlinear analysis has been poor. The need for foundries to supply accurate large-signal bias-dependent models is as important as the need for an efficient, general-purpose nonlinear analysis and optimization program. By introducing the concept of a load-pull tuner for characterization of large-signal performance of active devices (see Section 8.2), and the Microwave Harmonica program, a program based on the harmonic balance method (Compact Software, Inc.), both needs are now being met and improved MMIC designs are effected.

BIBLIOGRAPHY

Pengelly, Raymond, Compact Software's *Transmission Line News*, Vol. 1, No. 2, September 1987, pp. 4–5.

8.2 YIELD-DRIVEN DESIGN

The ability to produce low-cost MMICs in large quantities will be significantly enhanced by the microwave and millimeter wave integrated circuit (MIMIC) program, sponsored by the Department of Defense. Not only has it provided manufacturers with the incentive to speed research and development, but it has created a need for a higher level of computer-aided design, with all operations from design to fabrication integrated into a single interactive package.

A missing link has been the ability to design circuits for maximum yield when element tolerances are considered. A major step toward realization of that goal was taken by Compact Software in its role as a member of the Raytheon/Texas Instruments MIMIC team. It performed tolerance analysis (yield optimization) and accurate assessments of the yield of MMICs across a wafer.

The technique is a significant departure from the traditional manner in which MMICs are designed, or for that matter, the way microwave circuits in general are designed and constructed. It incorporates a range of tolerances rather than using rigid values for circuit elements. Think of it as fine versus coarse tuning.

The current manner in which circuits are designed using microwave CAE strives for optimum performance at the expense of yield. The new approach strives for performance within an acceptable range that is compatible with optimum yield; that is, performance criteria are met while keeping cost down. It is this "real world" approach that is demanded by the requirements of the MIMIC program: high-volume low-cost circuits, with acceptable performance.

8.2.1 No Simple Task

Yield optimization is a complex task, and was previously unavailable to the MMIC designed. It is actually an old concept, having been proposed by John Bandler more than 10 years ago. It bears close scrutiny. The concept was first proposed for MMIC design by Robert A Pucel at the 1984 GaAs IC Symposium.

Pucel concluded that since the optimization routines used in all commercial CAE software do not include the variability of commercial passive and active components in their analysis, the process is better suited to the research and development environment rather than to commercial production, where manufacturing yield is important.

The present method of CAE cannot guarantee that the optimized design will yield the lowest spread in performance for the given range of tolerance. It is inadequate for MMIC design and will not readily lead to circuit designs or topologies with the highest design yield.

The variations in performance inherent in any device used in a microwave circuit generally require that "tweaking" take place after the circuit is fabricated. Tweaking is not convenient for MMICs, but even if it were, it would not be desirable. Ideally, to achieve the goal of acceptable performance as well as optimum yield, the customer should allow the manufacturer a range of performance to which the delivered part can conform and still meet system requirements.

Manufacturers must have in their CAE program a database of reliable information on the components used in the MMICs. This is especially critical for active devices, which are always the components with the highest degree of uncertainty. The database must include tolerance data, accompanying statistics, and probability distribution.

8.2.2   Rethinking Design

Yield optimization requires a change in the design process. Rather than optimizing a circuit for best performance using one set of device values, the tolerance data are used for optimization, in a process called "design centering," a phrase attributable to Bandler.

We can illustrate this process using a graph. For the sake of simplicity, let's assume that the range of acceptable performance (Fig. 8.8) allows only two element

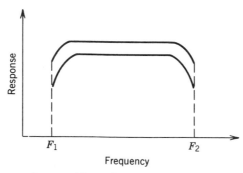

Figure 8.8   This range of acceptable performance represents a "window" whose limits represent acceptable performance from the customer's standpoint.

values (P_1 and P_2) to be varied. In the current CAE design process, only a fixed set of values can be used to optimize the circuit for maximum performance, with no regard for element tolerances.

It is more likely that more than one set of element values will be permissible if a "window" of acceptable performance is specified rather than a best-possible performance, thus allowing element tolerances. These sets of element values can be represented on a graph by defining the entire region of possibility (Fig. 8.9). This "element constraint region" consists of all possible element combinations that will yield performance within the acceptable range.

Taking this one step further, let's add tolerances to elements P_1 and P_2, and assume that they are distributed about some nominal value for each pair. This tolerance condition can be depicted by a rectangle (Fig. 8.10), within which is the tolerance of any pair of element values. The probability that any combination of elements values is in the area of the rectangle is inversely proportional to the area of the rectangle. By superimposing the tolerance rectangle over the element constraint region (Fig. 8.11) for some arbitrary choice of nominal values of P_1 and P_2, the number of circuits that will satisfy the performance window will be proportional to the fraction of the rectangle area that falls within the element constraint region.

8.2.3 Hitting the Mark

Thus, if the element values are chosen correctly, the design yield can be optimized more accurately. In the example shown (Fig. 8.12), the overlap in the tolerance rectangle and element constraint region is maximum. Of course, by loosening the performance window or reducing tolerances, 100% yield can be achieved (Fig. 8.13).

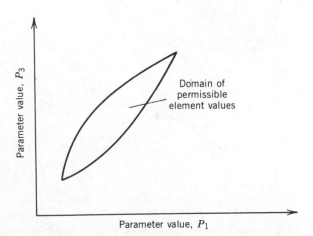

Figure 8.9 The permissible element values will provide performance defined as acceptable in Fig. 8.8.

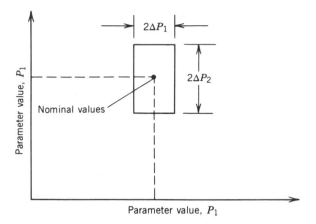

Figure 8.10 The rectangular area is the tolerances of circuit elements distribu.ed around a nominal valu⁻

This method, which could be called the "cost-driven" approach, is much more realistic than the "performance-driven" approach. It is intrinsically attuned to what the MMIC manufacturer will encounter in the real world, that is, assuming variability in device performance.

It is important to remember that design yield has been determined in advance of production. No wafers have been fabricated. With the cost of one pass through the foundry at approximately $50,000, this is an extraordinary saving in time,

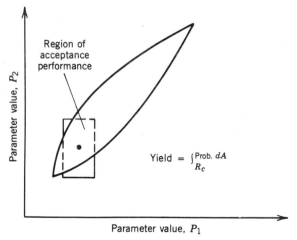

Figure 8.11 Yield is determined by superimposing the tolerance rectangle of Fig. 8.10 over the area of acceptable element values. The yield is proportional to the amount of rectangle that falls within the acceptable element value region.

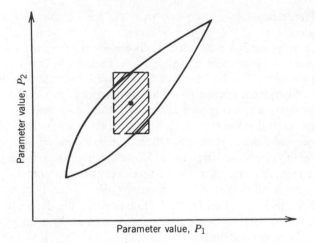

Figure 8.12 If element values are chosen carefully, yield can be maximized. This illustration represents that condition.

material, and processing costs. It is also possible to determine at this eary stage that the circuit topology chosen is inappropriate for production, which reaps additional benefits.

8.3 DESIGNING NONLINEAR CIRCUITS USING THE HARMONIC BALANCE METHOD

The use of nonlinear components such as bipolar transistors, GaAs FETs, and microwave diodes makes it necessary to predict large signal-handling perfor-

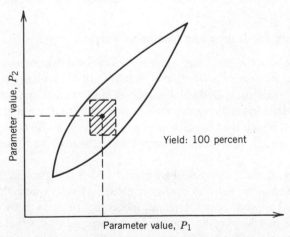

Figure 8.13 By relaxing the range of acceptance or tightening tolerance, 100% yield can be achieved.

mance. The traditional tools to do this were the SPICE approach and Volterra series expansion.

The SPICE program is a program operating solely in the time domain. SPICE is an outstanding workhorse for dc analysis as a function of bias and temperature and transient analysis. The drawbacks of SPICE are: (1) the lack of an optimizer; (2) the lack of distributed elements such as tee-junctions, crosses, and others; and (3) the slow execution speed related to the time-domain approach.

Another approach that has been tried is Volterra series expansion. This approach is a simulation where the actual computation time is somewhat independent of the values of the components used in a circuit. However, once the number of harmonics goes up, Volterra series expansion also becomes very time consuming. The Volterra series can be regarded as a nonlinear generalization of the familiar convolution integral. The Volterra series also has the limitation that the degree of nonlinearity must be mild, as the representation otherwise requires an intractably large number of details for adequate modeling.

The recently developed harmonic balance method avoids many of the time-consuming mathematical approaches mentioned previously. This method is a hybrid time- and frequency-domain approach which allows all the advantages of a time-domain device model, combined with the strength of the steady-state frequency-domain technique, to be presented in the lumped and distributed circuit elements in which the device is embedded. The time-domain model can be completely general, thus bypassing complicated determination of coefficients by curve fitting over different bias levels.

As introduced originally, the harmonic balance method is a "single-tone" method that cannot handle the more general case of nonlinearities such as mixers. To handle the multitone aspect, a modified harmonic balance method has been created and multidimensional Fourier transforms have been implemented. A variety of attempts have been made at accurate modeling of the field-effect transistor, including work by Madjar and Rosenbaum, Curtice, and Sussman-Fort.

8.3.1 Splitting the Linear and Nonlinear Portion

The key concept in the harmonic balance method is to take advantage of a linear program such as Super-Compact that handles all the microwave components accurately. The microstrip and stripline discontinuities are of greatest interest. This calculation is done in the frequency domain and thus is fast and efficient. An interface is required which then hands over the information to the nonlinear portion of the program, which uses the harmonic balance method, being computed in the time domain.

Figure 8.14 shows a complete FET model which is separated into an external portion, the parasitics, and the nonlinear model. If we assume that all elements marked with a Z are distributed elements, it becomes obvious that we need a microwave program that handles the transmission lines, a portion that can handle the lumped elements, and a program that can handle the nonlinearities. Figure 8.15 shows the separation between the frequency-domain and time-domain portions.

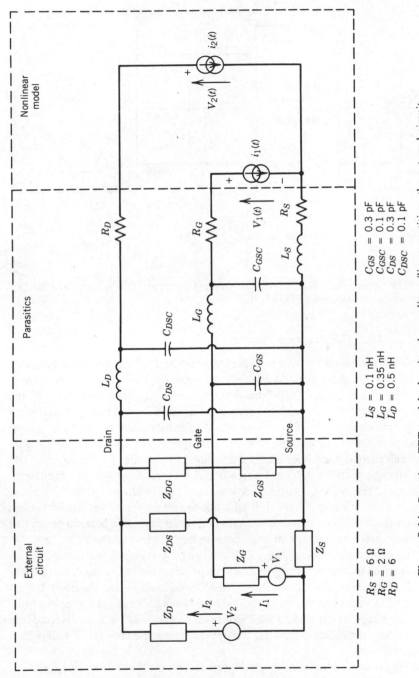

Figure 8.14 Complete FET model, showing the nonlinear/linear partition, the external circuit, and the parasitics used to model the NEC-72089.

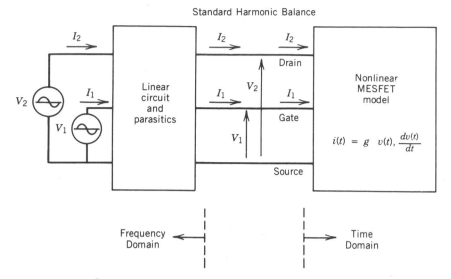

Figure 8.15 Partitioned MESFET circuit. Applied gate and drain voltages and relevant terminal voltages and currents are indicated.

8.3.2 How Does the Program Work?

For a fixed circuit topology (analysis case), the frequency domain is passed through only once; the admittance matrix of the linear subnetwork is computed and stored for subsequent use. Figure 8.16 shows the harmonic balance flowchart. In the time-domain path, the state-variable harmonics are first used to compute the corresponding time-domain waveforms. As mentioned earlier, these are fed to nonlinear device equipment to produce the time-domain device port voltages and currents. Voltage and current harmonics are then described by one- or two-dimensional fast Fourier transforms (FFTs) for the cases of single-tone and two-tone excitation, respectively. The voltage harmonics are used to generate "linear" current harmonics via the linear subnetwork admittance matrix. The two sets of current harmonics are finally compared to produce individual harmonic balance errors and a combined (global) harmonic balance error to be used in a convergence test.

In well-conditioned cases (e.g., FET circuits), a standard Newton–Raphson iteration may be used successfully as an update mechanism even though no starting-point information is available (i.e., if zero initial values are assumed for all harmonics). In such cases the harmonic balance errors are used via a simple perturbation mechanism to generate a Jacobian matrix. The latter is then inverted and applied to the error vector to generate the updated harmonic vectors. The algorithm is fast and accurate.

For circuits containing strongly nonlinear devices such as microwave diodes, a simple Newton iteration may sometimes fail to converge. To overcome this difficulty, Microwave Harmonica incorporates a second iteration scheme based on a

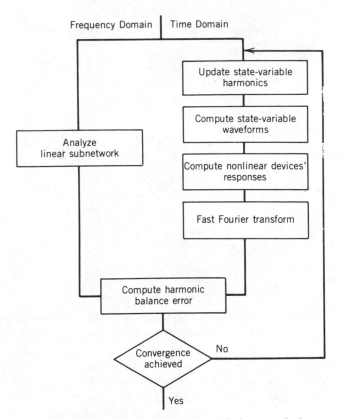

Figure 8.16   Flowchart of harmonic balance analysis.

variable metric algorithm (quasi-Newton iteration), which is slower although considerably more robust than the regular Newton method.

In ill-conditioned cases, the quasi-Newton iteration may be used to approach the required solution. After this has been done to a satisfactory extent, automatic switchover to Newton iteration takes place, so that the approach solution can quickly be refined to any desired accuracy.

Timing information pertaining to the analysis of a single-ended FET mixer is given in Fig. 8.17. When circuit optimization is requested, the algorithm flowchart is modified. Harmonic balance errors are computed in the same way, but now the variable circuit parameters are also updated and the linear subnetwork admittance is computed at each iteration. An objective function is defined as a combination of harmonic balance error and a contribution arising from the electrical specifications. Such an objective is then minimized by the variable metric algorithm until a minimum close enough to zero is reached. Circuit parameters and state-variables harmonics are updated simultaneously, thus avoiding the nesting of nonlinear analysis and circuit optimization loops. Microwave Harmonica is a general tool

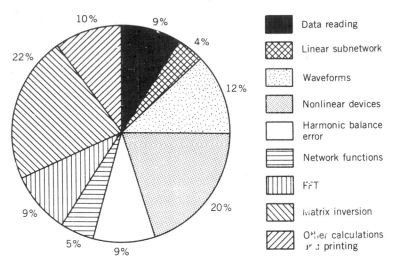

Figure 8.17   Analysis of a single-ended FET mixer. Overall computation time 12.9 s (VAX 8800).

using the harmonic balance method for microwave. The harmonic balance method is a generic mathematical approach and its use in a commercial CAD software is a first.

We mentioned earlier that there are cases where we mix linear and nonlinear components, and as the number of active devices and frequency points increase, the overall speed requirement and program size become an issue. At Compact Software we are looking at both PC-based and mainframe-based versions. In the case of the mainframe, we wanted to make sure that all mainframes that support Fortran 77 can handle the software and that it can be run on the largest machine (or rather, the fastest).

Figure 8.18 shows a plot diagram by which either a VAX or a Cray computer can be addressed. In the PC environment, a nice graphics interface has been developed to provide output information from the simulator. Figure 8.19 shows the FET curves, Fig. 8.20 shows the output waveforms as a function of time, and Fig. 8.21 shows the output level as a function of the input level for different harmonics.

Library Functions.   The program has a large library of components, both idealized and microstrip.

Idealized Components

Series-connected *RLC* one-port
Parallel-connected *RLC* one-port
Two- or three-port transformer
Voltage-dependent voltage source (including delay)
Current-dependent voltage source (including delay)

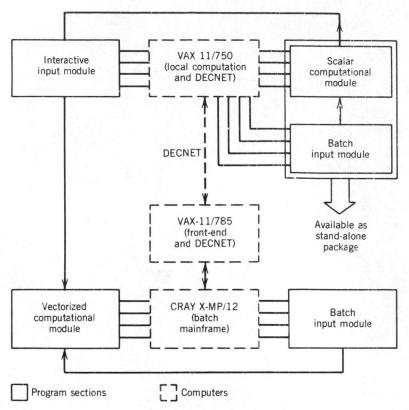

Figure 8.18 Block diagram of the software package and a typical computer system used to run it.

Voltage-dependent current source (including delay)

Current-dependent current source (including delay)

Loss-free TEM line section

Short-circuited loss-free TEM line stub

Open-circuited loss-free TEM line stub

Loss-free symmetric TEM coupled-line section

Microstrip Components

Fringing capacitance of microstrip open end

Parasitics of microstrip impedance step

Right-angle microstrip bend

Compensated right-angle microstrip bend

Microstrip tee-junction

Microstrip cross-junction

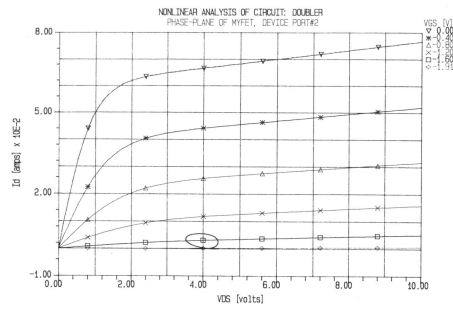

Figure 8.19 FET curves.

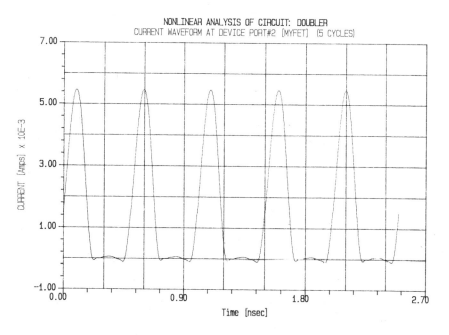

Figure 8.20 Output waveforms as a function of time.

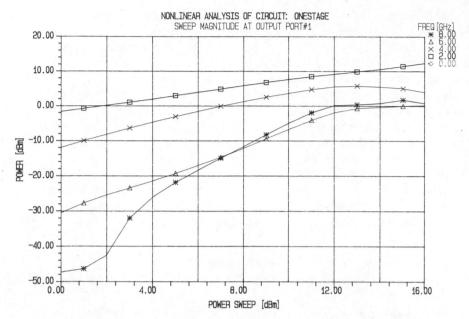

Figure 8.21 Output level as a function of input level for different harmonics.

Uniform microstrip section, lossy and dispersive

Short-circuited microstrip stub, lossy and dispersive

Open-circuited microstrip stub, lossy and dispersive, including open-end correction

Symmetric coupled-microstrip section, lossy and dispersive

Array of coupled microstrip lines of unequal widths and spacings, lossy and dispersive, using a simplified implementation of the spectral-domain approach

Rectangular microstrip resonator

Uniform microstrip section coupled to dielectric resonator

All microstrip components are described by state-of-the-art models. They are defined by means of geometrical data such as microstrip widths and lengths, and share a number of parameters related to their fabrication technology. These parameters, which may be input separately, include substrate thickness, dielectric constant, loss tangent, and roughness. Their values are taken into account in the calculations.

In addition, the program accepts any number of "measured" (i.e., a priori known) linear components described by a frequency-dependent impedance, admittance, or scattering matrix. Interpolation of input data is carried out by the program automatically whenever necessary. Finally, users can input any set of arbitrarily self-defined models to create their own technology-dependent libraries.

The description of the linear subnetwork comprises both the RF and bias circuits, including all free sources, in a unique circuit file. Any physical or electrical parameter may be selected as a variable to be optimized. A unique feature of this package is that the bias source voltages are also optimizable, so that the user can ask the program to choose the best bias conditions for the nonlinear devices.

The specification of design goals marks one of the essential differences between linear and nonlinear optimization. In the linear case, a network function is simply an algebraic consequence of the scattering parameters. On the other hand, to compute the performance of a nonlinear network, the actual steady-state regime must be completely known since the network functions can be obtained only from the voltage and current harmonics at the circuit ports. This increased complication is handled completely by the program; it requires no special efforts by the user. A menu of optimizable network functions is available. This includes all of the most common performance indexes of nonlinear circuits:

1. The output power from a given port at any harmonic
2. The spectral purity of the output signal at a specified harmonic from a given port
3. The return loss at any port connected with a free RF source
4. The power transfer efficiency from the dc bias sources to the output signal at a given port and harmonic
5. The transducer gain between given input and output ports at specified harmonics
6. The power-added efficiency between given input and output ports

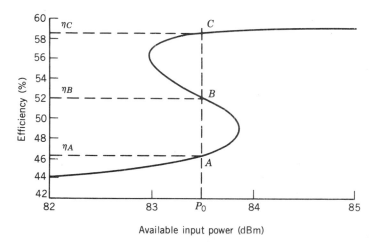

Figure 8.22 Hysteresis cycle of a microstrip parametric frequency divider.

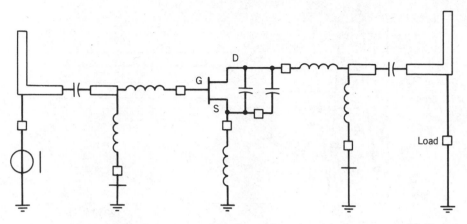

Figure 8.23 Schematic representation of an active microstrip frequency multiplier.

Any of these functions can be specified arbitrarily or simply monitored throughout the optimization. The program associated with each function identifies a set of operations on voltage and current harmonics at the relevant network ports, which is utilized to compute the required function automatically.

8.3.3 Examples

In conventional CAD programs, analysis and optimization were centered around a variation of components used to match active devices. Since these devices are

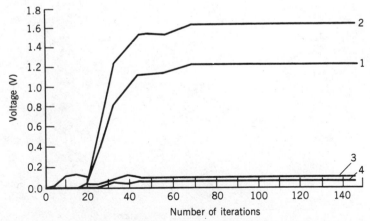

Figure 8.24 Harmonic balance analysis of the circuit in Fig. 8.23. Magnitudes of the drain voltage harmonics versus the number of iterations.

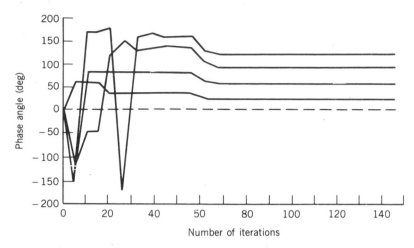

Figure 8.25   Harmonic balance analysis of the circuit in Fig. 8.23.

assumed to be linear and/or memoryless, optimization was independent of the level of operation. Microwave Harmonica takes into consideration nonlinear modeling and thus deals with the problem of optimizing the circuit not only as a function of frequency but also as a function of drive level. This required the development of totally new optimization techniques. A typical example is provided by nonlinear circuits having multiple operating points, such as the frequency divider, whose hysteresis cycle is displayed in Fig. 8.22. At an input power level P, this circuit

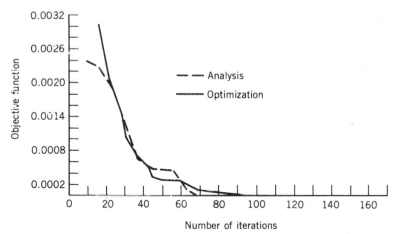

Figure 8.26   Dependence of the objective function on the number of iterations for an analysis and an optimization of the circuit in Fig. 8.23.

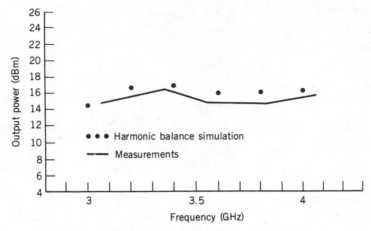

Figure 8.27 Output power of monolithic VCO.

will have three operating points, a, b, and c. Analyzing the circuit starting from zero harmonics will lead to point a. Constraining the analysis properly makes sure that all possible operating points can always be found in a short number of turns.

Another important application is the active microstrip frequency multiplier shown in Fig. 8.23. The harmonic balance analysis provides information about the magnitude of the drain voltage harmonics. The correct solution is found after approximately 70 iterations. Figures 8.24 and 8.25 show the magnitudes and phases of these harmonics, illustrating the fact that their proper values are reached after

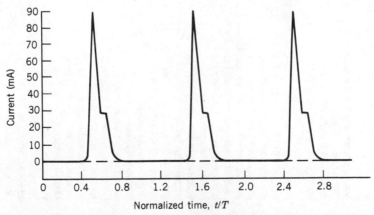

Figure 8.28 Conduction current waveform through the source varactor of a monolithic VCO.

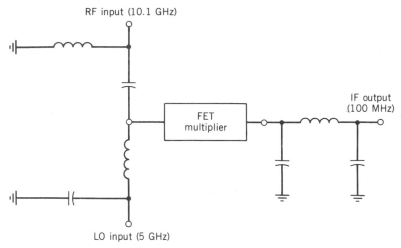

Figure 8.29   Schematic diagram of an FET harmonic mixer.

approximately 70 iterations. The rapid convergence for both analysis and optimization can be expressed as the error function shown in Fig. 8.26.

Many more applications are possible. Figure 8.27 shows a comparison of measured response versus the harmonic balance approach for a monolithic VCO. The conduction current waveforms through the source reactor of the VCO can be predicted (see Fig. 8.28). Finally, the FET harmonic mixer shown in Fig. 8.29

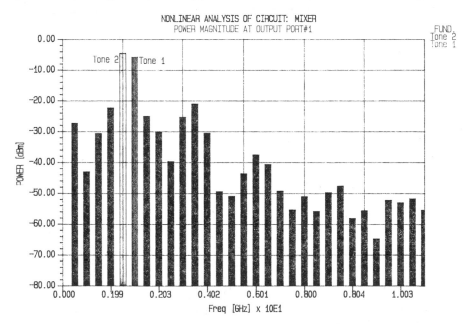

Figure 8.30   Drain voltage spectrum for the mixer in Fig. 8.29.

was analyzed. The resulting drain voltage spectrum for the mixer (Fig. 8.30) provides all the necessary information about its operation.

BIBLIOGRAPHY

Bandler, J. W., and C. Charalambous, "Theory of Generalized Least pth Approximation," *IEEE Transactions on Circuit Theory*, Vol. CT-19, 1972, pp. 287–289.

Charalambous, C. "Nonlinear Least pth Optimization and Nonlinear Programming," *Math. Programming*, Vol. 12, 1977, pp. 195–225.

D'Inzeo, G., F. Giannini, C. M. Sodi, and R. Sorrentino, "Method of Analysis and Filtering Properties of Microwave Planar Networks," *IEEE Transactions on Microwave Theory and Techniques*, Vol. MTT-25, July 1977, pp. 462–471.

Fletcher, R., "A New Approach to Variable Metric Algorithms," *Computer Journal*, Vol. 13, August 1970, pp. 317–322.

Gupta, K. C., R. Garg, and R. Chadha, *Computer-Aided Design of Microwave Circuits*, Artech House, Norwood, Mass., 1981, pp. 338–353.

Hammerstad, E., and O. Jensen, "Accurate Models for Microstrip Computer-Aided Design," *IEEE MTT-S International Microwave Symposium Digest*, Washington, D.C., May 1980, pp. 407–409.

Itoh, T., and R. Rudokas, "New Method for Computing the Resonant Frequencies of Dielectric Resonators," *IEEE Transactions on Microwave Theory and Techniques*, Vol. MTT-25, January 1977, pp. 52–54.

Jansen, R. H., and M. Kirschning, "Arguments and Accurate Model for the Power-Current Formulation of Microstrip Characteristic Impedance," *Archiv Der Elektrischen Übertragung*, Vol. 37, 1983, pp. 108–112.

Kirschning, M., and R. H. Jansen, "Accurate Model for Effective Dielectric Constant of Microstrip with Validity Up to Millimeter-Wave Frequencies," *Electronics Letters*, Vol. 18, March 18, 1982, pp. 272–273.

Kirschning, M., and R. H. Jansen, "Accurate Wide-Range Design Equations for the Frequency-Dependent Characteristics of Coupled Microstrip Lines," *IEEE Transactions on Microwave Theory and Techniques*, Vol. MTT-32, January 1984, pp. 83–90.

Kirschning, M., R. H. Jansen, and N. H. L. Koster, "Accurate Model for Open-End Effect of Microstrip Lines," *Electronics Letters*, Vol. 17, February 1981, pp. 123–125.

Kirschning, M., R. H. Jansen, and N. H. L. Koster, "Measurement and Computer-Aided Modeling of Microstrip Discontinuities by an Improved Resonator Method," *IEEE MTT-S International Microwave Symposium Digest*, Boston, June 1983, pp. 495–497.

Komatsu, Y., and Y. Murakami, "Coupling Coefficient between Microstrip Line and Dielectric Resonator," *IEEE Transactions on Microwave Theory and Techniques*, Vol. MTT-31, January 1983, pp. 34–40.

Materka, A., and T. Kacprzak, "Computer Calculation of Large-Signal GaAs FET Amplifier Characteristics," *IEEE Transactions on Microwave Theory and Techniques*, Vol. MTT-33, Feb. 1985, pp. 129–135.

Menzel, W., and I. Wolff, "A Method for Calculating the Frequency-Dependent Properties of Microstrip Discontinuities," *IEEE Transactions on Microwave Theory and Techniques*, Vol. MTT-25, February 1977, pp. 107–112.

Mirshekar-Syahkal, D., and J. B. Davies, "Accurate Solution of Microstrip and Coplanar Structures for Dispersion and for Dielectric and Conductor Losses," *IEEE Transactions on Microwave Theory and Techniques*, Vol. MTT-27, July 1979, pp. 694–699.

Peterson, D. L., et al., "A GaAs FET Model for Large-Signal Applications," *IEEE Transactions on Microwave Theory and Techniques*, Vol. MTT-32, March 1984, pp. 276–281.

Rauscher, C., and H. A. Willing, "Simulation of Nonlinear Microwave FET Performance Using a Quasi-static Model," *IEEE Transactions on Microwave Theory and Techniques*, Vol. MTT-27, October 1979, pp. 834–840.

Rizzoli, V., C. Cecchetti, and A. Lipparini, "A General-Purpose Program for the Analysis of Nonlinear Microwave Circuits under Multitone Excitation by Multidimensional Fourier Transform," *Proceedings of the 17th European Microwave Conference*, Rome, September 1987.

Rizzoli, V. A. Lipparini, and E. Marazzi, "A General-Purpose Program for Nonlinear Microwave Circuit Design," *IEEE Transactions on Microwave Theory and Techniques*, Vol. MTT-31, September 1983, pp. 762–770.

Rohde, Ulrich L., "Harmonic Balance Method Handles Nonlinear Microwave CAD Problems," *Microwave Journal*, October 1987, pp. 203–210.

Scott, J. R., and R. A. Minasian, "A Simplified Microwave Model of the GaAs Dual-Gate MESFET," *IEEE Transactions on Microwave Theory and Techniques*, Vol. MTT-32, March 1984, pp. 243–248.

Watson, H. W. (ed.), *Microwave Semiconductor Devices and Their Circuit Applications*, McGraw-Hill, New York; 1969, pp. 164–165.

Willing, H. A., C. Rauscher, and P. De Santis, "A Technique for Predicting Large-Signal Performance of a GaAs MESFET," *IEEE Transactions on Microwave Theory and Techniques*, Vol. MTT-26, December 1978, pp. 1017–1023.

8.4 PROGRAMMABLE MICROWAVE TUNING SYSTEM

The Programmable Microwave Tuning System, developed by the David Sarnoff Research Center, provides a unique new measurement system that performs a variety of automatic microwave tuning and device characterization procedures.

8.4.1 The PMT System

The Programmable Microwave Tuner System (PMTS) is a computer-controlled state of-the-art method of dynamically testing microwave circuits under full-power conditions while automatically varying the load and/or source impedance. When used as an interface to the appropriate peripheral equipment (RF source, power meters, noise-figure meter, etc.), this system greatly simplifies active-device characterization by providing accurate and repeatable measurements. When coupled with a host computer running the PMT application software, the system can be directed optimally to seek either a specific impedance point that maximizes (or minimizes) a measurable parameter (e.g., peak output power, maximum efficiency, minimum noise) or a family of impedance points that satisfies a specific criteria (e.g., constant output power, constant efficiency, or a constant noise figure).

The PMTS provides an arbitrary impedance to circuits and measures their responses. This is accomplished by a combination of precision hardware and software—no network analyzer is required. The PMTS software combines equivalent-circuit models and sophisticated search algorithms to characterize the load and source pull characteristics of a device or circuit with respect to power, efficiency, and noise figure. It obviates the need to precharacterize at discrete impedance points using a network analyzer system.

8.4.2 Tuning Techniques

Maximum power transfer between an RF source and a load is obtained when the complex load impedance is the conjugate of the complex source impedance. If an active device is placed between the source and load, the input and output impedances must be conjugately matched simultaneously.

Before automatic vector network analyzers made impedance measurements routine, microwave engineers would place tuners before and after their circuit and simply "diddle" the tuners for maximum output power. After the tuners were adjusted properly, the impedance of each was determined by using a slotted line. This technique was repeated for other frequencies in the range of interest. In the case of linear small-signal circuits, such tuning procedures are no longer necessary. Modern CAD programs can synthesize matching networks analytically using the measured or modeled network parameters of the circuit (S, Y, or Z) parameters and the source and load impedances. Although tuners are no longer needed in the design of small-signal linear amplifiers, they are still required in the design of low-noise circuits, nonlinear circuits, and even large-signal linear circuits.

When designing low-noise-amplifier matching networks, the proper impedance needed for optimum input noise match is independent of the device's network parameters. It can be determined by placing a tuner on the input of the device and adjusting the tuner for the minimum noise figure. An indirect measurement technique can also be used in which the noise figure is measured for several different known impedances at the input of the device. These data are then used to calculate the noise parameters of the device: the minimum noise figure, the source impedance for minimum noise figure, and the equivalent noise resistance. These parameters allow the circuit designer to compute the noise figure for any arbitrary source impedance.

Large-signal designs also require terminating impedances which cannot be determined from the linear network parameters. The most effective technique for determining the required matching network is to place a tuner on the output of the device and tune it for maximum (or some target) power. Information on how the output power, efficiency, and distortion vary with load impedance and frequency is usually needed to design the optimum matching network. Such load-pull data require further tuner measurements.

8.4.3 The PMTS Approach

The most apparent problem when using a manual tuner is the enormous amount of time required to obtain results. Each data point requires manual adjustment of the tuner to obtain the desired performance, and then removal of the tuner and measurement of its impedance on a network analyzer. If the tuner's loss is taken into account, which should be done for accurate results, all four S parameters must be measured and the desired response (noise figure, output power, efficiency, etc.) recalculated. The results may not be correct even when this procedure for calculating loss is followed since the measured parameter cannot be corrected for tuner loss until after the tuner is measured. The PMT system overcomes such difficulties by incorporating the following components: (1) a family of programmable tuners, (2) a programmable tuner controller, and (3) application software that provides tuner control, optimization procedures, and data manipulation routines. Figure 8.31 shows the system hardware components.

Tuner. Tuners are available to cover the frequency range 400 MHz to 26.5 GHz. Each is built around a slotted transmission-line section fitted with two low-impedance elements (slugs) riding on the center conductor. The tuners are designed to achieve a minimum VSWR of approximately 10:1 over an octave bandwidth. High-precision stepper motors and zero-backlash ball screw assemblies provide extremely precise slug positioning anywhere along the line. PMTS incorporates an electrical model of the tuner in the control software. Each tuner is precisely characterized over its entire bandwidth/tuning range. The characterization data are stored in a microprocessor chip within each tuner and read on demand by the tuner controller.

Programmable Tuner Controller. PTC-8700 tuner controller provides a full-function instrument for controlling up to two programmable tuners and interacting with the host computer running the PMT application software. The front panel consists of a high-resolution alphanumeric display panel, a function/data-entry keyboard, and a joystick tuning control lever. The joystick is used for manual control of the tuner. In the automatic mode of operation, the PTC-8700 communicates with a host computer running the PMT software. System messages appear on the alphanumeric display and the front-panel keyboard provides for operator

Figure 8.31 PMTS system hardware components.

interaction. Keyboard functions include setting the operating frequency and measurement reference planes, as well as setting the tuner impedance.

PMT Software. The tuners and PTC-8700 act together in accordance with commands from the PMT application software and provide any arbitrary impedance value anywhere in a particular tuner's VSWR/frequency range. As its basic function, PMT translates any desired impedance request into tuner settings, and conversely, determines the impedance corresponding to any tuner setting. In addition to controlling the tuner itself, the software can measure the gain of a device using programmable power meters, determine the noise figure using a programmable noise figure meter, measure bias voltage and currents via a data acquisition unit, and provide graphical output on a CRT or plotter.

By combining the capability to control test instrumentation along with the ability to "synthesize" an impedance accurately, a variety of high-level measurement procedures can now be performed quickly and easily. The PMTS can determine the impedance that produces the maximum output power or can determine (and plot/display) the locus of all impedance points that provide a specified output power (constant power contours). Similar functions can be performed for efficiency and noise. The PMTS determines performance contours and operating points by

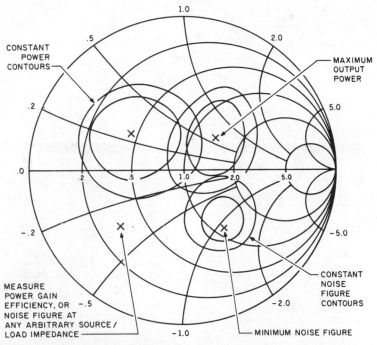

Figure 8.32 Graphical representation of the PMT measurement data includes constant power and noise-figure contours.

actually presenting the device with impedances, the selection of which is guided by the measured device response and the optimized search algorithms incorporated into PMT. Figure 8.32 shows the various types of measurement data that the PMT system can provide. Figure 8.33 depicts typical measurement system configurations for performing power measurements and noise characterization.

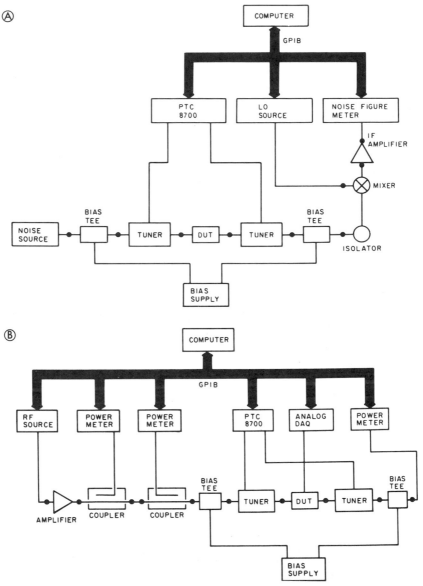

Figure 8.33 Block diagram of equipment configuration used for performing (*A*) power measurements, and (*B*) noise characterization.

BIBLIOGRAPHY

Schepps, Jonathan and Stewart Perlow, David Sarnoff Research Center, Inc., Compact Software's *Transmission Line News*, Vol. 2, No. 1, March 1988, pp. 4–5.

8.5 INTRODUCTION TO MMIC CONSIDERING LAYOUT EFFECTS

As the density of the circuits has increased, the effect of coupling has to be considered. A special program called LINMIC+ developed by Professor Jansen at the University of Duisburg can handle these cases and offers other important enhancements. LINMIC+ enables the computer-aided design of strip-type planar MICs and MMICs and incorporates analysis, sensitivity analysis, and a stable and efficient interactive optimization procedure. It accomplishes this by making direct, automated use of a very general, rigorous field-theoretical approach to the generation of design information for a wide class of structures up to high millimeterwave frequencies.

Its description is based on a fast, enhanced spectral-domain technique which computes the required design data in the form of multidimensional look-up tables. In analysis and optimization, these tables are used together with a fast interpolation method. This approach constitutes a shift away from analytical models, further extending the range of applicability and accuracy in circuit predictions (particularly at high frequencies). Circuit simulation and optimization in LINMIC+ are performed by using the layout's geometrical data and the electrical specifications of available commercial hybrid components; that is, the design parameters are physical quantities that are actually under the designer's control. Practical realizability of a design can be enforced by user-definable constraints. A user-controlled objective function for the design of MICs and MMICs can handle any of the complex network electrical quantities generated internally in up to nine user-selected formats (also mixed): real or imaginary, magnitude or phase. It can optimize any of these quantities to be greater, equal to, or less than a given specification. In addition, the noise figure and stability factor can be handled. Circuit performance specifications can be defined within and far outside a band of interest. Also, interactive control of the features of the optimization algorithm allows gradual change between maximally flat and equal ripple designs. Up to 60 parameters in a circuit can be varied during optimization.

In addition, LINMIC+ enables the automated generation and implementation of transistor models for very broadband applications from measured or data sheet *S* parameters. This allows for device simulation outside the frequency range of measurement.

As a consequence of its field theoretical program, the LINMIC+ package has the potential to describe internally a variety of microwave circuit structures which could never be described by analytical models. LINMIC+ is thus applicable for microstrip, stripline, suspended substrate, and multilayer MMIC transmission-line structures.

LINMIC+'s LCPACK can handle interdigital capacitors, multiturn square spiral inductors, and even planar spiral transformers and couplers at very high frequencies, on a substrate with or without passivation or second-level dielectric. Another option, named MELINE, allows for accurate characterization of a single strip meander line and parallel coupled meander structures. Of course, whenever accurate analytical models for microwave strip structures are available, they are implemented in LINMIC+ as an alternative. A variety of models are used to describe the frequency-dependent characteristics of single and coupled microstrips (accurate at frequencies well into the millimeter wave region). These models have been used in the major commercial CAD packages because of their high accuracy. An overall illustration of the LINMIC+ program is shown in Fig. 8.34.

8.5.1 Component and Interconnection Modules

The following elementary component modules are supplied with the LINMIC+ package:

SSTRIP   Single srip transmission lines, including microstrip and suspended substrate; up to 20 different user-definable substrate configurations/

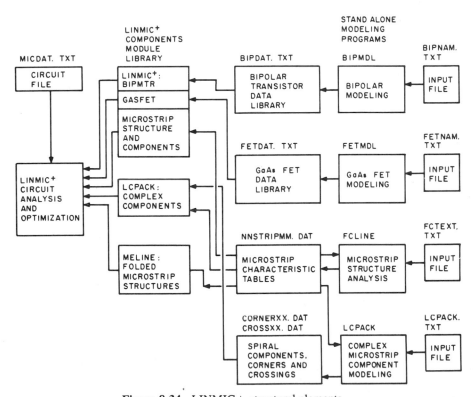

Figure 8.34 LINMIC+ structural elements.

transmission media, each controlled by an integer identifier (applies to all component modules)

CSTRIP	Section of symmetrically coupled strips, four-port
MSTRIP	Analog to SSTRIP; avoids parameter recomputation
NSTRIP	Section of N multiply coupled strips, $2N$-port ($N = 3, \ldots, 10$)
CPLPRT	Coupled strip section loaded with admittances at two ports
DFLINE	User-defined strip transmission line
OPSTUB	Transmission-line stub with ideal/nonideal open end
SHSTUB	Transmission-line stub with ideal/nonideal short-circuited end (or via hole short)
SCSTUB	Sector-shaped radial stub structure, open ended
CHBEND	Chamfered 90° strip transmission-line bend
STEPLN	Symmetrical/unsymmetrical impedance step, including line section
LJUNCT	Symmetrical/unsymmetrical loaded strip-type T-junction
TJUNCT	Symmetrical/unsymmetrical T-junction three-port
UNSGAP	Unsymmetrical gap between two end-to-end coupled strips
SECTOR	Radial strip structure extruding from 90° bend, two-port
DLBCAP	Chip capacitor inserted into a gap between two strips
RSCHIP	Chip resistor mounted over a gap between two strips
BIPMTR	Bipolar microwave transistor, chip or packaged
GASFET	Gallium arsenide microwave FET, chip or packaged
RLCPRT	Ideal network elements R, L, C or any shunt or series combination of these; includes the definition of strips to contact these elements
BRANCP	Branch guide coupler, including the coupling effects between the involved parallel strips
RATRCP	Rat-race ring coupler
SLANCP	Straight, interdigitated (Lange) coupler, using four strips
FLANCP	Folded, interdigitated (Lange) coupler, using four strips
BRSDCP	Broadside-coupled stripline-type directional coupler

Note that many of the modules show the more general case. For example, NSTRIP represents a total of 48 physically different structures; RLCPRT represents 28 different configurations.

Additional interconnection operations can be performed:

COMPNT	Repeated use and linking of stored components
SUBNET	Use of subnetworks as (super) components
SNLINK	Link between ports of the current subnetwork
TNLINK	Interconnections of ports associated with different subnetworks; final linking operation in a design
METLINE	extension modules are:

SNGMEA	Folded single-strip structures

CPLMEA   Parallel folded structures coupled to each other
LCPACK   extension modules are:

INDUCT   Rectangular MIC/MMIC spiral inductors
PLTRAN   Rectangular MIC/MMIC spiral transformers
INTCAP   Interdigitated capacitors
NSLACP   Four- and six-strip interdigitated coupler configurations

The component module used for data-bank handling and black-box circuit operations is

SCFILE   Reading, using, and writing S-, Y-, or Z-parameter files for components with $1, \ldots, 4$ ports in the LINMIC+ environment (compatible with Super-Compact file format)

A conventional procedure based on the segmentation approach reduces circuits to a number of capacitors, inductors, transmissions lines, rectangular inductors, or other components. For example, a circuit with a rectangular inductor can be simulated by using one of the better mathematical expressions, such as the one developed by Ingo Wolff. These mathematical models assume, however, no under-

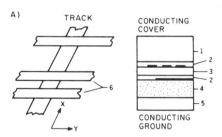

1-AIR 2-PASSIVATION 3-POLYIMIDE 4-SUBSTRATE
5-OPTIONAL DIELECTRIC MEDIUM 6-STRIPS (SPIRAL WINDINGS)

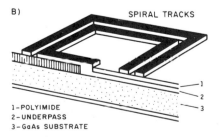

1-POLYIMIDE
2-UNDERPASS
3-GaAs SUBSTRATE

Figure 8.35 Spiral (inductor) tracks: (*A*) cross-sectional view illustrates multilayer composition; (*B*) underpass provides contact to inner turn.

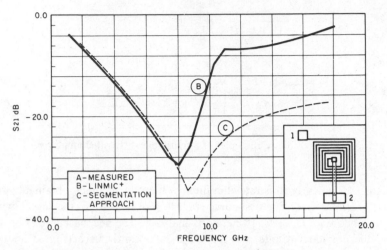

Figure 8.36 Multiturn inductor scattering parameter S_{21} frequency: Ⓐ values measured on HP-8510 analyzer; Ⓒ LINMIC+ predicted performance; Ⓒ predicted performance using the standard segmentation approach.

pass or air-bridge connection for inductors. LINMIC+ extends this analytic capability.

Figure 8.35 is a drawing of spiral tracks and a general structure. The underpass is used to connect the inner portion of the inductor to the outside. The inductor was built using seven turns, with a track and gap dimension of 12 μm and an inner turn on a 100×100 μm grid.

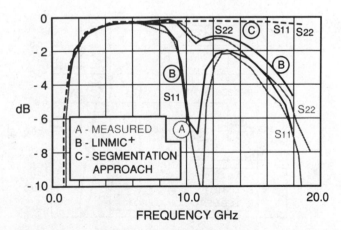

Figure 8.37 Multiturn inductor input and output scattering parameters S_{11} and S_{22} versus frequency: Ⓐ HP-8510 measured values; Ⓑ LINMIC+ predicted performance; Ⓒ predicted performance using the standard segmentation approach.

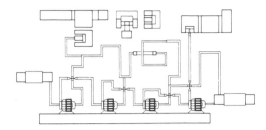

Figure 8.38   Calma plot of a four-stage distributed amplifier.

Figure 8.36 clearly illustrates the improvement in accuracy obtainable through the use of LINMIC+. The first curve (Fig. 8.36a) is a plot of the actual measured data obtained with an HP8510 network analyzer. The second curve (Fig. 8.36b) shows the predicted results of LINMIC+. The third curve (Fig. 8.36c) is the predicted curve obtained through the use of a popular PC-based software package.

Figure 8.37 is even more impressive. By looking at the input and output reflection coefficients, it becomes obvious that the measured data and those predicted by LINMIC+ are very close, while the dashed-line prediction by the PC product (Fig. 8.37) shows no difference between the input and output parameters. This apparent (and nonvalid) symmetry resulted from a disregard of the coupling between the tracks and the effect of the underpass.

Having understood this, it is even more interesting to analyze a four-stage amplifier for circuit simulation. Figure 8.38 is a Calma plot of an experimental four-stage traveling-wave amplifier which was used as a tool to verify the accuracy of the software. This Plessey amplifier was first generated on a standard simulator

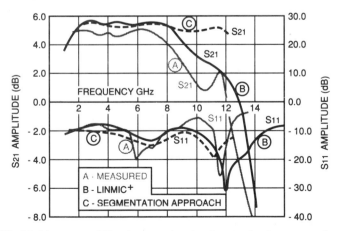

Figure 8.39   Multistage amplifier input and gain characterization versus frequency: Ⓐ HP-8510 measurements; Ⓑ LINMIC+ predictions; Ⓒ segmentation predictions relate to differences between measurements, and LINMIC+ predictions relate to use of statistical rather than actual measured data for the FET.

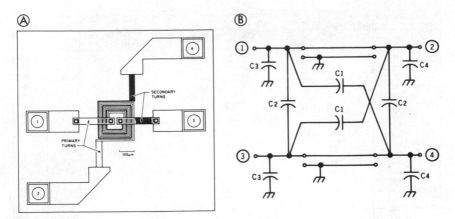

Figure 8.40 Transmission-line-based spiral transformer: (*A*) physical representation; (*B*) electrical equivalent circuit up to 18 GHz.

without taking into consideration the effect of coupling. Figure 8.39*c* shows the predicted performance using the segmentation approach.

Actual measurements on an HP8510 (Fig. 8.39*a*) show significantly different results; the magnitude of S_{21} drops much earlier than originally predicted. By using LINMIC+ and taking into consideration the coupling of the various transmission lines while using an enhanced model for the bends, the simulation prediction is very close (Fig. 8.39*b*). LINMIC+ can also handle exotic structures such as the transmission line–based spiral transformer shown in Fig. 8.40. LINMIC+ is a dedicated MMIC simulator and outperforms the segmentation approach for high-density MMIC designs.

BIBLIOGRAPHY

LINMIC+ User's Manual.

Jansen, R. H., R. G. Arnold, and I. G. Eddison, "A Comprehensive CAD Approach to the Design of MMICs Up to mm-Wave Frequencies," *IEEE Transactions on Microwave Theory and Techniques*, February 1988.

8.6 GaAs MMIC LAYOUT SOFTWARE

When used in conjunction with electrical and/or optical simulation software, graphics layout software assists the analog designer to develop GaAs MMIC and millimeter wave circuits and optoelectronic components. One particular package, the GaS STATION, utilizes mouse-driven menu-oriented inputs to pull up a main display and up to 12 different menus. The process is initiated by choosing from a main menu up to four separate menus from the following list:

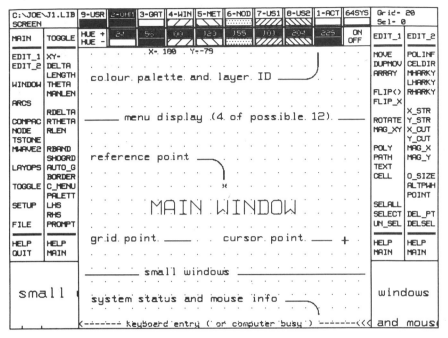

Figure 8.41   Maximum of four menus can be used with the main window.

EDIT1	ARCS	TSTONE	TOGGLE
EDIT2	COMPAC	MWAVE2	SETUP
WINDOW	NODE	LAYOPS	FILE

Figure 8.41 illustrates this with a sceen presentation that includes the MAIN and TOGGLE to the left and the EDIT1 and EDIT2 menus to the right of the main window.

8.6.1   Capabilities

GaS STATION is used to generate patterns. It utilizes what is known as the GDS-II data format, an industry standard, that allows for up to 64 layers (of processing) with choices from a 256-color palette. One of the powerful features of GaS STATION is that besides using a graphics input, it can pull from geometrically defined circuit files available in several leading microwave CAD packages. (e.g., the COMPAC and TSTONE menus listed above). A node capability is built into the software to help facilitate this interaction between the electrical circuit description and the geometrical layout.

Typical shapes that can be generated, such as circles, arcs, and sinusoidal and exponential functions, are illustrated by the distributed filter section in Fig. 8.42.

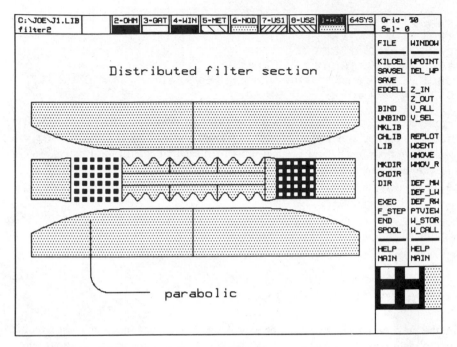

Figure 8.42 Distributed filter section requires parabolic shape generation.

Notice that in this case only two menus are called up to the right of the main window.

Command files can be run from GaS STATION, which enables the user to run a setup file to configure the system. Command files can also be used to store a record of the design session or as mentioned before, to generate or run Super-Compact or Touchstone files.

8.6.2 Example

Consider the "open-end effects circuit" shown in Fig. 8.43. The first half of it is delineated in the following circuit block descriptor:

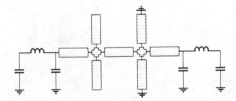

Figure 8.43 Open ends affect circuit.

```
CAP    1  0  C=1.2PF
IND    1  2  L=1NH
CAP    2  0  C=1.2PF
TRL    2  3  W=100UM P=400UM SUB1
CROS   3  4  5 6 W1=100UM W2=100UM W3=100UM W4=100UM SUB1
OST    4     W=100UM P=400UM SUB1
```

From the above, notice that a 400-μm-long 100-μm-wide dielectric factor 1 (SUB1) transmission line was placed between nodes 2 and 3. GaS STATION users can work with both metric and English units of measurement. In the English system the numbers can range from 0.1 to 100,000 μm, literally a ratio of 1 million:1. Appropriately, the TOGGLE menu includes a variety of relative and absolute angle and length indicators.

Figure 8.44 indicates another useful feature available from the software's command structure. A transmission line can be chamferred to maintain a constant impedance through a bend.

8.7 PRACTICAL DESIGN EXAMPLE

In the previous chapters we have been looking at both linear and nonlinear circuits, but in reality circuits are parts of systems. These systems incorporate various appli-

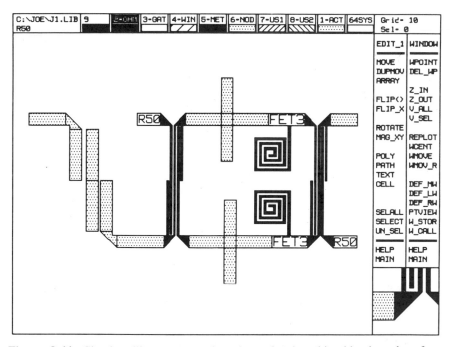

Figure 8.44 Circuit utilizes constant impedance bends achievable through software commands.

cations. In our opinion, a good example of how to integrate these things is a 4-GHz anticollision radar, consisting of a pulsed transmitter and a selective receiver.

8.7.1 The Design

Figure 8.45 shows a sketch of the receiver and transmitter portion. As can be seen, the receiver consists of an antenna, a three-element input filter, and a matching circuit for the input amplifier. The output of the amplifier shows a matching circuit and a dc decoupler capacitor which feeds into one arm of a branch-line coupler. The branch-line coupler is also driven from a dielectric resonator oscillator. The output of the branch-line coupler terminates in a video detector consisting of two diodes.

The transmitter has its own antenna, and a coupler in the output monitors the power level. For reasons of overall power, the oscillator uses a FET as the active device, and the gate voltage is pulsed to turn the transistor on and off.

8.7.2 The Elements

Figure 8.46 shows the insertion losses and return losses of adjacent planar antennas at various distances (1 m, 70 cm, and 50 cm). Figure 8.47 shows the radiation diagram of a planar antenna in the E and H field with a height of 3.14 mm and a dielectric constant of 2.33. Figure 8.48 shows the pattern of a slightly different antenna with $H = 1.57$ m. Here the antenna impedance is 88 Ω, versus 52 Ω in Fig. 8.47.

8.7.3 The Input Filter

The input filter was designed using common requirements for bandpass filters using parallel coupled lines. The frequency was 4.3 GHz with a 200-MHz bandwidth. The filter response is shown in Fig. 8.49. Figure 8.50 shows the list of an equivalent model of a parallel modeled resonator bandpass filter. Comparison of the CAD simulation and measurements indicates excellent tracking. The response curve shown in Fig. 8.51 is very close to measurements done with the network analyzer.

8.7.4 The Dielectric Resonator

The dielectric resonator for the oscillator can be configured in several ways. Figures 8.52 to 8.54 show the resonant behavior of the dielectric resonator in different configurations. A circuit simulation again provides good agreement between measurement and prediction.

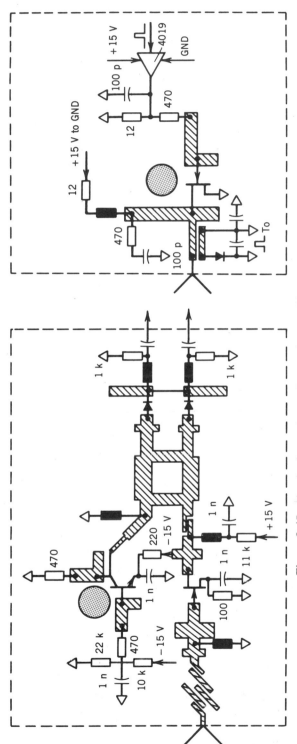

Figure 8.45 Sketch of a receiver and transmitter used in an anticollision radar.

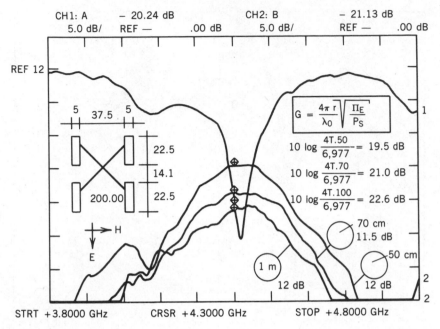

Figure 8.46 Insertion losses and return losses of adjacent planar antennas at different distances.

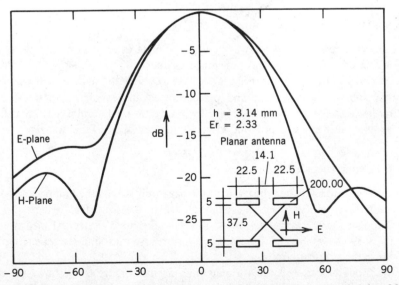

Figure 8.47 Radiation diagram of a planar antenna in the E&H field with a height of 3.14 mm and a dielectric constant of 2.33.

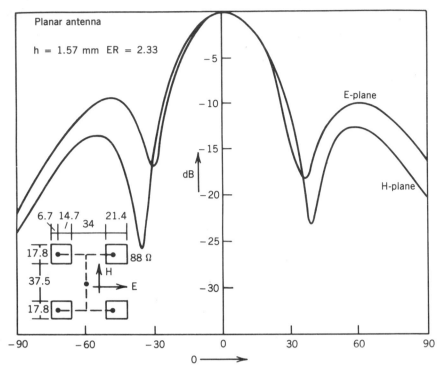

Figure 8.48   Pattern of a slightly different antenna with $H = 1.57$ m. Here the antenna impedance of 88 versus 52 Ω in the preceding example.

8.7.5   The Branch-Line Coupler

The branch-line coupler has always caused a lot of interest in modeling, because it consists of a T-junction and transmission lines as shown in Fig. 8.55. As the device consists of four tees, it is a good test to verify the accuracy of the model. Comparison of measurement and prediction again reveals close tracking and resemblance (see Figs. 8.56 to 8.58). Typically, in CAD tools, it is important to run as many test cases as possible to verify the accuracy of the models.

8.7.6   Other Circuit Elements

The dc separation between the chain of the preamplifier and the branch-line coupler also requires some verification. Although we recommend that readers use their CAD tools to model those elements, we have provided the actual measurement. Figure 8.59 shows the insertion loss and return loss for a quarter-wave coupler as shown in the original circuit.

The diode detector requires a matching circuit. Figure 8.60 shows the return loss of the HP diode HP5082-2217 for various currents through a 1-kΩ resistor at

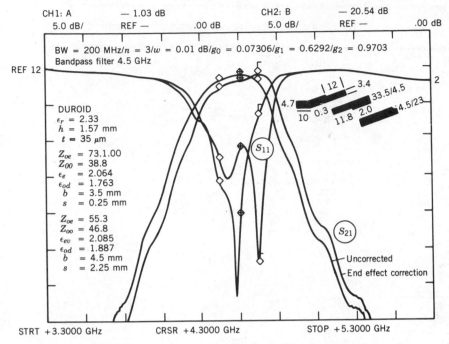

CH1: A   — 1.03 dB   CH2: B   — 20.54 dB
  5.0 dB/   REF —   .00 dB   5.0 dB/   REF —   .00 dB

BW = 200 MHz/n = 3/w = 0.01 dB/g_0 = 0.07306/g_1 = 0.6292/g_2 = 0.9703
Bandpass filter 4.5 GHz

REF 12

DUROID
ϵ_r = 2.33
h = 1.57 mm
t = 35 μm

Z_{oe} = 73.1.00
Z_{00} = 38.8
ϵ_e = 2.064
ϵ_{od} = 1.763
b = 3.5 mm
s = 0.25 mm

Z_{oe} = 55.3
Z_{00} = 46.8
ϵ_{ev} = 2.085
ϵ_{od} = 1.887
b = 4.5 mm
s = 2.25 mm

S_{11}

S_{21}

Uncorrected
End effect correction

STRT +3.3000 GHz   CRSR +4.3000 GHz   STOP +5.3000 GHz

Figure 8.49 Structure and measured response of 4.3 GHz band+pass filter.

the particular mechanical configuration. The purpose of this was to determine the right matching combination. The transmitter requires a power monitor. Figure 8.61 shows insertion loss and return loss for a quarter-wave microstrip coupler as a function of frequency.

8.8 CAD APPLICATIONS

The following circuit files from the application notes section of the Super-Compact PC manual serve as a good introduction to the use of linear CAD tools and demonstrate the power of modern CAD microwave tools.

APP1: Single-Stage Amplifier (Figs. 8.62, 8.63). Application Note 1 shows a combination of several useful features in the Super-Compact PC program. The main purpose is to illustrate the design of a single-stage amplifier using microstrip. In conjunction with pure circuit analysis, the TRL program was used. Optimization techniques and the use of the DATABANK are also demonstrated. TRL is used to determine the line width of the microstrip with a certain characteristic impedance (Z_0 = 50 Ω in this case). This step is important since you can improve the quality factor (Q) by selecting impedances around 70 Ω, or you can omit discontinuities

```
*******************************************************************
*                                                                 *
* EQUIVALENT MODEL OF A PARALLEL-COUPLED RESONATOR BANDPASS FILTER *
*                                                                 *
*******************************************************************
O1:3.5MM
O2:4.4862MM
BLK
     CPL  1  2  3  4   W=O1   S=0.25MM   P=12.697MM  SUB1
     OPEN  2   W=O1   SUB1
     OPEN  4   W=O1   SUB1
     CPL  3  5  6  7   W=O2   S=2.1595MM  P=12.404MM  SUB1
     OPEN  5   W=O2      SUB1
     OPEN  7   W=O2      SUB1
A:2POR  1  6
END
BLK
     CPL  1  2  3  4   W=O1   S=0.25MM   P=11.800MM  SUB1
     OPEN  2   W=O1   SUB2
     OPEN  4   W=O1   SUB2
     CPL  3  5  6  7   W=O2   S=2.1595MM  P=11.700MM  SUB1
     OPEN  5   W=O2      SUB2
     OPEN  7   W=O2      SUB2
B:2POR  1  6
END
BLK
     A  1  6
     A  8  6
C:2POR  1  8
END
BLK
     B  1 2
     B  3 2
D: 2POR 1 3
END
FREQ
     STEP  3.3GHZ   5.3GHZ   0.1GHZ
END
OUT
     PRI  C  S
     PRI  D  S
     PRI  A  S
     PRI  B  S
END
DATA
SUB1:  MS  H=1.57MM  ER=2.33  TAND=0.001  MET1=CU  35UM
SUB2:  MS  H=1.57MM  ER=2.33
END
```

Figure 8.50 Super-Compact circuit description of bandpass filter shown in Fig. 8.49.

by always selecting the same impedance. Optimization is used to obtain the highest possible gain (MS21) and the lowest possible reflection (MS11) in order to meet the specifications of the circuit. Because the SCOMPACT DATABANK contains most of the common transistor data, it comes in very handy for the user. All the available information about the device will be linked to the circuit as long as it is defined in the DATA block in the following format:

HAMP : HPMCC FILE=\BANK01\HPM.FLP

where HAMP = label of the user-defined black box, HPMCC = name of the device in the databank, and BANK01 = directory where the databank can be found HPM.FLP = filename for the databank

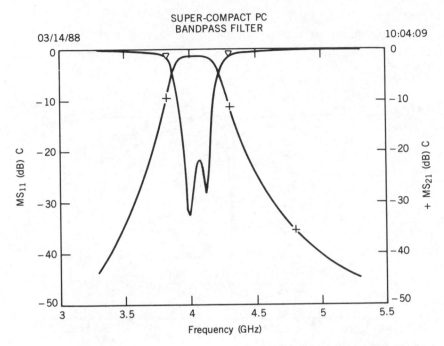

Figure 8.51 Calculated response of 4.3-GHz bandpass filter.

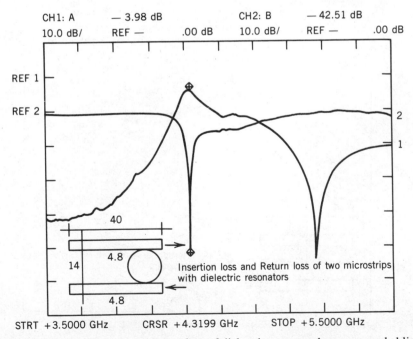

Figure 8.52 Insertion loss and return loss of dielectric resonator between coupled lines.

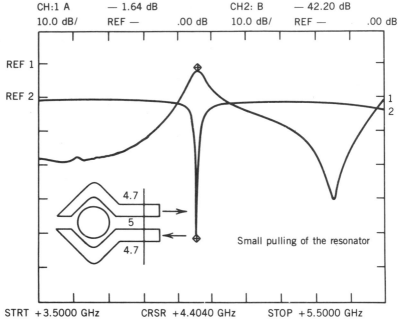

Figure 8.53   Insertion loss and return loss for different transmission line configurations.

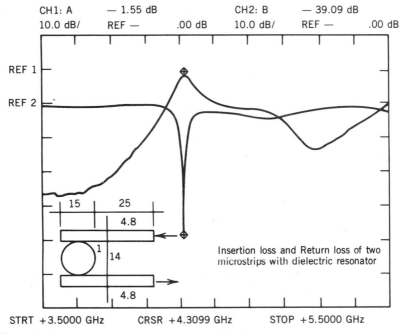

Figure 8.54   Insertion loss and return loss with different coupling to the resonator.

```
*   BRANCH-LINE DIRECTIONAL COUPLER (BRNLINE2.CKT)
WW: 7.6MM
LW: 14.5MM
WN: 4.7MM
LN: 9.80MM
BLK
    T:   TEE 1  5 9 W1=WN W2=WW W3=WN SUB
    WL:  TRL 5  6     W=WW   P=LW   SUB
    NL:  TRL 9 12     W=WN   P=LN   SUB
    T    2  6 10
    NL 10 11
    T    4  8 12
    WL   8  7
    T    3  7 11
PHYS: 4POR 1 2 3 4
END
FREQ
    STEP 3.3GHZ 5.5GHZ   0.2GHZ
    4.465GHZ
END
OUT
    PRI PHYS S
END
DATA
    SUB: MS H=1.57MM ER=2.33 MET1=CU 35UM
END
```

Figure 8.55 Super-Compact circuit file listing of branch-line directional coupler.

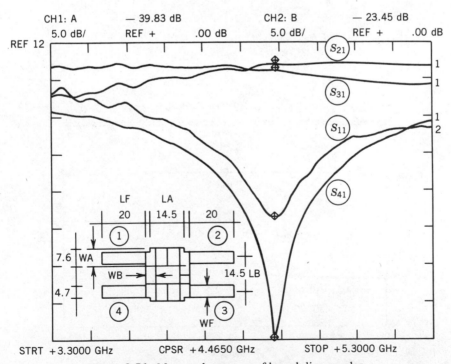

Figure 8.56 Measured response of branch-line coupler.

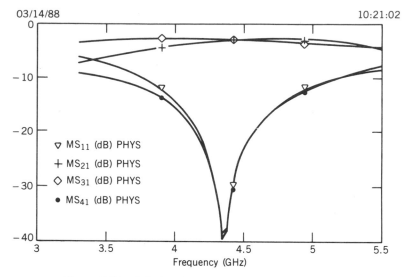

Figure 8.57   Calculated response of branch-line coupler.

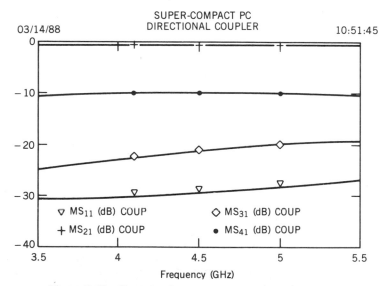

Figure 8.58   Coarse and response of branch-line coupler.

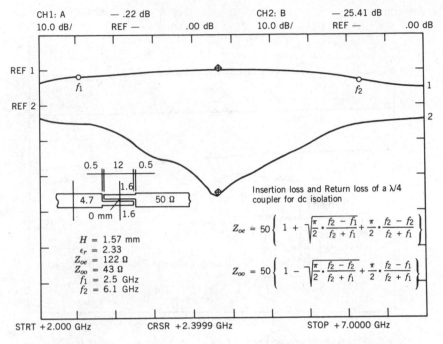

Figure 8.59 Insertion loss and return loss of control wavelength coupler.

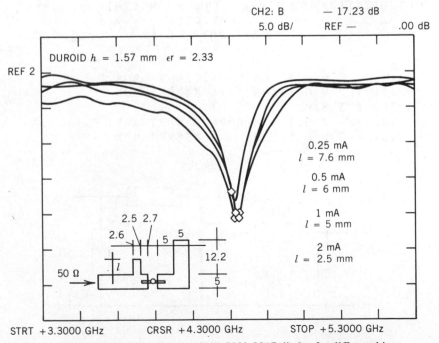

Figure 8.60 Return loss of HP-5082-2217 diodes for different bias.

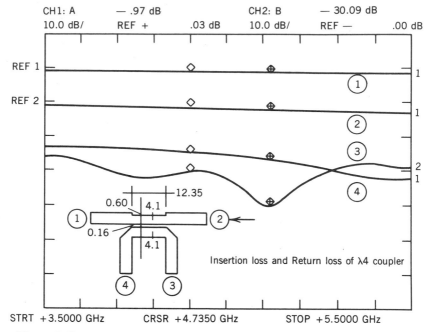

Figure 8.61   Insertion loss and return loss of quarter-wave coupler at the output.

```
*OPTIMIZATION OF A SINGLE STAGE MICROSTRIP AMPLIFIER
*                    (APP1.CKT)
*                BY ANTHONY W. KWAN

LAD
     TRL 1 0 W=0.221MM P=?4.97MM?  SUB
     TRL 1 2 W=0.576MM P=?4.838MM? SUB
     TWO 2 3 HAMP
     TRL 3 4 W=0.576MM P=?4.838MM? SUB
     TRL 4 0 W=0.576MM P=?4.838MM? SUB
```

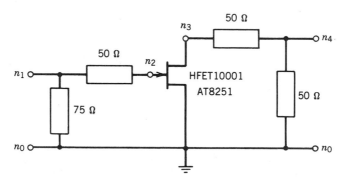

Figure 8.62   Schematic of single-stage amplifier.

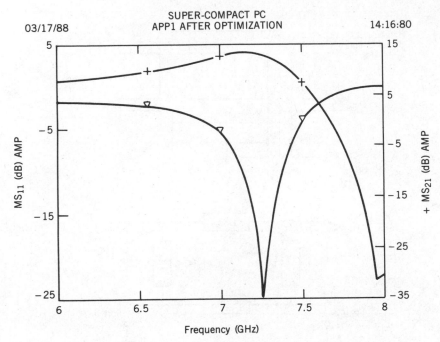

Figure 8.63 Frequency response of amplifier of Fig. 8.62.

```
AMP:2POR 1 4
END
FREQ
    STEP 6GHZ 8GHZ 0.1GHZ
END
OUT
   PRI AMP SK
END
OPT
  AMP
     F=7.2GHZ 7.3GHZ  MS21 12DB GT
     F=7.2GHZ 7.3GHZ  MS11 -15DB LT
END
DATA
    HAMP:HPMCC                FILE=\BANK01\HPM.FLP
    SUB:MS H=.635MM ER=10.0 MET1=CU 15UM TAND=0.0001
END
```

APP2: 3-dB Branch-Line Coupler (Figs. 8.64 to 8.66). Application Note 2
shows how a physical model is optimized to an ideal electrical model. Before
optimization the physical model did not resonate at 10 GHz and did not have the
right values for any of the S parameters. Since the line width of the microstrip
does not change the frequency, the physical lengths were chosen as the parameters

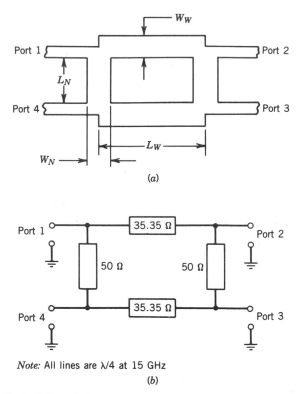

Note: All lines are λ/4 at 15 GHz

(b)

Figure 8.64   3-dB branch-line coupler: (a) physical model; (b) electrical model.

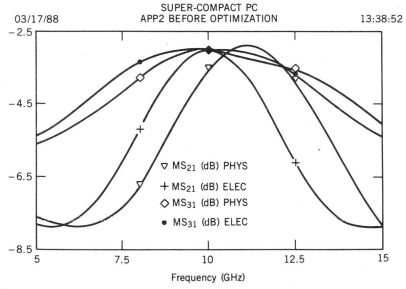

Figure 8.65   Frequency response of 3-dB branch-line coupler before optimization.

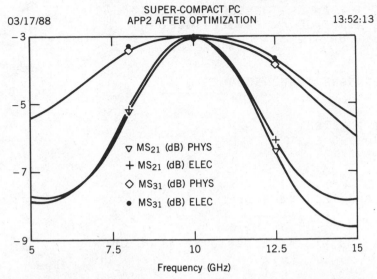

SUPER-COMPACT PC
APP2 AFTER OPTIMIZATION

Figure 8.66 Frequency response of 3-dB branch-line coupler after optimization.

to be optimized. After optimization the physical model characteristics came very close to the electrical model, especially the 3-dB bandwidth. In this case the goal is to come as close as possible to the response of the electrical network. Therefore, the complete set of S parameters is optimized using only one expression.

```
*3DB BRANCH-LINE DIRECTIONAL COUPLER (APP2.CKT)
*                 AFTER OPTIMIZATION

WW:  ?1.148MM?
LW:  ?2.74MM?
WN:  ?0.604MM?
LN:  ?2.84MM?
BLK
   T:   TEE 1   5 9 W1=WN  W2=WW W3=WN SUB
   WL:  TRL 5   6 W=WW   P=LW    SUB
   NL:  TRL 9  12 W=WN   P=LN    SUB
   T    2   6 10
   NL  10  11
   T    4   8 12
   WL   8   7
   T    3   7 11
PHYS: 4POR 1 2 3 4
END
BLK
   TRL 1 2 Z=35.35 E=90 F=10GHZ
   TRL 4 3 Z=35.35 E=90 F=10GHZ
```

```
   TRL 1 4 Z=50      E=90 F=10GHZ
   TRL 2 3 Z=50      E=90 F=10GHZ
ELEC: 4POR 1 2 3 4
END
FREQ
   STEP 5GHZ 15GHZ 500MHZ
END
OUT
   PRI PHYS S
   PRI ELEC S
END
OPT
   PHYS
F=7.5GHZ 12.5GHZ S=ELEC
END
DATA
   SUB: MS H=.635MM ER=10 MET1=AU 3UM
END
```

APP3: Edge-Coupled Microstrip Filter (Figs. 8.67, 8.68). Application Note 3 shows how to realize a microstrip filter with a center frequency of 6 GHz. The selected bandwidth is 65%. This particular bandwidth is not easy to realize using edge-coupled lines to build up the microwave filter. Nevertheless, the necessary values for the filter design are as follows:

Z_{0e}	Z_{00}	Gap (mm)	Width (mm)	Length (mm)
68.834	31.1655 ± 0.0217	4.4497		
	0.0583			
50.156	36.7023	0.0877	0.0881	4.3454
50.156	36.7023	0.0877	0.0881	4.3454
68.834	31.1655 ± 0.0217	4.4497		
	0.0583			

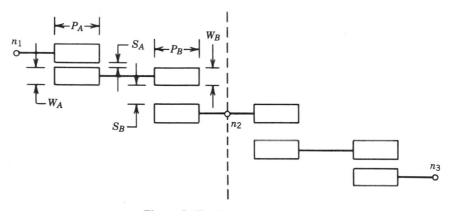

Figure 8.67   Edge-coupled filter.

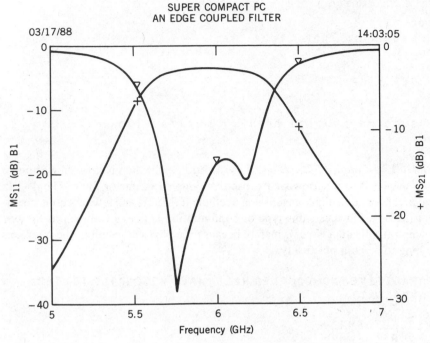

Figure 8.68 Couple frequency response of edge-coupled filter.

The actual design splits the symmetrical circuit into two pieces. First, one half of
the filter is defined in subcircuit $a1$. Then subcircuit $a1$ is used and connected with
itself again. The final circuit is called $b1$.

```
*Edge coupled filter with a bandwidth of 65% (APP3)
*and a center frequency of 6GHz
*using length for layout with open end effect

blk
      cpl  1  2  3  4  w=0.0583mm  s=0.0217mm  p=4.4281mm  sub
      cpl  3  5  6  7  w=0.0881mm  s=0.0877mm  p=4.3177mm  sub
      open  2          w=0.0583mm                          sub
      open  4          w=0.0583mm                          sub
      open  5          w=0.0881mm                          sub
      open  7          w=0.0881mm                          sub
a1:2por  1  6
end
*
blk
      a1   1  2
a1   3  2
b1:2por  1    3
end
```

```
*
freq
   step 5ghz 7ghz .1ghz
end
out
   pri b1 s
end
data
   sub: ms h=0.1mm er=12.9 met1=au 1um
end
```

APP4: End-Coupled Filter (Figs. 8.69, 8.70). Application Note 4 shows how to apply the GAP to a narrowband capacitive-coupled resonator filter. The 4-GHz bandpass filter has a high transmission coefficient (MS21) and low reflection coefficient (MS11). However, this type of configuration is not as common as the one shown in Application Note 3, mainly because the physical length of the filter gets too long to be built practically.

```
*CAPACITIVE END-COUPLED HALF-WAVE RESONATOR FILTER
*(APP4.CKT)
*
G1:?.50363MM?
G2:?.85663MM?
L1:?11.254MM?
L2:?4.4424MM?
L3:?4.3502MM?
WW1:   4.8MM
BLK
        TRL 1 2 W=WW1   P=L2  SUB2
        GAP 2 3 W=WW1    G=G1   SUB1
        TRL 3 4 W=WW1   P=L1   SUB2
        GAP 4 5 W=WW1    G=G2   SUB1
        TRL 5 6 W=WW1   P=L3   SUB2
        GAP 6 7 W=WW1    G=G2   SUB1
        TRL 7 8 W=WW1    P=L1   SUB2
        GAP 8 9 W=WW1    G=G1   SUB1
        TRL 9 10  W=WW1  P=L2 SUB2
```

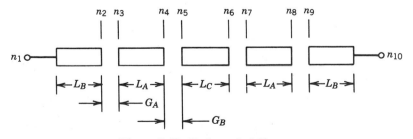

Figure 8.69 End-coupled filter.

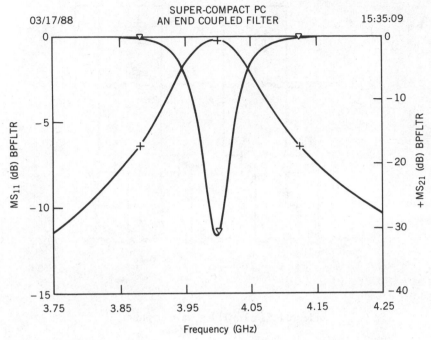

Figure 8.70 Frequency response of end-coupled filter.

```
BPFLTR:2POR 1   10
END
FREQ
    STEP 3.75GHZ 4.25GHZ 0.01GHZ
         4GHZ
END
OUT
    PRI BPFLTR SK
END
OPT
  BPFLTR F=3.98GHZ 4.02GHZ MS21 -.5DB GT
         F=4.1GHZ  5.00GHZ MS21 -15DB LT
         F=3.0GHZ  3.90GHZ MS21 -15DB LT
END
DATA
    SUB1: MS H=2.4MM ER=10.0
    SUB2: MS H=2.4MM ER=10.0 MET1=CU 35UM TAND=0.0001
END
```

APP5: Traveling-Wave Amplifier (Figs. 8.71 to 8.73). Application Note 5 shows the simulation of a traveling-wave amplifier (TWA). The distributed amplifiers provide a very flat gain slope and a bandwidth for several octaves. Today's TWAs are typically built as MMICs and the optimum number of active stages is about

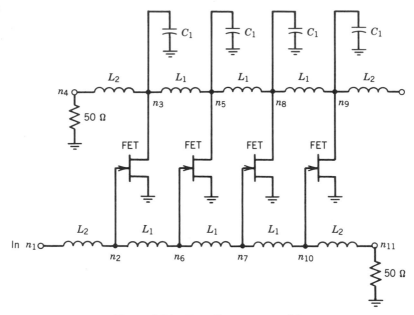

Figure 8.71   Travelling-wave amplifier.

four. The inductors shown compensate for the internal capacitors of the semiconductors. The active elements are simulated through simple voltage sources and the distributed transmission lines are approximated by the inductors.

```
*A TRAVELLING WAVE AMPLIFER (APP5.CKT)

L1:.5804NH
L2:.20778NH
C1:.05494PF
BLK
    CAP 1 0 C=0.25PF
    VCG 1 2 G=30MS R1=100E6 R2=100E6
    CAP 2 0 C=0.025PF
```

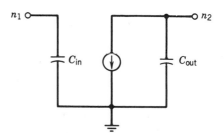

Figure 8.72   Equivalent FET model.

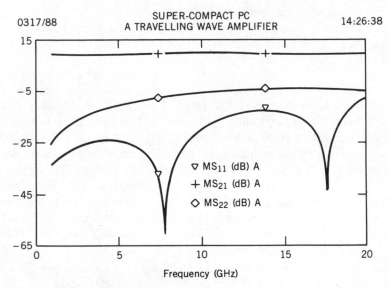

Figure 8.73 Frequency response of traveling-wave amplifier.

```
FET:2POR 1 2
END
BLK
     IND 1   2    L=L2
     FET 2   3
     IND 3   4    L=L2
     RES 4   0    R=50
     CAP 3   0    C=0.225PF
     IND 2   6    L=L1
     IND 3   5    L=L1
     FET 6   5
     IND 5   8    L=L1
     CAP 5   0    C=C1
     IND 6   7    L=L1
     FET 7   8
     IND 7   10   L=L1
     CAP 8   0    C=C1
     IND 8   9    L=L1
     FET 10  9
     IND 10  11   L=L2
     RES 11  0    R=50
     CAP 9   0    C=C1
     IND 9   12   L=L2
A:2POR 1    12
END
FREQ
     STEP 1GHZ 20GHZ 0.5GHZ
```

```
END
OUT
    PRI FET S
    PRI A   S
END
```

APP6: Voltage-Controlled Oscillator (Figs. 8.74, 8.75). Application Note 6
shows a 7.5-GHz voltage-controlled oscillator (VCO). This design takes advantage of the tweak feature of Super-Compact PC to illustrate the variation in
frequency by changing the tuning diode. The capacitance, which determines the
oscillating frequency, is varied to simulate the voltage change. However, as the
oscillating frequency varies, the impedance (Z11) changes as well. Therefore, the
output power will decrease if the load impedance is kept constant to 50Ω. Seven
curves are shown on the impedance plot, having initial capacitance 25 pF and final
capacitance 50 pF with a step of 5 pF. The use of the databank, one feature of
Super-Compact PC, is also demonstrated.

```
*                          A VCO DESIGN
*                 USING THE TWEAK FEATURE OF SUPER-
*                          COMPACT PC
*                       BY ANTHONY KWAN

BLK
    RES 1 2 R=10000
    CAP 1 0 C=220PF
    RES 2 0 R=22000
    CAP 2 0 C=220PF
    TWO 2 3 4   BIPL
    RES 1 4 R=3300
    CAP 4 5 C=?15.603PF?
*THIS CAP BELOW IS BEING TWEAKED TO ADJUST THE
*OSCILLATING FREQUENCY
    CAP 3 5 C=?40.783PF?
```

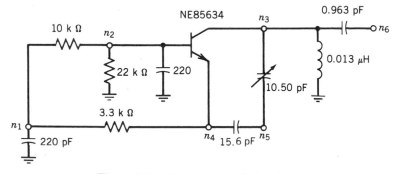

Figure 8.74   Voltage-controlled oscillator.

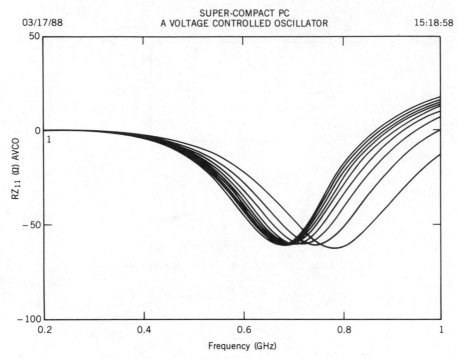

Figure 8.75 Optimized negative resistance for a VCO.

```
    IND 3 0 L=?.01298UH?
    CAP 3 6 C=?.96307PF?
AVCO:1POR 6
END
FREQ
    STEP 200MHZ 1GHZ 25MHZ
END
OUT
    PRI AVCO Z
    PRI AVCO SK
END
OPT
    AVCO F=775MHZ RZ11=-50
END
DATA
    BIPL:NECZW FILE=\BANK01\NEC.FLP
END
```

APP7: *Modeling a Microwave Transistor (Figs. 8.76 to 8.80).* Application Note
7 illustrates how optimization can be used to obtain a device model. The schematic
diagram shows a FET model in which six components, those through which an
arrow are drawn, are to be optimized. The goal of optimization is to choose values

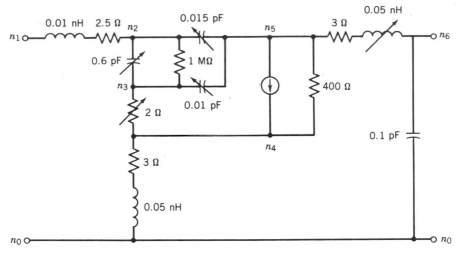

Figure 8.76 Equivalent circuit of a microwave transistor.

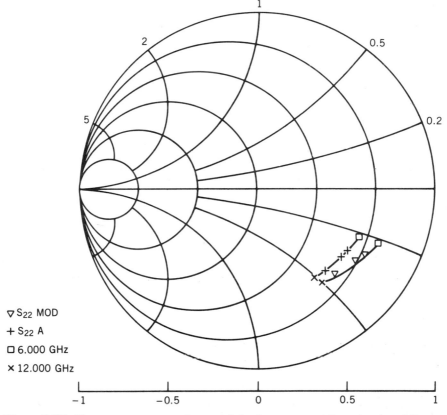

▽ S_{22} MOD
+ S_{22} A
□ 6.000 GHz
× 12.000 GHz

Figure 8.77 Frequency response of a modeled microwave transistor showing difference for S_{22}.

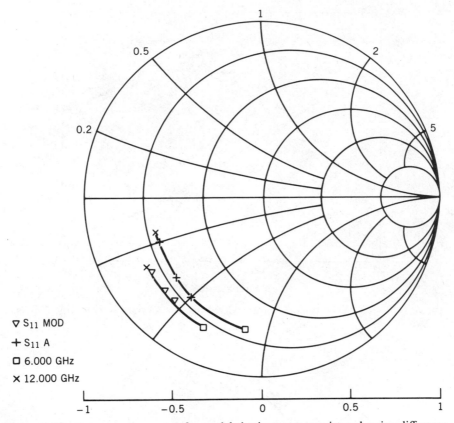

Figure 8.78 Frequency response of a modeled microwave transistor showing difference for S_{11}.

for these components such that computed S parameters for the model fit a set of measured S parameters. The model is named circuit a and is optimized subject to

$$S = Q1$$

as specified in the OPT section.

```
*MODELLING A MICROWAVE TRANSISTOR (APP7.CKT)

BLK
  SRL 1 2 R 2.5 L .01NH
  CAP 2 3 C ?.84279PF?
  RES 3 4 R ?3.1794?
  SRL 4   R 3    L  .05NH
  CAP 2 5 C ?.01448PF?
  SRL 5 6 R 3    L ?.39433NH?
  CAP 6   C ?374.72E-9PF?
```

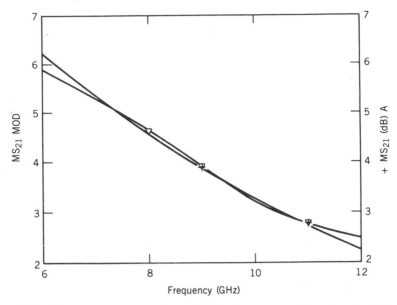

Figure 8.79   Frequency response of a modeled microwave transistor showing difference for S_{21}.

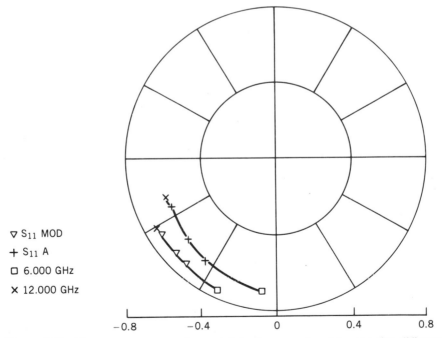

▽ S_{11} MOD
+ S_{11} A
□ 6.000 GHz
✕ 12.000 GHz

Figure 8.80   Frequency response of a modeled microwave transistor showing difference for S_{12}.

```
* NOTE GM IS IN MHOS
  VCG 2 5 3 4 G 0.05 R1 1E6 R2 400 F 1E18 T 5E-12
  CAP 3 5 C ?.10032PF?
MOD: 2POR 1 6
END
BLK
    TWO 1 2   Q1
A:2POR 1   2
END
FREQ
  6GHZ 7GHZ 8GHZ 9GHZ 10GHZ 11GHZ 12GHZ
END
OUT
  PRI MOD SK
  PRI A   SK
END
OPT
*
* MATCH COMPUTER MODEL TO MEASURED DATA
* ORDER OF WEIGHTS IS 11, 12, 21 AND 22
*
  MOD S=Q1 W 10 10 10 10
  TERM .01
END
DATA
*   DATA FOR HPMPP WITH PARASITIC LEAD INDUCTANCES
Q1:S
  6GHZ 0.732  -96.3 1.973 99.3 0.051 69.8 0.642 -26.0
  7GHZ 0.699 -110.9 1.835 88.2 0.058 70.8 0.629 -30.3
  8GHZ 0.673 -124.4 1.708 78.4 0.066 73.2 0.618 -34.8
  9GHZ 0.644 -136.9 1.570 69.8 0.073 77.3 0.613 -39.5
 10GHZ 0.622 -147.5 1.449 61.7 0.081 82.7 0.610 -44.9
 11GHZ 0.616 -154.8 1.378 53.5 0.093 87.7 0.605 -51.1
 12GHZ 0.623 -160.1 1.332 44.6 0.108 90.7 0.602 -58.1
```

APPENDIX A
Derivation of the Stability Factor*

The derivation of (1.138) was given by Carson [4.3] but the proof is repeated here for completeness. The graphical solution for stability from Fig. 1.29b is

$$|C_G| - r_G > 1 \tag{A.1}$$

Using (1.135) and (1.136) gives

$$\left| \frac{\left| S_{22} - DS_{11}^* \right| - \left| S_{12} S_{21} \right|}{\left| S_{22} \right|^2 - \left| D \right|^2} \right| > 1 \tag{A.2}$$

which becomes

$$\left\| S_{22} - DS_{11}^* \right| - \left| S_{12} S_{21} \right\|^2 > \left\| S_{22} \right|^2 - \left| D \right|^2 \right|^2 \tag{A.3}$$

We square and group terms to obtain

$$2 \left| S_{12} S_{21} \right| \left| S_{22} - DS_{11}^* \right| < \left| S_{22} - DS_{11}^* \right|^2 + \left| S_{12} S_{21} \right|^2 - \left\| S_{22} \right|^2 - \left| D \right|^2 \right|^2 \tag{A.4}$$

Using the identity

$$\left| S_{22} - DS_{11}^* \right|^2 = \left| S_{12} S_{21} \right|^2 + (1 - \left| S_{11} \right|^2)(\left| S_{22} \right|^2 - \left| D \right|^2) \tag{A.5}$$

we square (A.4) and combine terms to give

$$(\left| S_{22} \right|^2 - \left| D \right|^2)^2 \left\{ [(1 - \left| S_{11} \right|^2) - (\left| S_{22} \right|^2 - \left| D \right|^2)]^2 - 4 \left| S_{12} S_{21} \right|^2 \right\} > 0 \tag{A.6}$$

---

*The material in this appendix is largely adapted from R. S. Carson, *High Frequency Amplifiers*, Wiley, New York, 1975, pp 192–196.

which finally results in

$$2\left|S_{12}S_{21}\right| < 1 - \left|S_{11}\right|^{2} - \left|S_{22}\right|^{2} + \left|D\right|^{2} \tag{A.7}$$

$$k = \frac{1 - \left|S_{11}\right|^{2} - \left|S_{22}\right|^{2} + \left|D\right|^{2}}{2\left|S_{12}\right|\left|S_{21}\right|} > 1 \tag{A.8}$$

This is the stability factor used to describe the unconditional stability of a two-port.

APPENDIX B
An Important Proof Concerning Stability

It is not obvious that the inequality

$$\frac{B_1}{2|C_1|} > 1 \tag{B.1}$$

implies that the stability factor k is greater than unity. The proof follows.

Substituting the definitions of B_1 and C_1 from (1.168) and (1.167) into (B.1) gives

$$\frac{1 + |S_{11}|^2 - |S_{22}|^2 - |D|^2}{2|S_{11} - S_{22}^* D|} > 1 \tag{B.2}$$

$$1 + |S_{11}|^2 - |S_{22}|^2 - |D|^2 > 2|S_{11} - S_{22}^* D| \tag{B.3}$$

Using the identity

$$|S_{11} - S_{22}^* D|^2 = |S_{12} S_{21}|^2 + (1 - |S_{22}|^2)(|S_{11}|^2) - |D|^2) \tag{B.4}$$

and squaring both sides of (B.3) gives

$$(1 - |S_{22}|^2 + |S_{11}|^2 - |D|^2)^2$$
$$> 4|S_{12} S_{21}|^2 + 4(1 - |S_{22}|^2)(|S_{11}|^2 - |D|^2) \tag{B.5}$$

Rearranging terms yields

$$[(1 - |S_{22}|^2) + (|S_{11}|^2 - |D|^2)]^2 - 4(1 - |S_{22}|^2)$$
$$\cdot (|S_{11}|^2 - |D|^2) > 4|S_{12} S_{21}|^2 \tag{B.6}$$

Now using the identity

$$(a + b)^2 - 4ab = (a - b)^2 \tag{B.7}$$

will give

$$[(1 - |S_{22}|^2) - (|S_{11}|^2 - |D|^2)]^2 > 4|S_{12}S_{21}|^2 \qquad \text{(B.8)}$$

which can be written

$$[(1 - |S_{11}|^2 - |S_{22}|^2 + |D|^2)]^2 > 4|S_{12}S_{21}|^2 \qquad \text{(B.9)}$$

Taking the square root of both sides yields

$$1 - |S_{11}|^2 - |S_{22}|^2 + |D|^2 > 2|S_{12}S_{21}| \qquad \text{(B.10)}$$

which completes the proof.

APPENDIX C
S-Parameter Expressions Relevant to the Design of RF and Microwave Amplifiers*

Load Stability Circle

Center

$$C_L = \frac{S_{11} D^* - S_{22}^*}{|D|^2 - |S_{22}|^2} \tag{C.1}$$

Radius

$$R_L = \left| \frac{S_{12} S_{21}}{|D|^2 - |S_{22}|^2} \right| \tag{C.2}$$

Generator Stability Circle

Center

$$C_G = \frac{S_{22} D^* - S_{11}^*}{|D|^2 - |S_{11}|^2} \tag{C.3}$$

Radius

$$R_G = \left| \frac{S_{12} S_{21}}{|D|^2 - |S_{11}|^2} \right| \tag{C.4}$$

*The material in this appendix is largely adapted from P. L. D. Abrie, *The Design of Impedance Matching Networks for Radio-Frequency and Microwave Amplifiers*, Artech House, Norwood, Mass., 1985.

where

$$D = S_{11} S_{22} - S_{12} S_{21} \tag{C.5}$$

Stability Factor

$$k = \frac{1 - |S_{11}|^2 - |S_{22}|^2 + |S_{11} S_{22} - S_{12} S_{21}|^2}{2 |S_{12} S_{21}|} \tag{C.6}$$

Maximum Operating and Available Power Gain of an Inherently Stable Transistor and the Corresponding Terminations

$$G_{\text{ma}} = \left| \frac{S_{21}}{S_{12}} \right| (k - \sqrt{k^2 - 1}) \tag{C.7}$$

$$\Gamma_{Gm} = C_1^* [B_1 - (B_1^2 - 4|C_1|^2)^{1/2}]/(2|C_1|^2) \tag{C.8}$$

$$\Gamma_{Lm} = C_2^* [B_2 - (B_2^2 - 4|C_2|^2)^{1/2}]/(2|C_2|^2) \tag{C.9}$$

where

$$B_1 = 1 + |S_{11}|^2 - |S_{22}|^2 - |D|^2 \tag{C.10}$$

$$B_2 = 1 + |S_{22}|^2 - |S_{11}|^2 - |D|^2 \tag{C.11}$$

$$C_1 = S_{11} - S_{22}^* D \tag{C.12}$$

$$C_2 = S_{22} - S_{11}^* D \tag{C.13}$$

Circle of Constant Mismatch for a Voltage Generator (Generator Reflection Parameter Γ_G) Terminated in a Passive Load (Reflection Parameter Γ_L)

Transducer Power Gain

$$G_T = \frac{[1 - |\Gamma_L|^2][1 - |\Gamma_G|^2]}{|1 - \Gamma_G \Gamma_L|^2} = \frac{1}{M_c} \tag{C.14}$$

Circle

Center

$$C_M = \frac{\Gamma_G^* G_T}{1 - |\Gamma_G|^2 (1 - G_T)} \tag{C.15}$$

Radius

$$R_M = \frac{(1 - |\Gamma_G|^2)\, \sqrt{1 - G_T}}{1 - |\Gamma_G|^2\, (1 - \breve{G}_T)} \qquad \text{(C.16)}$$

Constant-Operating-Power-Gain Circle

Center

$$C_w = \frac{g_w\, (S_{22}^* - D^* S_{11})}{1 + g_w(|S_{22}|^2 - |D|^2)} \qquad \text{(C.17)}$$

Radius

$$R_w = \frac{(1 - 2\, k\, |S_{12}\, S_{21}|\, g_w + |S_{12}\, S_{21}|^2\, g_w^2)^{1/2}}{1 + g_w\, (|S_{22}|^2 - |D|^2)} \qquad \text{(C.18)}$$

where

$$g_w = G_w/|S_{21}|^2 \qquad \text{(C.19)}$$

and

$$G_w = \frac{|S_{21}|^2\, (1 - |\Gamma_L|^2)}{1 - |S_{11}|^2 + |\Gamma_L|^2\, (|S_{22}|^2 - |D|^2) - 2\, \mathrm{Re}\, (C_2 \Gamma_L)} = G \qquad \text{(C.20)}$$

Constant-Available-Power-Gain Circle

Center

$$C_{\mathrm{Av}} = \frac{g_{\mathrm{Av}}\, (S_{11}^* - D^*\, S_{22})}{1 + g_{\mathrm{Av}}\, (|S_{11}|^2 - |D|^2)} \qquad \text{(C.21)}$$

Radius

$$R_{\mathrm{Av}} = \frac{(1 - 2\, k\, |S_{12}\, S_{21}|\, g_{\mathrm{Av}} + |S_{12}\, S_{21}|^2\, g_{\mathrm{Av}}^2)^{1/2}}{|1 + g_{\mathrm{Av}}\, (|S_{11}|^2 - |D|^2)|} \qquad \text{(C.22)}$$

where

$$g_{\mathrm{Av}} = G_{\mathrm{Av}}/|S_{21}|^2 \qquad \text{(C.23)}$$

and

$$G_{Av} = \frac{|S_{21}|^2 (1 - |\Gamma_G|^2}{1 - |S_{22}|^2 + |\Gamma_G|^2 (|S_{11}|^2 - |D|^2) - 2 \, \text{Re} \, (C_1 \Gamma_G)} \qquad (C.24)$$

Constant-Gain Circles for a Transistor with $S_{12} = 0$

Transducer Power Gain

$$G_{T-u} = G_1 \, |S_{21}|^2 \, G_2 \qquad (C.25)$$

where

$$G_1 = \frac{1 - |\Gamma_G|^2}{|1 - \Gamma_G S_{11}|^2} \qquad (C.26)$$

$$G_2 = \frac{1 - |\Gamma_L|^2}{|1 - \Gamma_L S_{22}|^2} \qquad (C.27)$$

Constant-G_1 Circle

Center

$$C_{G1} = \frac{g_1 \, S_{11}^*}{1 - |S_{11}|^2 \, (1 - g_1)} \qquad (C.28)$$

Radius

$$R_{G1} = \frac{\sqrt{1 - g_1} \, (1 - |S_{11}|^2)}{1 - |S_{11}|^2 \, (1 - g_1)} \qquad (C.29)$$

where

$$g_1 = G_1/G_{1-\text{max}} = G_1 \, (1 - |S_{11}|^2) \qquad (C.30)$$

Constant-G_2 Circle

Center

$$C_{G2} = \frac{g_2 \, S_{22}^*}{1 - |S_{22}|^2 \, (1 - g_2)} \qquad (C.31)$$

Radius

$$R_{G2} = \frac{\sqrt{1 - g_2}\,(1 - |S_{22}|^2)}{1 - |S_{22}|^2\,(1 - g_2)} \tag{C.32}$$

where

$$g_2 = G_2/G_{2-\text{max}} = G_2\,(1 - |S_{22}|^2) \tag{C.33}$$

Constant-Noise-Figure Circle

Center

$$C_F = \frac{\Gamma_{0n}}{1 + N} \tag{C.34}$$

Radius

$$R_F = \frac{\sqrt{N^2 + N\,(1 - \Gamma_{0n})}}{1 + N} \tag{C.35}$$

where

$$N = |1 + \Gamma_{0n}|^2\,(F - F_{\text{min}})/(4r_n) \tag{C.36}$$

$$r_n = R_n/Z_0 \tag{C.37}$$

$$F = F_{\text{min}} + \frac{4\,Rn}{Z_0}\,\frac{|\Gamma_G - \Gamma_{0n}|^2}{|1 + \Gamma_{0n}|^2\,(1 - |\Gamma_G|^2)} \tag{C.38}$$

Equivalent Generator Reflection Parameter and Transducer Power Gain for a Given Constant-Operating-Power-Gain circle

$$\Gamma_{G-\text{OUT}} = \Gamma^*_{L-\text{opt}} \tag{C.39}$$

[Refer to (C.9)]

$$G_T = \frac{A_w}{2}\,(\sqrt{1 + 4/A_w^2} - 1) \tag{C.40}$$

where

$$A_w = \frac{|C_w|^2}{R_w^2\,|\Gamma_{G-\text{OUT}}|^2}\,[1 - |\Gamma_{G-\text{OUT}}|^2]^2 \tag{C.41}$$

Equivalent Load Reflection Parameter and Transducer Power Gain for a Given Constant-Available-Power-Gain Circle

$$\Gamma_{L-\text{IN}} = \Gamma_{G-\text{opt}}^{*} \tag{C.42}$$

[Refer to (C.8)]

$$G_T = \frac{A_{\text{Av}}}{2}\left(\sqrt{1 + 4/A_{\text{Av}}^2} - 1\right) \tag{C.43}$$

where

$$A_{\text{AV}} = \frac{|C_{\text{AV}}|^2}{R_{\text{AV}}^2 \, |\Gamma_{L-\text{IN}}|^2}\left[1 - |\Gamma_{L-\text{IN}}|^2\right]^2 \tag{C.44}$$

Equivalent Load Reflection Parameter and Transducer Power Gain for a Given Constant-Noise-Figure Circle

$$\Gamma_{L-\text{IN}} = \Gamma_{G-\text{opt}-F}^{*} = \Gamma_{0n}^{*} \tag{C.45}$$

[Refer to (C.38)]

$$G_T = \frac{A_F}{2}\left(\sqrt{1 + 4/A_F^2} - 1\right) \tag{C.46}$$

where

$$A_F = \frac{|C_F|^2}{R_F^2 \, |\Gamma_{L-\text{IN}}|^2}\left[1 - |\Gamma_{L-\text{IN}}|^2\right]^2 \tag{C.47}$$

APPENDIX D
Passive Microwave Elements

D.0 INTRODUCTION

The rapid expansion of microwave utilization derives principally from advances in two major areas: microwave semiconductors and the technology of microwave integrated circuits. The passive microwave elements technology has made use of the achievements in semiconductor fabrication and also has new branches, such as monolithic elements (discussed in Section D.4) and lumped elements (which are especially suitable for monolithic MICs and are discussed in Section D.1).

Since the analysis of transmission, such as microstrip lines and CPW is becoming more accurate and effective, CAD plays a more and more important role in microwave technology. The new analysis and modeling methods are given where we introduce the distributed elements (Section D.2) and discontinuities (Section D.3).

In the final section (Section D.5) we introduce several special-purpose elements. Dielectric resonators as an example, have been widely used in recent years, since the development of the dielectric resonator with high Q-factor.

The theory and design of passive microwave elements have been reported in numerous articles widely scattered in the technical literature, but there is no single comprehensive description available. This appendix is intended to fill that gap.

D.1 LUMPED ELEMENTS

The lumped elements can be used for definition in the microwave band if the sizes of lumped elements can be made much smaller than the wavelength. Computer-aided design of circuits using lumped elements requires complete and accurate characterization of thin film lumped elements of microwave frequency. Differences between distributed and lumped elements with MIC circuits are primarily to provide physical support and isolation between the various elements, whereas for MICs using distributed elements, most of the energy is stored or propagates within the substrate.

Resistor (Thin Film, etc.)

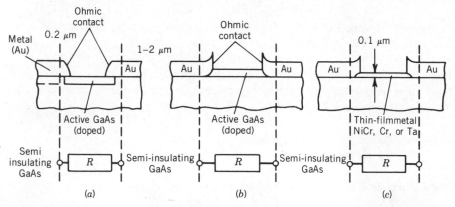

Figure D.1 Resistor types for GaAs MMICs: (a) implanted semiconductor resistor (surface resistance, $R_F > 20\ \Omega$); (b) mesa semiconductor resistance ($R_F > 3\ \Omega$); (c) vapor deposited or sputtered thin-film resistance ($3\ \Omega\ R_F < 100\ \Omega$).

Description: Resistors can be produced as semiconductor resistors (Fig. D.1a, b) or as thin-film resistors with resistive layers of NiCrTa or Cr.

References: D.1, pp. 70–71
D.2, pp. 108–121

Capacitor (Thin Film, etc.)

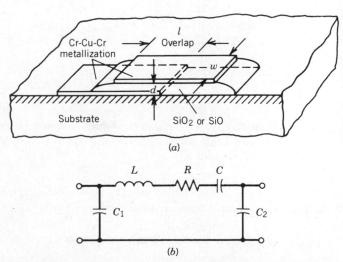

Figure D.2 (a) Configuration of an overlay capacitor; (b) Equivalent circuit for the capacitor in (a).

Description: With the dielectric layer (or insulator layer) between the two metal plates, we have a metal dielectric–metal capacitor.

Parameters: Overlap l, width of the top and bottom plates for capacitor W, and thickness of dielectric or insular layer d.

Equivalent Circuit: R represents losses in the capacitor; parasitics are represented by a series inductance and fringing capacitance C_1 and C_2 due to the ground. C, the capacitor, is much bigger than C_1 or C_2.

References: D.3, pp. 216–217
D.2, pp. 118–121

Bond Wire

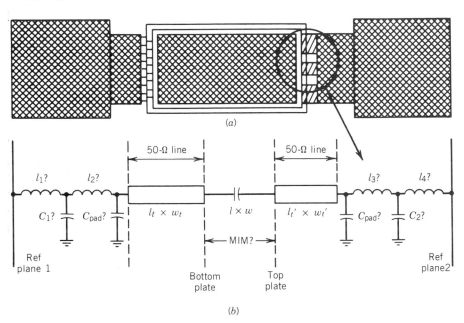

(a)

(b)

Figure D.3 Schematic of the circuit which includes bond wires. The elements shown with "?" are optimized. MIM has the distributed model incorporated in it. $(l_1 + l_2)$, $(l_3 + l_4)$ are the total inductances for the bond wires.

Description: The bond wires in Fig. D.3a have the equivalent circuit shown in Fig. D.3b. The analysis method is transmission-line theory.

Parameters: Length and width of the bond wires are required.

Equivalent Circuit: L1, L2, L3, and L4 are the inductors for the bond wires, C_1 C_2, and C_{pad} are the capacitors for the bond wires.

Reference: D.5.

Diodes (Beam Lead, etc.)

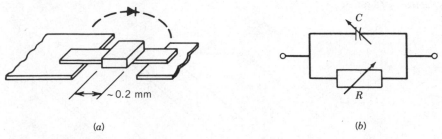

(a)   (b)

Figure D.4 Hybrid elements: (*a*) diode chip; (*b*) equivalent circuit.

Description: The beam-lead diode connected to two ends of microstrip exists in a form that can be built into MICs.

Equivalent Circuit: The variables C and R depend on the working condition of the diode, such as the bias voltage.

Reference: D.1, pp. 56–57

D.2 DISTRIBUTED ELEMENTS

Transmission Lines

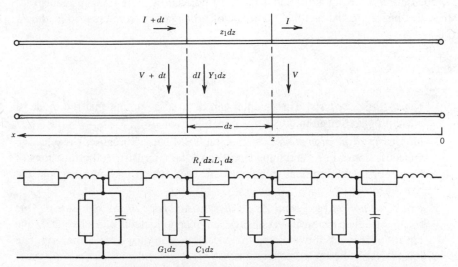

Figure D.5 Transmission line and its equivalent circuit.

The transmission line and its equivalent circuit are shown here. The three transmission-line examples are microstrip line, coplanar waveguide, and grounded coplanar waveguide, which are discussed below.

Microstrip Line

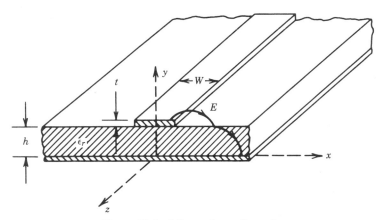

Figure D.6 Microstrip configuration.

Structure Parameters: W is the width of the microstrip, h is the thickness of the dielectric layer, ϵ_r is the dielectric permittivity, and T is the thickness of the microstrip.

Brief Description: Microstrip line is the most popular of these transmission structures, due mainly to the fact that the mode of propagation on microstrip is almost TEM.

Methods of Solution

1. *Quasi-static analysis.* This method can be used assuming that the mode of wave propagation in microstrip is pure TEM.
2. *Microstrip dispersion model.* As the non-TEM behavior causes the effective dielectric constant ϵ_{re} and impedance Zo of the microstrip to be functions of frequency, semiempirical techniques are used which take into account the non-TEM nature.
3. *Exact evaluation of ϵ_{re} and Zo_n; full-wave analysis of the microstrip.* One has to introduce time-varying electric and magnetic fields and solve the wave equation. Instead of evaluating the capacitance in quasi-static analysis, one has to determine the propagation constant.

Reference: D.6, pp. 4–5, 20, 43

Coplanar Waveguide (CPWC)

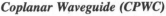

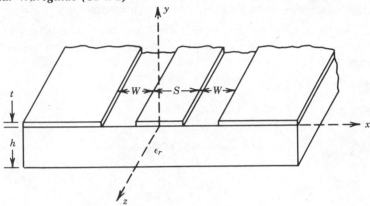

Figure D.7 Coplanar waveguide.

Parameters: Coplanar waveguide (all the conductors are in the same plane) consists of a center strip with two ground planes located parallel to and in the plane of the strip, and with W in between, the width of the center strip is S and the thickness of the conductor and dielectric layer (with ϵ_r) are t and h separately.

Methods: Coplanar lines have been studied using *quasi-static approximation* as well as *full-wave analysis*. A quasi-static analysis of these transmission lines was carried out using conformal mapping and with the assumption that the dielectric substrate is thick enough to be considered infinite. For commonly used thicknesses this assumption is valid for large values of the dielectric constant. A modification of the method studied takes the finite thickness of the dielectric substrate into consideration. The effect of enclosure on the characteristics of CPW has been determined using the finite difference method. A full-wave analysis of coplanar lines which provides information regarding the frequency dependence of phase velocity and characteristic impedance has been carried out by using Galerkin's method in the spectral domain by the variational method and by nonuniform discretization of integral equations.

Reference: D.6, p. 260

Grounded Coplanar Waveguide (GCPW)

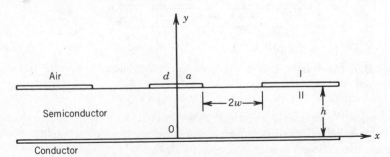

Figure D.8 Grounded coplanar waveguide.

Parameters: The difference between grounded coplanar waveguide and coplanar waveguide is the extra conductor plane under the dielectric layer.

Methods: As one kind of coplanar lines, GCPW can be studied using quasi-static approximation as well as full-wave analysis. The spectral domain method [D.7] has been proved to be an effective method.

Reference: D.7

Coupled Lines

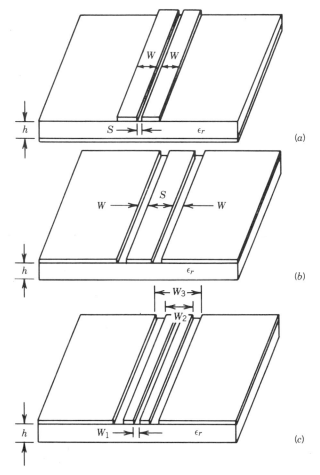

Figure D.9 Configurations of (*a*) coupled microstrip lines; (*b*) coupled slotlines; and (*c*) coupled coplanar waveguides.

Description: A coupled-line configuration consists of two transmission lines parallel to each other and in close proximity. Because of the coupling of support, there are two different modes of propagation; that is, they share their own characteristic impedances.

Parameters: The configuration for the coupled lines is shown in Fig. D.9, with the parameters labeled in the figure.

Methods

1. *Even- and odd-mode method.* Valid for description of symmetrical coupled lines. Wave propagation along a coupled line is expressed in terms of two modes, corresponding to an even or an odd symmetry about a plane which can, therefore, be replaced by a magnetic or electric wall for the purpose of analysis.

2. *Coupled-mode approach.* This method is quite general and is applicable to asymmetric coupled lines. Also, the wave propagation is expressed in terms of the modes of propagation on individual uncoupled lines modified by the coupling because of mutual capacitances and inductances.

3. *Graph transformation technique.* This technique uses Richard's transformation and allows the coupled-line structures to be treated in exactly the same manner as lumped networks.

4. *Congruent transformation technique.* This approach is powerful for establishing coupled-line properties when there are large number of lines coupled together.

Reference: D.6, pp. 303–305

Waveguides (Coax, Rectangular)

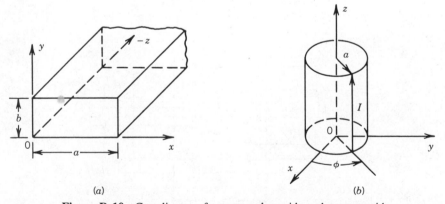

(a)   *(b)*

Figure D.10 Coordinates of a rectangular guide and a coax guide.

Description: In general, a waveguide consists of a hollow metallic tube of a rectangular or circular shape used to guide an electromagnetic wave. Waveguides are used principally at frequencies in the microwave range; inconveniently large guides would be required to transmit radio-frequency power at longer wavelengths. At frequency range X band from 8.00 to 12.0 GHz, for example, the U.S. standard rectangular waveguide WR-90 has an inner width of 2.286 cm (0.9 in.) and an inner height of 1.016 cm (0.4 in.); but its outside dimensions are 2.54 cm (1 in.) wide and 1.27 cm (0.5 in.) high.

In waveguides the electric and magnetic fields are confined to the space within the guides. Thus no power is lost through radiation, and even the dielectric loss is negligible, since the guides are normally air-filled. However, there is some power loss as heat in the walls of the guides, but the loss is very small.

It is possible to propagate several modes of electromagnetic waves within a waveguide. These modes correspond to solutions of Maxwell's equations for the particular waveguides. A given waveguide has a definite cutoff frequency for each allowed mode. If the frequency of the impressed signal is above the cutoff frequency for a given mode, the electromagnetic energy can be transmitted through the guide for that particular mode without attenuation. Otherwise, the electromagnetic energy with a frequency below the cutoff frequency for that particular mode will be attenuated to a negligible value in a relatively short distance. *The dominant mode in a particular guide is the mode having the lowest cutoff frequency.* It is advisable to choose the dimensions of a guide in such a way that for a given input signal, only the energy of the dominant mode can be transmitted through the guide.

Method

1. The desired waveguide quotients are written in the form of either rectangular or cylindrical coordinate systems suitable to the problem at hand.
2. Apply the boundary condition.
3. Get the partial differential equation and solve them by proper method (mathematical methods).

Reference: D.8, pp. 104–105

Taper

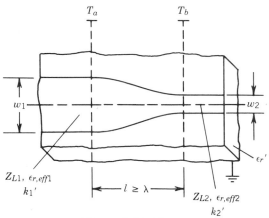

Figure D.11 Taper.

Description: The taper with the conductor width W_1 and W_2 has the effect that S_{11} and S_{22} can be maintained as low as required by choice of shape and length.

Method: The planar waveguide model, including high-order modes, can be used to analyze the the structure.

Reference: D.1, pp. 281–282

Air Bridge and Via

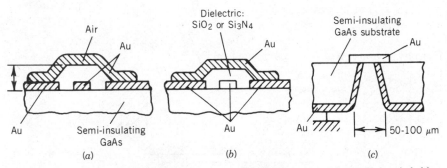

Figure D.12 Connecting elements for MMICs: (a) air bridge ("via"); (b) dielectric bridge; (c) ground through contact (via hole ground, "via hole").

Description: Connectors are important components for MMICs. To connect separated metallized areas, low-capacitance air bridges (air bridges, via) or dielectric bridges, or ground through contact (via a hole ground "via hole") are used. The dimensions and parameters are labeled in Fig. D.12.

Reference: D.1, pp. 70–71

Wrap

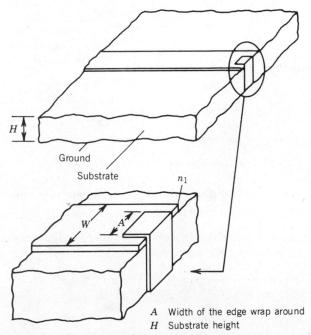

A   Width of the edge wrap around
H   Substrate height

Figure D.13 Wrap.

Description: Wrap around ground in microstrip.

Equivalent circuit: Wrap can be modeled by a parallel inductor and a resistor.

Reference: D.1, p. 306
D.9, pp. 4–136

Coupler (Lange, Rat-Race, Branch-Line)

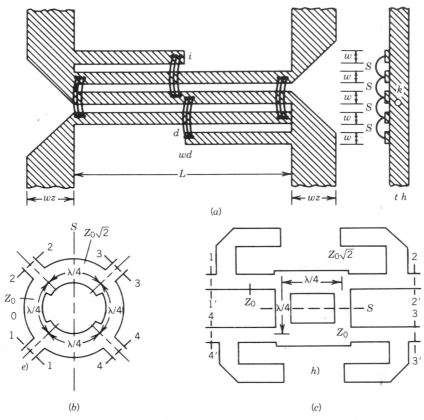

Figure D.14 Couplers.

Description: There are two kinds of couplers. Directional couplers and hybrid couplers (such as rat-race and branch-line)

Analysis Method: Even- and odd-mode theory can be used to analyze the couplers even though the couplers are not symmetrical in structure.

References: D.1, p. 14
D.2, pp. 154–162, 383–390
D.10, pp. 171–176
D.11, pp. 775–842

D.3 DISCONTINUITIES

The microstrip or other planar transmission-line discontinuities can be analyzed by the quasi-static or full-wave method. The latter will be more accurate.
The following analysis methods are discussed in detail and in two categories: quasi-static and full-wave analysis. These methods are applicable for such discontinuities as gap step, bend, tee, cross, slit, and open. After the discussion of methods, the equivalent models for each case (gap, step, etc.) are given.

Methods

1. *Quasi-static analysis.* This involves calculations of static capacitances and low-frequency inductances. The equivalent circuit is derived from these results. To consider the dispersion, a waveguide-type dynamic analysis taking dispersion into account is carried out.

   To calculate the capacitances, the following methods are effective: (a) matrix version method, (b) variational method, (c) Galerkin's method in the spectral domain, and (d) use of line sources with charge reversal.

   To calculate inductances using the quasi-static method, we may use the fundamental Maxwell equations and get the inductances expression in closed form.

2. *Fullwave analysis.* Based on the planar waveguide model, the Galerkin method in FTD and the contour integral method can be used.

References: D.1, pp. 189–202
              D.2, pp. 31–47
              D.6, pp. 107–193
              D.12, pp. 43–60

Gap

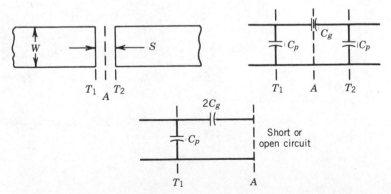

Figure D.15 Representation of a gap in microstrip and its equivalent circuit.

Step

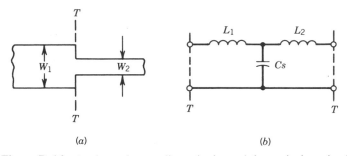

(a)   (b)

Figure D.16   A microstrip step discontinuity and the equivalent circuit.

Bend

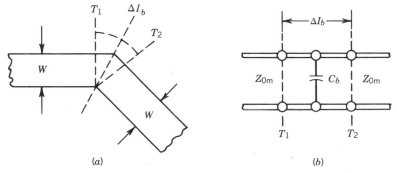

(a)   (b)

Figure D.17   Geometry and equivalent circuit of a microstrip bend.

Tee

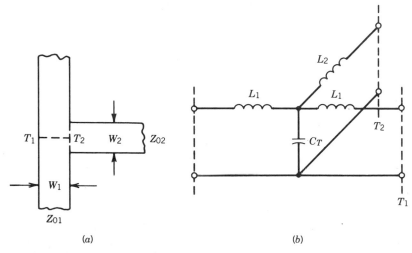

(a)   (b)

Figure D.18   Geometry and equivalent circuit of a microstrip tee junction.

Cross

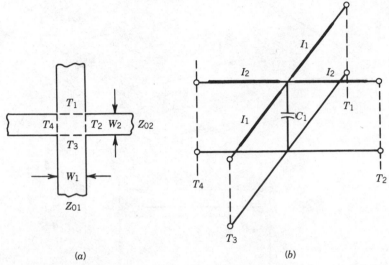

(a)   (b)

Figure D.19 Geometry and equivalent circuit of a microstrip cross junction.

Slit

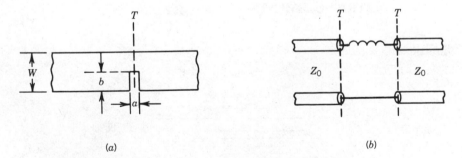

(a)   (b)

Figure D.20 Geometry and equivalent circuit of a microstrip slit.

Open

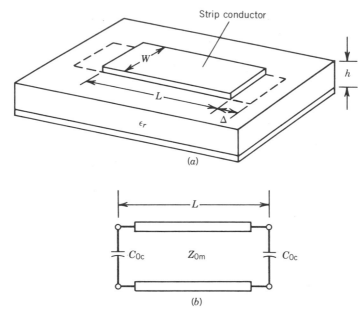

(a)

(b)

Figure D.21 Configuration for calculation of microstrip open-end capacitance and its equivalent circuit.

D.4 MONOLITHIC ELEMENTS

Interdigital Capacitor

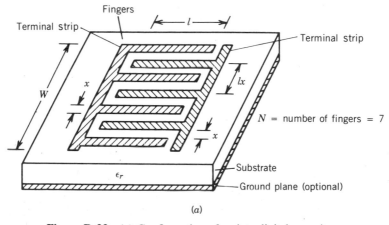

N = number of fingers = 7

(a)

Figure D.22 (a) Configuration of an interdigital capacitor.

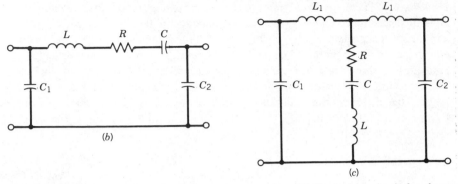

Figure D.22 (*b*) Equivalent circuit for series mounting; (*c*) Equivalent circuit for shunt mounting.

Methods: The capacitance between two sets of digits in interdigital structure is found by using the capacitance formula for the odd mode in coupled microstrip lines, with the ground plane spacing tending to an infinitely large value.

References: D.2, pp. 383–390
            D.3, p. 217

Interdigital Rectangular and Spiral Inductor

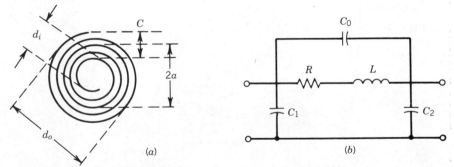

Figure D.23 (*a*) Configuration of a spiral inductor; (*b*) Equivalent circuit for a spiral inductor.

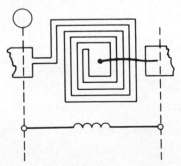

Figure D.24 Interdigital rectangular inductor layout and the equivalent circuit.

Description: The equivalent circuit for the spiral configuration does not consist of an inductance alone. There are associated parasitics in the form of self-capacitance and interturn capacitance, C_0, as well as the shunt fringing capacitances C_1 and C_2 due to the effects of ground. The equivalent circuit for a spiral inductor, including parasitics, is shown in Fig. D.23*b*. The series resistance R accounts for the loss. The typical range of values for parasitic elements for a spiral with diameter in the range 1.0 to 5.0 mm on an alumina substrate are as follows: C_0 is nearly 0.15 pF, C_1 ranges from 0.1 to 0.2 pF, C_2 ranges from 0.05 to 0.1 pF, and Q at 4 GHz ranges from 80 to 100.

References: D.3, p. 211
D.10, p. 160

Thin-Film Capacitor

See "Capacitor" in Section D.1.

Thin-Film Resistor

See "Resistor" in Section D.1.

Reference: D.2, pp. 253–260, 326–347

Interdigital transformer

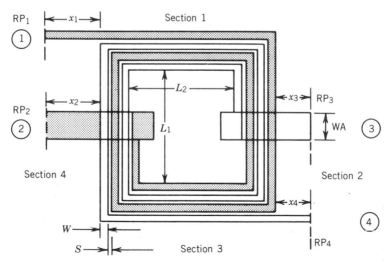

Figure D.25 Geometry parameters for the module PLTRAN describing the general transformer geometry.

Reference: D.15; D.16

Underpass/Overpass

See "Air Bridge and Via in Section D.2.

D.5 SPECIAL-PURPOSE ELEMENTS

Yig (Yttrium Iron Garnet, Y3Fe6012)

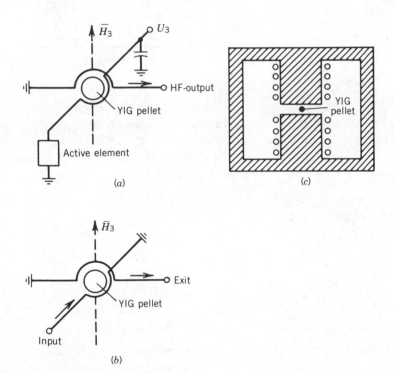

Figure D.26 (*a*) Diagram of a YIG band-pass filter; (*b*) YIG-tuned oscillator; (*c*) magnetic circuit for premagnetization.

Description: It has a very high unload resonator Q-factor Q_0 up to 110,000 and can be used at its ferromagnetic resonance as a resonator to tune oscillators and filers. The resonant frequency of the YIG element can be linearly changed over a wide range, by intensity of the magnetic bias field, and therefore, the element can be used to electrically tune oscillators and filters.

Reference: D.10, p. 212

Dielectric Resonator

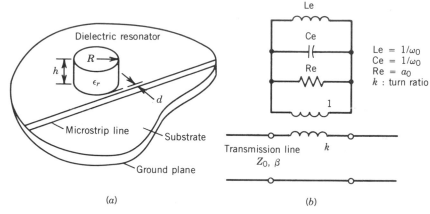

(a)   (b)

Figure D.27

Description: Dielectric resonators, offering high-Q cavity performance in microwave integrated circuits, are widely used in filter-stabilized oscillators, discriminators, and so on. In Fig. D.27, a dielectric resonator is placed beside the transmission line, and the equivalent circuit is shown.

Methods: Fundamental electromagnetic theory, other methods in reference book.

Reference: D.13

Gyrator

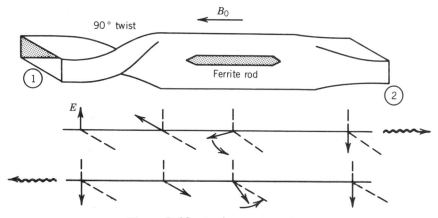

Figure D.28   A microwave gyrator.

Description: A gyrator is defined as a two-port device that has a relative difference in phase shift of 180° for transmission from port 1 to port 2 as compared with the phase shift for transmission from port 2 to port 1. *A gyrator may be obtained by employing the nonreciprocal property of Faraday rotation.* Figure D.28 illustrates a typical microwave gyrator. It consists of a rectangular guide with a 90° twist connected to a circular guide, which in turn is connected to another rectangular guide at the other end. The two rectangular guides have the same orientation at the input ports. The circular guide contains a thin cylindrical rod of ferrite with the ends tapered to reduce reflections. A static axial magnetic field is applied so as to produce 90° Faraday rotation of the TE_{11} dominant mode in the circular guide. Consider a wave propagating from left to right. In passing through the twist the plane of polarization is rotated by 90° in a counterclockwise direction. If the ferrite produces an additional 90° of rotation, the total angle of rotation will be 180°, as indicated in Fig. 6.28. For a wave propagating from right to left, the Faraday rotation is still 90° in the same sense. However, in passing through the twist, the next 90° of rotation is in a direction to cancel the Faraday rotation. Thus, for transmission from port 2 to port 1, there is no net rotation of the plane of polarization. The 180° rotation for transmission from port 1 to port 2 is equivalent to an additional 180° of phase shift since it reverses the polarization of the field. It is apparent, then, that the device just described satisfies the definition of a gyrator.

Reference: D.14, pp. 300–301

Circulator

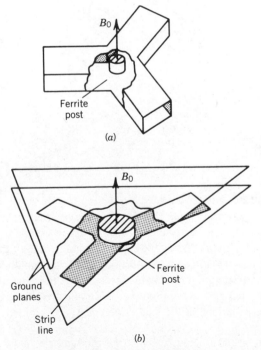

Figure D.29   Three-port circulators. (*a*) Waveguide version; (*b*) strip-line circulator.

Description: A circulator is a multiport device that has the property that a wave incident in port 1 is coupled into port 2 only, a wave incident in port 2 is coupled into port 3 only, and so on. The ideal circulator is also a matched device; that is, with all ports except one terminated in matched loads, the input impedance of the remaining port is equal to the characteristic impedance of its input line, and hence presents a matched load.

Reference: D.14, pp. 304–305

Isolator

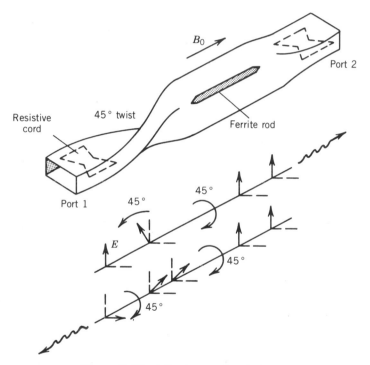

Figure D.30 A Farday-rotation isolator.

Description: The isolator, or uniline, is a device that *permits unattenuated transmission from port 1 to port 2 but provides very high attenuation for transmission in the reverse direction.* The isolator is often used to couple a microwave signal generator to a load network. It has the great advantage that all the available power can be delivered to the load and yet reflections from the load do not get transmitted back to the generator output terminals. Consequently, the generator sees a matched load, and effects such as power output variation and frequency pulling (change in frequency), with variations in the load impedance, are avoided.

Reference: D.14, p. 301

REFERENCES

D.1 Reinmat K. Hoffmann, *Handbook of Microwave Integrated Circuits*, Artech House Microwave Library, Norwood, Mass., 1987.

D.2 Jeffrey Frey and Kul Bhasin, *Microwave Integrated Circuits*, Artech House Microwave Library, Norwood, Mass., 1986.

D.3 K. C. Gupta, R. Garg, R. Chadha, "Computer Aided Design of Microwave Circuits," Artech House Inc., Norwood, Mass., 1981.

D.4 Jyotip Mondal, "An Experimental Verification of a Simple Distributed Model of MIM Capacitors for MMIC Applications," *IEEE Transactions on Microwave Theory and Techniques*, Vol. MTT-35, No. 4, April 1987.

D.5 J. P. Mondal, "An Experimental Verification of a Simple Distributed Model of MIM Capacitor for MMIC Applications," *IEEE Transactions on Microwave Theory and Techniques*, Vol. MTT-35, April 1987.

D.6 K. C. Gupta, *Microstrip Lines and Slot Lines*, Artech House Microwave Library, Norwood, Mass., 1979.

D.7 Y. C. Shik and T. Itoh, "Analysis of Conducted Backed Coplanar Waveguide," Electronics letters 10th June 1982, Vol. 18, No. 12, pp. 538–540.

D.8 S. Y. Liao *Microwave Devices and Circuits*, 2nd ed., Prentice-Hall, Englewood Cliffs, N.J., 1985.

D.9 Super-Compact Mainframe 1.91 Version.

D.10 Erich Pehl, *Microwave Technology*, Artech House Microwave Library, Norwood, Mass., 1985.

D.11 G. Matthaei, L. Young, E. M. T. Jones, *Microwave Filters, Impedance-Matching Networks and Coupling Structures*, McGraw-Hill, New York, 1964.

D.12 Takanor Okoshi, *Planar Circuits for Microwaves and Lightwaves*, Series in Electrophysics, Springer-Verlag, New York, 1984.

D.13 Darko Kajfez, *Dielectric Resonators*, Artech House Microwave Library, Norwood, Mass., 1986.

D.14 R. E. Collin, *Foundations for Microwave Engineering*, Physical and Quantum Electronics Series, McGraw-Hill, New York, 1966.

D.15 Linmic + User Manual Compact Software

D.16 User Manual Super Compact; Compact Software.

APPENDIX E
Spectral-Domain Analysis

Strictly speaking, the microstrip does not support pure TEM mode; hybrid TE and TM modes are present. At millimeter frequencies, the effects of longitudinal components of both the electric and magnetic fields become more important, and these effects cannot be described in terms of static capacitances and inductances. For accurate design, a full-wave analysis is needed to take into account the effects of longitudinal field components. Several rigorous methods are available for analysis of microstrip, such as the spectral domain method [E.1, E.2], equivalent waveguides model method [E.3, E.4], mode-matching method [E.5, E.6], finite difference method [E.7, E.8], finite element method [E.9, E.10], and so on. One of the most popular methods for analysis microstrip, including multilayer microstrip, is the spectral domain immittance approach in conjunction with Galerkin's method. In this approach an immittance matrix for the structure is derived from the combination of equivalent transmission lines in the transverse direction. Instead of lengthy field formulations, the matrix elements can be obtained almost by inspection. The propagation constant at a given frequency is extracted from algebraic equations that relate the Fourier transform of electric fields at the dielectric–air interface to those of the current on the microstrip. The advantages of this approach are its numerical simplicity and efficiency, and the accuracy of the solution may be improved systematically by increasing the number of basic functions. In addition, dispersion characteristics for the high-order modes can be generated.

The process of the spectral domain immittance approach is illustrated briefly as follows. Assume that the microstrip is infinitesimally thin and that its conductivity is infinite. The fields propagate in the z-direction according to $e^{-j\beta z}$. As shown in Fig. E.1, on the strip the tangential fields E_x and E_z have to be zero. If J_z and J_x represent the unknown currents on the microstrip, and G_{xx}, G_{xz}, G_{zx}, and G_{zz} are the Green's functions, which contain unknown propagation constant β, the following coupled integral equations can be derived:

$$\int \left[G_{xx}(x - x', h) J_x(x', h) + G_{xz}(x - x', h) J_z(x', h) \right] dx'$$
$$= E_x(x, h) = 0 \quad |x| \le w/2 \tag{E.1}$$

$$\int \left[G_{zx}(x - x', h) J_x(x', h) + G_{zz}(x - x', h) J_z(x', h) \right] dx'$$
$$= E_z(x, h) = 0 \quad |x| \le w/2 \tag{E.2}$$

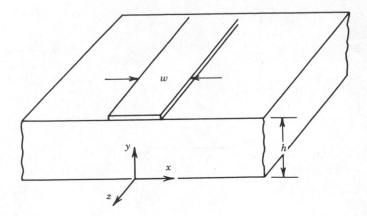

These equations may be solved provided that the Green's functions G_{xx}, etc. are given. The procedure is straight forward, but for the inhomogeneous structure these questions are not available in close forms. In addition, a significant amount of computer time may be required to solve the coupled integral equations. However, in the spectral domain, the coupled integral equations above become algebraic equations. To this end, taking the Fourier transform of (E.1) and (E.2), this results in

$$\tilde{G}_{xx}(\alpha, h)\, \tilde{J}_x(\alpha, h) + \tilde{G}_{xz}(\alpha, h)\, \tilde{J}_z(\alpha, h) = \tilde{E}_x(\alpha, h) \qquad \text{(E.3)}$$

$$\tilde{G}_{zx}(\alpha, h)\, \tilde{J}_x(\alpha, h) + \tilde{G}_{zz}(\alpha, h)\, \tilde{J}_z(\alpha, h) = \tilde{E}_z(\alpha, h) \qquad \text{(E.4)}$$

where the Fourier transform is defined by

$$\tilde{F}(\alpha) = \int_{-\infty}^{\infty} F(x)\, e^{j\alpha x}\, dx \qquad \text{(E.5)}$$

The algebraic equations contain four unknowns: $\tilde{J}_x$, $\tilde{J}_z$, $\tilde{E}_x$ and $\tilde{E}_z$. However, $\tilde{E}_x$ and $\tilde{E}_z$ will be eliminated later in the solution process based on Galerkin's procedure. The Green's functions $\tilde{G}_{xx}$, etc. are found by using the following equivalent transmission lines in the y direction. First, it is recognized that from the definition of the inverse Fourier transform, the field along the y direction is given by

$$Ey(x, y)\, e^{-j\beta z} = \frac{1}{2\pi} \int_{-\infty}^{\infty} \tilde{E}y(\alpha, y)\, e^{-j(\alpha x + \beta z)}\, d\alpha \qquad \text{(E.6)}$$

and similarly for $H_y(x, y)$. From this equation it is found that the field components are the superposition of inhomogeneous (in the y direction) waves propagating in the direction of θ from the Z axis, where $\theta = \cos^{-1}[\beta/(\sqrt{\alpha^2 + \beta^2})]$. By taking this into account, a new coordinate system (u, y, v) is chosen. In this coordinate system u is taken along the wave propagation direction, and v is transverse to the

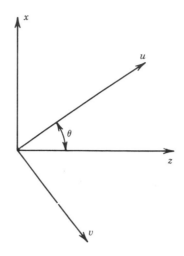

u and x axes (see Fig. E.2). The transformation from the u, v coordinate system to the x, y coordinate system is given by

$$u = x \sin \theta + z \cos \theta \qquad (E.7)$$

$$v = z \sin \theta - x \cos \theta \qquad (E.8)$$

The plane waves are now decomposed into TE-to-y (H_y, H_u, E_v) and TM-to-Y (E_y, H_v, E_u). The u and v axes are oriented such that if a current component J_u existed, it would generate only TM fields, and similarly, a J_u current component would generate only TE fields. Using this concept, an equivalent transmission-line circuit of TE and TM fields can be used to model the microstrip structure and is shown in Figure E.3. The characteristic admittances in each region are given by

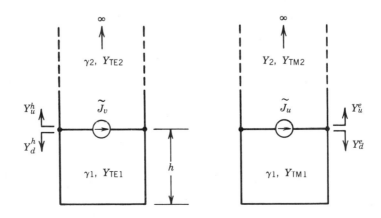

$$Y_{\text{TE}i} = -\frac{\tilde{H}_u}{\tilde{E}_v} = \frac{\gamma_i}{j\omega\mu_0} \qquad i = 1, 2 \qquad\qquad \text{(E.9)}$$

$$Y_{\text{TM}i} = \frac{\tilde{H}_v}{\tilde{E}_u} = \frac{j\omega\epsilon_0\epsilon_i}{\gamma_i} \qquad i = 1, 2 \qquad\qquad \text{(E.10)}$$

where $\gamma_i = \sqrt{\alpha^2 + \beta^2 - \epsilon_i K^2}$ is the propagation constant in the y direction and ϵ_i is the dielectric constant in the other region, respectively; α is the Fourier transform variable; β is the unknown propagation constant; and K is the free-space wave number. Note that section 1 of the transmission line corresponds to the dielectric layer of the microstrip, and section 2 represents the air above the microstrip. The metal strip is represented by the current densities $\tilde{J}_u$ and $\tilde{J}_n$. All of the boundary conditions are incorporated into the equivalent circuits. The ground plane is represented by a short circuit and the radiation condition is expressed by an infinitely long transmission line. The magnetic field discontinuity at the conductor interface is represented by the equivalent current sources. The fields $\tilde{E}_v$ and $\tilde{E}_u$ at the interface $y = h$ is related to the current sources via

$$\tilde{J}_u(\alpha, h) = Y^e \tilde{E}_u(\alpha, h) \qquad\qquad \text{(E.11)}$$

$$\tilde{J}_v(\alpha, h) = Y^h \tilde{E}_v(\alpha, h) \qquad\qquad \text{(E.12)}$$

Y^e and Y^h are the input admittances looking into the equivalent circuits at $y = h$ and are given by

$$Y^h = Y_u^h + Y_d^h \qquad\qquad \text{(E.13)}$$

$$Y^e = Y_u^e + Y_d^e \qquad\qquad \text{(E.14)}$$

where Y_u^e and Y_d^e are input admittances looking up and down at $y = h$ in the TM equivalent circuit, and Y_u^h and Y_d^h are those in the TE circuit and are given by

$$Y_d^e = Y_{\text{TM}1} \coth(\gamma_1 h) \qquad\qquad \text{(E.15)}$$

$$Y_u^e = Y_{\text{TM}2} \qquad\qquad \text{(E.16)}$$

$$Y_d^h = Y_{\text{TE}1} \coth(\gamma, h) \qquad\qquad \text{(E.17)}$$

$$Y_u^h = Y_{\text{TE}2} \qquad\qquad \text{(E.18)}$$

The fields in the u, v coordinate system are linearly related to the x and z components via the coordinate transform relations (E.7) and (E.8). Similarly, $\tilde{J}_x$ and $\tilde{J}_z$ are the superposition of $\tilde{J}_u$ and $\tilde{J}_v$. When the transformation relations are used, the Green's functions in (E.3) and (E.4) are given by

$$\tilde{G}_{xx}(\alpha, h) = \frac{\sin^2\theta}{Y^e} + \frac{\cos^2\theta}{Y^h} \qquad\qquad \text{(E.19)}$$

$$\tilde{G}_{xz}(\alpha, h) = \left(\frac{1}{Y^e} - \frac{1}{Y^h}\right) \sin \theta \cos \theta \qquad (E.20)$$

$$\tilde{G}_{zz}(\alpha, h) = \frac{\sin^2 \theta}{Y^h} + \frac{\cos^2 \theta}{Y^e} \qquad (E.21)$$

$$\tilde{G}_{zx}(\alpha, h) = \tilde{G}_{xz}(\alpha, h) \qquad (E.22)$$

The next step is to expand the current densities on the strip in terms of linearly independent basis functions to approximate the existing current distribution. The current densities are expressed by

$$\tilde{J}_x(\alpha, h) = \sum_{1}^{N} e_n \tilde{J}_{xn}(\alpha, h) \qquad (E.23)$$

$$\tilde{J}_z(\alpha, h) = \sum_{1}^{M} d_m \tilde{J}_{zm}(\alpha, h) \qquad (E.24)$$

where the $\tilde{J}_{xn}$ and $\tilde{J}_{zm}$ are basis functions, and c_n and d_m are unknown amplitude coefficients. Substituting the expansion functions into (E.3) and (E.4), the following equations are obtained:

$$\tilde{E}_x(\alpha, h) = \tilde{G}_{xx}(\alpha, h) \sum_{n=1}^{N} C_n \tilde{J}_{xn}(\alpha, h) + \tilde{G}_{yz}(d, h) \sum_{m=1}^{M} d_m \tilde{J}_{zm}(\alpha, h)$$

$$(E.25)$$

$$\tilde{E}_z(\alpha, h) = \tilde{G}_{zx}(\alpha, h) \sum_{n=1}^{N} C_n \tilde{J}_{xn}(\alpha, h) + \tilde{G}_{zz}(\alpha, h) \sum_{m=1}^{M} d_m \tilde{J}_{zm}(\alpha, h)$$

$$(E.26)$$

The final step in this process is to take the inner product of (E.25) and (E.26) with the known basis functions J_{xn} and J_{zm}. The tangential electric fields, which are zero on the strip, are multiplied by the currents, which are nonzero on strip. According to Parseval's theorem,

$$\int_{-\infty}^{\infty} F(x)G(x)\, dx = \frac{1}{2\pi} \int_{-\infty}^{\infty} \tilde{F}(\alpha)\tilde{G}(\alpha)\, d\alpha \qquad (E.27)$$

This results in a set of equations whose left side identically equals to zero. The following set of homogeneous matrix equations are obtained

$$\sum_{n=1}^{N} K_{pn} e_n + \sum_{m=1}^{M} K_{pm} d_m = 0 \qquad p = 1, 2, \cdots, N \qquad (E.28)$$

$$\sum_{n=1}^{N} K_{qn} C_m + \sum_{m=1}^{M} K_{qm} d_m = 0 \qquad q = 1, 2, \cdots, M \qquad (E.29)$$

where a typical matrix element is given by

$$K_{pm} = \int_{-\infty}^{\infty} \tilde{J}_{xp}(\alpha, h)\, \tilde{G}_{xz}(\alpha, h)\, \tilde{J}_{zm} d\alpha \qquad (E.30)$$

Once the integration is performed, the only unknown term in the K elements is the propagation constant. It is obtained by arranging the equation in matrix form,

Table E.1

Method	Features	Disadvantages
Spectral domain method	One of the most popular methods for infinitesimally thin conductors on multilayer structures Closed-form expressions for Fourier-transformed Green's functions Numerical efficiency	Cannot handle thick conductor structures For tight coupling the number of basic functions becomes large; would involve convergent problems
Finite difference method	Mathematical preprocessing is minimal Can be applied to a wide range of structures	Numerically inefficient precautions must be taken when the method is applied to an open-region problem Need layer computer storage for accurate solution
Finite element method	Similar to the finite difference method Has variational features in the algorithm and is more flexible in the application.	Developed to solve very large matrix equation Numerically inefficient Existence of so-called spurious (unphysical) zeros
Mode-mating method	Typically applied to the problem of scattering at the waveguide discontinuity Often used to solve enclosed planar structures, including metal thickness effects	Several different formulations possible, all theoretically equivalent; however, they may be different numerically Precautions must be taken on relative convergence for some problems
Equivalent waveguide model	Very useful method for analysis of microstrip discontinuity problem	

$$[K(\beta)] \begin{bmatrix} C_n \\ d_m \end{bmatrix} = 0 \qquad \text{(E.31)}$$

where K is a matrix of dimension of $(m + n) \times (m + n)$. For nontrivial solution, the determiner of the K matrix must be zero.

$$\text{Det}[K(\beta)] = 0 \qquad \text{(E.32)}$$

Solving this equation, the propagation constant is obtained. Next, the unknown coefficients c_n and d_m can be calculated by substituting β into (E.31). The propagation constant may be a complex number. The real part represents the wave number in the direction of propagation. The imaginary part is connected with the attenuation of the amplitude in the medium.

The characteristic impedance of the microstrip can be obtained by the power-current definition

$$Z_0 = \frac{2P_{\text{avg}}}{|1|^2} \qquad \text{(E.33)}$$

where P_{avg} expresses the average transmitted power and Z_0 represents the current in the propagation direction. Both can be calculated after β is determined.

The limitation of the spectral domain imittance method is that it cannot handle a finite metabolization thickness problem. For analyzing the dispersion characteristics of microstrip with finite metabolization thickness, the other full-wave analysis methods mentioned above may be used. A very brief summary of some popular methods is given in Table E.1.

REFERENCES

E.1 T. Itoh and R. Mittra, "Spectral-Domain Approach for Calculating the Dispersion Characteristics of Microstrip Lines," *IEEE Transactions on Microwave Theory and Techniques*, Vol. MTT-21, July 1973, pp. 496–499.

E.2 T. Itoh "Spectral Domain Imittance Approach for Dispersion Characteristics of Generalized Printed Transmission Line," *IEEE Transactions on Microwave Theory and Techniques*, Vol. MTT-28, July 1980, pp. 733–736.

E.3 I. Wolff, G. Kempa, and R. Mehran, "Calculation Method for Microstrip Discontinuities and T-Junctions," *Electronics Letters*, Vol. 8, 1972, pp. 177–179.

E.4 G. Kompa and R. Mehran, "Planar Waveguide Model for Calculating Microstrip Components," *Electronic Letters*, Vol. 11, 1975, pp. 459–460.

E.5 R. Mittra and S. W. Lee, *Analytical Techniques in the Theory of Guided Waves*, Macmillan, New York, 1971.

E.6 T. S. Chu, T. Itho, and Y.-C. Shih, "Comparative Study of Mode Matching Formulations for Microstrip Discontinuity Problems," *IEEE Transactions on Microwave Theory and Techniques*, Vol. MTT-33, 1985, pp. 1018–1023.

E.7 J. Hornsby and A. Gopinath, "Numerical Analysis of a Dielectric Loaded Waveguide with a Microstrip Line—Finite Difference Methods," *IEEE Transactions on Microwave Theory and Techniques*, Vol. MTT-17, 1969, pp. 684–690.

E.8 H. E. Green, "The Numerical Solution of Some Important Transmission-Line Problems," *IEEE Transactions on Microwave Theory and Techniques*, Vol. MTT-5, 1965, pp. 676–692.

E.9 P. Daly, "Hybrid-Mode Analysis of Microstrip by Finite Element Method," *IEEE Transactions on Microwave Theory and Techniques*, Vol. MTT-19, No. 2, 1973, pp. 19–25.

E.10 A. F. Thomson and A. Gopinath, "Calculation of Microstrip Discontinuity Inductances," *IEEE Transactions on Microwave Theory and Techniques*, Vol. MTT-23, No. 8, 1975, pp. 648–655.

APPENDIX F
Noise Program for Microwave Bipolar Transistors

The following is a short program to calculate the noise performance of a bipolar transistor based on the equivalent circuit shown in Fig. 3.5. It has some correction factors and the base time constant depends heavily on the device. A noise plot will follow using these equations.

```
10 CLS
20 pi = 4 * ATN(1)
30  '
40 PRINT "
*****************************************************************
*********"
50 PRINT "                      test for bip noise models .  "
55 PRINT "               copyright 1989 (R)  compact software
"
56 PRINT "               model includes parasitics"
57 PRINT "               good for chip evaluation  "
58 PRINT "               version 8 05 89"
60 PRINT "
*****************************************************************
*******"
70 PRINT
130 PRINT "enter Ic (mA)";
140 INPUT IC: IC = IC * .001
150 PRINT "enter Rb";
160 INPUT rbb
170 PRINT "enter Ft (Ghz) ";
180 INPUT ft: ft = ft * 1E+09
190 PRINT "enter hFE";
200 INPUT hfe
205 PRINT "enter emitter lead inductance (in nh)";
206 INPUT lb: lb = lb * 1E-09
207 PRINT "enter base lead inductance (in nH)";
208 INPUT le: le = le * 1E-09
210 a0 = hfe / (1 + hfe)
240 PRINT "enter operating frequency (GHz) ";
245 INPUT f0: f0 = f0 * 1E+09
250 VT = .026
260 re = VT / IC
270 fe = 3 * ft: fb = 3.8 * ft
280 cte = 1 / (2 * pi * re * fe)
285 c1 = (1 + (f0 / fb) ^ 2)
290 a = (c1 * (1 + (f0 / fe) ^ 2) - a0) * 1 / a0
300 XOPT = c1 * 2 * pi * f0 * cte * re ^ 2 / a / a0 - 2 * pi
```

```
* f0 * (lb + le)
320 rb = rbb
330 ropt1 = rb ^ 2 - XOPT ^ 2 + c1 * re * (2 * rb + re) / a0
/ a
335 IF ropt1 < 0 THEN
            PRINT "warning inductance too large, not enough
gain "
            STOP
            END IF
336 ropt = (ropt1) ^ .5
340 FMIN = a * (rb + ropt) / re + (c1 / a0): FMIN = 10 *
LOG(FMIN) / LOG(10)
350 RN = rb * (1 + 1 / hfe) + re * (1 + (rb / re) ^ 2 * (1 /
hfe + (f0 / fe) ^ 2))
380 RNN = RN / 50
390 PRINT "xopt = "; INT(XOPT * 100) / 100
400 PRINT "ropt ="; INT(ropt * 100) / 100
410 PRINT "Fmin (dB) ="; INT(FMIN * 100) / 100; "    ";
420 PRINT "rnn   ="; INT(RNN * 100) / 100; "     "; " Rn = ";
INT(RN * 100) / 100
430 RP = ropt + XOPT ^ 2 / ropt
440 XP = XOPT + ropt ^ 2 / XOPT
460 PRINT "Cp (pF ) = "; INT(1 / 2 / pi / XP / f0 * 1E+12 *
100) / 100; "    ";
470 PRINT "Rp = "; INT(RP * 100) / 100
480 RDENOM = ((ropt + 50) ^ 2 + XOPT ^ 2)
490 REGAMMA = (ropt ^ 2 + XOPT ^ 2 - 50 ^ 2) / RDENOM
500 IMGAMMA = (100 * XOPT) / RDENOM
510 PRINT "regamma ="; INT(REGAMMA * 100) / 100; "  imgamma
="; INT(IMGAMMA * 100) / 100
520 MGAMMA = SQR(REGAMMA ^ 2 + IMGAMMA ^ 2)
530 pgamma = ATN(IMGAMMA / REGAMMA) * 180 / pi
540 PRINT "magnitude (gammaopt)= "; INT(MGAMMA * 100) / 100;
550 IF REGAMMA < 0 THEN pgamma = pgamma + 180
560 PRINT "phase (gammaopt)= "; INT(pgamma * 100) / 100; "
deg"
570 GOTO 240 'NEXT F0
580 END
```

Noise simulation of the Avantek transistor AT41435 at 8V, 25 mA using previously shown equations and all parasitics. Figure F1 shows minimum noise figure and equivalent noise resistor. Figure F2 shows the gamma opt magnitude and phase for the same transistor as a function of frequency.

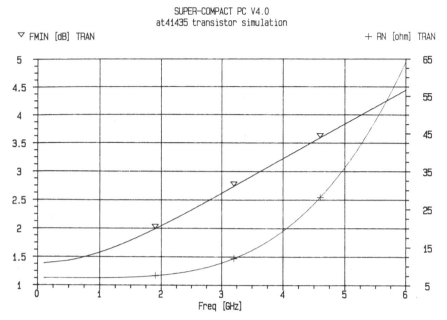

Figure F.1   Noise parameters of AT41435 from Avantek, F_{min} and R_n.

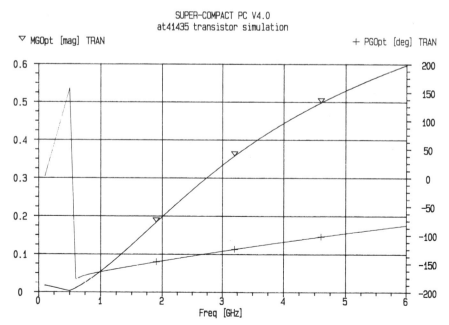

Figure F.2   Noise parameters of AT41435 from Avantek, Γ_{on}, magnitude and phase.

APPENDIX G
Noise Modeling of the Field-Effect Transistor

It is somewhat more difficult to derive expressions for the FET in a similar fashion as we have been able to do here. The paper by Fukui [3.49, 3.50] and subsequent more detailed papers by Pucel [3.45, 3.46] are the best foundation for these noise models. The most recent paper by Cappy [3.53] deals with the expressions for the HEMT.

Based on Fukui we obtain

$$F_{\min} = 1 + 0.016*f*Cgs*\text{sqrt}\left((R_g + R_s)/\text{gm}\right)$$

$$R_{\text{opt}} = 2.2*\left[1/(4\text{ gm}) + R_g + R_s\right]$$

$$X_{\text{opt}} = 160/fCgs$$

$$R_n = F(I_{\text{dss}}, I_d, f)/\text{gm}$$

After some additional consideration of external components and manufacturing differences we can write the following: [3.51]

```
CLS  :  '5-05-89   7:55
pi = 4 * ATN(1)
PRINT "
****************************************************************
**********"
PRINT "              copyright 1989 (R)  compact software "
PRINT "                 model includes parasitics"
PRINT "                    version 5 5 89"
PRINT "                      NOISE_FITTER "
PRINT "mes analyis"
PRINT
PRINT "enter gm   (mS) ";
INPUT gm
PRINT "enter Ids/Idss ratio in %"
INPUT i
PRINT "enter Rg ";
INPUT rg
PRINT "enter Rs ";
INPUT rs
PRINT "enter Ri ";
INPUT ri
PRINT "enter Cgs (pf) ";
INPUT cgs
cd:
```

```
PRINT "enter process codes (avt=.7,tach=1,trq=.8,ti=.5) ";
INPUT cd: IF cd < .3 GOTO cd
GO:
PRINT "operating frequency (GHz) ";
INPUT f0:
RN = .8 / (gm / 1000) * EXP(i * 2.2 * .01) ' if gm at idss
ropt = 12.2 * cd * (1000 / gm / 4 + rg + rs + ri) / (2 * pi
* f0 * cgs)'rohde
xopt = 160 / f0 / cgs            'fukui
FMIN1 = 1 + .016 * f0 * cgs * SQR((rg + rs) / gm * 100)
FMIN1 = 10 * LOG(FMIN1)
PRINT "Fmin (dB) ="; INT(FMIN1 * 100) / 100
RNN = RN / 50
PRINT "rn  ="; INT(RNN * 100) / 100

PRINT "xopt ="; INT(xopt * 100) / 100
PRINT "ropt ="; INT(ropt * 100) / 100
    GOSUB 1
END
1 :
    rp = ropt + xopt ^ 2 / ropt
    xp = xopt + ropt ^ 2 / xopt
    cp = 1 / 2 / pi / xp / f0 * 1E+12 / 1E+09
    PRINT "Cp (pF ) = "; INT(cp * 100) / 100; "   ";
    PRINT "Rp = "; INT(rp * 100) / 100
    RDENOM = ((ropt + 50) ^ 2 + xopt ^ 2)
    REGAMMA = (ropt ^ 2 + xopt ^ 2 - 50 ^ 2) / RDENOM
    IMGAMMA = (100 * xopt) / RDENOM
PRINT "regamma ="; INT(REGAMMA * 100) / 100; "  imgamma =";
INT(IMGAMMA * 100) / 100
    MGAMMA = SQR(REGAMMA ^ 2 + IMGAMMA ^ 2)
    pgamma = ATN(IMGAMMA / REGAMMA) * 180 / pi
PRINT "magnitude (gammaopt)= "; INT(MGAMMA * 100) / 100;
    IF REGAMMA < 0 THEN pgamma = pgamma + 180
    IF pgamma > 180 THEN pgamma = -360 + pgamma
PRINT "phase (gammaopt)= "; INT(pgamma * 100) / 100; " deg":
GOTO GO
RETURN
END
```

REFERENCE

[G1] Ulrich L. Rohde, Anthony M. Pavio, Robert A. Pucel, "Accurate Noise Simulation of Microwave Amplifiers Using CAD," *Microwave Journal*, December 1988, pp. 130–141.

Compact Software "Super Compact PC 4.0 Now Includes Nodal Noise Analysis Capability," *Transmission Line News*, (3), September 1988.

Index